A Consumer's Guide to Archaeological Science

MANUALS IN ARCHAEOLOGICAL METHOD, THEORY AND TECHNIQUE

Series Editors:

Charles E. Orser, Jr., *Illinois State University*, Normal, Illinois
Michael B. Schiffer, *University of Arizona*, Tucson, Arizona

For more information on this series, visit: http://www.springer.com/series/6256

Mary E. Malainey

A Consumer's Guide to Archaeological Science

Analytical Techniques

 Springer

Mary E. Malainey
Department of Anthropology
Brandon University
Manitoba, Canada

ISSN 1571-5752
ISBN 978-1-4419-5703-0   e-ISBN 978-1-4419-5704-7
DOI 10.1007/978-1-4419-5704-7
Springer New York Dordrecht Heidelberg London

Library of Congress Control Number: 2010934257

Printed on acid-free paper

Springer is part of Springer Science+Business Media (www.springer.com)

Preface

There is growing interest in the application of analytical techniques to archaeological materials; however, many archaeologists and students of archaeology have little or no background in the natural sciences. The purpose of this book is to explain archaeological science to archaeologists. Analytical techniques are presented in a manner that the average North American archaeologist can understand. Sample requirements, potential benefits, and limitations of each approach are outlined so that an archaeologist can be a better informed consumer. It is not intended to train archaeological scientists.

The work is written with the assumption that the reader is familiar with archaeological terminology and methodology; in this respect, it is more suitable for senior undergraduate- and graduate-level students. It can also serve as a reference guide for academic researchers and cultural resource management archaeologists interested in employing these techniques, researchers supervising students who employ them, or those who simply want to better understand their application.

The information presented should enable an archaeologist to understand and critically evaluate

- the suitability of various analytical techniques to address particular archaeological questions;
- the data generated through the application of these techniques;
- the validity of archaeological interpretations made on the basis of the data.

The book is divided into four parts, Basic Science, Applications, Materials, and Instrumentation. Basic Science consists of six chapters covering the concepts from chemistry, physics, and the biological sciences upon which the analytical techniques are based. This scientific primer may be required by some readers to fully understand the rationale and theory behind sample selection, processing, and analysis. Descriptions of atoms, elements, molecules, and their properties are given in Chapter 1. Electromagnetic radiation, Chapter 2, is particularly relevant to the spectroscopic techniques covered in Chapters 34–36, and 39. The principles of radioactive decay, described in Chapter 3, are applicable to the dating techniques described in Chapters 8 and 9 and analytical techniques described in Chapters 32 and 39. Not surprisingly, the stable isotopes described in Chapter 4 are relevant to

Chapter 13. The organic compounds described in Chapter 5 are those analyzed in Chapters 8 and 13 (collagen), 14 (lipids), 15 (blood and other proteins), and 16 (DNA). The topics covered in Chapter 6 are broadly relevant to the study of organic compounds as well.

Part II, Applications, includes chapters describing principles, procedures, and issues related to the most common applications of scientific methods to archaeological materials. After a brief examination of methodological considerations (Chapter 7), several dating techniques are considered. Radiocarbon dating is presented in Chapter 8; other radioactive decay-based dating techniques, uranium series, potassium–argon and argon–argon, and fission track are described in Chapter 9. Trapped charge dating techniques, which include thermoluminescence, optically stimulated luminescence, and electron spin resonance dating, are considered together in Chapter 10. Amino acid racemization, obsidian hydration, cation-ratio dating, and archaeomagnetism are presented together in Chapter 11. General issues related to provenance studies are discussed in Chapter 12. The analysis of the stable isotopes of archaeological interest, carbon, nitrogen, strontium, lead, oxygen, and hydrogen, is described in Chapter 13. Lipid residue analysis (Chapter 14), blood and protein analysis (Chapter 15), and ancient DNA and the polymerase chain reaction (Chapter 16) are described in the final chapters of Part II.

Case studies involving the analysis of specific material types are presented in Part III. The analyses of the fundamental constituents of archaeological materials are described separately from the analysis of residues introduced through use. The materials discussed include pottery (Chapter 17), flaked and ground stone tools (Chapter 18), bone and teeth (Chapter 19), cultural rock (Chapter 20), organic artifacts (Chapter 24), paint, pigment, and ink (Chapter 25), metal and glass (Chapter 26), plant remains (Chapter 27), matrix and other environmental deposits (Chapter 28), and other materials (Chapter 29). The analysis of food residues from pottery (Chapter 21) and other artifacts (Chapter 22) is discussed separately from non-food residues (Chapter 23).

Sampling considerations, sample introduction, and the analytical techniques employed, together with descriptions of instrument components, are presented in Part IV. Although the material is presented in lay language, the reader should be familiar with the relevant concepts presented in Part I, Basic Science.

This work is different from other books on archaeometry with respect to target audience and scope. It is specifically for archaeologists without a strong background in science rather than archaeological scientists. Those wishing more advanced treatments should consult Pollard and Heron (1996), Ciliberto and Spoto (2000), Pollard et al. (2007), and volumes in *The Advances in Archaeological and Museum Science Series*, edited by Martin J. Aitken, Edward V. Sayre, and R. E. Taylor. Archaeological science topics in this text include dating techniques and the analysis of organic and inorganic materials, which makes this text broader in scope than Henderson (2000) but narrower than Brothwell and Pollard (2001). In contrast with Lambert (1997), case studies only represent one part of this text, Part III. The basic scientific knowledge required for understanding archaeological science (Part I), the rationale and issues surrounding analytical approaches (Part II), and the instrumentation (Part IV) are also emphasized. In addition, the majority of case studies consider the analysis of New World archaeological materials.

My personal experience equipped me to take on the task of writing this book. Before transferring to anthropology, I was enrolled in an undergraduate honours chemistry program. As a senior undergraduate, I used instrumental neutron activation analysis to examine the trace element composition of Precontact period Aboriginal pottery. As a graduate student, I received training in both nuclear magnetic resonance spectroscopy and mineralogical thin section analysis. My doctoral research included the processing and analysis of archaeological and experimental residues and the potential plant and animal food sources using gas chromatography. I continue to analyze archaeological residues extracted from a variety of materials in my laboratory using gas chromatography with mass spectrometry. In addition, I have several years of experience as a field archaeologist in both academic research and cultural resource management settings, during which I obtained age estimates for archaeological materials using radiocarbon and thermoluminescence dating. I first taught a senior undergraduate and graduate level course in archaeological science to archaeology students with social science backgrounds in 1999.

I am incredibly grateful to the many people who encouraged me to complete this project, especially James M. Skibo, Michael B. Schiffer, Charles E. Orser, Teresa Krauss, Katherine Chabalko, and LuAnn Wandsnider. The text benefited from the comments of many people who took the time to review particular parts, chapters, or sections including Joan Buss, Wulf A. Gose, Karen G. Harry, Darren Johnston, Anne Keenleyside, W. J. (Jack) Rink, Henry P. Schwarcz, Orin C. Shanks, and Dongya Y. Yang. I am especially indebted to Krisztina L. Malisza and Timothy Figol who read the entire manuscript and aided in the compilation of the references. Timothy Figol drafted all figures, except Fig. 11.3. I am responsible for any errors that still exist.

Manitoba, Canada   Mary E. Malainey

Contents

Part I
Basic Science

This section covers concepts from the physical and life sciences upon which the analytical techniques are based and data interpretations are made. It is presented so that readers will be better able to more fully understand the rationale and theory behind sample selection, processing, analysis, and data interpretation. Most scientific principles are applicable to more than one analytical technique. Relevant topics and data were selected and condensed from Block et al. (1990), Connelly et al. (2005), Leigh et al. (1998), Lido (2006), Panico et al. (1993), Wieser (2006), and several textbooks in general and organic chemistry, biochemistry, biology, and physics including Halliday and Resnick (1981), Halliday et al. (2005), Kimball (1984, 1994), Lehninger (1970), Lehninger et al. (1993), Mortimer (1986), Nelson and Cox (2008), Smallwood and Alexander (1984), Solomons (1980), and Solomons and Fryhle (2004).

Chapter 1
Elements, Atoms, and Molecules

Matter and the Atom

The general term, matter, can be applied to all material of interest to archaeologists. Most materials are combinations of many pure substances. A pure substance is one that consists of only one type of element or only one type of molecule. Elements are the simplest substances because they consist of only one type of atom. There are a limited number of elements that form the building blocks of all matter in the universe. The properties of elements and atoms are described below.

Some analytical techniques only analyze elements and samples must be purified before analysis. Radiocarbon dating examines only the carbon atoms and the sample is converted to carbon dioxide or graphite prior to analysis. Stable isotope analysis of archaeological materials most often measures elemental carbon, nitrogen, strontium, oxygen, and/or lead. In other cases, specific elements within a mixture are targeted for analysis. For example, atomic absorption measures the abundance of a single element within a sample by detecting energy specific to that element. By using analyzers capable of monitoring emissions of different frequencies simultaneously, techniques such as inductively coupled plasma and instrumental neutron activation analysis can determine the abundance of dozens of elements at the same time.

Compounds are substances composed of more than one element in fixed proportions. The particular combination of elements is called a molecule of that compound. Molecules and chemical bonding are described in more detail below. Organic materials are most often analyzed as molecules rather than elements. Carbon, hydrogen, oxygen, and nitrogen are the major constituents of all organic compounds, so elemental analysis is of limited use. In general, only one class of organic material is targeted for analysis at a time. The targeted material, such as lipids, DNA, amino acids, or proteins, is isolated from other substances using established chemical and/or physical processes. Chromatographic techniques are often used to separate specific compounds, which can then be identified through comparisons with reference material. Separated compounds can then be analyzed using mass spectrometry, which uses the mass of whole molecules (as molecular ions) or their fragments to more precisely identify them.

M.E. Malainey, *A Consumer's Guide to Archaeological Science*, Manuals in Archaeological Method, Theory and Technique, DOI 10.1007/978-1-4419-5704-7_1, © Springer Science+Business Media, LLC 2011

Provenance studies often use the specific combination of pure substances in a mixture to characterize the source of the material. This approach works well in situations where the mixture is homogeneous or uniform throughout. Heterogeneous mixtures vary in composition and consist of parts that are physically distinct.

The Atom

Atoms are the smallest particles that retain the unique characteristics of an element. Molecules are combinations of atoms that maintain the unique characteristics of a compound. All atoms, except hydrogen, consist of negatively charged electrons surrounding a positively charged nucleus of protons and neutrons (Table 1.1). The hydrogen atom is composed of a single electron and a single proton.

Table 1.1 Subatomic particles. (Fundamental Physical Constants by Mohr, Peter J. and Barry N. Taylor (2006), in CRC Handbook of Chemistry and Physics 87th Edition 2006–2007, Editor-in-Chief David R. Lido, pp. 1–1 to 1–6, CRC Press, Taylor & Francis Group, Boca Raton, FL)

Particle	Atomic mass units	Charge
Electron	0.00054857990945	$1-$
Proton	1.00727646680	$1+$
Neutron	1.00866491560	0

Atoms and molecules are electrically neutral because they have equal numbers of protons and electrons. The positive charges balance the negative charges. The number of protons in the nucleus of a particular element does not change. If radioactive decay or another type of nuclear reaction alters the number of protons in a nucleus, the atom is transformed into a different element. On the other hand, electrons are mobile. As part of a chemical bond, they can be donated to, accepted from, or shared with another atom. The removal or addition of electrons is called ionization. This process transforms neutral atoms into positively charged ions called cations or negatively charged ions called anions.

Structure of the Atom

The nucleus of the atom is a dense cluster of protons and neutrons. It is located at the center of the atom and carries the positive charge. Since electrons are small compared to protons and neutrons, the bulk of the atom's mass is centralized. The area occupied by electrons is called the extranuclear space. Electrons move around the nucleus so rapidly, they cannot be precisely located. It is only possible to identify regions where the probability of finding electrons is highest. The rapid movement of these particles forms a negatively charged electron cloud, or charge cloud, around the nucleus. The extranuclear space is very large compared to the size of the subatomic particles, so an atom consists mainly of empty space. If the nucleus were the size of a tennis ball, the atom would have a diameter of over 1.5 km.

The Periodic Table of Elements

The periodic table provides a means of organizing elements by their atomic weights and properties (Fig. 1.1 and Table 1.2). In 1869, Dmitri Mendeleev noticed that as the atomic masses of elements increase certain properties recur. In his table, elements with similar physical and chemical properties were arranged in vertical columns called groups. The horizontal rows are called periods.

In the periodic table, each element is identified by a symbol consisting of one or two letters. The symbol was assigned by international agreement and is usually based on the English or Latin name for the element. The atomic number (Z) is the number of unit positive charges on the nucleus or simply the number of protons. Elements are electrically neutral so the atomic number is also equal to the number of extranuclear electrons. The mass number (A) is the total number of protons and neutrons in the nucleus of the atom. The number of neutrons in an atom can be determined by subtracting the atomic number from the mass number:

$$A - Z = \text{number of neutrons}$$

An atom is designated by the chemical symbol for the element with the atomic number of the element placed at the lower left and the mass number at the upper left:

$^{A}_{Z}$ Symbol $\quad ^{35}_{17}$Cl Chlorine atom with 17 protons and electrons and 18 neutrons

While the mass number of a particular atom is always a whole number, the atomic weight of an element shown in the periodic table is not. The atomic weight of an element is the weighted average of the atomic masses of its natural isotopes. There are two natural isotopes of chlorine: 75.781% have 17 protons and electrons and 18 neutrons; 24.22% have 17 protons and electrons and 20 neutrons. The mass of the lighter isotope in atomic mass units is 34.97 u, while 36.95 u is the mass of the heavier isotope. The atomic weight is the sum of the natural abundances of each natural isotope multiplied by its respective mass. In the case of chlorine, it is about 35.453 u.

Electron Configuration

Information about the configuration of electrons in an element is contained in the periods, or horizontal rows, of the periodic table. The relationship is straightforward for elements with atomic numbers from 1 to 18, which are found in the first three periods. Until recently, a Roman numeral and the letter A were used to indicate the group to which these elements belong. The Roman numeral indicated the number of electrons in the outermost or valence shell. Arabic numbers, applied consecutively across the periodic table from left to right, are now used to designate the groups. Noble gases are placed in group 0; the outermost shell or subshell of these elements is full, giving them a stable configuration of electrons.

metals ←———→ non-metals

1/IA	2/IIA	3/IIIB	4/IVB	5/VB	6/VIB	7/VIIB	8/VIII	9/VIII	10/VIII	11/IB	12/IIB	13/IIIA	14/IVA	15/VA	16/VI	17/VIIA	18/VIIIA
1 **H** 1.0079																	2 **He** 4.0026
3 **Li** 6.941	4 **Be** 9.0122											5 **B** 10.811	6 **C** 12.011	7 **N** 14.007	8 **O** 15.999	9 **F** 18.998	10 **Ne** 20.180
11 **Na** 22.990	12 **Mg** 24.305											13 **Al** 26.982	14 **Si** 28.086	15 **P** 30.974	16 **S** 32.065	17 **Cl** 35.453	18 **Ar** 39.948
19 **K** 39.098	20 **Ca** 40.078	21 **Sc** 44.956	22 **Ti** 47.867	23 **V** 50.942	24 **Cr** 51.996	25 **Mn** 54.938	26 **Fe** 55.845	27 **Co** 58.933	28 **Ni** 58.693	29 **Cu** 63.546	30 **Zn** 65.38	31 **Ga** 69.723	32 **Ge** 72.64	33 **As** 74.922	34 **Se** 78.96	35 **Br** 79.904	36 **Kr** 83.798
37 **Rb** 85.468	38 **Sr** 87.62	39 **Y** 88.906	40 **Zr** 91.224	41 **Nb** 92.906	42 **Mo** 95.96	43 **Tc** (98)	44 **Ru** 101.07	45 **Rh** 102.91	46 **Pd** 106.42	47 **Ag** 107.87	48 **Cd** 112.41	49 **In** 114.82	50 **Sn** 118.71	51 **Sb** 121.76	52 **Te** 127.60	53 **I** 126.90	54 **Xe** 131.29
55 **Cs** 132.91	56 **Ba** 137.33	57–71 *	72 **Hf** 178.49	73 **Ta** 180.95	74 **W** 183.84	75 **Re** 186.21	76 **Os** 190.23	77 **Ir** 192.22	78 **Pt** 195.08	79 **Au** 196.97	80 **Hg** 200.59	81 **Tl** 204.38	82 **Pb** 207.2	83 **Bi** 208.98	84 **Po** (209)	85 **At** (210)	86 **Rn** (222)
87 **Fr** (223)	88 **Ra** (226)	89–103 #	104 **Rf** (261)	105 **Db** (262)	106 **Sg** (266)	107 **Bh** (264)	108 **Hs** (270)	109 **Mt** (268)	110 **Ds** (281)	111 **Rg** (272)	112 **Uub** (285)	113 **Uut** (284)	114 **Uuq** (289)	115 **Uup** (288)	116 **Uuh** (291)		118 **Uuo** (294)

* 57 **La** 138.91 | 58 **Ce** 140.12 | 59 **Pr** 140.91 | 60 **Nd** 144.24 | 61 **Pm** (145) | 62 **Sm** 150.36 | 63 **Eu** 151.96 | 64 **Gd** 157.25 | 65 **Tb** 158.93 | 66 **Dy** 162.50 | 67 **Ho** 164.93 | 68 **Er** 167.26 | 69 **Tm** 168.93 | 70 **Yb** 173.05 | 71 **Lu** 174.97

89 **Ac** (227) | 90 **Th** 232.04 | 91 **Pa** 231.04 | 92 **U** 238.03 | 93 **Np** (237) | 94 **Pu** (244) | 95 **Am** (243) | 96 **Cm** (247) | 97 **Bk** (247) | 98 **Cf** (251) | 99 **Es** (252) | 100 **Fm** (257) | 101 **Md** (258) | 102 **No** (259) | 103 **Lr** (262)

Fig. 1.1 Periodic table of the elements. From Wieser, M. E. (2006) IUPAC Technical Report: Atomic Weights of the Elements 2005 "Pure and Applied Chemistry" 78(11):2051–2066

Table 1.2 Table of atomic weights (Adapted from Wieser (2006))

Element	Symbol	Atomic number	Atomic weight	Element	Symbol	Atomic number	Atomic weight
Actinium	Ac	89	(227.0277)	Europium	Eu	63	151.964
Aluminum	Al	13	26.9815386	Fermium	Fm	100	(257.0951)
Americium	Am	95	(243.0614)	Fluorine	F	9	18.9984032
Antimony	Sb	51	121.760	Francium	Fr	87	(223.0197)
Argon	Ar	18	39.948	Gadolinium	Gd	64	157.25
Arsenic	As	33	74.92160	Gallium	Ga	31	69.723
Astatine	At	85	(209.9871)	Germanium	Ge	32	72.64
Barium	Ba	56	137.327	Gold	Au	79	196.966569
Berkelium	Bk	97	(247.0703)	Hafnium	Hf	72	178.49
Beryllium	Be	4	9.012182	Hassium	Hs	108	(277)
Bismuth	Bi	83	208.98040	Helium	He	2	4.002602
Bohrium	Bh	107	(264.12)	Holmium	Ho	67	164.93032
Boron	B	5	10.811	Hydrogen	H	1	1.00794
Bromine	Br	35	79.904	Indium	In	49	114.818
Cadmium	Cd	48	112.411	Iodine	I	53	126.90447
Calcium	Ca	20	40.078	Iridium	Ir	77	192.217
Californium	Cf	98	(251.07196)	Iron	Fe	26	55.845
Carbon	C	6	12.0107	Krypton	Kr	36	83.798
Cerium	Ce	58	140.116	Lanthanum	La	57	138.90547
Cesium	Cs	55	132.905452	Lawrencium	Lr	103	(262.1097)
Chlorine	Cl	17	35.453	Lead	Pb	82	207.2
Chromium	Cr	24	51.9961	Lithium	Li	3	6.941
Cobalt	Co	27	58.933195	Lutetium	Lu	71	174.967
Copper	Cu	29	63.546	Magnesium	Mg	12	24.3050
Curium	Cm	96	(247.0704)	Manganese	Mn	25	54.938045
Darmstadtium	Ds	110	(271)	Meitnerium	Mt	109	(268.1388)
Dubnium	Db	105	(262.1141)	Mendelevium	Md	101	(258.0984)
Dysprosium	Dy	66	162.500	Mercury	Hg	80	200.59
Einsteinium	Es	99	(252.0830)	Molybdenum	Mo	42	95.94
Erbium	Er	68	167.259	Neodymium	Nd	60	144.242

Table 1.2 (continued)

Element	Symbol	Atomic number	Atomic weight	Element	Symbol	Atomic number	Atomic weight
Neon	Ne	10	20.1797	Selenium	Se	34	78.96
Neptunium	Np	93	237.0482	Silicon	Si	14	28.0855
Nickel	Ni	28	58.6934	Silver	Ag	47	107.8682
Niobium	Nb	41	92.90638	Sodium	Na	11	22.9897693
Nitrogen	N	7	14.0067	Strontium	Sr	38	87.62
Nobelium	No	102	(259.1010)	Sulfur	S	16	32.065
Osmium	Os	76	190.23	Tantalum	Ta	73	180.94788
Oxygen	O	8	15.9994	Technetium	Tc	43	97.9072
Palladium	Pd	46	106.42	Tellurium	Te	52	127.60
Phosphorus	P	15	30.973762	Terbium	Tb	65	158.92535
Platinum	Pt	78	195.084	Thallium	Tl	81	204.3833
Plutonium	Pu	94	(244.0642)	Thorium	Th	90	232.03806
Polonium	Po	84	(208.9824)	Thulium	Tm	69	168.93421
Potassium	K	19	39.0983	Tin	Sn	50	118.710
Praseodymium	Pr	59	140.90765	Titanium	Ti	22	47.867
Promethium	Pm	61	(144.9127)	Tungsten	W	74	183.84
Protactinium	Pa	91	231.03588	Ununbium	Uub	112	(285)
Radium	Ra	88	226.0254	Ununhexium	Uuh	116	(289)
Radon	Rn	86	(222.0176)	Ununoctium	Uuo	118	(294)
Rhenium	Re	75	186.207	Ununpentium	Uup	115	(288)
Rhodium	Rh	45	102.90550	Ununquadium	Uuq	114	(289)
Roentgenium	Rg	111	(272.1535)	Uranium	U	92	238.0289
Rubidium	Rb	37	85.4678	Vanadium	V	23	50.9415
Ruthenium	Ru	44	101.07	Xenon	Xe	54	131.293
Rutherfordium	Rf	104	(261.1088)	Ytterbium	Yb	70	173.04
Samarium	Sm	62	150.36	Yttrium	Y	39	88.90585
Scandium	Sc	21	44.955912	Zinc	Zn	30	65.409
Seaborgium	Sg	106	(266.1219)	Zirconium	Zr	40	91.224

The Bohr Model of the Hydrogen Atom

In 1913, Niels Bohr proposed a model for the configuration of electrons of an atom that successfully explained the experimental observations of the hydrogen atom. In this case, hydrogen gas was heated to a high temperature and the light it emitted passed through a prism. The prism separated the light according to its constituent wavelengths. The Bohr model explained the positions of the lines of the hydrogen spectrum (Fig. 1.2).

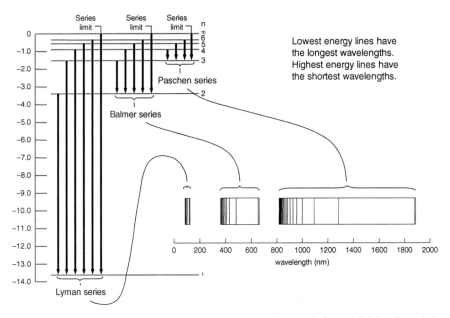

Fig. 1.2 Lines of the hydrogen spectra ("Fundamentals of Physics', Second Edition Extended, David Halliday and Robert Resnick, copyright 1981, Reproduced with permission of John Wiley & Sons, Inc)

Bohr suggested that, prior to heating, the distance at which the negatively charged electron orbited the positively charged nucleus was as small as possible. An electron located in this area was in the K level or $n = 1$ shell. This represents the lowest possible energy condition and ground state configuration of the hydrogen atom. When the gas was heated, electrons absorbed energy and moved into spherical orbits with large diameters, thus were farther from the nucleus. Atoms in this high-energy condition are referred to as being in an excited state. In order to return to a lower energy state, the atom must release the excess energy. Electrons rid themselves of the excess energy by emitting a photon of light and then dropping to a lower orbit.

Bohr suggested that spectral lines of hydrogen with the shortest wavelengths were due to emissions from electrons dropping from outer orbits to the innermost orbit possible, i.e., the K level or $n = 1$ shell. These spectral lines, called the Lyman series, arose from electrons that lost the greatest amount of energy and returned to the ground state condition. The next grouping of spectral lines, called the Balmer series, was due to electrons dropping down to the next orbit, designated the L level or $n = 2$ shell, situated farther way from the nucleus. Less energy was emitted so the wavelengths of Balmer series lines were longer than those of the Lyman series. The third grouping of spectral lines, called the Paschen series, resulted from electrons falling from outer orbits to the M level or $n = 3$ shell, situated even farther from the nucleus. Wavelengths of Paschen series lines are even longer because electrons dropping to the M level emit less energy.

Bohr believed that each orbit occurred at a fixed distance from the nucleus and calculated the energy of an electron on the basis of the orbit it occupied. He also calculated the distance from the nucleus of each orbit. In the hydrogen gas experiment, heating caused many electrons to jump to higher orbits while others simultaneously fell to lower ones. The positions of individual spectral lines of the emitted light corresponded to the difference in energy between discrete orbits.

Heisenberg's Uncertainty Principle

While Bohr's model explained the spectra of hydrogen, it could not account for the line spectra of other elements. Other researchers found that only a few electrons were actually distributed in spherical shells and, furthermore, they were impossible to precisely locate. According to Heisenberg's uncertainty principle, it is not possible to simultaneously know the location and momentum of an electron-sized particle. This is because the methods used to determine the location of such a tiny particle will produce a change in its speed and direction.

Because of its rapid movement around the nucleus, the charge of the electron is spread out, forming a charge cloud. The probability of finding an electron is highest where the density of the charge cloud is greatest. For an electron at ground state, the charge cloud is densest near the nucleus. It is not possible to say that an electron will occur at a distance from the nucleus corresponding to a particular shell, but this is the region where the electron is most likely to be found. For hydrogen, the probability of finding a ground state electron is highest when the distance from the nucleus equals the value Bohr calculated for the $n = 1$ shell.

Electron Configurations of Other Elements

The electron configuration of each element is unique. Many analytical techniques make use of these differences to determine the elemental composition of a sample. The ground state distribution of electrons in other elements can be quite complex. The maximum number of electrons in a given shell increases with its distance from the nucleus and is equal to $2n^2$. The $n = 1$ shell can contain no more than two electrons ($2 \times 1 \times 1 = 2$), $n = 2$ can contain eight ($2 \times 2 \times 2 = 8$), $n = 3$ can contain $18 (2 \times 3 \times 3 = 18)$, $n = 4$ can contain $32 (2 \times 4 \times 4 = 32)$, and so on. The element, Lawrencium, has 103 electrons distributed across seven shells at ground state. Rather than orbits of fixed distances from the nucleus, it is only possible to identify areas around the nucleus where an electron has a high probability (90%) of occurring.

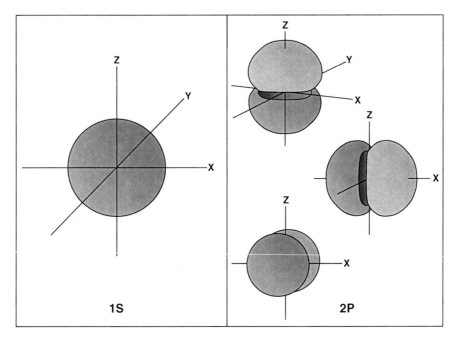

Fig. 1.3 *s* and *p* orbitals

Areas where electrons have high probability of occurring are called orbitals and can be spherical or non-spherical (Fig. 1.3). Their level number and their shape are used to identify different orbitals; for example, the spherical orbital of the fourth shell is 4*s*. Every shell has a spherical *s* orbital that contains a maximum of two electrons. All, except the first shell, have *p* orbitals in which up to six electrons occur in three paired dumbbell-shaped lobes which are of equal distances from the nucleus along the *x*-, *y*-, and *z*-axes. All, except the first two shells, have *d* orbitals, in which up to 10 electrons occur in five paired lobes along and between the *x*-, *y*-, and *z*-axes. All except the first three have *f* orbitals where a maximum of 14 electrons occur in 7 pairs; all except the first four have *g* orbitals where up to 18 electrons can occur in 9 pairs.

All electrons in a given subshell have the same energy. In the absence of a magnetic field, they appear identical. If one is applied, however, differences arise due to the slightly different orientations of the orbitals with respect to the magnetic field. In spectral studies, a single emission line can be split into many lines when a magnetic field is applied. This is referred to as the Zeeman effect.

The order in which orbitals are filled depends upon their relative energies: lowest energy orbitals are filled first. This is known as the Aufbau principle. Although the third shell can contain a total of 18 electrons, after the 3*s* and 3*p* orbitals are completed, two electrons move into 4*s* orbitals before the 3*d* orbitals are filled. In fact, the electron configuration of argon (Ar), with orbitals 1*s*, 2*s*, 2*d*, 3*s,* and 3*p*

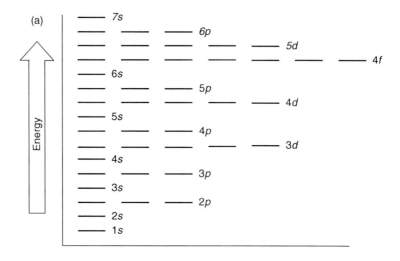

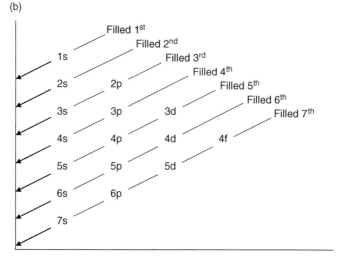

Fig. 1.4 Relative energy and filling order of atomic orbitals ((**a**) Reprinted from T.W. Graham Solomons, Figure 1.22, "Organic Chemistry", Second Edition, copyright 1980, with permission of John Wiley & Sons, Inc. (**b**) K. Malisza)

filled, is so stable that the element is in group 0. The order in which orbitals are filled by electrons is shown in Fig. 1.4.

The distribution of electrons is governed by Hund's rule of maximum multiplicity. Two electrons do not occupy an orbital until other orbitals of equal energy are each occupied by one electron. Each orbital of a subshell has the same energy and so each receives one electron before electron pairs are formed. A spinning electron creates a magnetic field; it acts like a little bar magnet with a north and south pole. The spins of unpaired electrons in the orbitals of a subshell are parallel. It can only be paired with an electron with the opposite spin, so that the magnetic moments cancel. This is outlined in the Pauli exclusion principle.

Unpaired electrons in atoms, ions, or molecules affect how materials react to a magnetic field. Paramagnetic substances contain unpaired electrons and are drawn into magnetic fields. All electrons in diamagnetic substances are paired; they are weakly repelled by a magnetic field.

The Division Between Metals and Non-metals

Most elements in the periodic table are metals; only those on the rightmost side are non-metals and a step-like division exists between them. Boron, silicon, arsenic, tellurium, astatine and all elements to the right of them in their respective periods are non-metals. Non-metals are poor conductors, not lustrous, and are brittle as solids. Metals are good conductors of heat and electricity and have a characteristic luster. Metals are malleable, which means they can be pounded flat, and ductile, so that they can be drawn out into a wire.

Chemical Bonding

Atoms of different elements combine in different ways to produce compounds. Chemical bonds form between atoms and the distribution of electrons between them changes. Noble gases in group 0 do not readily react with other elements. Very reactive elements include alkali metals, those with only one valence electron (group 1, formerly IA), and halogens, those with one electron short of a stable configuration (group 17, formerly VIIA). The three fundamental types of bonding are ionic, covalent, and metallic.

Ionic bonding occurs between a metal and a non-metal. In general, metals have relatively few valence electrons while the outer shells of non-metals are almost complete. In ionic bonds, electrons are transferred from the metal to the non-metal. The atoms of the element that loses electrons become positively charged ions. The atoms of the other element gain electrons and become negatively charged ions. Electrostatic attraction between the oppositely charged ions holds them in a crystal. Positively charged ions are known as cations; negatively charged ions are called anions.

Ionic bonding enables non-metals to gain enough valence electrons to attain a stable electron configuration. Metals lose their valence electrons, which eliminates the incomplete shell. A valence bond, or Lewis (dot), structure consists of the symbol of an element with its valence electrons represented by dots. These are useful for illustrating changes in the distribution of electrons that occur during chemical bonding.

Sodium (Na), a group 1 element, has one valence electron and chlorine (Cl), which is in group 17, requires one electron to complete a shell (Fig. 1.5). An ionic

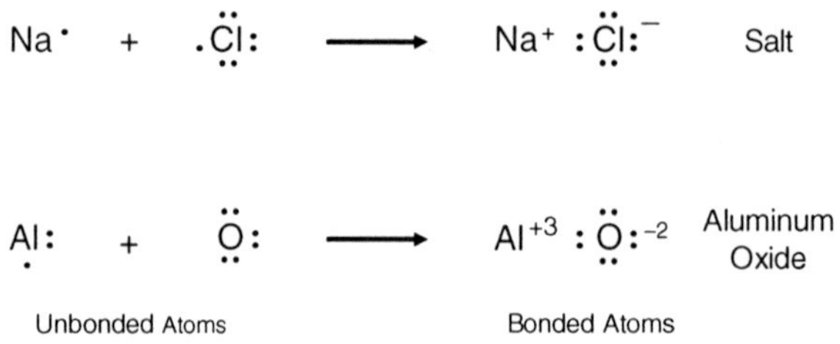

$$Na \cdot \quad + \quad \cdot \ddot{\underset{..}{Cl}}: \quad \longrightarrow \quad Na^+ : \ddot{\underset{..}{Cl}}:^- \quad \text{Salt}$$

$$Al: \quad + \quad \ddot{\underset{.}{O}}: \quad \longrightarrow \quad Al^{+3} : \ddot{\underset{..}{O}}:^{-2} \quad \begin{array}{l}\text{Aluminum}\\\text{Oxide}\end{array}$$

Unbonded Atoms   Bonded Atoms

bond is formed when sodium transfers its valence electron to chlorine and the Na^+ cation is attracted to Cl^- anion. Sodium chloride, NaCl, which is table salt, does not occur as a single molecule but forms a crystal in which the ratio of sodium to chlorine atoms is 1:1.

Aluminum is a group 3 (formerly IIIA) element that forms an ionic bond with the group 16 (formerly VIA) element oxygen (Fig. 1.5). Each aluminum atom loses its three valence electrons and each oxygen atom gains two. The aluminum oxide, Al_2O_3, crystal has a ratio of two aluminum to three oxygen atoms and is electrically neutral.

Occasionally, an atom that is chemically similar may be used in place of another in the crystal. Strontium and calcium are both group 2 (formerly group IIA) elements; strontium is only slightly larger than calcium. During the formation of gypsum, plagioclase feldspar, and carbonate minerals, strontium will be substituted for calcium at times. Likewise, rubidium can be used in place of potassium in alkali feldspars and micas; both are group 1 elements.

In covalent bonds, atoms share electrons to mutually attain stable electron configurations. This type of bond occurs between non-metals; the element hydrogen, which is technically a metal, forms covalent bonds with non-metals and itself as H_2. A single covalent bond consists of a pair of electrons shared by two atoms. The strength of the bond comes from the attraction of the positively charged nuclei to the negatively charged electron cloud of the bond. The covalent bonds between the atoms of water, ammonia, and methane are depicted in Fig. 1.6.

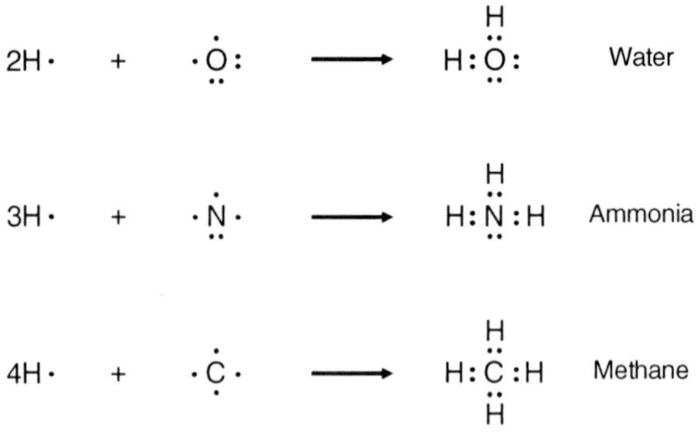

$$2H \cdot \quad + \quad \cdot \ddot{\underset{..}{O}}: \quad \longrightarrow \quad \overset{\textstyle H}{H : \ddot{\underset{..}{O}}:} \quad \text{Water}$$

$$3H \cdot \quad + \quad \cdot \dot{\underset{..}{N}} \cdot \quad \longrightarrow \quad \overset{\textstyle H}{H : \ddot{\underset{..}{N}} : H} \quad \text{Ammonia}$$

$$4H \cdot \quad + \quad \cdot \dot{\underset{.}{C}} \cdot \quad \longrightarrow \quad \begin{array}{c} H \\ H : \ddot{C} : H \\ H \end{array} \quad \text{Methane}$$

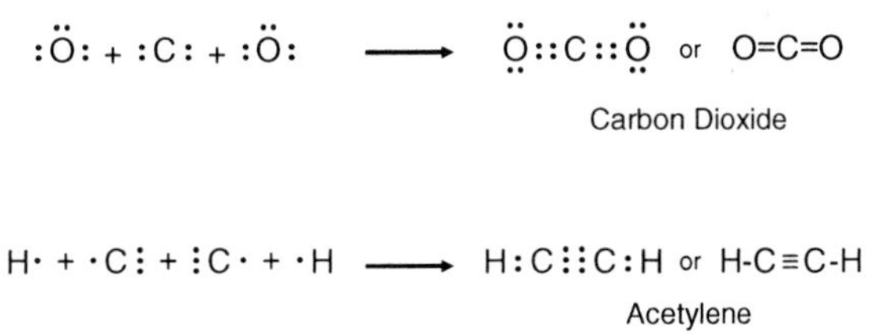

Fig. 1.7 Lewis dot structures showing covalent bond formation of carbon dioxide and acetylene

$$:\overset{..}{O}: + :C: + :\overset{..}{O}: \longrightarrow \overset{..}{O}::C::\overset{..}{O} \text{ or } O=C=O$$

Carbon Dioxide

$$H\cdot + \cdot C: + :C\cdot + \cdot H \longrightarrow H:C::C:H \text{ or } H\text{-}C\equiv C\text{-}H$$

Acetylene

Double covalent bonds, where two pairs of electrons are shared, and triple covalent bonds, where three pairs of electrons are shared, can also form (Fig. 1.7). Double and triple bonds can be very unstable. Acetylene, which contains a triple bond, is used in welding because it produces a very hot flame when it reacts with oxygen. Unsaturated fatty acids contain double bonds between adjacent carbon atoms and multiple double bonds occur in polyunsaturated fatty acids.

Transition Between Ionic and Covalent Bonding

The bonds that form between atoms in most compounds are intermediate between pure ionic and pure covalent. Ionic bonds have characteristics of a covalent bond when the positively charged ion (cation) attracts and deforms the electron cloud of the anion. This is most pronounced when a large anion, with outer electrons far from the nucleus, bonds with a small cation with a high positive charge.

Pure covalent bonds only occur in molecules formed from identical species, e.g., H_2, O_2, Cl_2. When atoms of different elements are joined by a covalent bond, the electrons are not shared equally. The density of the electron cloud of the bond will be greater around elements with the greater attraction for electrons, or electronegativity (Fig. 1.8). The lack of symmetry of the electron cloud around the two nuclei causes the region of the bond with the highest electron density to have a partial negative charge ($\delta-$). The other end of the bond develops a partial positive charge ($\delta+$). This type of bond is called a polar covalent bond. While the bond has positive and negative poles (or a dipole), the molecule is neutral.

The types of bonds between individual atoms in the molecule affect the solubility of a substance in water. Water is polar because the attraction of electrons by oxygen is much greater than that by hydrogen; other polar substances readily dissolve in water. The difference in electronegativities between individual atoms of nonpolar compounds is not large. Non-polar substances are water insoluble but generally dissolve well in solvents with low polarity, such as chloroform, benzene, and carbon tetrachloride.

metals ← | → non-metals

1/IA	2/IIA	3/IIIB	4/IVB	5/VB	6/VIB	7/VIIB	8/VIII	9/VIII	10/VIII	11/IB	12/IIB	13/IIIA	14/IVA	15/VA	16/VIA	17/VIIA	18/VIIIA
1 **H** 2.20																	2 **He**
3 **Li** 0.98	4 **Be** 1.57											5 **B** 2.04	6 **C** 2.55	7 **N** 3.04	8 **O** 3.44	9 **F** 3.98	10 **Ne**
11 **Na** 0.93	12 **Mg** 1.31											13 **Al** 1.61	14 **Si** 1.90	15 **P** 2.19	16 **S** 2.58	17 **Cl** 3.16	18 **Ar**
19 **K** 0.82	20 **Ca** 1.00	21 **Sc** 1.36	22 **Ti** 1.54	23 **V** 1.63	24 **Cr** 1.66	25 **Mn** 1.55	26 **Fe** 1.83	27 **Co** 1.88	28 **Ni** 1.91	29 **Cu** 1.90	30 **Zn** 1.65	31 **Ga** 1.81	32 **Ge** 2.01	33 **As** 2.18	34 **Se** 2.55	35 **Br** 2.96	36 **Kr**
37 **Rb** 0.82	38 **Sr** 0.95	39 **Y** 1.22	40 **Zr** 1.33	41 **Nb** 1.6	42 **Mo** 2.16	43 **Tc** 2.10	44 **Ru** 2.2	45 **Rh** 2.28	46 **Pd** 2.20	47 **Ag** 1.93	48 **Cd** 1.69	49 **In** 1.78	50 **Sn** 1.96	51 **Sb** 2.05	52 **Te** 2.1	53 **I** 2.66	54 **Xe** 2.60
55 **Cs** 0.79	56 **Ba** 0.89	57 **La** 1.10	72 **Hf** 1.3	73 **Ta** 1.5	74 **W** 1.7	75 **Re** 1.9	76 **Os** 2.2	77 **Ir** 2.2	78 **Pt** 2.2	79 **Au** 2.4	80 **Hg** 1.9	81 **Tl** 1.8	82 **Pb** 1.8	83 **Bi** 1.9	84 **Po** 2.0	85 **At** 2.2	86 **Rn**
87 **Fr** 0.7	88 **Ra** 0.9	89 **Ac** 1.1															

Lanthanides	58 **Ce** 1.12	59 **Pr** 1.13	60 **Nd** 1.14	61 **Pm**	62 **Sm** 1.17	63 **Eu**	64 **Gd** 1.20	65 **Tb**	66 **Dy** 1.22	67 **Ho** 1.23	68 **Er** 1.24	69 **Tm** 1.25	70 **Yb**	71 **Lu** 1.0
Actinides	90 **Th** 1.3	91 **Pa** 1.5	92 **U** 1.7	93 **Np** 1.3	94 **Pu** 1.3	95 **Am** (1.3)	96 **Cm** (1.3)	97 **Bk** (1.3)	98 **Cf** (1.3)	99 **Es** (1.3)	100 **Fm** (1.3)	101 **Md** (1.3)	102 **No** (1.3)	103 **Lr**

Fig. 1.8 Known electronegativities of elements (Data from Lido (2006, pp. 9–83))

Nomenclature

Rules have been established for naming compounds, that is, substances that contain two or more different elements. A binary compound consists of two elements. If the bond between them is covalent, the less electronegative element appears first in the name (Fig. 1.8). If the bond is ionic, the name of the metal appears first. For all binary compounds, the ending *-ide* is substituted for the usual ending of the element appearing second. The names of a few ternary compounds, consisting of three elements, also end in *-ide*, including cyanides (NaCN – sodium cyanide) and hydroxides (NaOH – sodium hydroxide).

When hydrogen-containing compounds dissociate in water, the ending *-ic acid* is substituted for the usual ending of the element that appears last in the name:

hydrogen chloride (HCl) in water becomes hydrochloric acid;
hydrogen fluoride (HF) in water becomes hydrofluoric acid;
hydrogen sulfide (H_2S) in water becomes hydrosulfuric acid;
hydrogen sulfate (H_2SO_4) in water becomes sulfuric acid.

The combination of certain elements and oxygen can result in the formation of anions. These anions appear in a wide variety of substances, including those categorized as carbonates, silicates, phosphates, and sulfates (Table 1.3). These compounds are metal salts with the metal cation forming the alkali and the anion derived from an acid. For example, the carbonate ion (CO_3^{2-}) is the anion of carbonic acid (H_2CO_3).

Table 1.3 Examples of complex anions containing oxygen. "Data from Leigh et al. 1998, Table 4.1, "Principles of Chemical Nomenclature: A Guide to IUPAC Recommendations", with permission of Wiley-Blackwell"

Formula	Name	Formula	Name
$C_2H_3O_2^-$	Acetate	OH^-	Hydroxide
AsO_3^{3-}	Arsenite	NO_3^-	Nitrate
CO_3^{2-}	Carbonate	NO_2^-	Nitrite
ClO_3^-	Chlorate	ClO_4^-	Perchlorate
ClO_2^-	Chlorite	MnO_4^-	Permanganate
CrO_4^{2-}	Chromate	PO_4^{3-}	Phosphate
CN^-	Cyanide	SiO_4^{4-}	Silicate
$Cr_2O_7^{2-}$	Dichromate	SO_4^{2-}	Sulfate
ClO^-	Hypochlorite	SO_3^{2-}	Sulfite

Hydrogen Bonds

The intermolecular attractions of certain hydrogen-containing compounds are unusually strong and called hydrogen bonds. They occur when hydrogen is covalently bonded to elements that strongly attract electrons and are small in size. Oxygen, nitrogen, fluorine, and chlorine are the most highly electronegative elements; however, chlorine only forms weak hydrogen bonds because of its larger size. Small, highly electronegative elements strongly attract the bonding electrons and leave hydrogen atoms with a significant $\delta+$. This causes the hydrogen atom to act like a bare proton. The hydrogen atom of one molecule and a pair of unshared electrons of the electronegative atom of another molecule are mutually attracted.

Effective hydrogen bonds occur in compounds such as NH_3, H_2O, and HF. Hydrogen bonds cause the boiling points of these compounds to be higher than expected because additional energy is required to separate molecules in liquid state. The melting points and viscosities of these compounds are also high.

Hydrogen bonding is strongest in water because both pairs of electrons of the oxygen atom attract hydrogen atoms of other molecules. Hydrogen bonding is responsible for the unusual properties of water. Covalent and hydrogen bonds cause water to be organized in a tetrahedral structure. These bonds become rigid when water is in solid phase (ice), which causes the density of ice to be lower than that of liquid water. Hydrogen bonding is also the reason why water acts as such an effective solvent, especially for certain compounds containing O, N, and F.

Hydrogen bonding is very important in determining the structure and properties of molecules in living systems. Hydrogen bonding is responsible for connecting the two strands of the DNA molecule and causes it to twist into a double helix. Fibrous proteins, such as those found in hair and muscle, are organized in bundles held together by hydrogen bonds.

Chemical Equations

By definition, a mole (mol) is the amount of a substance that contains the same number of chemical units as the number of atoms in exactly 12 grams (g) of ^{12}C. This number, called Avogadro's number, is 6.022×10^{23}. If you have a mole of any element, it weighs its atomic mass in grams. Returning to the earlier example, 1 mol of ^{17}Cl weighs 34.97 g and 1 mol of ^{18}Cl weighs 36.95 g but 1 mol of Cl weighs 35.45 g, the number in the periodic table that takes into account the natural abundances of each isotope.

When considering compounds, 1 mol of substance consists of 6.022×10^{23} molecules. A mole of any compound has a mass equal to its molecular weight in grams. The molecular weight of a compound is the sum of the atomic weights of its constituent atoms. The molecular weight of water (H_2O) is 18 [(2 H × 1) + (1 O × 16)]; so 1 mol of water has a mass of 18 g. This approach is used to determine the molecular weight of compounds that form both covalent and ionic bonds, even though ionic substances occur as crystalline structures rather than molecules.

The composition of substances that form molecules is expressed in a molecular formula. A molecule of butane contains four atoms of carbon and 10 atoms of hydrogen and is expressed as C_4H_{10}. One mole of butane contains 48 (4 × 12) g of carbon and 10 (10 × 1) g of hydrogen. In terms of percentage composition, butane is about 17% hydrogen and 83% carbon.

The chemical formula for ionic substances is the simplest whole number ratio of atoms in a compound and called its empirical formula. While it is necessary to express the formula of ionic substances in this manner, it can also be used for covalent compounds. For example, the empirical formula of butane is C_2H_5; there are 2.5 times as many hydrogen atoms as carbon atoms in butane.

Cellular Energy

Cells require energy to function and this energy usually comes from the breakage of chemical bonds. The amount of energy released during this process is equal to the energy required to break the bond. The first and second laws of thermodynamics are followed. First, the energy is conserved; it is neither created nor destroyed but can be transformed. Second, the amount of disorder or entropy in the universe increases. Plants and microorganisms capable of photosynthesis acquire energy from the sun; other organisms acquire energy from the nutrients in their food.

Reactions occur in living organisms at relatively constant temperature and pressure, which makes them quite easy to predict and understand. Certain reactions are spontaneous while others require the addition of energy from the environment. The sum of chemical reactions that occur within the cells of an organism is called its metabolism. Plants and animals are capable of storing energy for future use in highly energetic chemical bonds. An important energy storage process is phosphorylation, which involves the linking together of phosphate groups. Adenosine monophosphate (AMP) is a ribonucleic acid (RNA) nucleotide consisting of the purine base, adenine, bonded to the 5-carbon sugar, ribose, and an inorganic phosphate (P_i) group (Fig. 1.9). Addition of a second P_i group produces adenosine diphosphate (ADP); addition of a third P_i group produces adenosine triphosphate (ATP). Bonds between AMP and each additional inorganic phosphate group are highly energetic because they must overcome the electrostatic repulsion between negatively charged oxygen atoms. When energy is required for a chemical reaction to proceed, ATP can be converted to ADP and one P_i or AMP and two P_i.

Enzymes may serve as catalysts to start and control certain metabolic reactions that occur in cells. Enzymes facilitate reactions and may form temporary bonds with molecules to ensure a reaction proceeds, but are not changed by the process.

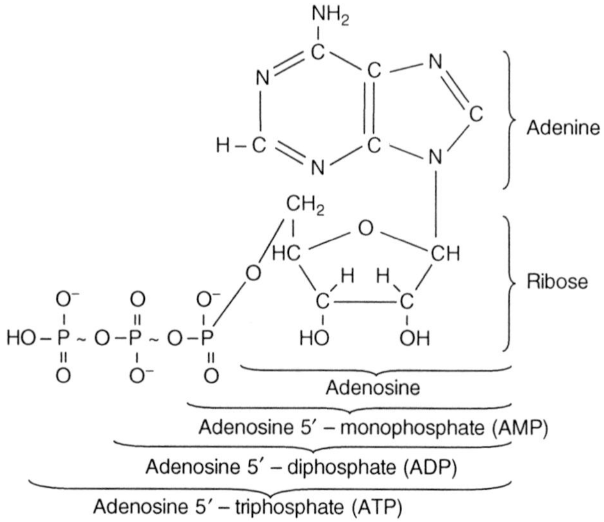

NH$_2$
|
N=C—C—N
| | \\
H—C=N—C—N   C
| |
Adenine

CH$_2$
|
HC   O
\\   CH
O
H H
C———C
| |
HO—P~O—P~O—P   HO   OH
O   O   O⁻
HO   OH

Adenosine

Adenosine 5′ – monophosphate (AMP)

Adenosine 5′ – diphosphate (ADP)

Adenosine 5′ – triphosphate (ATP)

Fig. 1.9 Structural formulas showing the relationship between adenosine and its 5′ phosphates

Electrical Conductivity

Elements of the periodic table (Fig. 1.1) are divided into two large groups, metals and non-metals. As noted earlier, electrical conductivity is a property of metals; one explanation is that their outermost electrons are not tightly bound to the atom. These "free electrons" are able to move through a solid carrying electrical charges; this is the "free electron model" of conductivity.

The band-gap pattern of atoms provides an explanation for the behavior of conductors, insulators, and semiconductors. According to the band theory model, the outermost electrons of an atom occupy either a valence band or a conduction band and an energy gap exists between the two (Fig. 1.10). Only electrons in conduction bands are available to carry electrical charges; those in valence bands are not. All of the outermost electrons of insulators occur in the valence band and the energy gap between it and the conduction band is very large. External energy, in the form of heat or an applied electrical field, is insufficient to move electrons from the valence to the conduction band. The conduction band remains empty, so the substance cannot conduct electricity. The configuration of conductors is different in that the outermost electrons are in a valence band that also serves as a conduction band and no energy gap exists between them.

The semiconductors share properties with both insulators and conductors. At ground state, a semiconductor resembles an insulator in that electrons are only located in the valence band, the conduction band is empty and an energy gap exists between them. The difference is that the energy gap between the conduction and outermost valence band in a semiconductor is relatively small. At ground state, a semiconductor cannot conduct electricity but only a small amount of energy is

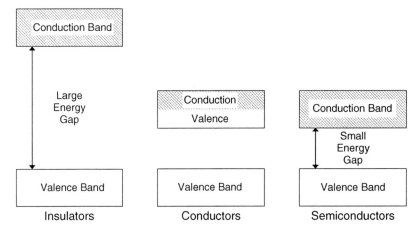

Fig. 1.10 The band-gap patterns of insulators, conductors, and semiconductors ("Fundamentals of Physics", Second Edition Extended, David Halliday and Robert Resnick, copyright 1981, Reproduced with permission of John Wiley & Sons, Inc.)

required to move electrons into the conduction band transforming it into a conductor. When an excited electron moves into the conduction band, a "hole" that behaves like a positive charge is created in the valence band.

Semiconductors are of archaeological interest because their properties enabled the development of trapped charge dating techniques. At ground state, crystalline materials, such as quartz, feldspar, chert, and tooth enamel, are semiconductors with unoccupied conduction bands. Bursts of energy from the external environment cause valence electrons to temporarily move to the conduction band; after the excess energy is diffused, they return to the valence band. Occasionally, electrons dropping back to the valence band become trapped in flaws in the crystal structure of a mineral and remain at an intermediate energy level. Semiconductors that formed about the time of human occupation, such as tooth enamel, or were returned to ground state as a result of human activity, such as minerals in fired pottery, fire-cracked rock, or heat-treated lithics, gradually accumulate trapped electrons over time. The number of trapped electrons depends on the energy available in the natural environment from cosmic radiation and the decay of radioactive isotopes. By measuring the number of trapped electrons and estimating the annual radiation dose to which the semiconductor was exposed, it is possible to calculate the time that has passed since the material was last at ground state.

Chapter 2
Electromagnetic Radiation

Bohr demonstrated that information about the structure of hydrogen could be gained by observing the interaction between thermal energy (heat) and the atom. Many analytical techniques used today follow the same principle. Atoms or molecules can absorb energy, become excited, and then emit the excess energy in order to return to their original ground state configuration. High-energy impacts with charged particles and induced nuclear reactions are also used to excite sample atoms, but the most common approach is to employ electromagnetic radiation. The electromagnetic radiation absorbed or emitted provides information about elemental composition, molecular configuration, or other characteristics about the sample.

Wavelength, Frequency, and Energy

Electromagnetic radiation (EM) is described in terms of its wavelength, frequency, or energy. All electromagnetic energy travels at the speed of light, c, which is 2.998×10^8 m/s, so wavelength (λ) and frequency (v) are inversely related: $c = \lambda v$. Long waves have a low frequency and short waves have a high frequency (Fig. 2.1). The wavelength and frequency also indicate the energy of the wave. The relationship between wavelength and energy, E, is described by the equation, $E = hc/\lambda$, where h is Planck's constant ($h = 6.625 \times 10^{-34}$ Joule-seconds or J s) and c is the speed of light. By replacing the constants h and c with their respective values, we see that $E = 1.986 \times 10^{-25}$ Joule-meters or J m/λ. An inverse relationship exists; electromagnetic radiation with shorter wavelengths is more energetic. The relationship between energy and frequency is given by the equation, $E = hv$, where h is Planck's constant. A direct relationship exists; electromagnetic radiation with a higher frequency is more energetic.

The Electromagnetic Spectrum

The electromagnetic spectrum has been divided into many sections; familiar names are assigned to the different ranges (Fig. 2.2). Gamma rays and X-rays are types of

M.E. Malainey, *A Consumer's Guide to Archaeological Science*, Manuals in
Archaeological Method, Theory and Technique, DOI 10.1007/978-1-4419-5704-7_2,
© Springer Science+Business Media, LLC 2011

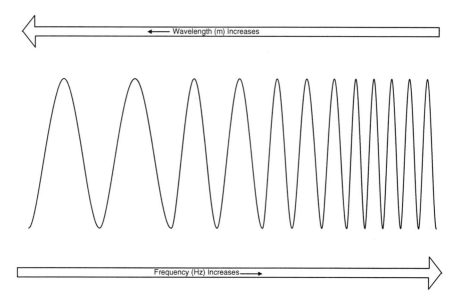

Fig. 2.1 The relationship between wavelength and frequency (Original Figure by M.E. Malainey and T. Figol)

EM radiation with the shortest wavelength, highest frequencies, and, consequently, the highest energy. At the other end of the spectrum are the low energy waves. In the order of increasing energy, these include radio waves, television waves, microwaves, and radar waves. The energy we know as light is the middle of the spectrum. The lowest energy light is infrared, followed by visible and ultraviolet. The human eye is capable of discriminating between EM waves of different energy in the visible range between about 400 and 700 nm in wavelength. The spectrum of visible light appears as the different colors: red, orange, yellow, green, blue, and violet. Within the visible range, red light has the lowest energy and violet the highest.

Energy Absorption and Emission by Atoms and Molecules

As outlined in Chapter 1, electrons are found at various discrete energy levels within an atom. Electrons are very small particles and the electronic energy levels are widely spaced. Electrons occupying the energy levels closest to the nucleus have the lowest energy; as one moves to the outer energy levels, electrons have progressively higher amounts of energy. Certain types of EM radiation are absorbed by the atoms or molecules in a sample and move them to a higher energy or excited state. Atoms or molecules in an excited state emit excess energy in order to return to ground state. The excess energy can be lost by the production of heat or the emission of radiation. Spectroscopic techniques involve monitoring the energy absorbed, emitted, or the behavior of excited molecules.

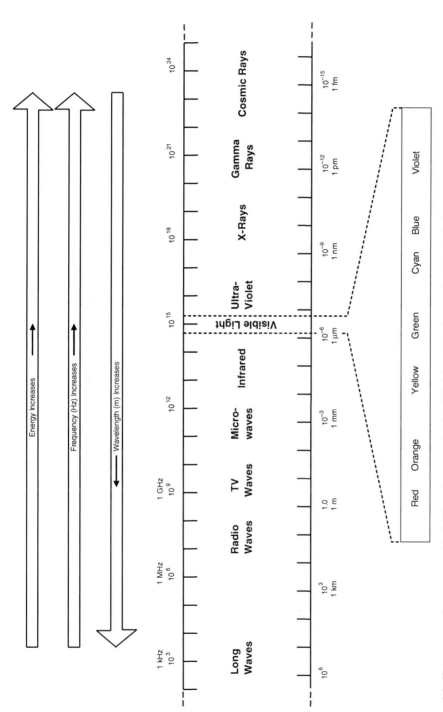

Fig. 2.2 Electromagnetic spectrum with visible region expanded (Modified from "Fundamentals of Physics", Second Edition Extended, David Halliday and Robert Resnick, copyright 1981, Reproduced with permission of John Wiley & Sons, Inc.)

Several of the analytical techniques described in Part IV either manipulate a sample with EM energy or expose it to conditions that cause its constituent atoms to emit EM energy. With spectroscopic techniques, electromagnetic radiation, $h\nu$, is directed at atoms or molecules. Under certain conditions, some radiation is absorbed which is called the photoelectric effect. The effect of this absorption varies with the types of atoms or molecules present, the type of radiation, and the environment in which it occurs. With nuclear magnetic resonance (NMR) spectroscopy, the absorption of radio frequency energy in a strong magnetic field causes the nuclei of certain sample atoms to flip. Electron spin resonance (ESR) dating involves the application of microwave energy to a crystalline sample in a strong magnetic field, which causes trapped electrons to resonate. Several spectroscopic techniques utilize EM radiation in the infrared (IR), visible (VIS), or ultraviolet (UV) range to characterize a sample. Molecular bonds absorb light in the UV and visible range; IR radiation causes molecular bonds to vibrate.

Inductively coupled plasma-atomic emission spectroscopy (ICP-AES), optical emission spectroscopy (OES), and atomic absorption spectroscopy (AA) all involve heating the sample and monitoring characteristic light emitted or absorbed by excited atoms. Instrumental neutron activation analysis (INAA or NAA) determines the elemental composition of a sample on the basis of the characteristic gamma rays emitted by artificially created radioisotopes.

X-radiography uses the differential absorption of high-energy X-rays by elements to image the interior of an object. X-ray diffraction (XRD) uses the characteristic interference pattern of X-rays produced by its crystal structure to identify specific minerals. In X-ray fluorescence (XRF), whether it be energy-dispersive (ED) or wavelength-dispersive (WD) spectrometry, high energy X-rays excite sample electrons which then return to ground state by fluorescing X-rays characteristic of that element. Characteristic X-rays emitted by elements following high-energy impacts with electrons can be monitored with either scanning electron microscopy with ED XRF or electron microprobe analysis. High-energy impacts with protons or α-particles also cause the emission of X-rays and gamma rays characteristic of elements in a sample when using proton/particle-induced X-ray emission (PIXE) or proton/particle-induced gamma ray emission (PIGE).

Chapter 3
Radioactive Isotopes and Their Decay

Each element has a specific number of protons but the number of neutrons can vary to form different isotopes. Multiple isotopes exist for all elements, except sodium, beryllium, and fluorine. The balance between the number of protons and neutrons in its nuclei determines the stability of an isotope (Fig. 3.1). When the ratio of protons to neutrons is appropriate for the size of its nucleus, the isotope is stable. If the ratio of protons to neutrons is either too low or too high, the isotope will spontaneously change into a different atom with a more stable configuration. Unstable isotopes are radioactive; the process by which they achieve more stable configurations is called radioactive decay.

For lighter elements, with atomic numbers of 40 or less, the number of neutrons in stable isotopes is approximately equal to the number of protons. For example, the ratio of protons to neutrons in stable isotopes of carbon, ^{12}C and ^{13}C, is 1:1 and 1:1.17, respectively. The proton to neutron ratio in radioactive carbon, ^{14}C, is 1:1.33. Elements with atomic numbers greater than 40 are stable with proton to neutron ratios that progressively increase to a maximum of about 1.5. None of the heaviest elements in the periodic table can achieve stable nuclear configurations. The largest stable isotope is an isotope of bismuth, ^{209}Bi, with a proton to neutron ratio of 1:1.52. All uranium isotopes are radioactive; the most common, ^{238}U with a proton to neutron ratio of 1:1.59, decays to lead, ^{206}Pb with a proton to neutron ratio of 1:1.51.

In addition to the proton–neutron ratio, the absolute numbers of these subatomic particles affect isotope stability. The majority of stable isotopes, almost 160, have even numbers of protons and neutrons. About 50 stable isotopes have an even number of protons and an odd number of neutrons; about the same have an odd number of protons and an even number of neutrons. Very few stable isotopes have an odd number of both protons and neutrons.

Emissions

The transformation of a radioactive isotope into one with a stable configuration involves the emission of particles and energy. Natural radioactive isotopes can

M.E. Malainey, *A Consumer's Guide to Archaeological Science*, Manuals in
Archaeological Method, Theory and Technique, DOI 10.1007/978-1-4419-5704-7_3,
© Springer Science+Business Media, LLC 2011

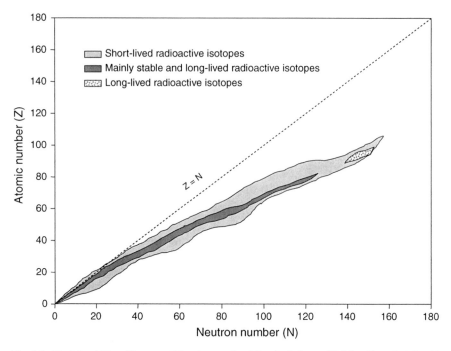

Fig. 3.1 Nuclei stability of isotopes ("Fundamentals of Physics", Second Edition Extended, David Halliday and Robert Resnick, copyright 1981, Reproduced with permission of John Wiley & Sons, Inc)

Table 3.1 Characteristics of common radioactive decay products. Data from "Alpha, Beta and Gamma Radioactivity, Essays on Radiochemistry No. 3, http://www.rsc.org/images/essay3_ tcm18-17765.pdf – Reproduced by permission of The Royal Society of Chemistry (Radiochemistry Group)"

Product	Symbol	Charge	Mass
Alpha particle	α	2+	4.0015
Beta particle (Negative emission)	β	1−	0.0054858 (~0)
Positron+	β^+	1+	0.0054858 (~0)
Gamma ray	γ	0	0

undergo three different types of emissions: alpha (α), beta (β), and gamma (γ) rays (Table 3.1).

Alpha emissions involve the ejection of alpha particles, which consist of two protons and two neutrons, giving them a charge of 2+ and an atomic mass number of 4. This is identical to the nucleus of a helium atom. Only the heaviest radioactive isotopes, those with atomic numbers greater than 82 and masses greater than 209, regularly decay through the emission of alpha particles. The loss of two protons transforms the radioactive element with atomic number Z into the element with atomic number Z–2.

Energy is released when a radioactive isotope decays through alpha emission. Much of the decay energy is transferred to the alpha particle, which is ejected from a radioactive substance at a speed of about 16,000 km/s. The recoil of the nuclei caused by ejection of the alpha particle and the emission of gamma radiation accounts for the remaining decay energy.

From a subatomic perspective, an alpha particle is very large and cannot travel far in air or through matter. In order to obtain an age estimate using a trapped charge dating technique, it is necessary to estimate the annual radiation dose to which a sample is exposed. Because these large particles travel such short distances, removing the outermost surface of the sample can eliminate the external contribution from alpha radiation.

Beta emissions are equivalent to electrons; they have a negative charge but almost no mass. Beta emissions are associated with radioactive isotopes that have an excess of neutrons. To ease the imbalance, a neutron is converted into a proton by emitting a negative charge. The mass of the nuclei produced is unchanged but the formation of a proton increases the atomic number by one. Radioactive ^{14}C decays through beta emissions into nitrogen:

$$^{14}_{6}\text{C} \rightarrow {}^{14}_{7}\text{N} + {}^{0}_{-1}\text{e}$$

The emission of gamma rays is also associated with this type of radioactive decay.

The velocity of beta emissions is approximately 130,000 km/s and they can travel longer distances in air or through matter than an alpha particle. Unless several millimeters of the exterior surface can be removed, the annual contribution of external beta radiation must be considered for trapped charge dating.

Positrons ($^{0}_{+1}\text{e}$) are analogous to negative beta emissions in that they have almost no mass, but carry a positive charge instead. Some artificially produced radioactive isotopes emit positrons; natural isotopes do not. If there are insufficient neutrons in the nucleus, a proton will emit a positron and convert to the neutral particle. The mass of the nuclei is unchanged but the loss of the proton decreases the atomic number by one.

An electron capture reaction is a third type of beta decay and also involves the loss of a proton. In this case, an inner shell, or k-shell, electron is captured by the nucleus; the negatively charged electron combines with a positively charged proton to produce a neutron.

Gamma rays are a type of highly energetic electromagnetic radiation with very short wavelengths. Gamma rays do not carry a charge but their emission enables a nucleus to expend excess energy. As noted above, any excess decay energy from an alpha particle emission is emitted as gamma radiation. Gamma ray emissions are also associated with both the formation and the decay of radioactive nuclei produced by neutron irradiation. As outlined below, by studying these gamma ray emissions, elements in a sample can be identified and quantified. This is the basis of instrumental neutron activation analysis (INAA) (see Chapter 32).

Radioactive Decay

Unlike many other reactions, the rate of radioactive decay (or activity) is not affected by temperature, only by the number of radioactive isotopes in the sample. Consequently, the decay rate is highest in newly formed radioactive material and decreases over time as the number of remaining radioactive isotopes, N, is reduced. As an example, if we have 1000 atoms of radioactive material with an initial decay rate of 5 disintegrations each second, when 500 atoms (one-half of the original material) remain the decay rate will have dropped to 2.5 disintegrations each second. When 250 atoms (one-quarter of the original material) remain the decay rate will be 1.25 disintegrations each second. The decay constant, λ, for a particular radioactive isotope is calculated by dividing the rate of radioactive decay, R, by the number of radioactive isotopes, N, in a sample:

$$\lambda = R/N \tag{3.1}$$

The decay constant for the isotope in the example is 0.005/s. Each isotope has a specific decay (or disintegration) constant.

When using the decay of radioactive carbon to date an artifact, we do not know the number of ^{14}C atoms that were initially in the sample or its initial decay rate but we can measure the current rate of decay. In order to obtain a radiocarbon age estimate for the sample, it is necessary to rearrange Equation (3.1) as follows:

$$R = \lambda N \tag{3.1a}$$

The decay rate of five atoms each second when there were 1000 atoms of radioactive material changes as soon as the next atom decays. For this reason, calculus is required to describe radioactive decay mathematically. Using calculus, R can be expressed as the number of atoms lost through disintegration in a time interval, $-\Delta N$, divided by the time interval, Δt:

$$R = -\Delta N/\Delta t \tag{3.2}$$

By substituting Equation (3.2) into Equation (3.1a) and using calculus, one can show that

$$\log (N_0/N) = \lambda t/2.303 \tag{3.3}$$

where N_0 is the number of radioactive isotopes present at time zero. This equation includes a base-10 logarithm; usually radioactive decay relationships are expressed using natural logarithms (see Chapters 8). The activity, in beta decay counts per minute per gram, can be used as a measure of the radioactive isotope remaining in the sample. It is possible to estimate the age of an artifact by comparing the activity of the sample to the activity of a modern standard. The activity of the sample provides a measure of N; the activity of the modern sample is used as N_0. This is a simple explanation of how a conventional radiocarbon date is obtained.

A feature commonly used to describe a radioactive isotope is its half-life, which is the time required for one-half of the material to decay. After one half-life, the number of atoms in a sample, N, is one-half the original amount, or $\frac{1}{2} N_0$. The half-life, $t_{1/2}$, of an isotope depends upon its decay constant and it can be shown that

$$t_{1/2} = 0.693/\lambda \tag{3.4}$$

For the isotope in the example

$$t_{1/2} = 0.693/0.005/\text{s}$$
$$t_{1/2} = 139 \text{ s}$$

The half-lives of radioactive isotopes are known to vary from fractions of a second to billions of years. The decay curve for radioactive isotopes is depicted in Fig. 3.2. Regardless of the length of its half-life, only a small fraction of the original amount of any radioactive isotope remains after about ten half-life periods have elapsed and the decay rate is no longer constant.

As outlined in Chapter 8, Libby calculated the half-life of ^{14}C to be 5568 years, which gives ^{14}C a decay constant of 1.24×10^{-4} per year using Equation (3.4). We can calculate t, the age estimate, with Equation (3.3) using the activities of the sample and a modern standard as measures of N and N_0, respectively. If a wooden artifact submitted for radiocarbon dating has an activity of 7.00 decay counts per minute per gram of carbon and the modern standard decays at a rate of 15.3 disintegrations per minute per gram of carbon, the logarithm of the activities is log (15.3/7.00)

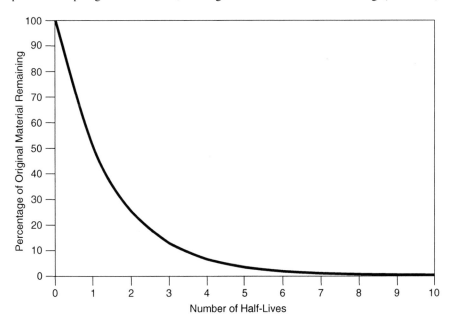

Fig. 3.2 Radioactive decay curve (Original Figure by M.E. Malainey and T. Figol)

or log (2.19), which is 0.340. An age estimate for the artifact is calculated as follows:

$$\log(N_0/N) = \lambda t/2.303$$
$$0.340 = 0.000124t/2.303$$
$$t = (0.340)(2.303)/0.000124$$
$$t = 6315 \text{ years}$$

Determining the Rate of Radioactive Decay

Instruments that detect the energy from emissions are used to measure the decay rate of radioactive isotopes. Liquid scintillation counters contain a chemical or mixture of chemicals that fluoresce (emit a flash of light) when they absorb beta particles. The photon of light produces a signal that is amplified and recorded by a counting device. Geiger–Müeller counters contain tubes of argon gas; the energy from radioactive decay emission removes an electron and creates positively charged argon ions. The ions enable a pulse of electric current to flow through a circuit, which operates a counter or device that makes a clicking sound. The amount of radiation emitted by a substance per unit of time, or its activity, is generally expressed in curies; one curie is 3.70×10^{10} disintegrations per second.

The Uranium Decay Series

The radioactive decay of a naturally occurring isotope usually changes the number of protons in its nucleus, which is its atomic number. Radioactive isotopes of one element are transformed into stable isotopes of another. The decay of an atom of radioactive carbon (^{14}C) to a stable nitrogen (^{14}N) atom is a one-step process involving the emission of a single β particle. In other cases, the nucleus of the decay product, called its daughter, is also unstable and undergoes radioactive decay. If the parent decays to many different radioactive daughters before a stable isotope forms, the multiple step process is called a decay series. The decay of ^{238}U to ^{206}Pb involves 14 steps and includes several α- and β-particle emissions (Fig. 3.3). Branching occurs in several places where it is possible for different radioactive daughter products to form, but the pathways always rejoin and one decay product is always preferred over the alternative. For example, ^{218}Po usually undergoes α-decay (99.97%) but β-decay (0.03%) is known to occur.

Uranium is soluble in water but its daughters are not. In environments such as cave formations, uranium is often deposited by water with newly formed calcite and a new decay series is established. Over a long period of time, all daughter isotopes in the decay series form and their activities, or decay rates, match that of the parent. This is the law of secular equilibrium. At equilibrium, the amount of each member

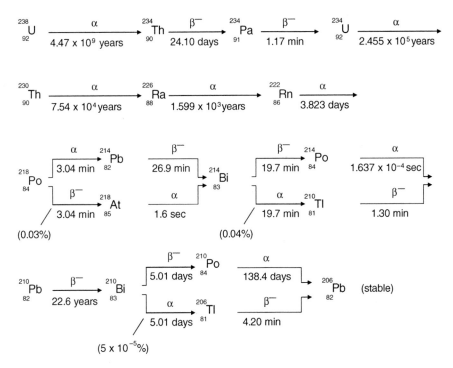

Fig. 3.3 Uranium decay series ("Data from Lido (2006)")

remains essentially constant because the rate at which it is produced matches the rate at which it decomposes.

Uranium series dating (Chapter 9) uses ratios between two daughters or between the parent uranium and a daughter to estimate the time since the decay chain was established. Uranium–thorium dating compares ^{234}U and ^{230}Th. Unless they were deposited at the same time, it may also be possible to compare ^{234}U to ^{238}U. If the sample is very old, the lead end product, ^{206}Pb, can be compared to the ^{238}U parent. If sufficient quantities of the less abundant uranium isotope, ^{235}U, are present in the sample, it can be compared to protactinium, ^{231}Pa.

Nuclear Reactions and Particle Acclerators

While radioactive decay naturally transforms one unstable isotope into a different stable isotope, it is also possible to artificially convert a stable isotope into an unstable, radioactive isotope. The process usually involves directing a projectile at the nucleus of a target element, which results in the formation of a different isotope and the emission of a subsidiary particle. Protons, neutrons, and alpha particles can be used as projectiles, as well as certain ions. These types of reactions are classified according to the projectile employed and the subsidiary particle ejected.

The nucleus of a target carries a strong positive charge that repels protons, alpha particles, and positively charged ions. In order to overcome the electrostatic repulsion and enable these types of reactions to proceed, it is necessary to use a particle accelerator. The charge on different segments of a particle accelerator is alternated between positive and negative to control forces of electrostatic attraction and repulsion. A particle is attracted into an oppositely charged section of the accelerator; as it exits the charge is reversed so that the particle is repulsed outward and attracted into the next segment. A cyclotron accelerates particles traveling outward in a spiral pattern between two oppositely charged D-shaped segments. A linear accelerator acts on particles as they move through a series of progressively longer tubes. In either case, the accelerated particles hit the target nuclei with tremendous force as they exit the instrument.

Neutrons are electrically neutral; consequently they are not repelled by positively charged nuclei. The type of nuclear reaction depends upon the energy of the collision between neutron and the target. The force of the impact between a fast neutron and the target will cause the ejection of an alpha particle or a proton. The atomic number of the isotope produced is different from that of the target. Another option is to reduce the energy of the fast neutrons by passing them through a moderator consisting of paraffin, graphite, oxygen, hydrogen, or deuterium. The target incorporates the slow, or thermal, neutron through a neutron-capture reaction. A radioactive isotope of the target is produced and prompt gamma radiation is emitted. Until these emissions die down, the sample is highly radioactive or "hot." Instrumental neutron activation analysis (Chapter 32) involves monitoring decay emissions from the target after a cooling period. When the newly formed radioactive isotope decays, both beta particles and gamma rays are emitted. Instrumental neutron activation analysis uses the frequency or energy of the gamma radiation to identify elements present in a sample; the intensity of the gamma radiation indicates the amount of that element present.

Chapter 4
Stable Isotopes

The stable (non-radioactive) isotopes commonly used in archaeological analysis and their distributions in the natural environment are described in this chapter. The properties of the stable isotopes of an element are almost identical, but some processes favor one isotope over another. This results in end products with isotopic ratios that differ from the starting material, or fractionation. Some of these processes are temperature dependent; typically, biological fractionation is not. The mixing of two isotopically distinct reservoirs will also change the ratio of isotopes. How their distributions in different types of archaeological materials are used to interpret diet, mobility patterns, climate, and provenience is described in Chapter 13.

Carbon

Two of the three isotopes of carbon are stable, ^{12}C and ^{13}C; ^{12}C is almost 100 times more abundant than ^{13}C. The movement of carbon throughout different parts of the environment is depicted in the carbon cycle (Fig. 4.1). Inorganic carbon is present in the atmosphere as carbon dioxide (CO_2) and carbon monoxide (CO). It occurs in soil, carbonate rocks, such as limestone and chalk, and fossil fuels, such as oil, natural gas, and coal, and is dissolved in water as bicarbonate, HCO_3^-. Carbon is a key component of living organisms. Carbon is present in all organic molecules, including carbohydrates, fats, and proteins.

Photosynthesis

Inorganic carbon dioxide is incorporated into living organisms through the process of photosynthesis. Green plants, and certain microorganisms that contain chlorophyll, carotenoids, and other pigments, transform inorganic carbon dioxide and water into organic sugars. Because all other molecules in the process are recycled, the photosynthesis reaction simplifies to

$$6CO_2 + 6H_2O \rightarrow C_6H_{12}O_6 + 6O_2$$

M.E. Malainey, *A Consumer's Guide to Archaeological Science*, Manuals in Archaeological Method, Theory and Technique, DOI 10.1007/978-1-4419-5704-7_4, © Springer Science+Business Media, LLC 2011

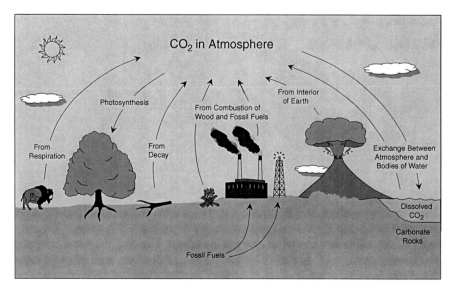

Fig. 4.1 The carbon cycle (Adapted from SILVER BURDETT BIOLOGY © 1981 Pearson Education, Inc. or its affiliates. Used by permission. All rights reserved.)

This pathway of photosynthesis was first described by Melvin Calvin and Andrew A. Benson in 1950 and is referred to as the Calvin or Calvin–Benson cycle. It illustrates photosynthesis in plants native to temperate environments, fruits, vegetables, trees, and Near Eastern crops. This process also occurs in aquatic plants, which produce the oxygen used by fish and mollusks.

During the first phase of photosynthesis, light energy is harnessed to produce molecules required for the next phase. When chlorophyll or other pigments absorb sunlight, an electron is raised to a higher energy level and the pigment molecule becomes excited. This energy and water are used to convert adenosine diphosphate (ADP) to adenosine triphosphate (ATP). The reduced form of nicotinamide adenine dinucleotide phosphate (NADPH) is also produced from nicotinamide adenine dinucleotide phosphate (NADP); oxygen is a by-product (Fig. 4.2). These reactions require energy harnessed from sunlight and are referred to as light reactions.

Carbon fixation does not require sunlight; instead, these dark reactions utilize energy stored in ATP and NADPH (Figs. 4.3 and 4.4). The ATP donates a phosphate group to ribulose phosphate, producing ribulose diphosphate. This compound incorporates, or fixes, CO_2 and the resulting products receive another phosphorus group from ATP. The reaction between these molecules and NADPH ultimately leads to the production of fructose and glucose. Glucose can form other sugars (e.g., sucrose), starches, and cellulose; all of the ADP, NADP, and ribulose phosphate are recovered. The intermediate products in the Calvin cycle, 3-phosphoglyceric acid (PGA), 1, 3-diphosphoglyceric acid (DPGA), and 3-phosphoglyceraldehyde (PGAL), have three carbon atoms. The plants that undergo photosynthesis in this manner are called C_3 plants. Carbon dioxide enters, and oxygen and water

Reactions that Require Energy from Sunlight Absorbed by Chlorophyll

$$2H_2O \xrightarrow{\text{Light Energy}} 4e^- + 4H^+ + O_2$$

$$2NADP^+ + 4e^- + 4H^+ \longrightarrow 2NADPH + O_2$$

which simplifies to (or net reaction):

$$2H_2O + 2NADP^+ \xrightarrow{\text{Light Energy}} 2NADPH + 2H^+ + O_2$$

Light Energy Storage

$$ADP + P_i \left(\begin{array}{c} \text{Inorganic} \\ \text{phosphate} \end{array} \right) \xrightarrow{\text{Light Energy}} ATP$$

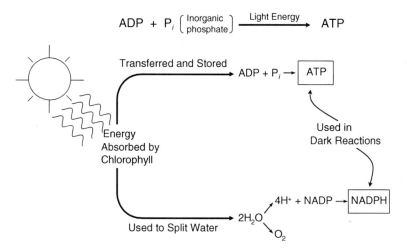

Fig. 4.2 Light reactions of the Calvin cycle (Original Figure by M.E. Malainey and T. Figol)

vapor exits, the plant through pores in the leaves called stromata. The entire C_3 photosynthetic process is summarized in Fig. 4.5.

Photosynthesis is different in certain plants native to hot environments, including corn, sugar cane, millet, and sorghum. Marshall Hatch and Rodger Slack described one C_4 pathway followed by these plants in the 1960s (Fig. 4.6); two others have since been recognized. Preliminary carbon fixation occurs near the surface of the leaf in mesophyll cells with phosphoenolpyruvate (PEP). The intermediate compounds produced, oxaloacetate and malate, have four carbon atoms each. The malate is transferred to bundle-sheath cells, where the carbon dioxide is recovered and regular C_3 photosynthesis occurs. The advantage of C_4 is that photosynthesis can occur even when carbon dioxide concentrations are very low. Plants reduce water loss by closing their stomata but still acquire enough CO_2 for the process.

$\textcircled{1}$ Fixation

CO_2 participates in condensation reactions with the ribulose
diphosphate acceptor

3 ribulose diphosphate + $3CO_2$ $\longrightarrow$ 6 Phosphoglycerate
molecules

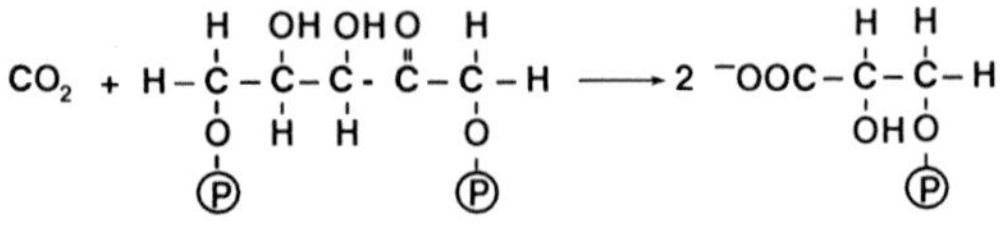

$$CO_2 \; + \; H-\overset{\overset{\displaystyle H}{|}}{\underset{\underset{\displaystyle \textcircled{P}}{|}}{\underset{O}{C}}} - \overset{\overset{\displaystyle OH}{|}}{\underset{\underset{\displaystyle H}{|}}{C}} - \overset{\overset{\displaystyle OH}{|}}{\underset{\underset{\displaystyle H}{|}}{C}} - \overset{\overset{\displaystyle O}{\|}}{C} - \overset{\overset{\displaystyle H}{|}}{\underset{\underset{\displaystyle \textcircled{P}}{|}}{\underset{O}{C}}} - H \; \longrightarrow \; 2 \; {}^{-}OOC - \overset{\overset{\displaystyle H}{|}}{\underset{\underset{\displaystyle OH}{|}}{C}} - \overset{\overset{\displaystyle H}{|}}{\underset{\underset{\displaystyle \textcircled{P}}{|}}{\underset{O}{C}}} - H$$

$\textcircled{2}$ Reduction

6 ATP and 6 NADPH (produced in light reactions) are used
to manufacture 6 glyceraldehyde phosphate (3 carbon sugar)
ADP, $NADP^+$ and inorganic phosphate (P_i) are recovered.

$\textcircled{3}$ Regeneration of CO_2 acceptor

3 ATP is used to manufacture 3 ribulose diphosphate
from 5 glyceraldehyde molecules.

The sixth glyceraldehyde phosphate molecule combines
with another to produce one glucose molecule.

Fig. 4.3 Carbon fixation or dark reactions of the Calvin cycle. (Original Figure by M.E. Malainey and T. Figol)

In addition to cellulose and starches, plants synthesize lipids and protein from the carbohydrates produced by photosynthesis. Plants are food producers; any organism that eats plants is termed a primary consumer. Carnivores that eat these primary consumer herbivores are secondary consumers. Tertiary consumers eat secondary consumers and so on. Each link in the food chain is termed a trophic level. After plants and animals die, fungi and certain bacteria process the tissue and release inorganic CO_2 and ammonia, NH_3, back into the environment; fire releases carbon dioxide and carbon monoxide into the atmosphere.

Under some circumstances, carbon-containing organic molecules do not decay but move into depositories or "carbon sinks." Terrestrial organisms may become incorporated into thick deposits of peat or over millennia form coal, oil, or natural gas. Carbon dioxide in the atmosphere is also exchanged with dissolved carbon dioxide in water. If the level of bicarbonate, HCO_3^-, is high, limestone and other carbonate-rich sediments may precipitate. Coral, sponges, and mollusks secrete calcium carbonate forming reefs, spicules, and shells that degrade slowly and may even become incorporated into limestone deposits.

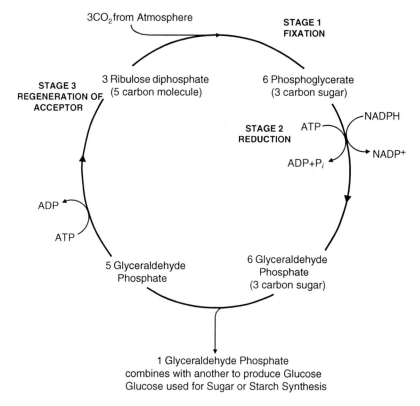

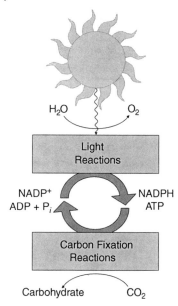

Fig. 4.4 The Calvin or Calvin–Benson cycle of C$_3$ photosynthesis

Fig. 4.5 Summary of C$_3$ photosynthesis (From Principles of Biochemistry, 2e by A. L. Lehninger, D. L. Nelson, and M. M. Cox. © 1993 by Worth Publishers. Used with the permission of W. H. Freeman and Company.)

Fig. 4.6 The Hatch–Slack
pathway of C$_4$ photosynthesis
(From Principles of
Biochemistry, 2e by A. L.
Lehninger, D. L. Nelson, and
M. M. Cox. © 1993 by Worth
Publishers. Used with the
permission of W. H. Freeman
and Company.)

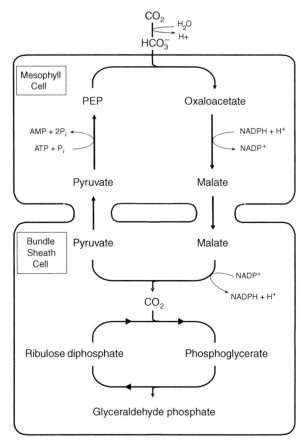

Isotopic Fractionation of Carbon

Stable isotopes of carbon occur in the atmosphere as CO_2. For every 100 atoms of
carbon, 99 will be ^{12}C and one will be ^{13}C. When incorporated into plant tissue
through photosynthesis, isotopic fractionation occurs and the ratio of ^{13}C to ^{12}C
changes significantly. When a herbivore eats plant material, the ratio of ^{13}C to ^{12}C
of its tissue is slightly different from the plants it consumes. The ratio of ^{13}C to
^{12}C of the tissue of the carnivore that eats the herbivore is slightly different from the
herbivore tissue. Instead of discussing these changes with respect to the atmosphere,
the ratio of ^{13}C to ^{12}C in a sample is compared to an international standard enriched
in ^{13}C, PDB or Vienna PDB; Pee Dee belemnite is a limestone from South Carolina.
Differences in the ratios between samples, including atmospheric carbon dioxide,
are very small and usually negative. Differences are presented as the value per mil
(‰), which is per thousand, rather than percent (%), which is per hundred.

Plants more readily use the lighter $^{12}CO_2$ molecule during photosynthesis so isotopic fractionation occurs. The first step of the Calvin cycle, carbon fixation with ribulose diphosphate, proceeds much faster with $^{12}CO_2$ than $^{13}CO_2$. Since more ^{12}C is incorporated into the plant, the average ratio of ^{13}C to ^{12}C in C_3 plants is very different from the atmosphere. Photosynthetic processes are also affected by temperature in that plants growing at higher latitudes have different ratios of ^{13}C to ^{12}C than those growing closer to the equator. By contrast, the ratio of ^{13}C to ^{12}C in C_4 plants is much closer to the ratio in the atmosphere. Crassulacean acid metabolism (CAM) photosynthesis of succulent plants produces intermediate values because they may follow C_3 photosynthetic processes during the day, C_4 photosynthetic processes at night, or a combination. For this reason, the ratio of ^{13}C to ^{12}C in CAM plants that fix CO_2 during the day resembles those of C_3 plants, while those that primarily undergo CO_2 fixation at night have ratios of ^{13}C to ^{12}C similar to C4 plants.

In stable carbon isotope analysis, the ratio between ^{13}C to ^{12}C measured in an animal bone can be used to determine its trophic level and diet. Stable isotope analysis can help identify the source of organic residues recovered from pottery, tools, and other artifacts. Stable isotope analysis of carbon containing rocks and minerals can be used in provenance studies.

Nitrogen

Nitrogen is a component of amines, amides, and amino acids, the building blocks of protein (see Chapter 5). The most common isotope of nitrogen is ^{14}N, but ^{15}N also occurs in nature. While the atmosphere is largely composed (79%) of molecular nitrogen, N_2, few organisms are able to use this form. Most can only utilize nitrogen that has combined with other elements or has been fixed (Fig. 4.7). Fixation is the process by which molecular nitrogen is converted into nitrates and other compounds. A small portion of nitrogen undergoes fixation as a result of lightning. The energy causes molecular nitrogen to break apart so the individual atoms can react with oxygen to form nitrates that contain the NO_3^- ion.

Certain species of bacteria and blue-green algae are capable of nitrogen fixation. Legumes (peas, beans, and alfalfa) are known as nitrogen-fixing plants, but genus *Rhizobium* bacteria that form symbiotic relationships with the plants actually perform the process. The bacteria invade the root hairs of the plant and form nodules where nitrogen fixation occurs. Both the plant and the surrounding soil are fertilized by the contribution of nitrates. Alone, neither the legume nor the rhizobia is capable of nitrogen fixation. Other types of nitrogen-fixing bacteria, such as *Azotobacter* and *Clostridium*, live freely in the soil.

Ammonia (NH_3), ammonium nitrate (NH_4NO_3), and urea [$(NH_2)_2CO$] can also be applied as fertilizer to provide fixed nitrogen for plant growth. Ammonia can be artificially synthesized for industrial purposes. Molecular nitrogen can combine with hydrogen, derived from a hydrocarbon such as petroleum or natural gas, but

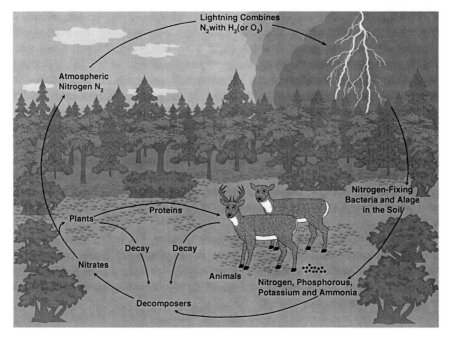

Fig. 4.7 The nitrogen cycle (Adapted from SILVER BURDETT BIOLOGY © 1981 Pearson Education, Inc. or its affiliates. Used by permission. All rights reserved.)

high temperatures and pressures are required and a catalyst must be present for the reaction to proceed.

Strontium

There are four isotopes of strontium: ^{88}Sr with an abundance of more than 82%; ^{86}Sr with an abundance of about 10%; ^{87}Sr with an abundance of approximately 7%; and ^{84}Sr with an abundance of 0.6%. Strontium-87 is the β-decay product of radioactive rubidium-87.

Strontium atoms are approximately the same size and carry the same 2+ charge as calcium. During the formation of minerals such as calcite, dolomite, gypsum, plagioclase feldspar, and apatite, strontium can be used in place of calcium. The ratio between ^{87}Sr and ^{86}Sr provides information about the environment in which these calcium-containing minerals formed. In general, recently formed basaltic rocks have the lowest ^{87}Sr/^{86}Sr values and very old granitic rocks have the highest. The ^{87}Sr/^{86}Sr value of the ocean is essentially homogeneous, but has varied over geological time. Fertilizers and air pollution contribute strontium to the modern environment.

This geochemical signature can be obtained from both inorganic materials and organic substances with a mineral component. The technique lends itself to diverse archaeological applications; it has been used in provenance studies of Greek marble and to trace the migration patterns of animals and humans by examining the ratio of ^{87}Sr to ^{86}Sr in bones and teeth.

Lead

There are four stable isotopes of lead, ^{204}Pb, ^{206}Pb, ^{207}Pb, and ^{208}Pb. While ^{204}Pb is not the product of any known radioactive decay series, the other isotopes are radiogenic. Lead-206 results from the decay of ^{238}U (half-life = 4.5 billion years), ^{207}Pb is produced by the decay of ^{235}U (half-life = 0.7 billion years), and ^{208}Pb is the stable daughter product of ^{232}Th (half-life = 14 billion years). Archaeological investigations using lead isotope analyses most often relate to provenance studies of metal artifacts. Lead is usually present in ore bodies; the abundance of each isotope is related to its age and the abundance and half-life of each parent radionuclide. As described in Chapter 13, the utility of lead isotope analysis for the study of artifact provenance has been recently questioned due to ore deposit inhomogeneity and likelihood of metal mixing/recycling in the past. Regardless, this data can be used to examine trade patterns and metal technology.

Lead actually occurs in most environments at low levels. The amount and isotopic ratios of the lead in animal bone can be used to assess its contribution from natural sources, or background. This data can be used to demonstrate the presence and source of lead contamination. A wide variety of cultural sources may directly or indirectly expose an individual to lead. Lead can be absorbed from food in direct contact with lead or a substance, such as metal, glass, glazed ceramics, or paint, which contains lead.

Oxygen and Hydrogen

There are three stable isotopes of oxygen, ^{16}O, ^{17}O, and ^{18}O, the natural abundances of which are 99.757, 0.038, and 0.205%, respectively. Oxygen isotope analysis involves comparisons between the two more abundant species, ^{16}O and ^{18}O. There are two stable isotopes of hydrogen, ^{1}H and ^{2}H (or D for deuterium), with natural abundances of 99.9885 and 0.0115%, respectively. The ratios of these isotopes in a particular region at a particular time depend upon the movement of water or the water cycle. During evaporation, water undergoes a phase change from liquid to a gas. Fractionation occurs because the lighter isotopes, ^{16}O and ^{1}H, are more likely to evaporate than the heavier isotopes, ^{18}O and ^{2}H. Consequently most water that falls as precipitation, which is called meteoric water, is depleted in both ^{18}O and ^{2}H and enriched in ^{16}O and ^{1}H. The relationship between the ratio of stable oxygen isotopes, ^{18}O/^{16}O or δ^{18}O, to stable hydrogen isotopes, ^{2}H/^{1}H or δ^{2}H or δD, is $\delta D = 8\delta^{18}O + 10$ (Craig 1961).

The ratios between isotopes of oxygen and hydrogen provide information about the climate on a regional and global basis. The isotopic value of meteoric water in a particular region depends on several factors including latitude, temperature, altitude, distance from the coast, and rate of precipitation. Consequently, both isotopes can be used in provenance studies. When the amount of water lost through evaporation is approximately equal to the amount that returns to oceans via rivers and streams, the ratio between the heavy and light isotopes in the ocean is constant.

Oxygen isotope analysis can be used to track global changes in temperature through the resulting impact on the water cycle. When the global temperature drops, more water that falls as precipitation becomes trapped in icesheets, glaciers, and snowcaps and less runs off the land and returns to the ocean. During extended periods of cold, ocean water becomes progressively more enriched in ^{18}O as more of the water containing the lighter oxygen isotope is trapped in ice. When the global temperature increases, more ice melts, run-off increases, and $\delta^{18}O$ decreases as water enriched in ^{16}O returns to the ocean.

Chapter 5
Organic Compounds

The term, organic compounds, is applied to materials that contain carbon and are associated with living organisms. Carbon atoms form strong covalent bonds to other carbon atoms and to hydrogen, oxygen, nitrogen, and sulfur. Families of organic compounds are classified on the basis of their structures (Table 5.1). For example, alkanes are a family of organic compounds whose members contain only carbon and hydrogen atoms linked by single bonds. The numbers of carbon and hydrogen atoms in the compounds vary but the names of all members share the same suffix -ane (Table 5.2).

All members of a family of organic compounds share similar features; often this is a particular arrangement of atoms called a functional group. As these functional groups are the sites of most chemical reactions, members of an organic compound family share similar properties. By replacing a single hydrogen atom in any member of the alkane family with a hydroxyl group, –OH, the compound becomes a member of the alcohol family. The characteristics of organic compound families of particular interest to archaeologists are described in this section.

The common structure of members of the same organic compound family can be described using a general formula consisting of the functional group and the symbol R. The symbol represents the remainder of the molecule, which may be an alkyl group formed by removing a hydrogen atom from an alkane. The general formula for organic compounds in the alkane family is R–H; the general formula for alcohols is R–OH. The standard chemical name of many organic compounds is a combination of the alkane name, without the final "e," and an organic family suffix (Figure 5.1).

Alkyl group names, such as methyl, ethyl, and butyl, apply only when hydrogen is removed from the last carbon in the chain. Removal of hydrogen from an inner carbon in the chain creates a branched structure (Figure 5.1). The term *iso-* is used when a methyl group is attached to the second carbon in a three to six carbon chain. One of the most commonly used chemicals with this prefix, *iso*-octane, does not actually follow the established naming system; its correct name is 2,2,4-trimethylpentane.

Carbon atoms attached to two other carbons are referred to as secondary carbons; a *sec*-group contains a carbon attached to a methyl group and a carbon

M.E. Malainey, *A Consumer's Guide to Archaeological Science*, Manuals in Archaeological Method, Theory and Technique, DOI 10.1007/978-1-4419-5704-7_5, © Springer Science+Business Media, LLC 2011

Table 5.1 Important families of organic compounds. Reprinted from Leigh et al. 1998, Table 3.2 Some Important Compound Classes and Functional Groups, *Principles of Chemical Nomenclature: A Guide to IUPAC Recommendations*, with permission of Wiley-Blackwell

	Family											
	Alkane	Alkene	Alkyne	Alcohol	Ether	Aromatic compounds	Aldehyde	Ketone	Carboxylic acid	Ester	Amine	Amide
Specific example	CH_3CH_3	$CH_2{=}CH_2$	$HC\equiv CH$	CH_3CH_2OH	CH_3OCH_3	(benzene ring)	$\underset{CH_3CH}{\overset{O}{|}}$	$\underset{CH_3CCH_3}{\overset{O}{|}}$	$\underset{CH_3COH}{\overset{O}{|}}$	$\underset{CH_3COCH_3}{\overset{O}{|}}$	CH_3NH_2	$\underset{CH_3CNH_2}{\overset{O}{|}}$
IUPAC name	Ethane	Ethene or ethylene	Ethyne or acetylene	Ethanol	Methoxy-methane	Benzene	Ethanal	Propanone	Ethanoic acid	Methyl ethanoate	Methan-amine	Ethanamide
Common name	Ethane	Ethylene	Acetylene	Ethyl alcohol	Dimethyl ether	Benzene	Acetal-dehyde	Acetone	Acetic acid	Methyl acetate	Methyl-amine	Acetamide
Functional group	C–H and C–C bonds	$\diagdown C{=}C \diagup$	$-C{\equiv}C-$	$-\overset{|}{\underset{|}{C}}-OH$	$-\overset{|}{\underset{|}{C}}-O-\overset{|}{\underset{|}{C}}-$	Aromatic ring	$-\overset{O}{\overset{\|}{C}}-H$	$-\overset{|}{C}-\overset{O}{\overset{\|}{C}}-\overset{|}{C}-$	$-\overset{O}{\overset{\|}{C}}-OH$	$-\overset{O}{\overset{\|}{C}}-O-\overset{|}{C}-$	$-\overset{|}{\underset{|}{C}}-N\diagdown^{|}_{|}$	$-\overset{O}{\overset{\|}{C}}-N\overset{|}{-}$

chain. A carbon attached to three other carbons is called a tertiary carbon; a *tert*-group consists of a carbon attached to three methyl groups. In general, the boiling point of many organic compounds increases with chain length. The boiling point of a compound containing a branched structure will be lower than a straight chain compound with the same number of carbons. The name of branched compounds is based on the longest continuous chain of carbon atoms. For alkanes, carbons are

Table 5.2 Examples of unbranched alkanes. Reprinted from T.W. Graham Solomons, Table 3.2, "Organic Chemistry", Second Edition, copyright 1980, with permission of John Wiley and Sons

Name	Number of carbons	Name	Number of carbons
Methane	1	Heptadecane	17
Ethane	2	Octadecane	18
Propane	3	Nonadecane	19
Butane	4	Eicosane	20
Pentane	5	Heneicosane	21
Hexane	6	Docosane	22
Heptane	7	Tricosane	23
Octane	8	Triacontane	30
Nonane	9	Hentriacontane	31
Decane	10	Tetracontane	40
Undecane	11	Pentacontane	50
Dodecane	12	Hexacontane	60
Tridecane	13	Heptacontane	70
Tetradecane	14	Octacontane	80
Pentadecane	15	Nonacontane	90
Hexadecane	16	Hectane	100

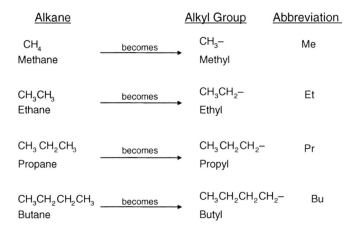

Alkane		Alkyl Group	Abbreviation

CH_4
Methane   —becomes→ CH_3-
Methyl   Me

CH_3CH_3
Ethane   —becomes→ CH_3CH_2-
Ethyl   Et

$CH_3 CH_2 CH_3$
Propane   —becomes→ $CH_3 CH_2 CH_2-$
Propyl   Pr

$CH_3CH_2CH_2CH_3$
Butane   —becomes→ $CH_3CH_2CH_2CH_2-$
Butyl   Bu

Alkyl groups are formed by removing a terminal hydrogen from an alkane.

$CH_3 CH_2 CH_2 CH_2-$

Butyl (or n -butyl)

$$CH_3CHCH_2- \quad (CH_3)$$

Isobutyl

$$CH_3CH_2CH- \quad (CH_3)$$

sec–Butyl (or s-butyl)

$$CH_3-C- \quad (CH_3, CH_3)$$

tert -Butyl (or t -butyl)

Four different alkyl groups contain four carbon atoms.

Fig. 5.1 Nomenclature of alkyl groups (Data from Leigh et al. 1998, "Principles of Chemical Nomenclature: A Guide to IUPAC Recommendations", with permission of Wiley-Blackwell)

numbered from the end of the chain closest to the branch and used to identify its location.

The chemical structure of specific organic compounds can be depicted in different ways (Fig. 5.2). Covalent bonds can be represented by dots or dashes. Condensed formulas show which atoms are bonded to each carbon. Bond line formulas represent the carbon skeleton of the molecule and location of atoms other than hydrogen; proper bond angles are shown. Each carbon atom forms four bonds. The bond angle of single bonds is 109.5°, which places the carbon atom at the center of a tetrahedral structure. If a carbon atom forms one double and two single bonds, bond angles are about 120°, and they are in the same plane. This places the carbon atom at the center of an equilateral triangle. The bond angle of a carbon atom forming one triple and one single bond is 180°, producing a linear arrangement. Full three-dimensional aspects of organic molecules can be illustrated using circle and line structures or dashline and wedges.

Type	Example	Description
Dot	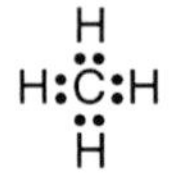 H H:C:H H	Shows pairs of shared electrons of covalent bonds.
Dash	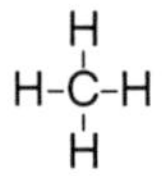 H H–C–H H	Single bonds are depicted as a single dash.
	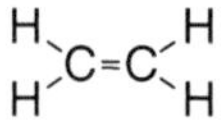 H H ＼C=C／ H H	Double bonds are depicted as two dashes.
	H – C≡C– H	Triple bonds are depicted as three dashes.
Condensed	$CH_3CHClCH_3$	Consists of every carbon atom and the atoms to which it is bonded.
Bond line	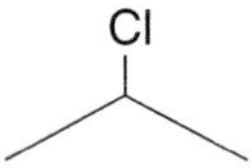 Cl	Shows only the carbon skeleton. Hydrogen atoms are assumed to be present.
3D circle and line	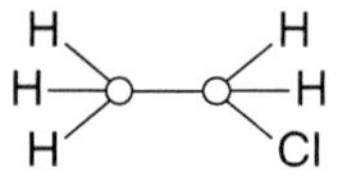 H H H–○ ○–H H Cl	Carbon atoms are depicted as a circle to which other atoms are bonded, forming tetrahedrons.
3D dash line and wedge	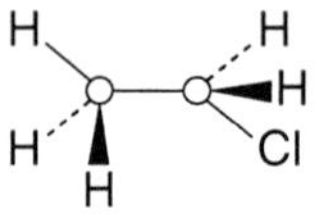 H H H–○ ○◄H H Cl	Dashed line projects behind plane of paper. Line projects on the plane of paper. Wedge projects out of the plane of paper.

Fig. 5.2 Structural representations of organic molecules

Important Organic Compound Families

Alkanes, Alkenes, and Alkynes

Organic compounds in these families are called hydrocarbons because they consist of only carbon and hydrogen atoms (Fig. 5.3). The primary source of alkanes is petroleum. Alkanes are compounds in which the carbon atoms are connected by only single bonds. If at least one pair of carbon atoms in a hydrocarbon is connected by a double bond, it is called an alkene. Naturally occurring alkenes include

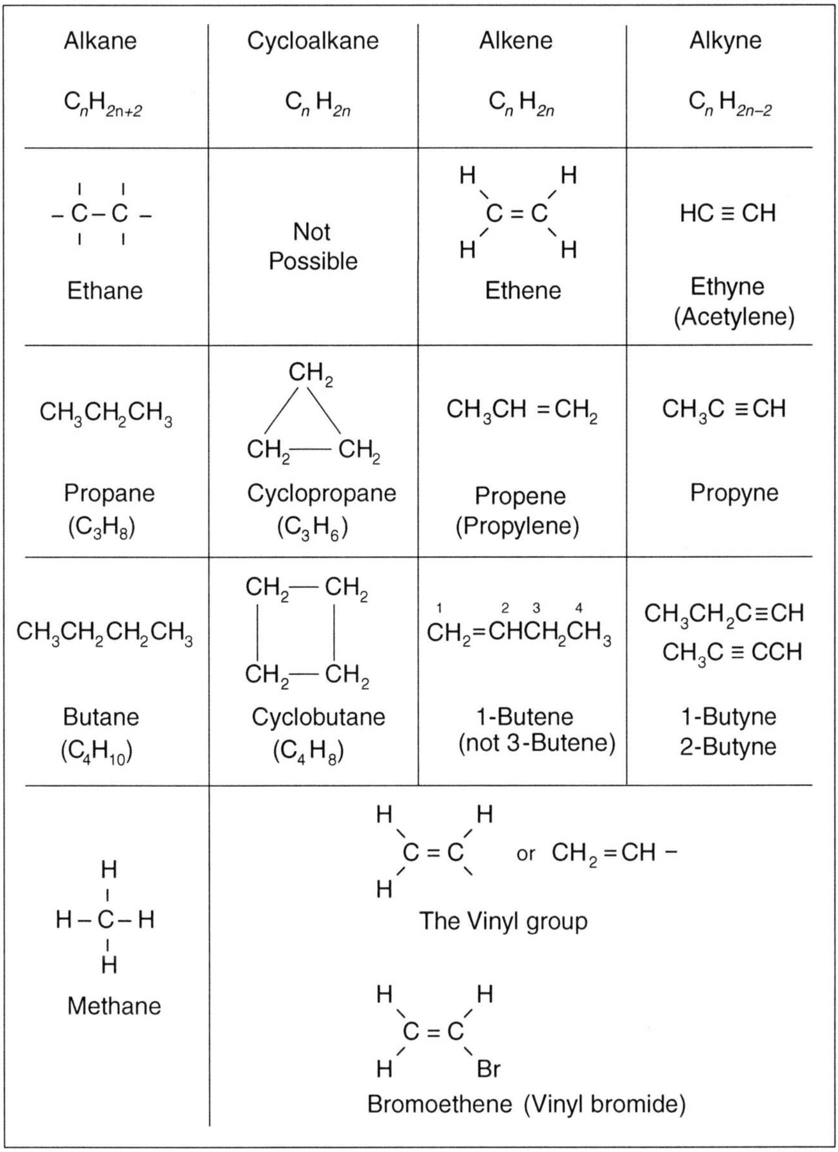

Alkane	Cycloalkane	Alkene	Alkyne
C_nH_{2n+2}	C_nH_{2n}	C_nH_{2n}	C_nH_{2n-2}
$-\overset{\mid}{\underset{\mid}{C}}-\overset{\mid}{\underset{\mid}{C}}-$ Ethane	Not Possible	$\overset{H}{\diagdown}\overset{}{\underset{\diagup}{C}}=\overset{}{\underset{H}{\overset{\diagdown}{C}}}\overset{H}{}$ Ethene	$HC \equiv CH$ Ethyne (Acetylene)
$CH_3CH_2CH_3$ Propane (C_3H_8)	$\overset{CH_2}{\diagup \diagdown}$ $CH_2 \!-\! CH_2$ Cyclopropane (C_3H_6)	$CH_3CH = CH_2$ Propene (Propylene)	$CH_3C \equiv CH$ Propyne
$CH_3CH_2CH_2CH_3$ Butane (C_4H_{10})	$CH_2\!-\!CH_2$ $\mid \qquad \mid$ $CH_2\!-\!CH_2$ Cyclobutane (C_4H_8)	$\overset{1}{CH_2}=\overset{2}{CH}\overset{3}{CH_2}\overset{4}{CH_3}$ 1-Butene (not 3-Butene)	$CH_3CH_2C\equiv CH$ $CH_3C \equiv CCH$ 1-Butyne 2-Butyne
H $\mid$ $H-C-H$ $\mid$ H Methane	$\overset{H}{\diagdown}\overset{}{\underset{\diagup}{C}}=\overset{}{\underset{H}{\overset{\diagdown}{C}}}\overset{H}{}$ or $CH_2 = CH-$ The Vinyl group $\overset{H}{\diagdown}\overset{}{\underset{\diagup}{C}}=\overset{}{\underset{H}{\overset{\diagdown}{C}}}\overset{H}{Br}$ Bromoethene (Vinyl bromide)		

Fig. 5.3 Examples of alkanes, alkenes, and alkynes

compounds known as terpenes and terpenoids, which are important components of essential oils. The term alkyne is applied to hydrocarbons containing a triple bond between a pair of carbon atoms. Hydrocarbons can be arranged in a straight or branched line or in a ring, forming cycloalkanes or cycloalkenes.

Alcohols and Ethers

Alcohols are organic compounds with a hydroxyl (–OH) group attached to a carbon atom (Fig. 5.4). They can be considered as derivatives of water, with an alkyl group replacing a hydrogen atom. Alcohols are named by replacing the -e of the corresponding alkane with -ol. Methanol (or methyl alcohol or wood alcohol), the simplest alcohol, is commonly found in automobile antifreeze and highly toxic. The combination of methanol and chloroform is often used to extract lipids from archaeological materials. Ethanol (grain alcohol) is present in alcoholic beverages. It can be made by the fermentation of sugars in grains and fruit juices. Isopropyl alcohol is commonly used as rubbing alcohol. Very short chain alcohols are completely miscible with water; but the solubility of alcohols decreases as the hydrocarbon chain lengthens. Long-chain alcohols are insoluble in water and components of plant waxes. Those unique to certain plant groups (families, genera, or species) can serve as biomarkers.

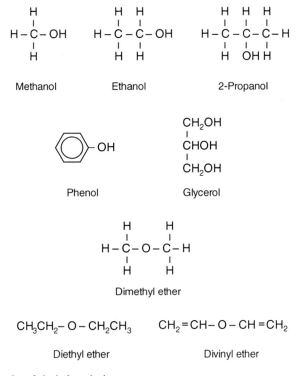

Fig. 5.4   Examples of alcohols and ethers.

Phenols are compounds that have a hydroxyl group attached to one aromatic ring; hence it is both an alcohol and an aromatic compound. Phenols consisting of a hydroxyl group attached to a cluster of two or more aromatic rings (called a polycyclic benzenoid ring) are called naphthols or phenanthrols. The compound, phenol or benzenol, is formed when a hydroxyl group is substituted on a benzene ring. Phenol is an important industrial chemical used in the production of plastics and many other commercial products.

Glycerol is a molecule with one hydroxyl group attached to each of its three carbon atoms. Glycerol forms the backbone of triacylglycerols and is released during the production of soap (saponification).

Hydrogen bonding occurs between hydroxyl groups of alcohol and phenol molecules. For this reason, the boiling points of alcohols and ethers are higher than compounds of similar molecular weight that do not form hydrogen bonds (see Chapter 1).

Ethers are organic compounds with two alkyl groups linked together by an oxygen atom (Fig. 5.4). For this reason, they can be considered di-alkyl derivatives of water. Common names are frequently used; these consist of both groups attached to the oxygen followed by *ether*. Proper chemical names are applied to complex ethers. Diethyl ether and divinyl ether have both been used as surgical anesthetics.

Aromatic Compounds

Most aromatic compounds, or arenes, contain a six-carbon ring unit. The parent of all these aromatic compounds is benzene, C_6H_6 (Fig. 5.5). Benzene is highly unsaturated, having the equivalent of three double bonds. Rather than acting like three isolated double bonds, the electrons are equally shared among the six carbons. This charge delocalization gives the benzene ring an unexpectedly high degree of stability. Instead of undergoing addition reactions, another atom or functional

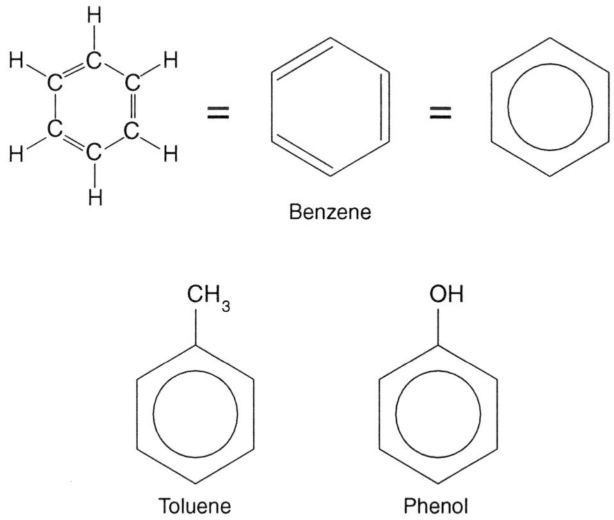

Benzene

Toluene

Phenol

Fig. 5.5 Examples of aromatic compounds

group generally replaces a hydrogen atom on the ring in a substitution reaction. Any compound that contains ring structures with charge delocalization is classed as an aromatic compound.

The name "aromatic" suggests these compounds are fragrant. Naphthalene, which consists of two benzene rings fused together, is the chemical with the distinct odor used in mothballs. Benzene also has a distinct odor but is highly toxic; this carcinogenic solvent should only be handled in a fume hood. Toluene, with a methyl group substitution, and phenol, with a hydroxyl group substitution, are widely used in the production of plastics. Two amino acids, phenylalanine and tyrosine, contain the benzene ring; another, tryptophan, contains a benzene ring fused to a five-membered ring. Purine and pyrimidine are heterocyclic aromatic rings that contain both carbon and nitrogen atoms. Derivatives of these compounds appear in deoxyribonucleic acid (DNA), ribonucleic acid (RNA), nicotinamide adenine dinucleotide phosphate (NADP), and its reduced form (NADPH).

Aldehydes and Ketones

An aldehyde is an organic compound with an oxygen atom double bonded to the last or terminal carbon in the chain; a hydrogen atom is also bonded to the terminal carbon (Fig. 5.6). The carbon–oxygen double bond (–C$=$O) is called the carbonyl functional group. Simple aldehydes are named by replacing the *-e* of the

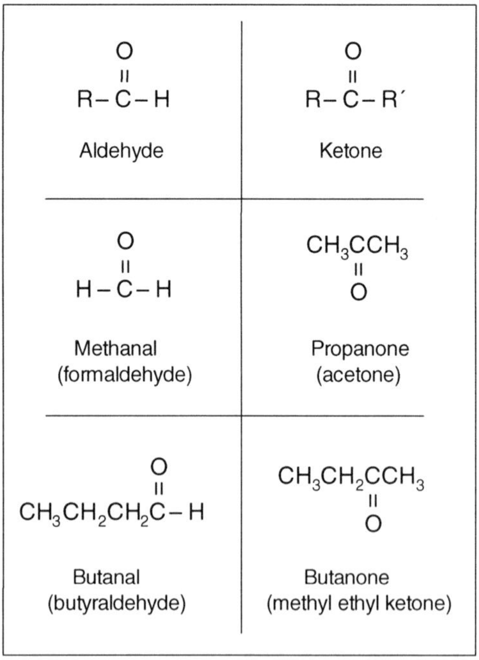

O ‖ R– C – H Aldehyde	O ‖ R– C – R′ Ketone
O ‖ H – C – H Methanal (formaldehyde)	CH$_3$CCH$_3$ ‖ O Propanone (acetone)
O ‖ CH$_3$CH$_2$CH$_2$C – H Butanal (butyraldehyde)	CH$_3$CH$_2$CCH$_3$ ‖ O Butanone (methyl ethyl ketone)

Fig. 5.6 Examples of aldehydes and ketones

corresponding alkane with *-al*. The simplest aldehyde, methanal, is commonly known as formaldehyde.

Ketones are formed when the carbonyl group is in the middle of a carbon chain (Fig. 5.6). Simple ketones are named by replacing the *-e* of the corresponding alkane with *-one*. The simplest ketone, propanone, consists of three carbons with a central carbonyl group and is better known by its common name, acetone.

Carboxylic Acids and Esters

The carboxyl group consists of a hydroxyl (–OH) group attached to a carbonyl group, which is the carbon–oxygen double bond (Fig. 5.7). A carboxylic acid forms when the carboxyl group occurs at the end of a carbon chain. Carboxylic acids are named by replacing the *-e* of the corresponding alkane (or longest chain of a

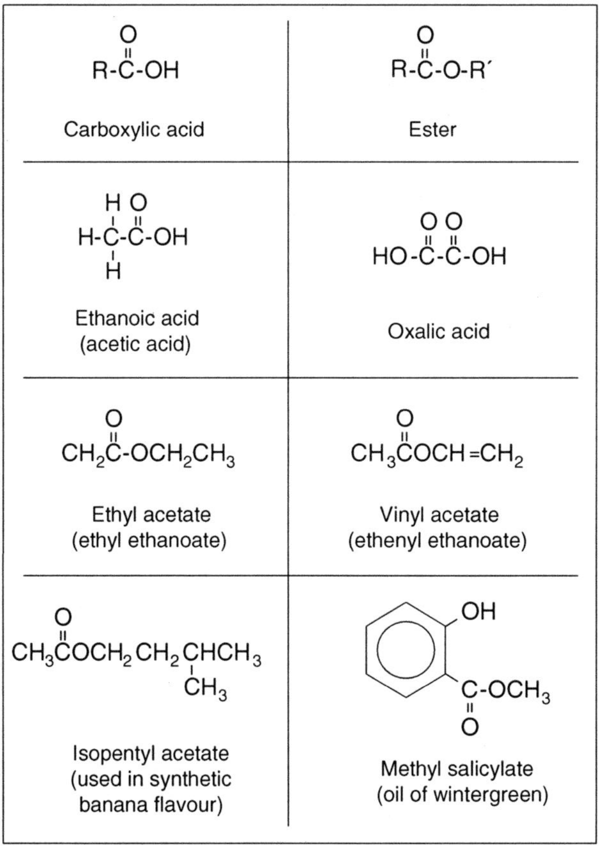

Fig. 5.7 Examples of carboxylic acids and esters

branched alkane) with *-oic acid*. The common name of methanoic acid is formic acid. Ethanoic acid is better known as acetic acid; vinegar is weak acetic acid. Oxalic acid, which is the standard for stable carbon analysis, is a dicarboxylic acid consisting of two carboxyl groups bonded together.

Carboxylic acids with between 4 and 36 carbons are called fatty acids. Carboxylic acids are also the parent group of esters, amides, and a variety of other related compounds. Salts of carboxylic acids occur when a metal ion replaces the hydrogen of the hydroxyl group and forms an ionic bond with oxygen. Sodium and potassium salts of long chain carboxylic acids are major ingredients of soap.

Esters are very similar to carboxylic acids but, instead of a hydroxyl group, an alkyl group is attached to the oxygen (Fig. 5.7). Esterification is the process of synthesizing an ester from a carboxylic acid and an alcohol in the presence of an acid; water is a by-product of the reaction. The names of esters consist of the alcohol ending with *-yl* followed by the acid ending in *-ate* or *-oates*. Esters often have pleasant odors and many different natural and synthetic esters are used as food flavorings (Fig. 5.6).

Amines and Amides

Amines and amides are organic compounds that contain nitrogen. A primary amine, or alkyl amine, is formed when the nitrogen is bonded to a terminal carbon and two hydrogen atoms (Fig. 5.8). These are named by combining the alkyl group with *-amine*. In a secondary amine, the nitrogen is attached to one hydrogen atom and two alkyl groups. The nitrogen in tertiary amines is attached to three alkyl groups. The names of secondary and tertiary amines designate the alkyl group names; the prefixes di- and tri- are used if identical groups appear. A number of medically and biologically important compounds are amines, including adrenalin, serotonin, and certain vitamins. Histamine, which produces symptoms associated with colds and allergies, is a toxic amine. Nicotine, morphine, and cocaine are special types of amines called alkaloids.

Amides contain a carbonyl group (C=O) attached to nitrogen (Fig. 5.9). When derived from a carboxylic acid, the *-ic* (or *-oic*) *acid* portion of the name is replaced by *-amide*. Substitutions on nitrogen are prefaced by *N-* (for a single substitution) or *N,N-* (for a double substitution).

Lipids

Lipids are compounds that generally have low solubility in water but can be extracted from plant and animal tissues with non-polar organic solvents, such as chloroform, methanol, and ether. A wide variety of organic compounds are categorized as lipids, including fatty (carboxylic) acids, triacylglycerols, steroids, waxes, terpenes, phospholipids, glycolipids, and prostaglandins. The first five have proven useful for identifying archaeological residues.

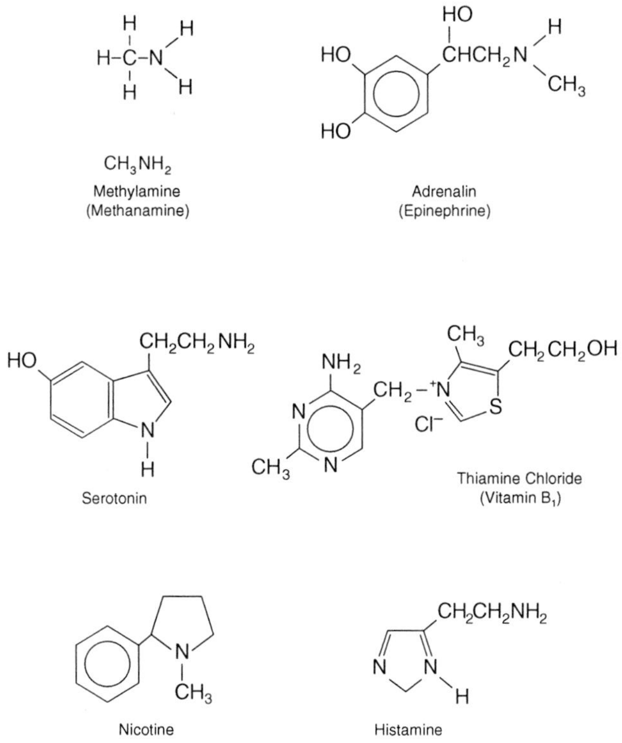

CH$_3$NH$_2$

Methylamine
(Methanamine)

Adrenalin
(Epinephrine)

Serotonin

Thiamine Chloride
(Vitamin B$_1$)

Nicotine

Histamine

Fig. 5.8 Examples of amines

Fig. 5.9 Examples of amides
(Reprinted from T.W. Graham
Solomons, Table 17.5,
"Organic Chemistry", Second
Edition, copyright 1980, with
permission of John Wiley &
Sons, Inc.)

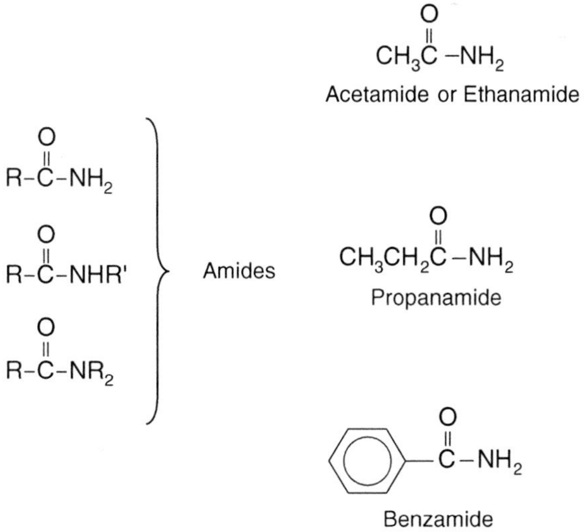

$$CH_3\overset{O}{\underset{||}{C}}-NH_2$$

Acetamide or Ethanamide

$$R-\overset{O}{\underset{||}{C}}-NH_2$$

$$R-\overset{O}{\underset{||}{C}}-NHR'$$ Amides

$$R-\overset{O}{\underset{||}{C}}-NR_2$$

$$CH_3CH_2\overset{O}{\underset{||}{C}}-NH_2$$

Propanamide

$$-\overset{O}{\underset{||}{C}}-NH_2$$

Benzamide

Fatty Acids and Triacylglycerols

Fatty acids are carboxylic acids with hydrocarbon chains ranging from 4 to 36 carbons in length; those with between 12 and 24 carbons are most common in plant and animal foods (Table 5.3). Fatty acids that do not contain double bonds are called saturated fatty acids; those that contain at least one double bond are referred to as unsaturated fatty acids because they have the potential to hold more hydrogen atoms. Those with only one double bond are mono-unsaturated, while those with multiple double bonds are polyunsaturated.

Table 5.3 Names of fatty acids. (From "Fatty Aicds: Physical and Chemical Characteristics" (1975), in CRC Handbook of Biochemistry and Molecular Biology, 3rd Edition, Lipids, Carbohydrates, Steroids, edited by Gerald D. Fasman, pp. 484–492, CRC Press, Cleveland, Ohio.)

Systematic name	Common name	Shorthand designation
Dodecanoic	Lauric	C12:0
Tetradecanoic	Myristic	C14:0
cis-9-Tetradecenoic	Myristoleic	C14:1
Pentadecanoic		C15:0
Hexadecanoic	Palmitic	C16:0
cis-9-Hexadecenoic	Palmitoleic	C16:1
Heptadecanoic	Margaric	C17:0
Heptadecenoic		C17:1
Octadecanoic	Stearic	C18:0
cis-9-Octadecenoic	Oleic	C18:1ω9
cis-11-Octadecenoic	Vaccenic	C18:1ω7
9,12-Octadecadienoic	Linoleic	C18:2
9,12,15-Octadecatrienoic	α-Linolenic	C18:3ω3
Eicosanoic	Arachidic	C20:0
cis-9-Eicosenoic	Gadoleic	C20:1
Docosanoic	Behenic	C22:0
cis-13-Docosenoic	Erucic	C22:1
Tetracosanoic	Lignoceric	C24:0
cis-15-Tetracosenoic	Nervonic	C24:1

The shorthand convention for designating fatty acids, Cx:$y\omega z$, contains three components. The "Cx" refers to a fatty acid with a carbon chain length of x number of atoms. The "y" represents the number of double bonds or points of unsaturation, and the "ωz" (where ω is omega, the last letter of the Greek alphabet) indicates the location of the most distal double bond on the carbon chain, i.e., closest to the methyl end. Thus, the fatty acid expressed as C18:1ω9 refers to a mono-unsaturated isomer with a chain length of 18 carbon atoms and a single double bond located 9 carbons from the methyl end of the chain. Similarly, the shorthand designation, C16:0, refers to a saturated fatty acid with a chain length of 16 carbons. Positions of double bonds can be designated with respect to the carboxyl group using the upper case form of the Greek letter delta, Δ. An 18-carbon fatty acid with double bonds located on the 9th, 12th, and 15th carbons from the carboxyl end can be indicated as $C18:3\Delta^{9,12,15}$.

Polyunsaturated fatty acids extracted from foods can have up to six double bonds. Research has shown that properties of unsaturated fatty acids are influenced by the position of double bonds with respect to the last carbon in the chain or methyl terminal. Fatty acids with a double bond occurring on the third carbon from the final, or omega, carbon are called omega-3 fatty acids. Those with a double bond on the ninth carbons from the methyl terminal are omega-9 fatty acids; omega-6 fatty acids also exist.

In nature, fatty acids are not usually found as isolated molecules called free fatty acids. Instead, they most often occur as parts of triacylglycerols, in which three fatty acids are bonded to a single glycerol molecule through ester linkages (Fig. 5.10). The fatty acids of simple triacylglycerols are identical; mixed triacylgycerols contain at least two different fatty acids. The terms, fat and oil, refer to the physical state of the triacylglycerols. Fats are solid at room temperature because their constituent triacylglycerols are mainly saturated fatty acids. About 55% of the fatty acids in beef tallow are saturated. Oils are liquid at room temperature because the triacylgycerols are mainly composed of unsaturated fatty acids. For example, about 80% of the fatty acids in olive oil are unsaturated. Oils are generally extracted from plants and fish. Polyunsaturated fatty acids readily react with oxygen in the presence of light at room temperature. Foods that contain fatty acids with multiple double bonds, such as fish, will quickly spoil and turn rancid if not refrigerated promptly.

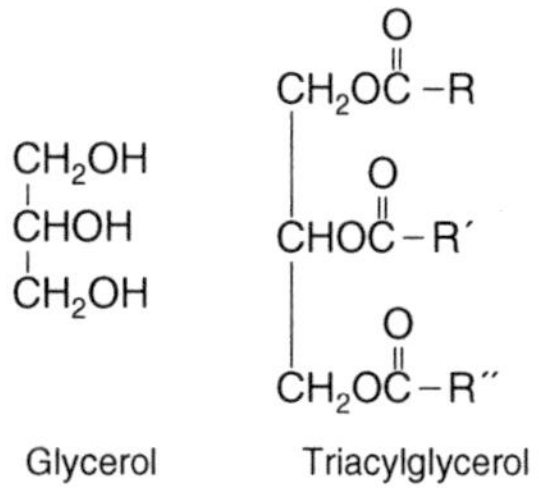

Fig. 5.10 The structure of triacylglycerols (Reprinted from T.W. Graham Solomons, Figure 21.1, "Organic Chemistry", Second Edition, copyright 1980, with permission of John Wiley & Sons, Inc.)

$$CH_2OH$$
$$CHOH$$
$$CH_2OH$$
Glycerol

$$\begin{array}{c} O \\ \| \\ CH_2OC-R \\ O \\ \| \\ CHOC-R' \\ O \\ \| \\ CH_2OC-R'' \end{array}$$
Triacylglycerol

Plants, animals, and other living organisms use fats and oils as stored forms of energy. Fats and oils are stored in the adipocyte cells of animals, forming adipose tissue, and in the seeds of plants. Fats and oils are the most concentrated forms of energy available. Gram for gram, the amount of energy released in kilocalories when an organism metabolizes triacylglycerols is more than twice that from either carbohydrates or proteins. The shortest fatty acids are found in milk and other dairy products; because the hydrocarbon chain is so short, these fatty acids are water soluble. Water solubility of fatty acids decreases with increased chain length.

Although fats and oils do not dissolve in water, the salts of carboxylic fatty acids are excellent soaps. Soap is prepared through a process called alkaline hydrolysis or saponification, which separates the individual fatty acids from the glycerol molecule (Fig. 5.11). Traditional soap manufacture involves boiling animal fat, often beef

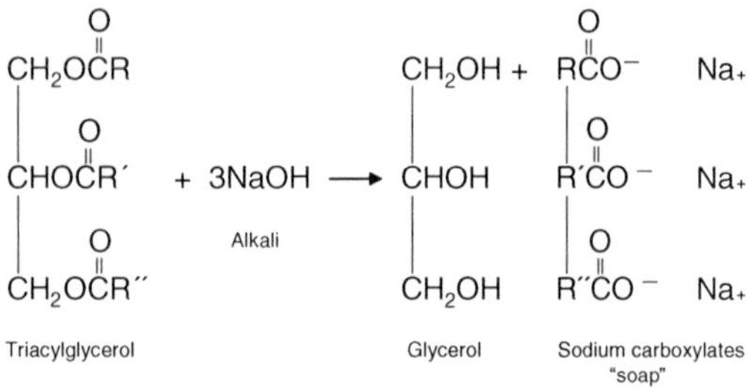

Fig. 5.11 Saponification of triacylglycerols (soap manufacture)

$$
\begin{array}{ccc}
\underset{\text{CH}_2\text{OCR}}{\overset{\overset{\displaystyle O}{\parallel}}{|}} & & \underset{\text{CH}_2\text{OH} +}{|}\quad \underset{\text{RCO}^-}{\overset{\overset{\displaystyle O}{\parallel}}{}}\quad \text{Na}_+ \\
\underset{\text{CHOCR}'}{\overset{\overset{\displaystyle O}{\parallel}}{|}} + 3\text{NaOH} \longrightarrow & & \underset{\text{CHOH}}{|}\quad \underset{\text{R'CO}^-}{\overset{\overset{\displaystyle O}{\parallel}}{}}\quad \text{Na}_+ \\
\underset{\text{CH}_2\text{OCR}''}{\overset{\overset{\displaystyle O}{\parallel}}{|}} & \text{Alkali} & \underset{\text{CH}_2\text{OH}}{|}\quad \underset{\text{R''CO}^-}{\overset{\overset{\displaystyle O}{\parallel}}{}}\quad \text{Na}_+ \\
\text{Triacylglycerol} & & \text{Glycerol}\quad \text{Sodium carboxylates} \\
& & \text{"soap"}
\end{array}
$$

tallow, with an alkali in water. Lye or wood ashes are typical sources of sodium hydroxide (NaOH) or potassium hydroxide (KOH), respectively. Once the reaction is completed, salt is added to cause solid soap to precipitate out while glycerol remains in solution.

Fats and oils do not dissolve in water unless soap is present. The non-polar hydrocarbon chain end of soap is hydrophobic (water fearing); the polar carboxyl end is hydrophilic (water loving). Soap molecules exist in water as clusters, called micelles, with the hydrocarbon end inward and the carboxyl end pointing out. The hydrocarbon ends dissolve into and then completely surround droplets of fats and oils. Individual micelles have negative charges at their surfaces and repel each other so the encapsulated fats and oils disperse (Fig. 5.12).

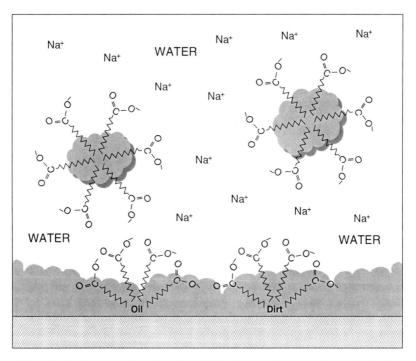

Fig. 5.12 Micelle formation (Reprinted from T.W. Graham Solomons, Figure 21.3, "Organic Chemistry", Second Edition, copyright 1980, with permission of John Wiley & Sons, Inc.)

Sterols

Sterols are structural lipids that are present in cell membranes. Steroids have a common nucleus called the perhydrocyclopentanophenanthrene ring system, which consists of four fused rings, three of which contain six carbons and one which has five (Fig. 5.13). Sterols are a special kind of alcohol that serve as precursors to a wide variety of products known as steroids, which have specific biological activities.

Fig. 5.13
Perhydrocyclopentanophenanthrene
ring system

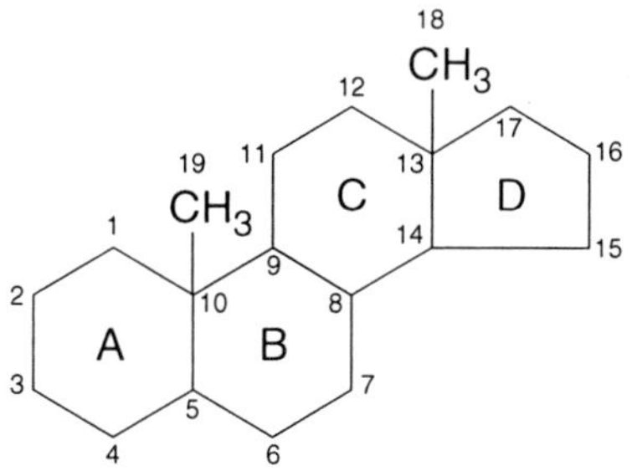

Cholesterol is the major sterol in animal tissue (Fig. 5.14); each person holds almost 250 g of cholesterol in their body tissues. Campesterol, stigmasterol, and sitosterol are found in plants; ergosterol occurs in fungi. Bacteria usually do not contain sterols. Sterols are used as biomarkers to identify the source of ancient residues.

Waxes

Waxes produced by living organisms consist of a long-chain alcohol connected to a long-chain fatty acid through an ester linkage (Fig. 5.15). The alcohol portion can contain between 16 and 30 carbons; the fatty acid chain can range from 14 to 36 carbons in length and be saturated or unsaturated.

Waxes act as protective barriers and resist decomposition. Birds, in particular, waterfowl, produce waxes to repel water and keep their feathers dry. Plants use waxes to both reduce water loss through evaporation and as a defense against parasites. The compositions of waxes produced by different organisms are distinct, which enables their use as biomarkers for the identification of ancient residues.

Terpenes and Terpenoids

The most important constituents of essential oils are terpenes and terpenoids. Essential oils are extracted from flowers and other plant materials by gently heating or steaming. Tar and pitches are removed from wood products by destructive distillation or pyrolyzation. The process involves heating wood in the absence of air until it has decomposed.

Fig. 5.14 The structure of cholesterol and stigmasterol

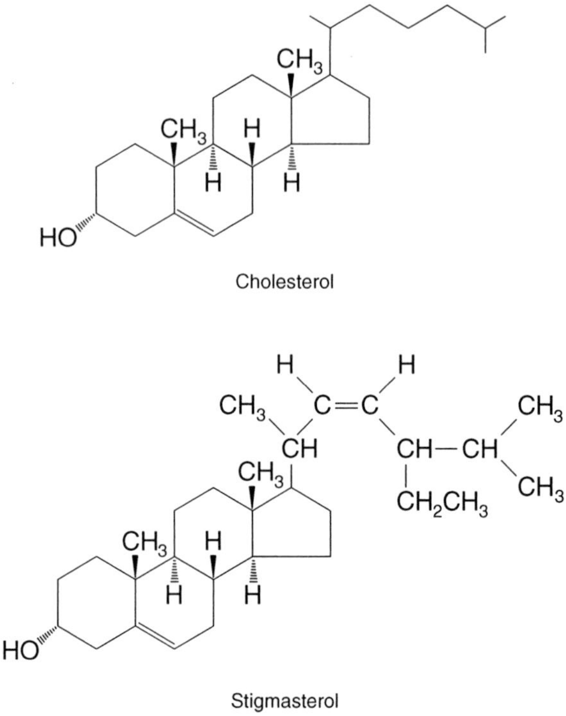

Cholesterol

Stigmasterol

Fig. 5.15 The general structure of waxes

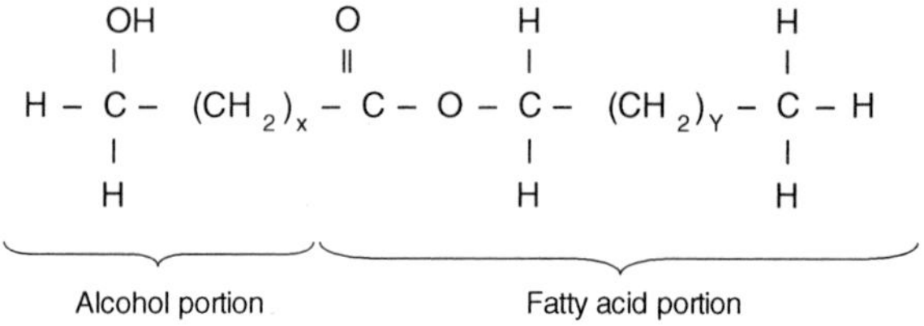

Alcohol portion   Fatty acid portion

These compounds are built up from two or more five-carbon structures called isoprene units (Fig. 5.16). Compounds with 10 carbons in two isoprene units are called monoterpenes; those with 15 carbons in three isoprene units are called sesquiterpenes. Diterpenes and triterpenes have 20 and 30 carbons in four and six isoprene units, respectively. The isoprene units of some terpenes are linked together to form ring structures. Terpenoids are terpenes that contain a hydroxyl group; menthol, which is extracted from peppermint, is a terpenoid. Essential oils have distinct odors and have been used for thousands of years in medicines and perfumes. Wood tars and pitches are used for waterproofing and as adhesives. The composition of terpenes and terpenoids is related to their plant source, which makes them effective biomarkers. Squalene, a triterpene, is the precursor of the important steroid, cholesterol.

Fig 5.16 The structure and examples of terpenes and terpenoids

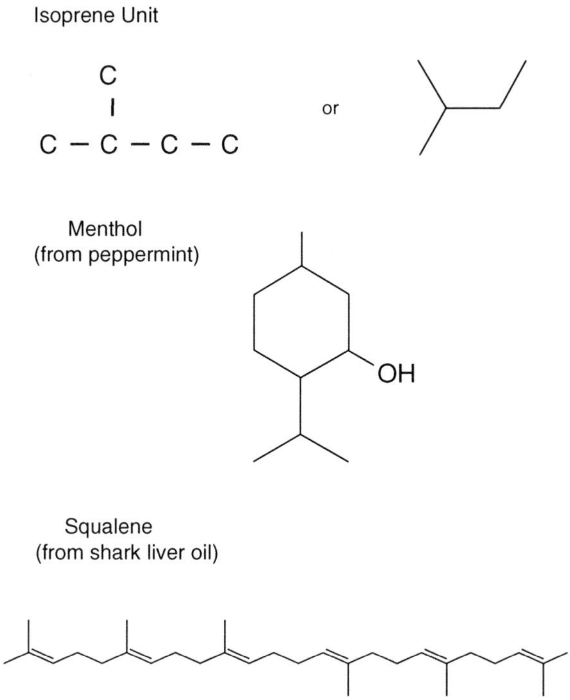

Isoprene Unit

or

Menthol
(from peppermint)

Squalene
(from shark liver oil)

Amino Acids and Proteins

Proteins are giant molecules thousands of times larger than carbohydrates or lipids. Each molecule is made up of individual amino acids linked like cars of a freight train. The properties of proteins are determined by their constituent amino acids and how they are arranged. Amino acids are compounds that have an amino group, a carboxyl group, and a hydrogen atom bonded to the same carbon atom, the α-carbon. The fourth group on the α-carbon is 1 of 20 side chains, called the R group, resulting in 20 different amino acids (Fig. 5.17).

In living organisms, the configuration of groups bonded to the α-carbon is the same. If the amino acid is oriented with the carboxyl group on top and the R group on the bottom, the amino group is on the left side of the α-carbon. These are referred to as L-amino acids, where L- is for levorotatory. This term is used because these amino acids are optically active and, when in solution, rotate plane-polarized light in a counter-clockwise direction or to the left (see Chapter 6). After an organism dies, some of their amino acids are transformed into their enantiomers (mirror images), D-amino acids. The D- is for dextrorotatory because these amino acids in solution rotate plane-polarized light in a clockwise direction or to the right. The rate at which this change occurs is different for each amino acid and the process stops when the number of D-forms of an amino acid equals the number of L-forms.

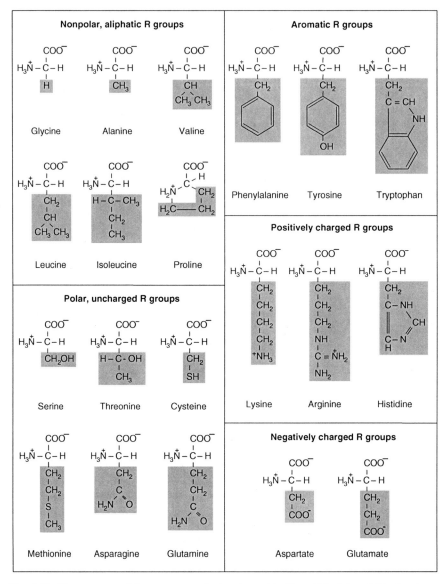

Fig. 5.17 The structure of the twenty standard amino acids (From Principles of Biochemistry, 2e by A.L. Lehninger, D.L. Nelson, and M.M. Cox. © 1993 by Worth Publishers. Used with the permission of W.H. Freeman and Company.)

Amino acids connect by dehydration sequences during which the acid group of one amino acid links to the amino group of another and a water molecule is lost. The process, called a condensation reaction, results in a covalent peptide bond (Fig. 5.18). Sequences of linked amino acids are called polypeptides or polypeptide

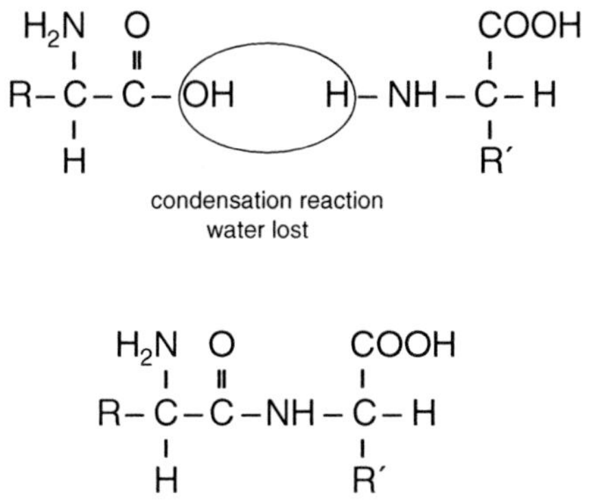

Fig. 5.18 Condensation reaction producing a peptide bond

$$
\underset{\text{H}}{\overset{\overset{\displaystyle \text{H}_2\text{N}}{|}}{\text{R}-\text{C}}}-\overset{\overset{\displaystyle \text{O}}{\|}}{\text{C}}-\text{OH} \qquad \text{H}-\text{NH}-\underset{\text{R}'}{\overset{\displaystyle \text{COOH}}{\overset{|}{\text{C}}}}-\text{H}
$$

condensation reaction
water lost

$$
\underset{\text{H}}{\overset{\overset{\displaystyle \text{H}_2\text{N}}{|}}{\text{R}-\text{C}}}-\overset{\overset{\displaystyle \text{O}}{\|}}{\text{C}}-\text{NH}-\underset{\text{R}'}{\overset{\displaystyle \text{COOH}}{\overset{|}{\text{C}}}}-\text{H}
$$

chains. A protein may consist of one or more polypeptide chains; those with multiple chains are called oligomeric. A hemoglobin molecule consists of four separate polypeptide chains.

Proteins can be divided into two categories, fibrous and globular. Fibrous proteins include water-insoluble collagen found in tendons, ligaments, and bones and α-keratin of hair, nails, horn, scales, feathers, and other tissues. Globular proteins are usually water soluble and have specific functions in cells. Globular proteins of archaeological interest include antibodies, serum albumin, hemoglobin, and myoglobin. Antibodies protect organisms from foreign substances; serum albumin has the function of transporting fatty acids. Both myoglobin and hemoglobin are involved in the transportation of oxygen.

The Structure of Proteins

Bond angles and forces of attraction or repulsion between different R groups on the amino acids determine the conformation of polypeptide chains within protein molecules. The sequence of amino acids bonded together in the polypeptide chains is referred to as the primary structure of the protein. The secondary structure refers to the spatial arrangement of the main polypeptide chain. It is primarily determined by the bond angles between adjacent atoms of amino acids forming the polypeptide. The secondary structure of fibrous proteins differs from that of globular proteins. Fibrous proteins occur in long strands or sheets. A common secondary structure of fibrous proteins is a "polypropylene rope-like" conformation, called the α-helix, of polypeptide chains. Their polypeptide chains are arranged as elongated helices around a straight screw axis. The twist of the α-helix in keratin is right handed while that of collagen is left handed. The fibrous proteins found in silk have a β-conformation where polypeptides are arranged side-by-side to form pleated sheets. The secondary structure of globular proteins is complex as they can incorporate many secondary structures into a single molecule.

The tertiary structure of proteins is due to interactions between the R groups, or side chains, of amino acids. Electrostatic attraction and repulsion and hydrogen bonding occur between amino acids of individual polypeptides and between adjacent polypeptide chains. In globular proteins, polypeptide chains are bent of folded to form compact structures. The general shape of these proteins is spherical or globular but the precise three-dimensional arrangement is specific to a particular globular protein. The manner in which the polypeptide chains of a globular protein are bent or folded to form the compact and tightly folded masses is referred to as its tertiary structure. Interaction between individual polypeptide chains of an oligomeric protein gives molecules their quaternary structure. Hemoglobin consists of four separate polypeptide chains closely linked through hydrogen bonds and electrostatic interactions between positively and negatively charged R groups.

The bonds that give proteins their primary and secondary structure are much stronger than the electrostatic attraction and hydrogen bonding responsible for their tertiary and quaternary structural organizations. The molecular structure of a native (or natural) protein is connected to how it functions in an organism or its biological activity. Most proteins are quite fragile and only retain their biological activity within a very limited range of temperature and pH. When subjected to heat or exposed to acids or bases, a protein can undergo denaturation (Fig. 5.19). The relatively weak linkages that produce the higher level structural organization of the molecule break. Denatured proteins have looser structures; most are insoluble and no longer able to function. The most visible effect of denaturation, decreased

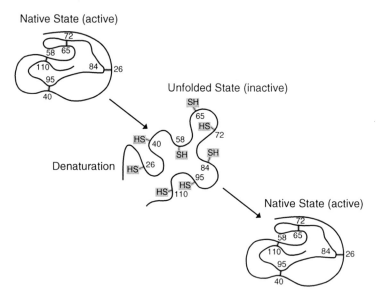

Fig 5.19 Tertiary structure of globular proteins: native state, denaturation, and renaturation (From Principles of Biochemistry, 2e by A.L. Lehninger, D.L. Nelson, and M.M. Cox. © 1993 by Worth Publishers. Used with the permission of W.H. Freeman and Company.)

solubility, is readily apparent in the case of egg white albumin. Egg white becomes white and hard upon heating. While this type of denaturation is irreversible, some proteins are able to slowly return to their native state if the denaturing conditions cease. Renaturation can occur in blood proteins, serum albumin and hemoglobin.

The study of protein and their constituent amino acids provide archaeologists with different data. Bone collagen is used for radiocarbon dating and stable isotope analyses for diet studies. Amino acid racemization, the degree to which particular amino acids have converted from their original L-forms into their D-forms is used as a dating technique. Identification of fresh blood to the family or species level is possible using immunological techniques and other forms of protein analysis.

Carbohydrates

Carbohydrates are products of plants that include sugars, starches, and cellulose. The simplest carbohydrates are monosaccharides, which are sugars that have names ending with the suffix -ose. In the presence of water and acid, more complex carbohydrates undergo a reaction called hydrolysis that produces monosaccharides. One mole of a disaccharide will produce 2 mol of monosaccharides; the hydrolysis of one mole of a trisaccharide will yield 3 mol of monosaccharides. Polysaccharides yield more than 10 mol of monosaccharides. Monosaccharides contain a carbonyl group, making them either aldehydes or ketones, and hydroxyl (–OH) groups are attached to other carbons. For this reason, carbohydrates can be described as hydroxy aldehydes or hydroxy ketones or compounds that yield hydroxy carbonyl compounds upon hydrolysis. Fructose and glucose are examples of monosaccharides. Sucrose, which is common table sugar, is a disaccharide. Plant starches and cellulose are examples of polysaccharides.

As outlined in Chapter 4, green plants produce carbohydrates during photosynthesis by the following general reaction:

$$xCO_2 + yH_2O + \text{sunlight} \rightarrow C_x(H_2O)_y + xO_2$$

Solar energy is captured by chlorophyll in the plants and converted to sugars and starches. When carbohydrates are eaten and digested by an animal, carbon dioxide, water, and energy are released. While some energy is converted to heat, some is stored as adenosine triphosphate (ATP) for later use. Animals also store energy as the carbohydrate, glycogen.

Sugars that contain nitrogen are components of ribonucleic acid (RNA) and deoxyribonucleic acid (DNA). Nucleosides have either a purine or a pyrimidine as the amino component and either D-ribose, in the case of RNA, or 2-deoxy-D-ribose, in the case of DNA, as the sugar component.

Deoxyribonucleic Acid

Deoxyribonucleic acid (DNA) is found in the cells of plants and animals. In cells with a nucleus, known as eukaryotic cells, 98% of its DNA is in the nucleus. The remaining 2% is in the mitochondria of plant and animal cells and the chloroplast found in the cells of plants and plant-like bacteria. DNA directs cellular functions; it contains the assembly instructions that living organisms use for the manufacture of protein.

James Watson and Francis Crick first proposed the double helix model of DNA in 1953. The DNA molecule is composed of two complementary strands. Each strand is made up of repeating subunits, called nucleotides, which consist of three covalently bonded chemical groups: a phosphate group, a five-carbon sugar, and a nitrogen base (Fig. 5.20). The phosphate group, triphosphate, and the sugar,

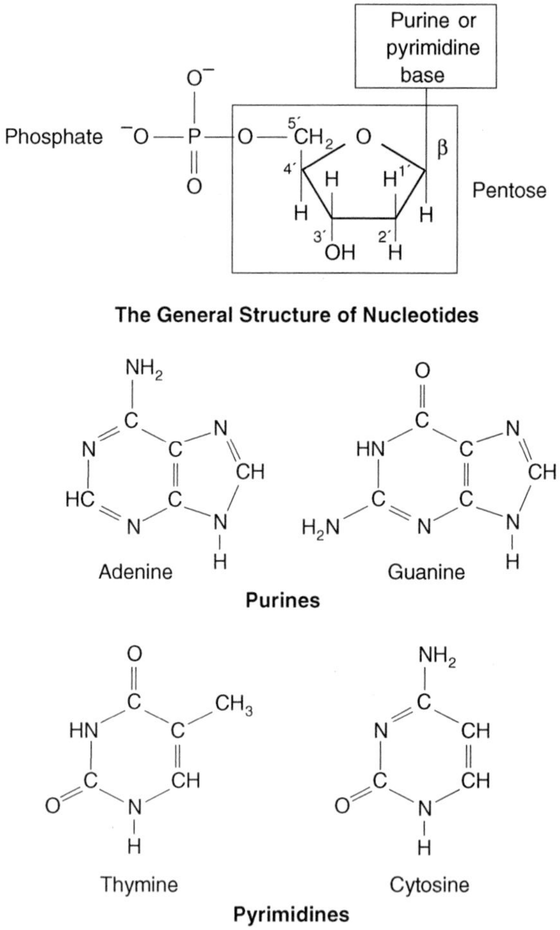

The General Structure of Nucleotides

Purines

Adenine   Guanine

Pyrimidines

Thymine   Cytosine

Fig. 5.20 The structure of nucleotides, purines, and pyrimidines

2-deoxy-D-ribose, are the same in every nucleotide, but four different nitrogen bases exist. The four bases of the amino-component are adenine (A) and guanine (G), which are purines, and cytosine (C) and thymine (T), which are pyrimidines. Each strand forms one-half of a DNA ladder, with phosphate and sugar forming the sides of the ladder and the nitrogen base making up one-half of the rung. Hydrogen bonds between nitrogen bases connect the two sides of the ladder together and cause the molecule to twist. The twisted nature of the two strands gives rise to the term, double helix.

Due to the manner in which DNA forms or breaks apart, the phosphate group is always found at one end of a strand and the sugar at the other. When describing a strand of DNA, the phosphate end is the 5′-end and the sugar is the 3′-end. The numbering designates the five carbons of the sugar. The nitrogen base is bonded to the first carbon, 1′ C. The hydroxyl group (OH−) of the sugar is bonded to the third carbon, 3′ C. The phosphate group is attached to the fifth carbon, 5′ C. When using shorthand notation, a DNA strand is described by the nitrogen bases present and specifies the ends as either 3′ or 5′ (Fig. 5.21).

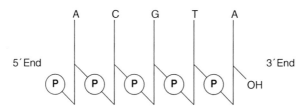

Fig. 5.21 Numbering the ends of DNA strands (From Principles of Biochemistry, 2e by A.L. Lehninger, D.L. Nelson, and M.M. Cox. © 1993 by Worth Publishers. Used with the permission of W.H. Freeman and Company.)

The four different nitrogen bases give rise to four different nucleotides (Fig. 5.20). The reason that DNA consists of complementary strands is due to the base pair rule. Although there are four nitrogen bases, the bonding between them is controlled. Adenine can only form a hydrogen bond with thymine and guanine only pairs with cytosine. The arrangement of base pairs on one strand must properly match the base pairs on the other, giving rise to complementary strands (Fig. 5.22). Because of the base pair rule, if the nucleotides of one strand are known, it is possible to deduce which nucleotides appear on the opposite strand.

DNA Replication

The operation of the base pair rule can be observed during DNA replication. DNA is capable of producing an exact copy of itself during a process directed, or catalyzed, by specialized enzymes called polymerases. The first step of replication is the separation of the DNA molecule into two strands by breaking the hydrogen bonds between the complementary bases (Fig. 5.23). This is similar to the unzipping of a zipper, except the bases are not intermeshed. Nucleotides from the environment

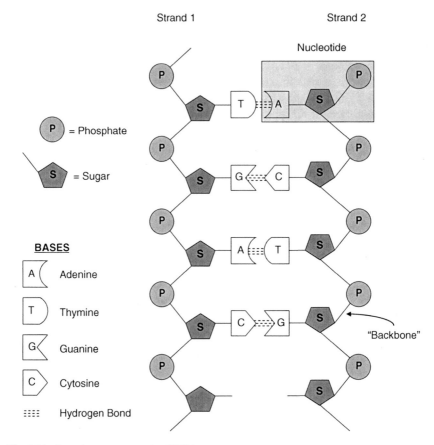

Fig. 5.22 Complementary strands of DNA

attach themselves to the exposed strands in accordance with the base-pairing rule. The final result will be two DNA molecules, identical to each other and to the original double helix.

In this manner, the two previously joined parental nucleotide chains serve as models or templates for the formation of a new strand of nucleotides.

Enzyme-Catalyzed DNA Synthesis

As mentioned above, polymerase enzymes must be present for DNA synthesis to occur. Polymerases cannot direct the formation of completely original strands; nucleotides can only be added to a pre-existing strand that serves as the template (Fig. 5.24). A primer must be in place on, and complementary to, the template to mark the starting point of DNA synthesis. The primer forms the first nucleotides of

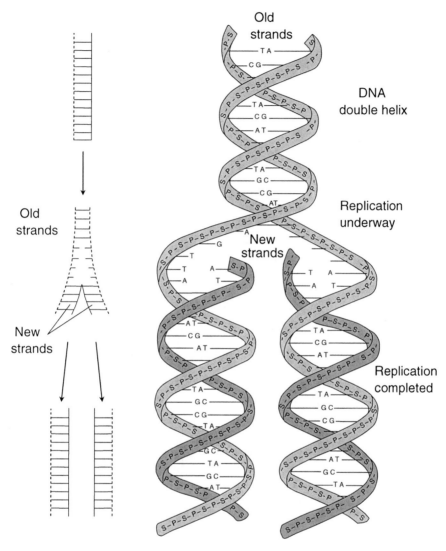

Fig. 5.23 DNA replication

the growing strand. All four different nucleotides must be present in the environment in the form of deoxynucleoside 5′ triphosphates (dNTPs). During the addition of nucleotides, phosphate binds to the O on the 3′ carbon of sugar and inorganic pyrophosphate is released. The new strand always grows in the 5′−3′ direction. The successive nucleotides of DNA are covalently bonded through phosphodiester linkages (Fig. 5.25). Strands consisting of 50 or fewer nucleotides are called oligonucleotides; longer strands are called polynucleotides.

In summary, natural DNA synthesis requires

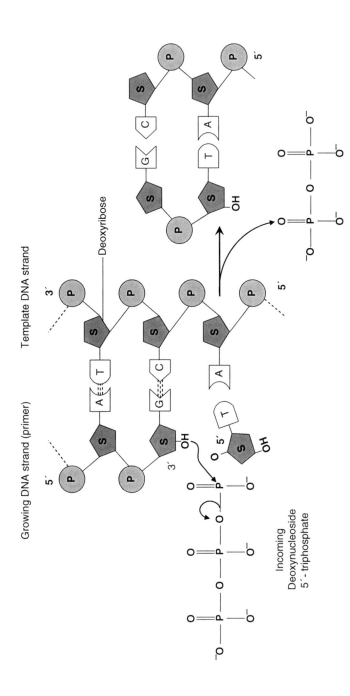

Fig. 5.24 Addition of nucleotides to growing strand (From Principles of Biochemistry, 2e by A.L. Lehninger, D.L. Nelson, and M.M. Cox. © 1993 by Worth Publishers. Used with the permission of W.H. Freeman and Company.)

Fig. 5.25 Phosphodiester
linkages in one strand of
DNA

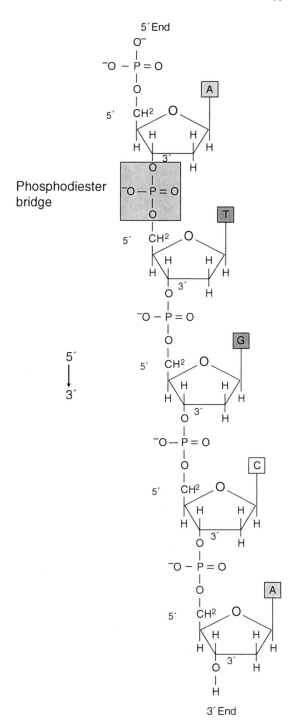

Phosphodiester
bridge

(1) Breakage of hydrogen bonds between two complementary strands of DNA;
(2) primers attached on the strands to mark the starting point of synthesis;
(3) the presence of all four varieties of nucleotides in the form of dNTPs: 2'-deoxyadenosine 5' triphosphate (dATP), deoxythymidine 5' triphosphate (dTTP), 2'-deoxyguanosine 5' triphosphate (dGTP), and deoxycytidine 5' triphosphate (dCTP);
(4) the presence of polymerase to direct the synthesis of DNA.

As described in Chapter 16, the study of ancient DNA was greatly facilitated by the development of the polymerase chain reaction (PCR). PCR mimics natural enzyme-catalyzed DNA replication and is used to detect and make millions of exact copies of a targeted segment of DNA. The area targeted for amplification depends upon the sample and data required by the researcher and may provide information on genus/species, kinship, gender, and evolutionary lineage.

Chapter 6
Other Useful Concepts

Acids and Bases

From a molecular perspective, acids are substances that donate protons (hydrogen nuclei or H^+). In the presence of strong acids, water acts as the base that accepts protons to produce hydronium ions (H_3O^+). Strong acids readily donate protons so the dissociation reaction goes to completion (Table 6.1). Bases include substances that have pairs of unshared, or loosely bound, electrons, which are capable of accepting protons. In the presence of strong bases, water acts as an acid that donates protons to form hydroxide ions (OH^-). Sodium hydroxide is a strong base; sodium and hydroxide ions are produced in water and the reaction goes to completion:

$$NaOH + H_2O \rightarrow Na^+ + OH^-$$

The strongest base that can form in an aqueous (water-based) solution is the hydroxide ion. The strongest acid that can form in an aqueous solution is the hydronium ion. The stronger the acid, the weaker its conjugate base; likewise, the stronger the base, the weaker its conjugate acid. Water is a very weak acid; its conjugate base (OH^-) is very strong. If stronger acids are required, they can be prepared in another solvent, such as liquid ammonia.

A strong acid in aqueous solution can be completely neutralized by a strong base in aqueous solution:

$$HCl + NaOH \rightarrow H_2O + Na^+ + Cl^-$$

In reality, the reaction actually involves only the hydronium and hydroxide ions; the sodium and chloride ions are spectator ions that do not participate.

Reactions of weak acids (or weak bases) in water do not go to completion. Instead, only a fraction of acid atoms donate their proton to water so fewer hydronium atoms form. An equilibrium point is reached where the number of reactions is equal to the number of reverse reactions that occur. The amount that reacts depends

M.E. Malainey, *A Consumer's Guide to Archaeological Science*, Manuals in Archaeological Method, Theory and Technique, DOI 10.1007/978-1-4419-5704-7_6, © Springer Science+Business Media, LLC 2011

Table 6.1 Acidity constants for various compounds. Reprinted from T.W. Graham Solomons, Table 2.2, "Organic Chemistry", Second Edition, copyright 1980, with permission of John Wiley and Sons

	Acid	Approximate K_a
	C_6H_{12}	10^{-45}
	CH_3CH_3	10^{-42}
	$CH_2=CH_2$	10^{-36}
	NH_3	10^{-34}
Increasing	$HC\equiv CH$	10^{-25}
Strength of	CH_3CH_2OH	10^{-18}
Acid	H_2O	10^{-16}
	CH_3COOH	10^{-5}
	CF_3COOH	1
	HNO_3	20
	H_3O^+	50
	HCl	10^7
	H_2SO_4	10^9
	H	10^{10}
	$HClO_4$	10^{10}
	$SbF_5 \cdot FSO_3H$	$>10^{12}$

upon the concentration of the acid. The following hypothetical reaction can be used to describe any acid dissolved in water:

$$HA + H_2O \rightarrow H_3O^+ + A^-$$

The equilibrium constant for this reaction is

$$K_a \text{ or } K_{eq}[H_2O] = \frac{[H_3O^+][A^-]}{[HA]}$$

Strong acids completely dissociate in water, which produces a large K_a value. Weak acids only partially dissociate, producing a small K_a value.

Not all acids are water soluble. Carboxylic acids with more than five carbons do not readily dissolve in water. As with other lipids, these fatty acids are soluble in organic solvents. Water-insoluble acids will dissolve in aqueous sodium hydroxide and react to form water-soluble sodium salts.

The pH Scale

A very small portion of pure water ionizes to produce H_3O^+ (sometime simplified to H^+) and OH^-. At 25°C, the ion product of water, or the water constant, K_w, is 1.0×10^{-14}. This can be expressed as follows:

$$1.0 \times 10^{-14} = [H^+][OH^-]$$

In pure water, the concentration of both ions is equal to 1.0×10^{-7} M. A strong acid in aqueous solution has a high concentration of hydronium ions and a low concentration of hydroxide atoms. The concentration of hydronium atoms in a strong base in aqueous solution is low and the concentration of hydroxide atoms is high.

The pH scale is a shorthand method of expressing the concentration of hydronium ions in a solution (Table 6.2). From the water constant, we know the concentration of hydronium will vary between 10^0 and 10^{-14} M. The pH of a solution is the negative common logarithm of the hydronium ion concentration:

$$pH = \frac{\log 1}{[H^+]} = -\log [H^+]$$

Table 6.2 The pH scale

pH	$[H^+]$	$[OH^-]$	
14	10^{-14}	10^0	
13	10^{-13}	10^{-1}	
12	10^{-12}	10^{-2}	Alkalinity
11	10^{-11}	10^{-3}	increases
10	10^{-10}	10^{-4}	
9	10^{-9}	10^{-5}	
8	10^{-8}	10^{-6}	
7	10^{-7}	10^{-7}	Neutral
6	10^{-6}	10^{-8}	
5	10^{-5}	10^{-9}	
4	10^{-4}	10^{-10}	
3	10^{-3}	10^{-11}	Acidity
2	10^{-2}	10^{-12}	increases
1	10^{-1}	10^{-13}	
0	10^0	10^{-14}	

In pure water, the hydronium concentration is 1.0×10^{-7} M,

$$[H^+] = 1.0 \times 10^{-7} \text{ M}$$
$$\log[H^+] = -7$$
$$pH = -\log [H^+] = 7$$

A strong acid in solution with a hydronium ion concentration of 1×10^{-1} M has a pH of 1. A strong base in solution with a hydronium ion concentration of 1×10^{-14} has a pH of 14.

Buffers

Many reactions are sensitive to pH so buffers are used to ensure the optimal level of acidity or alkalinity is maintained. Buffers are solutions that are capable of maintaining their pH at some fairly constant value, even with the addition of small amounts of acid or base. A buffer solution usually consists of a fairly high concentration of a weak acid and its salt in order to resist the effects of dilution.

Stereochemistry and the Polarimeter

Stereochemistry refers to the arrangement of atoms of a molecule in space. The particular way in which atoms of a compound are joined, or its connectivity, can greatly affect its properties. The term, isomers, is given to compounds that have the same composition, or molecular formula, but different three-dimensional structures. There are two major categories of isomers: structural (constitutional) isomers and stereoisomers. Structural isomers differ in the manner in which their atoms are joined, resulting in different constituents (Fig. 6.1). The chemical composition of a branched alkane may be the same as a straight chain alkane but the structure of their carbon chains is different; these are examples of chain isomers. A functional group may be bonded to one carbon in one isomer but the same functional group may be bonded to a different carbon in another, giving rise to positional isomers. In some cases, the arrangement of atoms in the molecules fundamentally changes the properties of the compound because different functional groups result. For example, an alcohol can have the same chemical formula as an ether; these are examples of functional group isomers.

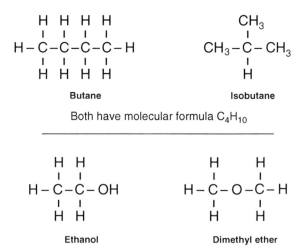

Fig. 6.1 Structural isomers

Fig. 6.2 Stereoisomers

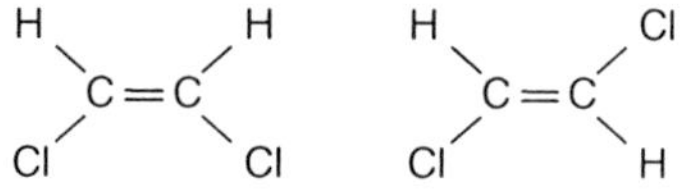

cis-1,2 -Dichloroethane   trans-1,2 -Dichloroethane

Stereoisomers have the same constituents and they are attached in the same order; however, the arrangement of the atoms in space is different. There are two categories of stereoisomers: diastereomers and enantiomers. Diastereomers are stereoisomers that are not mirror images of one another. This can easily occur in compounds that have multiple bonds, resulting in the possible formation of *cis*- and *trans*-isomers (Fig. 6.2). If groups are attached to the same side of a double bond, it is called the *cis*-isomer. If groups are attached to opposite sides of the double bond, it is called the *trans*-isomer. A *cis*-isomer is not a mirror image of the *trans*-isomer. Enantiomers are stereoisomers that are mirror images but cannot be superimposed. They occur in compounds that have chiral carbons. A chiral carbon is a carbon atom with four different groups attached to it. Any compound with one chiral carbon will have two enantiomers. Chiral objects have handedness, in the same way the left hand is the mirror image of the right hand but the two cannot be superimposed. This property can be demonstrated in 2-butanol, which has two enantiomers (Fig. 6.3).

While the basic physical properties, such as melting point, boiling point, and density, are different for structural isomers and diasteromers, they are the same for enantiomers. One way to separate enantiomers is to monitor their interactions with other chiral compounds. Since the spatial arrangements of atoms are not the same, the solubilities and reaction rates of enantiomers may differ. Reactions that result in the formation of only one set of stereoisomers are called stereoselective. If one set of stereoisomers, e.g., L-amino acids, react to produce a particular set of stereoisomers and the other set, e.g., D-amino acids, produce stereoisomerically different products, the reaction is stereospecific.

Another way to separate enantiomers is through the use of a polarimeter to observe their effect on plane-polarized light. Plane-polarized light has chiral properties so enantiomers are optically active. When a beam of plane-polarized light is passed through a solution containing only one enantiomer, the plane of polarization rotates. If the same light is passed through a solution containing only the other enantiomer, the plane of polarization will rotate exactly the same amount, but in the opposite direction. Samples containing equal concentrations of both enantiomers have no net effect on the plane of polarization; these are called racemic modifications or racemates.

Certain reactions can affect the optical activity of chiral compounds by inverting the configuration of some molecules. These reactions cause optically pure substances to undergo racemic modification. This process, called racemization, can occur among a small portion of the molecules, resulting in partial racemization. Complete racemization occurs when one-half of the molecules are converted into their inverse configuration and optical activity is lost.

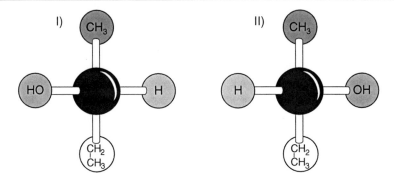

3D drawings of the 2-butanol enantiomers I and II.

Models of the 2-butanol enantiomers.

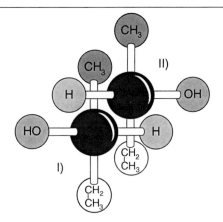

Unsuccessful attempt to superpose models of I and II.

Fig. 6.3 Enantiomers cannot be superimposed (Reprinted from T.W. Graham Solomons, Figure 8.3, "Organic Chemistry", Second Edition, copyright 1980, with permission of John Wiley & Sons, Inc.)

The number of chiral carbons present governs the number of stereoisomers possible for a chiral compound. The total number of stereoisomers cannot exceed 2^n, where n is the number of chiral carbons. A compound with one chiral carbon cannot have more than two stereoisomers; one with three chiral carbons cannot have more that eight (2^3 or $2 \times 2 \times 2$) stereoisomers.

The Polarimeter

Light beams consist of two mutually perpendicular oscillating fields, an electric field and a magnetic field. The planes in which these fields oscillate are also perpendicular to the direction the light beam is traveling or propagating. In ordinary, unpolarized light, electric and magnetic fields oscillate in all possible planes perpendicular to the direction of propagation. When ordinary light is passed through a polarizer, the electric field of the light that emerges will oscillate in only one plane and the magnetic field will oscillate in one perpendicular plane. This is referred to as plane-polarized light.

A polarimeter is used to measure the effect of optically active compounds, such as amino acids, on plane-polarized light (Fig. 6.4). The instrument consists of a light source, the beam of which is passed through a polarizer to generate plane-polarized light. A tube is positioned to hold the sample solution in the beam of light. Light emerging from the tube is passed through an analyzer, which is another polarizer, to determine the effect of the sample on the light beam. A scale measures the number of degrees the plane-polarized light has been rotated.

An optically active substance that rotates plane-polarized light in a clockwise direction, or to the right, is called dextrorotatary. A positive sign (+) is used to indicate this type of rotation and the substance can be called a D-enantiomer, or designated with letter R, for *rectus*, the Latin word for right. Rotation of plane-polarized light in the counter clockwise direction, or to the left, is termed levorotatory and indicated by a negative sign (−). These substances can be called L-enantiomers, or designated with the letter S, for *sinister*, the Latin word for left.

Plane-polarized light is not rotated by optically inactive samples because all possible orientations of molecules are present. The effect of molecules that rotate the light in one direction is cancelled by the effect of molecules rotating the light in the opposite direction. A solution containing equal amounts of both L-and D-enantiomers would be optically inactive, because the effects of all rotations are cancelled. These solutions are called racemic modifications or racemates.

Samples consisting of only one enantiomer, or optically pure, have the maximum effect on plane-polarized light. If any of the other enantiomer is present, the degree of rotation will be reduced. The optical purity of a sample is determined by comparing the effect of the sample, in degrees of rotation, to the effect of one that is optically pure:

$$\text{Optical purity} = \frac{\text{Observed rotation}}{\text{Specific rotation of the pure enantiomer}} \times 100\%$$

Chemical Reactions

Chemical reactions involve the breaking of existing chemical bonds and the making of new ones. In order for a reaction to occur, reacting substances must come into contact, or collide, with sufficient energy and in the proper orientation. An effective collision is one that enables the reaction to proceed. The frequency and energy of

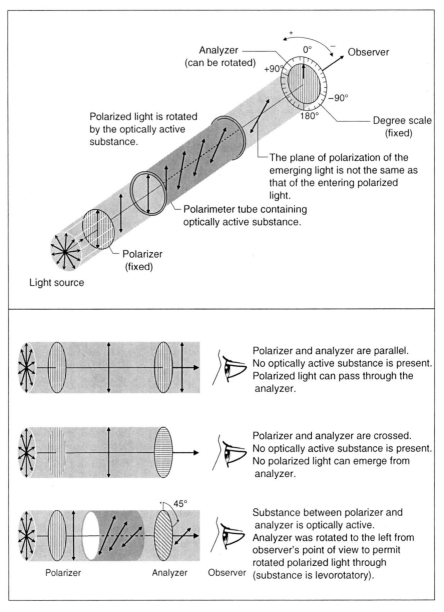

Fig. 6.4 The polarimeter and measurement of optical activity (Reprinted from John R. Holum, Figure 13.6, "Principles of Physical, Organic, and Biological Chemistry: An Introduction to the Molecular Basis of Life", copyright 1969, with permission of John Wiley & Sons, Inc)

collisions depend upon the concentrations of molecules and temperature. The number of collisions increases as the concentrations of reacting substances increase. Rates of reaction are usually highest when the concentrations of reactants are high. The actual effect of the concentration of reactants on reaction rate must be determined experimentally. Doubling the concentration of a reactant may double the reaction rate of one reaction and quadruple the reaction rate of another. In some cases, reaction intermediates form that are both produced and used in the course of the reaction. Intermediates are neither reactants nor products but their formation is required for the reaction to proceed. Under these circumstances, the formation of intermediates may limit the rate of the reaction.

Reacting substances must also collide with sufficient energy for a reaction to proceed. All molecules have negatively charged electron clouds. Slowly moving molecules may have insufficient energy to overcome the force of mutual electrostatic repulsion. A low-energy collision may simply cause the molecules to bounce off one another unchanged. Faster moving molecules have more energetic impacts so collisions are more likely to result in a reaction. The energy with which collisions occur increases with temperature. Molecules move more slowly at lower temperatures and faster at higher temperatures. Both the number of collisions and the number of effective collisions increase with temperature; consequently, a relatively small increase in temperature can greatly increase the reaction rate. In general, an increase of $10°C$ will cause the reaction rate to double. The energy required for a reaction to proceed is called the energy of activation.

The rate of chemical reaction may also be affected by the orientation of the molecules involved. A reaction may require a collision between specific parts of the molecules, such as functional groups. Depending on the complexity of their shapes, only a small portion of the collisions between reactants may occur with the molecules in the proper orientation. Since the orientation of the molecules is random, this is referred to as the probability factor. If the rate of reaction for one enantiomer is different from another when all other conditions are identical, one arrangement of atoms is preferential and gives rise to more effective collisions.

In organic chemistry, reduction reactions are those that increase the hydrogen content or decrease the oxygen content of a molecule. A reducing agent removes an oxygen atom from the molecule. Oxidation reactions are those that increase the oxygen content or decrease the hydrogen content of an organic compound. An oxidizing agent adds an oxygen atom to the molecule. In other words, reducing agents are oxidized and oxidizing agents are reduced.

Part II
Applications

Part II includes chapters on describing principles, procedures, and issues related to the most common applications of scientific methods to archaeological materials. After a brief examination of methodological considerations (Chapter 7), several dating techniques are considered. Radiocarbon dating is presented in Chapter 8; other radioactive decay-based dating techniques, uranium series, potassium–argon, and argon–argon, and fission track, are described in Chapter 9. Trapped charge dating techniques, which include thermoluminescence, optically stimulated luminescence, and electron spin resonance dating, are considered together in Chapter 10. Amino acid racemization, obsidian hydration, cation ratio dating, and archaeomagnetism are presented together in Chapter 11. The goal is to present these techniques in a straightforward and accurate manner; however, science-based dating techniques are very complex. The theoretical foundations of these approaches fall into the realm of calculus and physics and well beyond the scope of this text. Readers interested in pursuing these topics are further invited to consult general texts on quaternary dating methods and geochronology (Harper 1973; Mahaney 1984; Walker 2005) or works addressing specific approaches (Bourdon et al. 2003; Dalrymple et al. 1969; Dickin 2005; Eighmy and Sternberg 1990).

General issues related to provenance studies are discussed in Chapter 12. The analysis of the stable isotopes of archaeological interest, carbon, nitrogen, strontium, lead, oxygen, and hydrogen, is described in Chapter 13. Stable isotope analysis has broad and diverse archaeological applications ranging from metallurgy and geology to bioarchaeology and serves as a transition from the analysis of inorganic materials to organic materials. Lipid residue analysis (Chapter 14), blood and protein analysis (Chapter 15), and ancient DNA and the polymerase chain reaction (Chapter 16) are described in the final chapters of Part II. Foundations of these approaches lay in organic chemistry, biochemistry, and other biological sciences. The terminology presented with the descriptions is fairly basic but should enable an archaeologist to better understand and critically assess archaeological applications of these techniques.

Chapter 7
Methodological Considerations

In order to effectively and efficiently utilize archaeological science methods, a clear research problem must exist. General research questions often revolve around establishing the time of occupation, trade patterns, subsistence, settlement patterns, technologies, and site activities. These questions can be further refined into hypotheses which may be tested by applying specific analytical techniques to suitable artifacts. Key aspects of this process involve acknowledging (1) the limitations of the analytical technique and (2) assumptions made when assessing the suitability of the artifact.

Limitations of Analytical Techniques

Broadly speaking, most techniques discussed in this text provide compositional analyses of archaeological materials. An instrument only provides a measure of what is actually in a sample. If it was in proper working order and all the procedures regarding sample processing were followed, data generated in the course of analysis will reflect the composition of the sample submitted. However, an instrument cannot discriminate between constituents introduced by cultural processes from those introduced by natural processes. Even if a sample is free of contamination, diagenetic processes may have altered constituents introduced by cultural processes in ways that affect their interpretive value.

In general, procedures relating to the analysis of archaeological samples evolve over time. Sample selection guidelines change, processing procedures are modified, and new instruments are developed. It is not uncommon for concerns to be raised about the reliability of data that were obtained using techniques that were considered state of the art a few decades ago (Gale and Stos-Gale 1992; Pettitt et al. 2003). Even techniques that provide excellent data are not immune to change. The availability of instrumental neutron activation analysis (INAA) will decrease in the future because many research reactors may be decommissioned over the next 50 years (Kennett et al. 2002; Speakman and Glascock 2007).

M.E. Malainey, *A Consumer's Guide to Archaeological Science*, Manuals in Archaeological Method, Theory and Technique, DOI 10.1007/978-1-4419-5704-7_7, © Springer Science+Business Media, LLC 2011

Precision and Accuracy

Precision and accuracy of data are important considerations and discussed in virtually every analytical chemistry and instrumental analysis textbook. The material in this chapter is primarily based on Skoog and West (1982), Skoog et al. (1998), Willard et al. (1988), and Harris and Kratochvil (1981). Bishop et al. (1990) addressed these topics with respect to the construction of ceramic compositional databases. Precision refers to the ability to provide reproducible results; measures of precision can only be applied to samples processed and analyzed in exactly the same manner. One method of assessing precision is through replicate measurements on the same sample. From these measurements, an average and standard deviation can be calculated, which are then used to describe the confidence interval for the data (Fig. 7.1). Assuming the data set has a normal distribution, the true value will fall within the confidence interval defined by the average value plus and minus one standard deviation, which is the 1σ range, 68% of the time. The true value will fall within the confidence interval defined by the average value plus and minus two standard deviations, the 2σ range, 95% of the time. Small standard deviations and narrow confidence intervals are a feature of precise data. Even so, it is important to remember that at the 1σ range, the true value will fall outside the confidence interval 32% of the time.

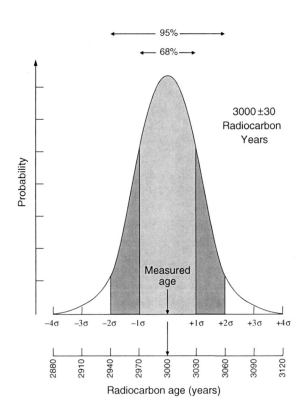

Fig. 7.1 Normal distribution (From LONGMAN SCIENCE BASED DATING IN ARCHAEOLOGY © 1990 Pearson Education, Inc. or its affiliates. Used by permission. All rights reserved.)

Accuracy is a measure of how close the value is to the actual, true, and correct value. Skoog et al. (1998:A1) note that "[s]trictly speaking, the only type of measurement that can be completely accurate is one that involves counting objects. All other measurements contain errors and give only an approximation of the truth." An accepted or consensus value of the quantity being measured may be provided for a standard. In this case, the difference between the value obtained and the accepted value is referred to as absolute error. When the difference is expressed as a percentage, parts per million (ppm), or parts per billion (ppb), it is referred to as relative error. Ideally, a measurement is made with high degrees of both accuracy and precision. It is possible to obtain a measure with a high degree of precision but a low degree of accuracy, however.

Sensitivity

Sensitivity refers to the detection limits of the instrument or, as Bishop et al. (1990:538) state with respect to compositional data, as "the minimal amount of an elemental concentration that can be detected given the experimental conditions." Sensitivity varies between instruments and between techniques. For example, the sensitivity of a mass spectrometer is given by its ability to discriminate ions of different mass, which is called its resolution (Chapter 31); an instrument with a resolution of 100,000 is more sensitive than an instrument with a resolution of 1000. Atomic absorption (AA) spectroscopy is more sensitive than optical emission spectroscopy, but with the introduction of the inductively coupled plasma (ICP) torch, the sensitivity of emission spectroscopy now surpasses AA (Chapter 34). Due in part to the higher sensitivity of ICP-MS (Chapters 31 and 34), some people are opting to use this technique to analyze the bulk composition of ceramics (Chapter 17) in place of INAA (Chapter 32). Instruments with very low detection limits are considered to be highly sensitive.

Sources of Error

Ensuring and maintaining the connection between the experimental results and the research question rests with both the archaeologist and the analyst. The archaeologist is largely responsible for selecting sufficient numbers of representative and suitable samples. The analyst is responsible for processing the samples in a manner that maximizes the accuracy and precision of the results. Either or both may be involved in the data processing but the archaeologist is generally responsible for interpreting the results and assessing their significance. Assumptions are inherently made with respect to the context and degree of alteration and/or contamination of the sample. Invalid assumptions lead to erroneous results. Critical analysis of analytical data is only possible if assumptions made regarding samples are acknowledged.

The Nature of Archaeological Samples

A number of factors affect the selection of material for analysis, including artifact availability, money, and time. The first factor, artifact availability, relates to the
nature and condition of the archaeological recoveries as well as the consent of those
with a vested interest in the material. While the effect of taphonomic processes
on archaeological faunal assemblages has been extensively considered (Hesse and
Wapnish 1985; Lyman 1994), all artifacts and materials with which they are associated are similarly affected. The compositions of artifacts recovered during the
course of an archaeological investigation were impacted by the decisions and activities of the former site occupants. The range of lithic materials at a site is related
to the intrinsic properties of the rock and the qualities the ancient inhabitants valued. Pottery is made from mixtures of materials; during pottery manufacture, one
or more clays from different sources may have been purified and then mixed with
one or more types of temper from different areas. The composition of metal artifacts may be even more complex. Refining processes, plating, formation of alloys,
and recycling can all cause a significant divergence between the compositions of
metal ores and the newly made metal objects. The initial composition of organic
residues can differ markedly from its source materials due to the effects of thermal
and oxidative degradation.

An artifact may be subject to a variety of alteration processing prior to and
after discard and burial. Exposure to sunlight, oxygen, microorganisms, and water
can cause chemical reactions that alter the composition of archaeological materials. These effects can be either apparent, resulting in corrosion or the formation of
patinas or carbonate deposits, or invisible, leading to protein denaturation, DNA
fragmentation, and lipid decomposition. The results of groundwater percolation,
proximity to radioactive substances, exposure to natural fires, soil acidity, to name a
few, all alter the composition of archaeological materials deposited in terrestrial
environments. Elements can be removed by leaching and demineralization processes, added through precipitation or modified through recrystallization. Materials
recovered from aquatic environments, either freshwater or marine, are subject to different processes, the effect of which also must be understood. The potential effects
of these and other factors with respect to various applications are described in the
balance of Part II; more specific effects on different material types are discussed in
the case studies appearing in Part III.

Artifact availability is also related to the value or significance of the material as perceived by the archaeologist, government regulators, collection managers,
museum curators, descendants of the previous site occupants, landowners, as well
as other stakeholders and interest groups. Archaeological resources are finite and
usually only a tiny portion of materials used by the previous inhabitants of a site
are preserved and recovered. Artifacts from burial contexts, connected with rituals, made from exotic material, or rarely recovered are less likely to be submitted
for analysis. If analyzed, minimally or non-destructive techniques are often applied.
The ability to determine the composition of whole or miniscule portions removed
or extracted from larger items can be advantageous, especially when valuable, rare,

or sacred objects are analyzed. The composition of a solid is not necessarily homogeneous; very small samples may not be representative of the whole. In particular, the composition of the surface may differ from the bulk of the solid. These factors are discussed in more detail in Chapter 30.

Archaeological investigations, particularly in the context of cultural resource management (CRM), frequently suffer shortages of both money and time. Lack of financial resources can limit the number and types of samples submitted for analysis. If an emphasis is placed on establishing the timing of occupation, few resources may remain to investigate other aspects of human behavior at the site and regional interaction patterns. The homogeneity and normality of potential source material and samples may be assumed rather than demonstrated. Inadequate numbers of samples may be analyzed to ensure the results are statistically significant. If materials initially obtained prove to be unsuitable for analysis, there may be insufficient resources to return to the site, collect new samples, and repeat the analytical procedure. Due to the abbreviated timeline between site discovery, assessment, mitigation, and reporting, there is often insufficient time to arrange for experts to visit a site and supervise sample collection. Expenses associated with obtaining such professional advice may not appear in a CRM budget.

The Nature of Laboratory Analysis

Sampling continues to be an issue in the laboratory because often only a portion of the material submitted is actually analyzed. While this affects all types of analysis, it is particularly relevant when non-destructive techniques are used, as only a miniscule amount of material is removed for compositional analysis. If a sample must be analyzed in solution, problems arise if some components do not dissolve or only partially dissolve.

Instrumental errors relate to the operation of the equipment during analysis. Voltage fluctuations or changes in temperature can affect the operation of an instrument. A leak can develop in the system; a component, such as a light, switch, circuit, or filament, may fail; a computer may crash. Instrumental errors can be detected by proper monitoring of the system and the running of standards. Human errors are those introduced by the analyst and can be minimized by double checking, experience with lab equipment, and by simply paying attention. Automated systems provide a degree of standardization rarely achievable by humans and can reduce the potential effect of these types of errors. Method errors arise during sample processing due to problems with reagents or reactions. These may be detected by processing a sample of known composition together with a batch of unknowns.

Conclusions

Compositional analysis provides a measurement of sample constituents. Unless the relationship between those constituents and activities at or connected to an archaeological site is empirically tested or otherwise clearly demonstrated, it is conjecture.

Each piece of archaeological information contributes to the overall picture; however, proof often remains elusive. Even under the most favorable conditions, archaeological data are incomplete and imperfect and it is important to understand these limitations. Multiple lines of evidence all pointing to the same conclusion strengthen an explanation. By amassing a solid body of evidence, a powerful case can be constructed in support of an archaeological interpretation.

Chapter 8
Radiocarbon Dating

A team of scientists led by Willard F. Libby developed radiocarbon dating; for this, Libby was awarded the 1960 Nobel Prize in chemistry. Today, it is the most widely used method of dating archaeological materials from the late Pleistocene and Holocene. The radiocarbon age of a sample is based on the amount of radioactive carbon remaining in it. Depending on how measurements are made, organic remnants of plants and animals that lived upward of 60,000 years ago are datable.

In this chapter, the principles of radiocarbon dating are presented, as well as the various procedures involved in calculating a radiocarbon age estimate. The relationship between radiocarbon ages and calendar dates is complex. By convention, laboratories use Libby's original calculation for the half-life of ^{14}C, which is off by more than 3%. The distribution of the isotopes of carbon in living organisms is not uniform so corrections for isotopic fractionation in samples other than wood or charcoal are necessary. A variety of factors cause fluctuations in the amount of radioactive carbon in the environment. Dates can be calibrated to account for fluctuations in atmospheric carbon over the last 25,000 years. Additional corrections must be made when dating the remains of marine organisms due to differences in radioactive carbon levels between the atmosphere and oceans. Appropriate adjustments are not available for all types of samples or all environments. Work continues to improve and refine the technique (Scott et al. 2004) but it is important to recognize that radiocarbon dates are age estimates. After all corrections and calibrations are made, the results are probabilities that an organism lived during one or more time ranges.

More information about radioactive isotopes and their decay is provided in Chapter 3; the theoretical basis of mass spectrometry and instrument components are described in Chapter 31. Detailed descriptions of radiocarbon dating can be found in general geochronology books and compilations, including those referred to in the introduction to Part II. The journal *Radiocarbon*, published by the Department of Geosciences at the University of Arizona, is openly available online through the university's institutional repository: www.radiocarbon.org. Several overviews of radiocarbon dating for an archaeological audience are available including Aitken (1990), Taylor (1997, 2000), Hedges (2000), and Bronk Ramsey (2008). A historical overview of radiocarbon dating in North America

M.E. Malainey, *A Consumer's Guide to Archaeological Science*, Manuals in
Archaeological Method, Theory and Technique, DOI 10.1007/978-1-4419-5704-7_8,
© Springer Science+Business Media, LLC 2011

can be found in R. E. Taylor (2000). The Radiocarbon Web-Info site supported by radiocarbon dating laboratories at the University of Waikato and University of Oxford is a valuable online resource (either www.c14dating.com or http://c14.arch.ox.ac.uk/embed.php?File=webinfo.html).

Carbon-14

Radiocarbon dating is based on the detection of the radioactive isotope of carbon, ^{14}C, in the remains of organisms. This isotope is formed in the upper atmosphere about 15 km above the ground. Cosmic radiation from space interacts with atmospheric gas molecules, generating neutrons. The newly formed neutrons, n, are highly reactive and combine with atmospheric nitrogen; the result is ^{14}C and a proton, p:

$$^{14}N + n \rightarrow {}^{14}C + p$$

Only a small amount of the total carbon in the world is radioactive; only one in a trillion (1×10^{12}) carbon atoms is ^{14}C. The stable isotope of carbon, ^{13}C, is much more common, forming 1% (1 in 100) of all carbon.

The movement of carbon through the environment depicted in Fig. 4.1 applies to both stable and radioactive isotopes because they are chemically equivalent. As outlined by Taylor (1997), within a few days of being formed, radioactive carbon reacts with atmospheric oxygen, O_2, to form heavy carbon dioxide, $^{14}CO_2$. A small portion, 1–2%, of carbon dioxide, becomes incorporated directly into terrestrial plants through photosynthesis and then indirectly into the herbivores and carnivores that consume them (see Chapter 13). About 85% of the carbon dioxide is dissolved in oceans and a portion becomes incorporated into carbonate compounds, such as calcite and dolomite, which occur in rocks such as limestone and marble. Living organisms use carbonate compounds to form calcareous skeletons and defense structures, such as mollusk shells, sponge spicules, and coral.

Using Radioactive Carbon to Date Organic Material

The nucleus of a ^{14}C atom contains six protons and eight neutrons, making the ratio 1 to 1.33. This represents a significant deviation from the ideal ratio of 1 to 1 for small atoms and causes the isotope to be unstable. The radioactive decay enables atoms to restore a favorable balance between protons and neutrons. In ^{14}C, this is achieved through a β-decay process during which a negative charge without mass (i.e., an electron) is emitted, causing one neutron to transform into a proton. The additional proton changes the atomic number of the element from six to seven, converting it from carbon to nitrogen. The atomic mass of the element is unchanged but with seven protons and seven neutrons, instead of six protons and eight neutrons, and the nucleus is now stable. The chemical equation for the radioactive decay of ^{14}C is

$$^{14}_{6}C \rightarrow ^{14}_{7}N + ^{0}_{-1}e$$

Libby calculated that ^{14}C had a half-life of 5568 years, with a mean lifetime of 8033 years. These values have since been refined; the accepted value for the half-life of ^{14}C is now 5730 and 8267 years for mean lifetime.

Living organisms constantly exchange carbon with the world around them so the levels of stable and radioactive carbon they contain closely match their environment. This exchange ceases when the organism dies so the ^{14}C that decays is not replenished. The difference between the amount in the environment and the amount in the remains of the organism increases over time and forms the basis of radiocarbon dating.

Different approaches are used to measure the amount of ^{14}C left in the sample. The earliest techniques relied on measuring the β-emission rate; accelerator mass spectrometry uses direct or ion-counting technology to determine the relative concentrations of carbon isotopes.

Counting β-Emissions

The rate of β-particle emissions is proportional to the amount of radioactive carbon in the sample. Consequently, the accuracy of this type of radiocarbon dating depends upon the ability to precisely count a sufficient number of β-emissions over a sufficiently long period of time. Samples containing more radioactive carbon will have higher emission rates than samples with less ^{14}C. If the sample size is small or so old that little radioactive carbon remains, the β-emission rate must be monitored for an extended period of time.

Since its development, techniques have improved to ensure detection of weak sample emissions and avoid inadvertent counting of cosmic radiation. The earliest method for counting β-emissions is known as the solid carbon technique. Organic samples were converted to carbon by burning; then the carbonized residue was painted on the inside of a modified Geiger counter. Placement inside the counter improved the detection of weak emissions and provided some protection from cosmic radiation.

Later, proportional gas counters were used to detect β-emissions. Samples were burned in oxygen to transform the carbon into carbon dioxide (CO_2) gas; after purification, methane (CH_4) or acetylene (C_2H_2) could be synthesized from it. Beta emissions from the CO_2 gas create bursts of ionization, which are counted as electrical pulses. Proportional gas counters were more precise because β-emissions could be discriminated from α-particles by pulse size. Lead was used to shield the instrument from cosmic radiation and anticoincidence detectors were used to further reduce the possibility of external contamination. Anticoincidence detectors monitor radiation pulses both within and outside of the instrument. They operate under the assumption that an emission from a sample is unlikely to occur at precisely the same time that cosmic radiation enters the lab. If an electric pulse occurs simultaneously inside the instrument and outside of it in the laboratory, it is assumed to be contamination and no count is registered.

Today, liquid scintillation counters detect β-emissions using compounds that fluoresce when exposed to ionizing radiation. Instead of registering electric pulses, liquid scintillation counters record flashes of light. Sample preparation is more elaborate. Acetylene is synthesized from the CO_2 produced by sample combustion; the acetylene is then converted to benzene (Tamers et al. 1961). The sample benzene, alone or together with toluene, is then mixed with a liquid scintillation "cocktail," usually a combination of PPO (2,5-diphenoloxazole) and POPOP [1,4-bis (5-phenyloxazol-2-yl) benzene]. Total volumes range between 0.3 and 20 ml. The flash of light from a β-particle emission is called a fluorescence event. The number of fluorescence events is proportional to the number of ^{14}C atoms that decay during the monitored period or radioactive decay events. These, in turn, are directly proportional to the number of ^{14}C atoms remaining in the sample.

Advances have reduced the sample size required; using micro techniques, less than 1 g is needed. A certain number of counts are required to obtain an acceptable degree of statistical precision. Several months of counting may be necessary to obtain ages that have an acceptable degree of precision from very old or small samples with low β-emission rates. An instrument monitors fluorescence events; sample and background vials are changed automatically during this period.

Dates obtained by monitoring β-emissions are conventional radiocarbon dates and may require correction for isotopic fractionation prior to calibration.

Accelerator Mass Spectrometry Dating

The introduction of accelerator mass spectrometry (AMS) represents a profound change in radiocarbon dating. Rather than using emission rates to estimate the amount of radioactive carbon present, the relative amount of ^{14}C atoms in the sample is determined. A mass spectrometer connected to a particle accelerator sorts the ions of carbon on the basis of mass and measures their relative concentrations. Using electrostatic tandem accelerator mass spectrometry is more expensive than decay counting, but there are several advantages. Typically AMS samples are 1000 times smaller than the size of those submitted for β-emission counting and samples weighing less than a milligram can be dated. Processing is generally simpler and analysis time for individual samples is very short, measured in minutes. Several days or months of decay counting would be required to obtain a radiocarbon age estimate with the same degree of precision achievable using AMS.

A diagram of an accelerator mass spectrometer is shown in Fig. 8.1; it generally consists of two or three magnets and an accelerator. The usual procedure requires

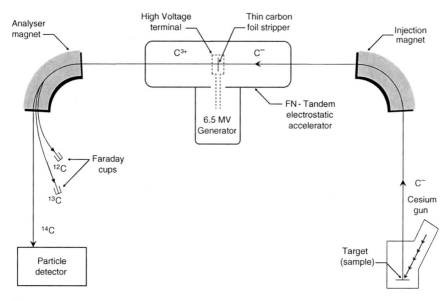

Fig. 8.1 Accelerator mass spectrometer (AMS) schematic (From LONGMAN SCIENCE BASED DATING IN ARCHAEOLOGY © 1990 Pearson Education, Inc. or its affiliates. Used by permission. All rights reserved.)

combustion of the sample to produce CO_2, which is then converted to graphite. The solid carbon is then placed into the ion source where it is bombarded with cesium ions under vacuum. Elements in the sample are converted into ions and then focused into a fast-moving beam with about 25 keV of energy. Carbon only forms negative ions; this prevents confusion with nitrogen, ^{14}N, which has a mass of 14 but does not form a negative ion.

The ion beam enters the first magnet called the injector magnet. This magnet acts like an ordinary mass spectrometer (see Chapter 31) and selects ions with an atomic mass of 14. Negative ions of all three carbon isotopes with atomic mass 14, ^{14}C, ^{13}CH, and $^{12}CH_2$, are directed into the high-voltage acceleration chamber. The positively charged high-voltage terminal pulls the negatively charged ions into the acceleration chamber where they collide with gas molecules or a thin carbon foil in the terminal's stripper canal. These high-energy collisions cause molecules to break apart and produce positively charged ions, either C^{3+} or C^{4+}. Positive carbon ions are repelled and accelerate out of the positively charged terminal and then pass through either one or two analyzer magnets that separate the carbon ions by mass.

The analyzer magnet operates on the principle that when an electrically charged particle moves in a magnetic field there is a force on it that causes its path to be curved. Paths of heavier particles have less curvature than lighter particles, causing them to separate. The curvature of ^{12}C ions will be the greatest; ^{14}C ions will have the straightest path. Measurements of two ratios, $^{13}C/^{12}C$ (δ ^{13}C) and

^{14}C/^{13}C (δ^{14}C), made on the sample and a modern standard are used to calculate the radiocarbon age estimate.

The entire process is very rapid because it is not necessary to wait for β-emissions to occur. AMS measurements on a sample can be made with a precision of ±1% in a matter of minutes; an extended period of counting β-emissions may be required to attain a similar degree of precision.

The practical limit for radiocarbon dating is 55,000–60,000 years for samples from which 3–7 g of carbon can be obtained. Age estimates can be obtained from large samples, consisting of more than 15 g of carbon that are as much as 75,000 years old. In the future, it may be possible to date much smaller and older samples using AMS dating.

Conventional Radiocarbon Ages

The conventional radiocarbon age (CRA) is a sample determined under a set of criteria established by Stuiver and Polach (1977). These include the use of Libby's half-life of 5568 years, use of appropriate modern standards, correction for sample isotopic fractionation, use of A.D. 1950 as 0 B.P., and assumption that radiocarbon levels in the different reservoirs have remained constant over time. These criteria assure uniformity of radiocarbon age calculations between different laboratories; however, they are also a source of error. We know that Libby's half-life is off by more than 3% and, as outlined below, fluctuations have occurred in radiocarbon reservoirs over time. Consequently, additional calculations must be made in order to convert a CRA to a calendar date. Radiocarbon ages obtained using decay counting techniques must also be corrected for isotopic fractionation.

Correcting Radiocarbon Dates for Isotopic Fractionation

When using decay counting techniques, the radiocarbon decay equation is given as

$$t = -8033 \ln (A_{sn}/A_{on})$$

where time, t, is the product of Libby's mean lifetime of ^{14}C, -8033, and the natural logarithm, ln, of the ratio between the activity of sample in counts per minute, A_{sn}, and the activity of modern standard in counts per minute, A_{on}.

Carbon-14 is heavier than other isotopes of carbon and not as readily incorporated into living organisms as the lighter carbon isotopes (see Chapters 4 and 13). Differences in the uptake of ^{14}C compared to ^{13}C and ^{12}C is termed isotopic fractionation. In animals, small species to species variations in the degree of isotopic fractionation relate to diet. The photosynthetic pathway of a plant determines how readily it uptakes carbon. Conventional radiocarbon ages given by laboratories are normalized to a common δ^{13}C value of $-25‰$ (per mill), which is appropriate for wood and charcoal. For other samples, it is necessary to measure the ^{13}C concentration in the sample as either a ^{14}C/^{13}C ratio or ^{13}C/^{12}C ratio. The former ratio is used for some AMS dates and the latter is used for other AMS dates and dates based on liquid scintillation counts. The isotopic fractionation of ^{14}C is about twice that of ^{13}C.

The per mil depletion of sample ^{14}C is determined as

$$\delta^{14}C = ((A_{sn}/A_{on}) - 1) \times 1000‰$$

then the normalized value of Δ^{14}C (D^{14}C) is calculated as

$$\Delta^{14}C = \delta^{14}C - (2 \delta^{13}C + 25)(1 + \delta^{14}C/1000)‰$$

and used to obtain a corrected conventional radiocarbon age where

$$\text{Corrected conventional radiocarbon age} = -8033 \ln (1 + \Delta^{14}C/1000)$$

Laboratories normalize radiocarbon dates to the base value of $-25.0‰$, the value of the Pee Dee belemnite carbonate standard. Dates on charcoal and wood do not require correction but the effects of isotopic fractionation can alter a radiocarbon age estimate on some materials by several hundred years. It is now common practice for laboratories to provide fractionation-corrected radiocarbon ages for each sample; however, it should be noted that procedures have changed over time. Some radiocarbon dating laboratories used an average stable isotope ratio value for the material class in the age calculation unless specifically requested to measure the precise value in the sample. Depending on the laboratory and when it was obtained, the age may be uncorrected. The program, CALIB, provides a means of obtaining a calibrated radiocarbon age for ^{14}C age estimates not previously normalized for δ^{13}C.

Carbon Exchange Reservoir

The total amount of carbon in the system is called the carbon exchange reservoir. The ratio of ^{14}C to non-radioactive carbon is approximately the same throughout the reservoir; however, significant exceptions arise from differences in circulation time, modern effects due to recent human activities, and long-term environmental trends.

Differences in Circulation Time in the Carbon Cycle

Living organisms constantly exchange carbon with the reservoir, so their internal levels of ^{14}C closely match that of the environment. Upon death and decomposition, the carbon in their bodies is returned to the environment. For most plants and animals, the turn around time is very rapid, ranging from a few days or months to a few hundred years. Chances are that radioactive carbon will be present when the last traces of an organism vanish.

The cycling time for carbon that enters certain other parts of the environment can be significantly longer (Fig. 8.2a). Dissolved carbonates remain in the depths of the ocean or trapped in glacial ice for thousands of years. Carbon atoms in the carbonates of limestone or marble, hydrocarbons in fossil fuels, and the carbon dioxide dissolved in magma are fixed for millions of years. Continual exchange does not occur with other parts of the carbon reservoir so the amount of radioactive carbon in these contexts is far less than that observed in living organisms. When deep ocean upwellings occur, surface water dissolves limestone, a volcano erupts or fossil fuels are burned, carbon depleted in ^{14}C is released into the environment. On the other hand, neutrons released when atomic weapons were detonated above ground resulted in the formation of high levels of ^{14}C (Fig. 8.2b). The effects of each of these factors are described below.

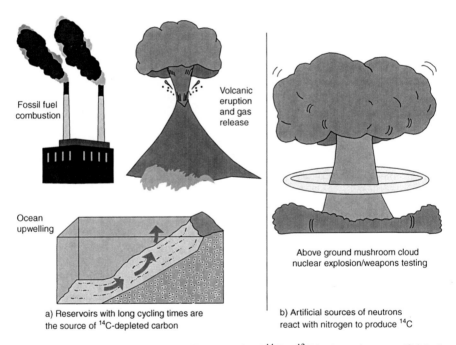

Fossil fuel combustion

Volcanic eruption and gas release

Ocean upwelling

Above ground mushroom cloud nuclear explosion/weapons testing

a) Reservoirs with long cycling times are the source of ^{14}C-depleted carbon

b) Artificial sources of neutrons react with nitrogen to produce ^{14}C

Fig. 8.2 Natural and human activities affect the ratio of ^{14}C to ^{12}C in the environment (Original Figure by M.E. Malainey and T. Figol)

Deep Ocean and Glaciers

Differences between the ^{14}C concentration of surface ocean water and the atmosphere cause shell and the remains of other marine organisms to date older than terrestrial charcoal formed at the same time. Subtracting 400 radiocarbon years from a sample gives the marine model age for shell. This represents the average discrepancy between terrestrial organisms and those in equilibrium with marine carbon reservoirs.

While carbon can remain in the atmosphere for several decades, carbon can move into the deep ocean and remain there for thousands of years. Any carbon dioxide trapped or dissolved in glacial ice may have been there since the Pleistocene. In the absence of constant exchange, the radiocarbon decays; water found in both the deep ocean and in glaciers is depleted in ^{14}C. Organisms that incorporate ^{14}C-depleted carbonates into their tissues will date older than expected.

Periodic upwellings bring up water from the deep ocean, which then mixes with surface ocean water in coastal environments. During the upwelling and for a time afterward, mollusks living in the region form their shells with ^{14}C-depleted carbonates. This creates regional differences between local and global ocean environments. Regional marine reservoir corrections are adjustments to the radiocarbon age of a sample to account for both short-term regional differences due to upwellings and global differences between the atmospheric and marine concentrations of radioactive carbon. Calibration curves can only be applied to the region for which they were constructed. For example, Deo et al. (2004) used paired charcoal and shell samples to develop a calibration curve for Puget Sound and the Gulf of Georgia for the past 3000 years.

Hard Water in Limestone Bedrock Regions

In non-marine environments, such as freshwater or brackish water lakes, the magnitude of the reservoir effect depends upon the level of total dissolved inorganic carbon (TDIC) and atmospheric CO_2. Areas with limestone bedrock pose a problem because ground and surface water slowly dissolve it. The water becomes enriched in very old carbonates and the ^{14}C-depleted carbon is then passed on to living organisms. This phenomenon, known as the freshwater reservoir effect (or the hardwater effect), results in radiocarbon dates that are older than expected. Geyh et al. (1998) found that the reservoir effect is not constant for a specific lake but subject to seasonal and temporal fluctuations related to changes in water depth.

Volcanic Eruptions and Gas Discharges

Molten rock deep beneath the Earth's surface is under such high pressure that carbon dioxide gas is dissolved in the magma. In the absence of active exchange with the atmosphere, only stable isotopes of carbon are retained. As it approaches the surface, this ^{14}C-depleted CO_2 separates from the parent magma to become a major constituent of volcanic gas. Expulsion of large quantities of carbon dioxide is usually associated with eruptions; but sudden releases between eruptions can also be significant. Plants growing on the slopes of volcanoes absorb the ^{14}C-depleted CO_2 during photosynthesis and, if dated, can appear to be several hundreds of years old.

Secular Variation

Taylor (1997) describes secular variation as any systematic variability in the ^{14}C time spectrum other than that caused by ^{14}C decay. The plot of the differences in ^{14}C activity over the last 9000 years is characterized by both short-term fluctuations in the form of small "wiggles," "kinks," or "warps" and a major downward trend (Fig. 8.3).

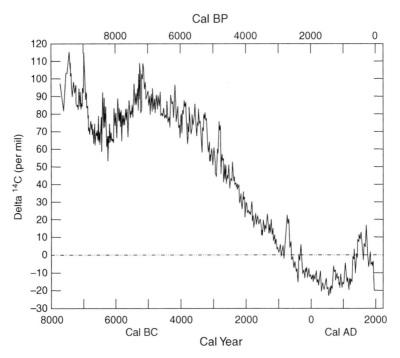

Fig. 8.3   Changes in carbon-14 activity in the atmosphere over the last 9000 years. (Reprinted with permission from M. Struiver, G.W. Pearson and T.F. Braziunas, 1986, Figure 1, "Radiocarbon" 28(2b):984.)

Short-Term Fluctuations or de Vries Effects

De Vries effects are short-term fluctuations in ^{14}C levels that have occurred throughout the Holocene. They are named after Hessel de Vries, who recognized that variability in the concentrations of ^{14}C meant that radiocarbon years could not be considered equivalent to calendar or solar years. Some changes in radioactive carbon can be attributed to natural fluctuations in sunspot activity. For example, rapid changes in solar magnetic intensity in the seventeenth century resulted in significant variability in the rate of ^{14}C production. Even larger perturbations have been identified between 8000 and 10,000 B.P. Recent human activity seems to be directly responsible for others.

Modern effects are recent fluctuations in the amount of ^{14}C in the environment that appear to result from human activity. These include the effects of burning fossil fuels and the atmospheric testing of atomic weapons. Hans Suess described Industrial or Suess Effects in 1955 as an explanation for older than expected dates on organic material since A.D. 1890. He suggested that the use of coal during the Industrial Revolution had diluted the amount of ^{14}C in the atmosphere. Increased reliance on coal began in the 18th century, when it was first used to fuel the steam engine. Later, petroleum products such as oil, gasoline, and natural gas were employed. These fossil fuels contain very old carbon and their combustion results in the emission of carbon dioxide and carbon monoxide depleted in ^{14}C. It is estimated that the decreased concentration of ^{14}C in the atmosphere causes "modern" material living since A.D. 1890 to date 2–2.5% older.

The atomic bomb, nuclear, or Libby effect is due to the atmospheric testing of nuclear weapons. The power of these weapons is from the nuclear fission of a heavy element, such as uranium. Splitting large nuclei produces a massive amount of thermal energy, gamma rays, and neutrons. The neutrons released by the detonation of atomic weapons react with nitrogen to produce ^{14}C; a single weapons test could produce several tonnes of radioactive carbon. Atmospheric testing of nuclear weapons doubled the amount of ^{14}C activity in terrestrial carbon bearing material. The amount of ^{14}C peaked in 1963 at 100% above normal levels but is decreasing as ^{14}C moves deep into the ocean. The effect of this ^{14}C enrichment is that terrestrial organisms of the nuclear age (>modern) can date up to 160% younger.

Taylor (1997, 2001) suggests that due to these recent and pronounced fluctuations in the amount of ^{14}C in the atmosphere, it is not possible to assign an unambiguous age to samples less than 300 years old except under very special circumstances.

Long-Term Trends

Dating methods based on annual accretions can provide an exact date for a specimen. Dendrochronological, or tree ring, dates on California sequoia, Douglas fir, bristlecone pine, Irish oak, and German oak are used to assess the accuracy of radiocarbon age estimates for samples almost 12,000 years old. Paired radiocarbon and uranium–thorium dates on marine corals have effectively doubled the length of the calibration curve (Fig. 8.4). In general, the deviation between radiocarbon ages and actual calendar dates increases with the age of the specimen but the cause is not fully understood. The divergence may be due to changes in the Earth's magnetic field, which affects the amount of cosmic radiation entering the atmosphere. Climate changes at the end of the last glacial period may also account for some of the deviations. Alternatively, both factors may be partially responsible.

At present, there is no consensus about ^{14}C levels prior to 25,000 years ago; studies involving a variety of chronometric techniques have produced conflicting results.

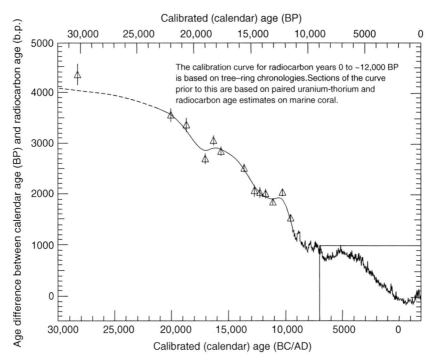

Fig. 8.4 Difference between radiocarbon and calibrated calendar ages over the last 30,000 years (Adapted from Struiver and Braziunas (1993) and Taylor et al. 1996; Reprinted with permission from M. Struiver and T.F. Braziunas, 1993, Figure 1, "Radiocarbon" 35(1):138; Reprinted from Quaternary Science Reviews 15(7), R.E. Taylor, R.E., Minze Stuiver, and Paula J. Reimer, Development and Extension of the Calibration of the Radiocarbon Time Scale: Archaeological Applications, page 661, Copyright 1996, with permission of Elsevier.)

Converting Radiocarbon Ages into Calendar Dates

Several of the assumptions related to the calculation of conventional radiocarbon ages are not valid. Libby calculated the half-life of ^{14}C to be 5568 years; this value differs from currently accepted value of 5730 ± 40 years by more than 3%. As a result, the older the conventional radiocarbon date, the more it deviates from the calendar date (Fig. 8.5). The other and more confounding problem is that atmospheric radiocarbon concentration has varied over time. For these reasons, the relationship between a radiocarbon age estimate and a calendar date is complex.

The process of converting a radiocarbon age estimate into calendar years is called calibration. Calibration compares the amount of the radioactive carbon remaining in a sample of unknown age to amounts measured in materials of known age. The calibration curve is based on the radiocarbon dates on known age trees and paired uranium–thorium and radiocarbon dates on marine corals (Stuiver and Braziunas 1993; Taylor et al. 1996). The trees that form the dendrochronological record reflect changes in atmospheric radiocarbon levels in their wood over the last 12,000 years.

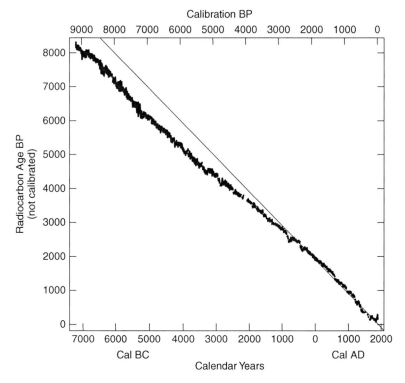

Fig. 8.5 Relationship between conventional radiocarbon age and calendar date (Reprinted from G.W. Pearson, 1987, Figure 1, "Antiquity" 61 (231):99 with permission from Antiquity Publications Ltd.)

The radiocarbon age estimate on a sample of unknown age is given as the mean date together with the error associated with its measurement (see Fig. 7.1). There is a 68% chance that the true radiocarbon age falls within the 1σ (sigma) range of the measured radiocarbon age (i.e.,±1 SD). There is a 95% chance that the true radiocarbon age falls within the 2σ range (i.e., +/− 2 SD) of the radiocarbon date.

For dates less than 12,000 years old, the process of calibration identifies tree rings of known age that are most likely to produce the same radiocarbon age as the sample of unknown age (Guilderson et al. 2005). This portion of the calibration curve is based on radiocarbon age estimates of 20-year-wide segments of tree rings presented at the 68% confidence interval. The radiocarbon age for the sample is considered at the 68 and 95% confidence interval. Calibrated dates are given as calendar date ranges and the probabilities that the sample date falls within them.

At the 1σ range, the sample with a radiocarbon age of 3000 ± 30 B.P. is associated with three calibrated age ranges (Table 8.1). Of these, the probability is highest (93.94%) that the calendar date falls within the calibrated age range of 1309–1208 cal B.C. At the 2σ range, there are two calibrated age ranges; there is a 91.7% chance the calendar date falls between 1319 and 1129 cal B.C.

Table 8.1 An example of radiocarbon age calibration at the 68 and 95% confidence intervals using the Calib 6.0 Radiocarbon Calibration Program by M. Struiver and P.I. Reimer

3000 ± 30 BP	Calibrated age ranges	Relative area under probability distribution
68.3% 1σ	cal BC 1309–1208	0.934
	cal BC 1202–1195	0.035
	cal BC 1139–1135	0.031
95.4% 2σ	cal BC 1374–1340	0.083
	cal BC 1319–1129	0.917

Guilderson et al. (2005) note that improved precision, for example, from ±40 to ±15 radiocarbon years, does not necessarily narrow the calibration age range. Several sections of the calibration curve are quite horizontal because the $^{14}C/^{12}C$ ratio fell at a rate equal to that of the radiocarbon decay. One of these "age plateaus" occurs at about 2450 radiocarbon years B.P. and lasts nearly 350 years (Fig. 8.6). The period of chronological ambiguity extends from about 750–400 B.C. and completely encompasses the "Golden Age of Greece" (Guilderson et al. 2005). By

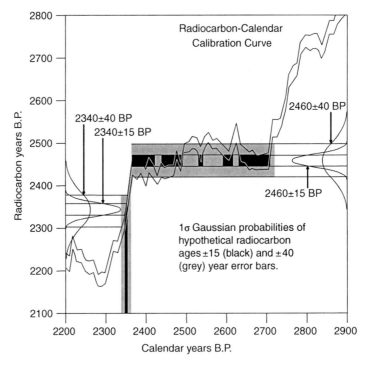

Fig. 8.6 Plateau in the curve increases the uncertainty associated with calibrated radiocarbon ages (From Guilderson et al. (2005). Reprinted with permission from AAAS)

contrast, a date of 2340 ± 40 radiocarbon years B.P. intercepts a nearly vertical portion of the calibration curve so the 1σ calibrated age range spans only a few decades. In similar situations, site chronology may be refined on the basis of stratigraphy and other sample data or by comparing the pattern of radiocarbon dates relative to the variation in the calibration curve or "wiggle-matching."

Sampling Considerations

While a radiocarbon age estimation can be obtained from a wide variety of materials, its quality ultimately depends upon the integrity of the sample. Radiocarbon laboratories analyze the organic component of material submitted. Proper selection of samples will increase the chance that the date obtained relates to the archaeological event under consideration. The material selected should be free of contaminants and as firmly associated with the targeted event as possible.

Charcoal and wood are suitable for radiocarbon dating; however, associating the material with the event in question may be problematic. The resulting age estimate pertains to the life of the tree, which may span decades or centuries. The relationship between the tree and the archaeological event in question may be uncertain. The wood appearing in a site may have been harvested from a tree that lived many years before the event occurred. Charcoal may have been deposited during a forest fire that burned decades after a site was abandoned. Bioturbation may cause the downward movement of small fragments of wood and charcoal.

Bone is largely a combination of the mineral hydroxyapatite and the protein collagen. Only bone collagen, which is the organic fraction, is capable of providing accurate radiocarbon age estimations and only if significant amounts of intact collagen are retained. Taylor (2001) suggests that age estimations on bones depleted in collagen have a high potential for producing anomalous values. Minimally, 5% of the original amount of collagen should be present; this would give a collagen yield of about 1%. Weber et al. (2005) suggest that it should be a standard practice to report the collagen yield whenever a radiocarbon age on bone is presented. It may be useful for archaeologists to adopt techniques used to assess bone diagenesis (see Chapter 19) because they can aid in the selection of samples suitable for radiocarbon dating.

Shell is often found in large quantities in coastal archaeological sites. Marine mollusks deposit the carbonates that build their shells during feeding; however, these carbonates are depleted in ^{14}C. The marine reservoir correction adjusts for differences in carbon circulation rates between oceans and the atmosphere. Marine carbonates produce radiocarbon ages that are an average of 400 years older than samples from terrestrial organisms that lived at exactly the same time.

The effects of coastal upwellings of deep water are restricted in area so the concentration of ^{14}C in the ocean is not uniform. For this reason, it is also necessary to establish regional marine reservoir corrections, ΔR values, for radiocarbon age estimates on marine shell. Stuiver and Braziuna (1993) presented coastal regional values for several different parts of the world. Compounding the problem is that the effects of upwellings may be of relatively short duration. Deo et al. (2004) found that increased upwelling offshore only occurred during the periods from 0 to 500 B.P. and again from 1200 to 3000 B.P. Only dates on Northwest Coast carbonates from these periods required an additional correction of 401 years.

Depending upon the diet of the organism, both the marine and the freshwater reservoir effect must also be considered when selecting bone for radiocarbon dating. Cook et al. (2001) obtained radiocarbon age estimates on five sets of human remains and terrestrial ungulate bones found in the same graves. Stable isotope analysis revealed the humans heavily relied on a freshwater aquatic diet. This resulted in radiocarbon ages that were on average 440 ± 45 years older than those on directly associated terrestrial animal bones. Cook et al. (2001) suggested that a simple linear relationship existed between the $\delta^{15}N$ value on the human bone and the age offset required to correct the date.

Additional adjustments are required for samples from the southern hemisphere. Levels of ^{14}C in the atmosphere south of the equator are not identical to levels in the northern hemisphere. The radiocarbon age of samples from the southern hemisphere should be reduced by 23 ± 3 radiocarbon years prior to calibration to account for this difference.

Sample Pretreatment

Samples submitted for radiocarbon dating are usually subjected to pretreatment to eliminate extraneous carbon. The pretreatment method depends upon the type of sample submitted. Most samples are first crushed and washed in deionized water. Charcoal and wood are usually given an "acid/alkali/acid" pretreatment. Hydrochloric acid (HCl) is used to dissolve carbonates and then secondary organic acids, such as humic acid, are removed with sodium hydroxide (NaOH); the final acid treatment serves to neutralize the alkali by removing carbon dioxide. Bone collagen is extracted by dissolving the mineral component with acid followed by NaOH to remove secondary organic acids. In the case of shell and other calcareous materials, secondary carbonate components are dissolved in HCl.

After pre-treatment, the carbon in clean charcoal, wood, and bone samples is converted into CO_2 by combustion. The carbon in shell and other carbonates are oxidized by acid hydrolysis. After purification, the CO_2 gas can be converted into acetylene or benzene for liquid scintillation counting or solid graphite for AMS.

Current Issues in Radiocarbon Dating

Less Destructive Sampling Techniques

While amounts of carbon required for an AMS dating is small, it still involves removal and destruction of a portion of the object. For this reason, it may be very difficult to obtain the ages of artifacts that are very small, rare, or considered sacred. The viability of alternative sampling techniques is under investigation. Plasma oxidation holds the promise of being a minimally or even non-destructive means of sampling organic artifacts (Chapter 24 and 25). By dating iron or steel artifacts with the carbon preserved in rust, material that would otherwise be discarded forms the sample and the integrity of the stable portion of the artifact is preserved (Chapter 26).

Judging the Quality of Radiocarbon Ages

The quality of radiocarbon age estimates has been reconsidered in terms of how closely the material being dated can actually be associated with a particular archaeological event. While it is the responsibility of the laboratory to accurately assess the amount of radioactive carbon in a sample; archaeologists have the responsibility of selecting appropriate samples. If the connection between the organic material submitted for dating and the archaeological event is dubious, the reliability of radiocarbon age estimates obtained from it is compromised. Problems surrounding the radiocarbon dating of organic material trapped in rock varnish are described in Chapter 11. Another concern involves the reliability and comparability of radiocarbon ages obtained by different laboratories, at different times and using different protocols. The method proposed by Pettitt et al. (2003) to evaluate the quality of radiocarbon age estimations on Old World materials is described below. Weber et al. (2005) examined the ability of radiocarbon dating to discriminate closely spaced events. Due to the uncertainty associated with radiocarbon age estimates, not all temporal issues can be resolved by simply dating more samples.

Assessing the Reliability of Radiocarbon Ages

An archaeological model of human behavior over time requires solid chronologies for the events it encompasses. In order to construct a model for Old World population movements between 20,000 and 8000 B.P., hundreds of radiocarbon age estimations from Europe, the Near East, and Africa must be used. As a first step, Pettitt et al. (2003) developed a method of assessing the reliability of individual age estimates. A variety of factors relating to the sample and laboratory procedures were evaluated including: certainty of association, sample context, type of material, sample size as a measure of resistance to migration, freedom from contaminants, laboratory protocols, calibration, and number of corroborating age estimates.

For each of nine attributes, a score between a high of 4 and a low of 0 was assigned to reflect the degree in confidence in its quality. Out of a possible 36 marks, 20 marks related to chronometry and 16 related to interpretation. A calibrated age obtained on bone collagen from a human-manufactured object with a maximum dimension greater than 10 cm and supported by five or more statistically identical ages would be awarded the highest possible score. Low scores were given to age estimates (1) obtained prior to 1970, (2) made on small samples that could migrate or on bulk samples due to the possibility of mixing, (3) lacking corroboration with other samples, (4) made on paleontological materials, and (5) made on samples poorly associated with cultural materials.

Pettitt et al. (2003) recommended that only radiocarbon ages with scores of 27 or higher out of 36 be used for modeling. Age estimations with scores between 10 and 26 should be used with caution. Those with scores of 9 or less were deemed unreliable.

Dating Closely Spaced Events

When faced with the prospect of needing to date closely spaced events, the standard archaeological procedure is to obtain an abundance of age estimations. Weber et al. (2005) investigated the question of whether radiocarbon dating bone could actually provide the degree of resolution that archaeologists want. Age estimations were obtained from a large hunter-gather cemetery on Lake Baikal, Siberia, utilized between about 2700 and 2000 B.C.

A total of 93 bones from 85 individuals were submitted for dating. Sufficient collagen was extracted from all but 6 bones; however, more than 5% of the original amount of collagen was retained in only 28 bones. Some bones with low collagen levels gave radiocarbon ages that were broadly accurate but an increase in variation occurred. Fine temporal resolution was not possible when the period of internment was short compared to the standard deviations associated with the radiocarbon ages or when they fell on a portion of the calibration curve that was flat or contained an inversion. Weber et al. (2005) noted that even large sets of AMS age estimations are "still beset by limitations and interpretive difficulties" inherent to the radiocarbon method in general, and specifically to the use of bone samples.

Conclusions

Radiocarbon dating revolutionized the practice of archaeology. Recent advances relate to sample processing techniques, increasing the precision of measurements, the extension of the calibration range, and refinement of the calibration curve. It is, however, necessary for archaeologists to recognize not only the potential, but also the limitations of the technique. A radiocarbon age is an estimate, or inference, of the age of a sample based on the amount of ^{14}C remaining in it. A number of factors affect the certainty with which these age estimates can be made. In cases where multiple radiocarbon ages are possible or overlap exists, relative dating techniques, such as sample ordering on the basis of stratigraphy or seriation, may provide clarification.

Chapter 9
Other Radioactive Decay-Based Dating Techniques

Uranium series dating, potassium–argon dating, argon–argon dating, and fission track dating are presented in this chapter. These techniques are widely employed by geologists because very old deposits can be dated but the number of archaeological applications is increasing. More detailed descriptions of these techniques can be found in general geochronology books and compilations, including those referred to in the introduction to Part II.

Uranium Series Dating

Uranium series dating has the potential to provide an age estimate for a variety of materials ranging from several hundred to several million years old. It is most often used to determine the age of Paleolithic cave deposits in the Old World; in this respect, the technique is probably under utilized, particularly in the New World. Theoretically, uranium series dating could be applied to any substance containing calcite, but some materials are superior because they are less likely to be contaminated. Recent advances in mass spectrometry have increased the degree of precision, reduced analysis time, and enabled age determinations on smaller and older samples. Uranium–thorium ages on marine coral have extended the radiocarbon calibration curve thousands of years beyond the limit of the dendrochronological master record. Techniques modeling uranium uptake and assessing the degree of environmental contamination have made it possible to obtain ages for less than ideal samples.

Materials suitable for uranium series dating and the analytical procedures involved are described in this section. In addition to the geochronology books mentioned in the introduction to Part II, the works of Bourdon et al. (2003) and Ivanovich and Harmon (1992) are thorough compilations of theoretical issues involving uranium series geochemistry and methodological issues of U-series dating. Schwarcz and Blackwell (1992), in the latter volume, focus on archaeological applications and discuss sampling considerations and methods of detecting contamination. Schwarcz (1997) and Latham (2001) present detailed summaries of

M.E. Malainey, *A Consumer's Guide to Archaeological Science*, Manuals in
Archaeological Method, Theory and Technique, DOI 10.1007/978-1-4419-5704-7_9,
© Springer Science+Business Media, LLC 2011

this technique for an archaeological audience. Examples of specific archaeological applications are presented in Part III.

Using Uranium as a Clock

Uranium is an element that occurs in nature as either ^{238}U or ^{235}U; both isotopes are radioactive. As outlined in Chapter 3, radioactivity is an element's response to a significant imbalance between the number of neutrons and protons in its nucleus. Both uranium isotopes decay through a long series of radioactive daughters to stable isotopes of lead: ^{238}U decays to ^{206}Pb and ^{235}U decays to ^{207}Pb. In the case of ^{238}U, ^{206}Pb is actually the 14th daughter in the decay chain (see Fig. 3.3). The half-life of ^{238}U is 4.5 billion years and that of ^{235}U is about 704 million years.

Uranium series dating uses the radioactive decay of uranium and/or its daughter isotopes to estimate the age of a sample. Uranium–thorium dating is most common and uses ^{234}U and ^{230}Th, which are the third and fourth daughter isotopes in the ^{238}U decay series, respectively. Comparisons between other uranium series isotopes have also been used as clocks. If the two isotopes were not deposited in the mineral at exactly the same time, ^{234}U can be compared to its parent, ^{238}U. Protactinium, ^{231}Pa, can be compared to its parent, ^{235}U, if the material has a very high concentration of uranium. In materials dating to the Middle Paleolithic and older, the lead end product, ^{206}Pb, can be compared to the ^{238}U parent.

Uncontaminated calcite that formed close to the time of a targeted archaeological event are ideally suited for dating. Uranium is water soluble and often deposited when calcite precipitates out of solution. Its daughter isotope, thorium, is much less soluble and does not occur in the newly formed mineral. In the absence of its daughter isotope, a state of non-equilibrium exists and a new multiple step uranium decay chain is established. Equilibrium is achieved when all daughter isotopes are present and their activities, or decay rates, are equal to that of the initial parent. In cases where the mineral is contaminated by thorium from the environment it is also possible to detect the degree of contamination and adjust the age accordingly.

The limits of the dating technique are governed by the sensitivity and precision of the detectors. The lower age limit corresponds to the point at which the daughter isotopes are detectable. The upper age limit corresponds to the equilibrium level when the ratios of the two isotopes are so close to one that samples give "infinite" ages. Comparisons between ^{234}U and ^{230}Th are possible for samples that formed between several hundred and about 500,000 years ago. When comparing ^{231}Pa to ^{235}U, the upper limit for age estimates is about 300,000 years.

In the case of uranium–thorium dating, changes in the ratios of the activities of ^{234}U and ^{230}Th over time are depicted in Fig. 9.1. The ratio of the activity of ^{230}Th to that of ^{234}U is shown on the Y-axis; the time in thousands of years (ka) since formations is shown on the X-axis. Young, uncontaminated samples contain ^{234}U but no ^{230}Th, so the initial ratio is zero; over time, the activity of ^{230}Th increases to virtually match that of ^{234}U. The difference between isotope activities becomes

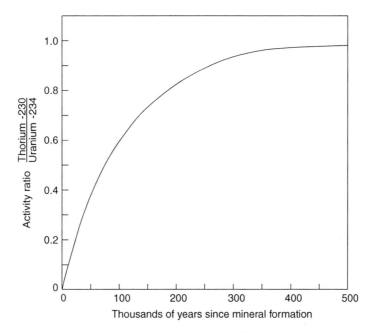

Fig. 9.1 Changes in the activity ratios between ^{230}Th and ^{234}U over time (From Chronometric Dating in Archaeology, Advances in Archaeological and Museum Science Vol. 2, 1997, Uranium Series Dating by H. P. Schwarcz, Figure 6.2, with kind permission of Springer Science and Business Media)

very small as the isotopes approach secular equilibrium, so only instruments capable of very precise measurements are able to estimate the age of materials older than 300, 000 years.

Measuring Radioactivity

The activity (or decay rate) of a radioisotope, R, can be determined by counting the number of radioactive disintegrations as either emissions of particles or energy, per minute per unit of weight of material. An alpha spectrometer can be used to monitor alpha particle emissions from the sample with an accuracy of $\pm$5–10%. The upper limit of uranium–thorium dating using alpha spectrometry is about 350,000 years. Gamma spectrometry can be used to count gamma ray emissions but instruments are not as sensitive and the shape of the sample must be considered; the accuracy of measurements is usually less than $\pm$10%.

As outlined in Chapter 3, the activity of a sample is directly related to the concentration of radioactive isotopes. When mass spectrometry is used to directly count the number of radioactive isotopes present, N, significantly more accurate measurements are possible. The calculation of combined equations (3.1) ($\lambda = R/N$) and (3.4) ($t_{1/2} = 0.693/\lambda$) is as follows:

$$R = \frac{0.693}{t_{1/2}} \times N$$

where $t_{1/2}$ is the half-life of the isotope. Thermal ionization mass spectrometry (Chapter 31) has an accuracy of $\pm 1\%$ and extends the upper limit of uranium–thorium dating to about 500,000 years. The upper limit may be further extended through the use of inductively coupled plasma-mass spectrometry (Chapter 34), which can discriminate isotopes on the basis of mass with an accuracy of 2–4% and has a detection limit between 0.1 and 10 ppb. These techniques measure atomic abundances that must be converted to activity ratios using Equations 3.1a and 3.2.

Sample Selection

This dating technique is most effectively applied to calcite that formed at, or close to, the time of an archaeological event. Ideal mineral samples contained uranium at the time of formation but were completely free of thorium and other daughter isotopes. Dating is straightforward if (1) the mineral was a closed system, that is, no uranium was gained or lost over time and (2) the mineral is free of contaminants so the source of daughter isotopes was limited to the decay of the parent uranium. Calcites that formed in caves or hot springs are particularly well suited for dating; biological materials from archaeological contexts, such as tooth enamel, bone, and shell, can be dated but these materials are often not closed systems. Biological materials are initially free of uranium and absorb it from the environment after death.

Uranium-containing calcites from deep within caves most closely match the ideal. Cave speleothems, such as stalactites, stalagmites, and flowstones, form when dissolved carbonates precipitate out of groundwater. Those that develop deep within caves are less likely to be contaminated by wind-blown sediments. Travertine, which precipitates at hot springs when the mineral-rich water reaches the surface and cools, is also suitable for dating. Other dateable materials include marl, a fine limestone, and caliche or calcrete, which forms in arid regions. These carbonates develop in soil and are prone to contamination. Non-biological minerals used to estimate the age of archaeological deposits may form prior to, during, or after, the archaeological event.

Certain biological materials can be dated; in fact marine corals provide excellent age estimates because they incorporate uranium into their skeletons during their lifetimes. Dating most other biogenetic materials, such as tooth enamel, bone, mollusk shell, and thick eggshell, is more difficult because they must be open systems. These materials do not contain uranium at the time of formation; it is absorbed after the animal dies. Although biogenetic materials tend to have more definite archaeological associations, the quality of the age estimates is usually inferior. Age calculations are greatly affected by how the timing and rate of the uranium uptake is modeled. As outlined below, by obtaining both uranium series and electron spin resonance (see Chapter 10) ages on a sample, uncertainty in the manner of uranium uptake can be reduced. If the correct uranium uptake model is identified, the methods should give identical age estimates.

Biogenetic samples submitted for uranium series dating should be free of contaminants, because uranium or its daughter isotopes are usually present in detritus. Samples that are dense, such as tooth enamel, are less likely to be affected by contamination than those that are porous, such as bone. Recrystallization must not have occurred, as the formation of new minerals resets the uranium series clock. Only ostrich and emu eggs have shells thick enough to provide a uranium series age. Mollusk shells composed of either pure calcite or pure aragonite are dateable but may give erroneous ages. Their porosity and susceptibility to recrystallization means that uranium series age estimates on bones can be older or younger than the probable age. Unless the uranium uptake model is established (see below), uranium series ages on all biogenetic materials, except coral, should be treated with caution.

Biogenetic Materials and Uranium Uptake

The two challenges posed by the dating of most biological materials are (1) establishing the uranium uptake model and (2) their open system behavior. With the exception of marine coral, biogenetic minerals only contain uranium absorbed from the environment; they must act as open systems in order to be dateable. If the materials continue to act as open systems, uranium can leave or enter it at any time. The introduction of new uranium will cause the sample to date too young; the loss of uranium will cause the sample to date too old. Bones are particularly susceptible to problems associated with open system behavior.

Two commonly discussed models of uranium absorption are early uptake and linear uptake. Under the early uptake (EU) model, uranium absorption occurs soon after deposition and then ceases. The production of daughter isotopes begins almost immediately, so the development of the decay series in biogenetic materials is similar to that observed in non-biological calcite of the same age. Calculations made using the EU model should be considered the minimum age of the sample. A linear uptake (LU) model assumes that uranium has been absorbed at a constant rate since deposition. New uranium is continuously added to the system so equilibrium can never be achieved. If it started slow then increased and stabilized over time, the absorption was supralinear. These uranium uptake models assume that the absorption of uranium begins shortly after the death and deposition of the organism. Erroneous ages on mollusk shells indicate that uranium absorption can occur long after deposition; Schwarcz (1997) suggests that all uranium series age estimations on this material should be treated with caution.

It is possible to avoid assumptions about uranium uptake by obtaining uranium series and electron spin resonance (ESR; Chapter 10) ages on a biogenetic sample. Uranium uptake is a factor in both methods so if early uptake is the correct model, the uranium series age on the sample will match the ESR age. If uranium uptake was linear or supralinear, the age determinations will differ but can be reconciled through the use of the equation:

$$U_t = U_0[t/T]^{p+1}$$

where p is an adjustable parameter indicating uranium uptake, T is the age of the sample, t is any time between 0 and T, U_t and U_0 are amounts of U at time t and the final measured value of uranium, respectively. In samples where the uranium entered the system in accordance with an early uptake model, the uranium series and ESR ages will agree when $p = -1$. If the uptake of uranium was linear, the ages will match when $p = 0$. If the uptake of uranium was some combination of early and linear uptake, the ages will agree at a value of p between -1 and 0. Values of p greater than one indicate the uptake of uranium was supralinear. The accuracy of the resulting age may be greatly improved when the method of uranium uptake is more firmly established.

Marine coral is different from other biological materials in that no assumptions about uranium uptake are necessary. Uranium is incorporated into its skeleton throughout the life of the coral. Uranium series ages for coral are being used to extend the calibration curve for radiocarbon dating into the late Pleistocene. The only assumption required for dating coral is that the $^{234}U/^{238}U$ ratio in seawater during the Pleistocene closely matches the current value of 1.14.

Sample Preparation

In order to obtain a uranium series age estimate, the sample is first dissolved in acid. A tracer or spike, consisting of known concentrations of artificial Th and U isotopes of different masses than those that occur naturally, is added to the sample. Ion exchange chromatography (see Chapter 33) is used to separate and purify the various isotopes. The ratios of the isotopes are determined using either emission (alpha or gamma) or mass spectrometry, corrected for variable yield and the age is calculated.

Sample Contamination

During calcite formation, detritus in the form of dust or mud can contaminate the deposit. If thorium is present in the detritus, the initial level of ^{230}Th in the sample will not be zero. Age estimates from contaminated samples can sometimes be corrected because they will contain both ^{230}Th and ^{232}Th and their activities will be approximately equal. The presence of ^{232}Th is an indicator of contamination because this primeval isotope occurs in nature; it is not the product of radioactive decay. If initial assumptions about the initial ratio of $^{230}Th/^{232}Th$ are accurate, the activity of ^{232}Th in the sample can be used to correct the level of contaminating ^{230}Th. The technique works well if the type of contamination is the same but the degree of contamination varies throughout the deposit being dated. Under these conditions, a systematic error exists which can be identified and corrected. Subsamples reflecting different degrees of contamination are collected from the deposit and analyzed. An isochron plot is made of the isotope ratios of $^{230}Th/^{232}Th$ vs. $^{234}U/^{232}Th$.

Values from the different subsamples will plot as a straight line, the slope of which is the ratio of ^{230}Th/^{234}U in uncontaminated calcite. A second isochron plot of ^{238}U/^{232}Th against ^{234}U/^{232}Th gives the ratio of ^{234}U/^{238}U. The age of the sample is determined by comparing the ^{230}Th/^{234}U ratio from the first plot to the ^{234}U/^{238}U ratio from the second plot (Fig. 9.2). However, the isochron method of correcting for contamination does not work well if all subsamples contain large amounts of limestone or if there is little isotopic variation among the subsamples.

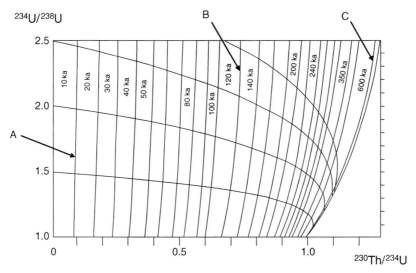

A) Samples that plot along the 10 ka Isochron are 10,000 years old.
B) Those that plot along the 120 ka Isochron are 120,000 years old.
C) Those that plot along the 600 ka Isochron are 600,000 years old.

Fig. 9.2 A graph showing the relationship between ^{230}Th/^{234}U and ^{234}U/^{238}U activity ratios in a closed system with no initial ^{230}Th (Reprinted from A. G. Latham, Figure 5.3., "Handbook of Archaeological Sciences", copyright 2001, with permission of Wiley-Blackwell)

Conclusions

Uranium series dating has most frequently been applied to Middle and Late Pleistocene materials, but it could be applied to materials that are only several hundred or a few thousand years old. Uncontaminated calcite deposited in caves, rock shelters, and natural springs usually contain uranium at the time of deposition but none of its daughter products. The extent to which a newly established decay chain has progressed toward secular equilibrium provides a means of dating matrix deposited before, during or after the time of human utilization. Ages obtained on samples that occur above or below the layer containing cultural materials can be used to set maximum or minimum ages of deposition of the archaeological layer. While "pure" material is desirable, methods have also been developed to detect the degree of contamination and correct the age estimates on samples containing detritus. For this reason, uranium series dating could provide age estimates on carbonate deposits that form in soil, such as caliche, within arid or semi-arid environments.

With the exception of marine coral, biological materials are free of uranium until it is absorbed from the environment after burial. The process of uranium absorption by the sample from the environment affects its calculated uranium series age. Subjecting samples to both uranium series and electron spin resonance dating can reduce the uncertainty associated with modeling uranium uptake. If the uptake of uranium corresponded to the early uptake model, both techniques will yield identical age estimates for the sample. If the uptake of uranium was linear or supralinear, the uranium series age will be younger, possibly much younger, than the ESR age on the same sample. In this case, it is necessary to calculate a new age for the sample based on the p value.

Potassium–Argon and Argon–Argon Dating

Potassium–argon (K–Ar) dating and argon–argon (Ar–Ar) dating are related techniques primarily used to estimate the age of geological deposits. K–Ar dating was developed in the mid-1950s; Ar–Ar dating was introduced in the following decade. Detailed explanations of these techniques appear in the geochronology books mentioned previously as well as Schaeffer and Zähringer (1966), Dalrymple and Lanphere (1969), and McDougall and Harrison (1988). Curtis (1975) and Walter (1997) present general summaries of the technique for archaeologists.

The Production of Radiogenic Argon

Both techniques involve the measurement of argon gas produced by the radioactive decay of potassium-40 (^{40}K), which has a half-life of 1.25 billion (10^9) years. Of the three naturally occurring isotopes, ^{40}K is the least common with an abundance of 0.012%. Potassium-40 usually decays to ^{40}Ca through the emission of β-particles; however, about 10.5% of ^{40}K decays to ^{40}Ar. As outlined by Walter (1997), ^{40}Ar is produced when ^{40}K undergoes an electron capture reaction during which an inner (k) shell electron is captured by the nucleus and combined with a positively charged proton to create a neutron. The single positive charge on the proton is cancelled by the single negative charge on the electron. The result is a neutral atom with the same atomic mass number ($Z = 40$) but the atomic number is reduced from 19 (potassium) to 18 (argon).

While potassium commonly occurs in rocks, only about 1 in 100,000 potassium atoms in a sample will decay to argon. Because argon is a gas, little is retained in molten rock; accumulation of gas commences with the onset of solidification. Both techniques are employed as a means of obtaining an age for volcanic material; for archaeological applications, contexts in which human activity coincides with volcanic activity are ideal. The major difference between the techniques rests on how sample components are measured. In K–Ar dating, the amounts of ^{40}K and ^{40}Ar are measured separately on different parts of the sample using different instruments. With Ar–Ar dating, potassium-39 is converted to argon-39 and the two argon isotopes are measured simultaneously on the same part of the sample. For this reason, Ar–Ar dating can be performed on single crystals, which are less vulnerable to contamination, and the internal consistency of the sample can be examined by obtaining age estimates on several crystals.

Very little argon gas is released from very young material or potassium-poor samples. The error associated with age estimates on Quaternary age, potassium-rich samples is about 10%.

Procedures for Obtaining a K–Ar Age

One of the most problematic aspects of K–Ar dating is that amounts of potassium and argon in the sample must be determined independently, which requires that the sample be divided. Because potassium is a solid, its concentration can be easily measured using a technique such as atomic absorption (Chapter 34). Before argon gas can be measured, it must first be collected from the sample without loss. One approach is to melt a small fragment of the rock, weighing only few grams, within the sample chamber and direct the liberated gas into a mass spectrometer (Chapter 31). An accurate age estimate can only be obtained if the two portions (aliquots) of the sample are completely homogeneous. For this reason, fine-grained lava and obsidian are more likely to produce reliable ages. It is also critical that the rock retains all radiogenic argon produced and that potassium is neither gained nor lost over time. This condition is met if the sample existed as a closed system with respect to both K and Ar.

Accurate K–Ar ages are easier to obtain from older and/or potassium-rich samples because the amount of argon produced is higher. If the sample is young or contains only small amounts of potassium, contaminants, such as argon from the atmosphere, can greatly affect the calculated age. Argon contamination is indicated by the presence of ^{36}Ar and correctable by subtraction because the ratio between ^{40}Ar and ^{36}Ar is known to be 295.5:

$$\text{Radiogenic } ^{40}\text{Ar} = ^{40}\text{Ar measured} - (295.5 \times ^{36}\text{Ar})$$

The age is calculated by substituting the measured values into the following equation:

$$t = 1/\lambda \ln \left[(\text{radiogenic } ^{40}\text{Ar}/^{40}\text{K}) \, (\lambda/\lambda_\varepsilon) + 1 \right.$$

The other variables in the equation are constants; λ is the specific decay constant of ^{40}K and $\lambda/\lambda_\varepsilon$, the fraction of ^{40}K that yields ^{40}Ar.

Procedures for Obtaining an Ar–Ar Age

Argon–argon dating involves the conversion of ^{39}K in the sample to ^{39}Ar and the simultaneous measurement of all argon isotopes with mass spectrometry. The whole sample is subjected to nuclear irradiation and neutron capture reactions convert a portion of the ^{39}K isotopes in the sample into ^{39}Ar, which does not occur in nature (Chapter 32). A small part of the sample, perhaps a single grain, is melted within the sample chamber; after purification, the gas released is directed into a mass spectrometer. The amount of radioactive ^{40}K in the sample is not directly determined, but ^{39}Ar provides a measure of K. The Ar–Ar age is calculated by comparing the ratio of ^{40}Ar to ^{39}Ar in the sample to the ^{40}Ar/^{39}Ar ratio in a standard of known age irradiated at the same time. Constants related to radioactive decay, the ratio of ^{40}K to ^{39}K and irradiation parameters are included in the ^{40}Ar/^{39}Ar age equation. A complicating factor is that argon can be produced when both calcium and ^{40}K are irradiated. In order to attain reliable ages, it is important to recognize and make corrections for these types of argon interferences.

Argon is released from the sample by a process known as laser fusion; the sample is heated incrementally (called step-wise heating) and the age of the gas liberated at each temperature can be determined. This represents a tremendous advantage as measurements are made on the same material at precisely the same location within the crystal lattice (i.e., single-crystal laser fusion), which eliminates the possible effects of sample inhomogeneity (Curtis 1975). The internal consistency of the subsample can be verified because the gas released at each step should be the same age. This permits the dating of weathered samples because gas from altered layers can be distinguished from unweathered cores. In addition, it is possible to recognize unreliable ages from samples that retained argon at the time of its solidification or lost a portion of the argon produced after its solidification.

Ar–Ar ages are more often precise than K–Ar ages because many replicate analyses can be made on a given sample (Chapter 7). The precision of the age of the sample as a whole is increased by averaging ages obtained on multiple subsamples. Grain-discrete laser-fusion is particularly useful if volcanic material is from a mixed context rather than a primary deposit because different minerals will likely yield different ages.

Fission Track Dating

Fission track dating examines the cumulative effects of damage that occurs when energy is released by spontaneous fission. When used as a clock, the resulting age corresponds to the last time the material was free of fission tracks, usually when it was newly formed. The density of tracks in the material and the rate at which they occur provide a mechanism for measuring time. The upper limit for calculating age depends upon the ability of the material to distinctly amass and retain the effects of damage. It is reached when new fission tracks obscure older ones. The technique is only briefly considered here; more extensive descriptions of the technique can be found in the geochronology books mentioned at the beginning of Part II.

Fission track dating was developed in the 1960s and is usually applied to glass or zircon minerals. It is more commonly employed in geological applications, but the data have been used to source archaeological obsidian (Bellot-Gurlet et al. 1999; see Chapter 18). The technique can be used to obtain an age estimate on volcanic material associated with early hominid remains, which enables comparisons between fission track age determinations and those obtained by potassium–argon dating. Fission tracks also occur in relatively young material with high uranium content.

Sample Requirements

Fission track dating can only be applied to material containing ^{238}U. Although it is a relatively rare occurrence, this isotope of uranium can decay by spontaneous fission. During spontaneous fission, the nucleus splits into two roughly equally sized fragments possessing a substantial amount of energy. The fission recoil causes damage to the crystal structure, producing a fission track. Fission tracks measure about 0.01 mm (10 μm) in length and have a small diameter.

The second requirement is that the material must have been free of fission tracks at some time in the past. An accurate fission track age can be determined if the targeted event corresponds to the last resetting or zeroing of the material. When the material contains fission tracks that predate the targeted event, the fission track age will be too old. If the material was exposed to conditions that obliterated fission tracks after the event of interest, the age will be too young.

Exposure to heat, or annealing, will obliterate fission tracks in crystalline materials. The mineral zircon must be exposed to a temperature of 800°C for at least 1 h to completely remove previous fission tracks. For glass, less time is required and the required exposure temperature is lower. Generally speaking, the time at which the mineral or glass was formed serves as an excellent zeroing event. This applies to both natural glass, which is obsidian, and glass manufactured by humans.

Procedure

Fission track dating is a three-step process. First, the tracks are enlarged so that they will be easier to count; this is achieved by etching them with either a strong acid or a strong base. Sodium hydroxide (NaOH) is used to enlarge fission tracks in zircon; hydrofluoric acid (HF) is used for obsidian and other types of glass. The fission tracks are less resistant to chemicals than undamaged areas. Other imperfections, such as air bubbles, are easily discriminated from fission tracks.

The second step is to count the number of tracks using an optical microscope with magnifying powers between 500 and 2500. The tracks in a small area, usually 1 cm × 1 cm, are counted. The number of tracks depends upon the age of the sample and the amount of uranium it contains. A very old sample with low uranium content may have relatively few fission tracks. Conversely, many fission tracks may appear in a young sample with high uranium content. This is especially true for leaded crystal or flint glass produced up until the early twentieth century. Lead oxide with uranium concentration exceeding 1% was added during the manufacturing process. Count densities of 100 tracks/cm^2 appear after only 10 years in these materials.

Age Calculation

The third step in calculating the age involves establishing the rate at which tracks appear. The rate of track formation is established experimentally by re-setting the clock and artificially inducing uranium fission, causing the production of new tracks. After the natural tracks are counted, the sample is annealed to erase all old tracks. The sample is then subjected to neutron irradiation (see Chapter 32), which induces thermal fission of ^{235}U. The new tracks that appear are counted and the sample age is calculated using the formula:

$$\text{Age} = N \times D_s/D_i \times 6 \times 10^{-8}$$

where D_s is the density of natural tracks, D_i is the density of induced tracks, and N is the neutron density of the irradiation.

Age Range and Potential Errors

Fission track dating is limited by the ability to accurately count tracks. The upper limit for track density is usually 10 million tracks/cm^2. Unstable tracks may fade with time and overlap between tracks produces counting errors. The lower limit is usually 100 tracks/cm^2. Results are usually considered statistically dubious if the track density is too low.

Chapter 10
Trapped Charge Dating Techniques

Thermoluminescence (TL), optically stimulated luminescence (OSL), and electron spin resonance (ESR) are all trapped charge dating techniques. These techniques use signals arising from electrons trapped in the crystalline structure of a sample to calculate the time since the traps were empty. For TL and OSL, the population of trapped charges is measured by the amount of light emitted by electrons released from their traps. Electrons are not evicted by ESR spectrometry; the strength of the signal emitted by trapped electrons provides a measure of the population size.

Trapped charge dating techniques can be applied to a variety of materials including quartz, feldspar, zircon, flint, and calcite formed by either geological processes (stalactite, stalagmite, and other types of speleothems) or biological processes (tooth enamel, mollusk shell, eggshell). Samples that were newly formed, heated to a high temperature or exposed to sunlight at the time of occupation, can be used to date an archaeological event. Under these conditions, previously accumulated electrons are released from some or all of the traps, effectively re-setting or zeroing the clock. The number of electrons trapped in the crystalline material and the rate at which they accumulate provide a mechanism for measuring time. The upper limit for calculating age is reached when all or most available traps are full.

More detailed information about the techniques is available from the previously mentioned geochronology texts. Archaeological applications of trapped charge dating techniques are considered in the following overviews: Aitken (1989, 1997); Feathers (1997, 2000); Grün (1997, 2000, 2001); Grün and Stringer (1991); Roberts (1997); Troja and Roberts (2000); Wintle (2008). Examples of specific archaeological applications are given in Part III.

Theoretical Considerations

Exposure to radiation energy causes electrons to temporarily move from a low-energy to a high-energy state. Occasionally, electrons become trapped in flaws that exist in the crystal structure of a mineral and remain at an intermediate energy level. The radiation that affects the electrons comes from both internal and external sources. Internal radiation is due to the decay of radioactive elements within the

M.E. Malainey, *A Consumer's Guide to Archaeological Science*, Manuals in
Archaeological Method, Theory and Technique, DOI 10.1007/978-1-4419-5704-7_10,
© Springer Science+Business Media, LLC 2011

sample itself; external radiation is from the environment in which the sample was situated. The number of trapped electrons in a sample is related to the degree of radiation exposure over time and the number of available traps in the material.

In order to obtain an age estimate, three variables must be determined: (1) the size of the population of trapped electrons; (2) the amount of radiation to which the sample was exposed each year, called the annual dose or natural dose rate; and (3) the sensitivity of the material to the effects of radiation. Highly sensitive material has many traps in which electrons can be held, increasing the chances of this event occurring. Material with a low sensitivity has few traps and a lower chance of electron entrapment. Given the same degree of radiation exposure, material with fewer traps will develop a smaller population of trapped electrons. The relationship between these variables is

$$\text{Age} = \text{signal from trapped electrons}/(\text{dose rate})(\text{sensitivity})$$

The sensitivity of the material is assessed by artificially irradiating it with fixed doses and examining the corresponding effect on signal strength. In this manner, the amount of artificial radiation required to produce a signal identical to the natural signal from an archaeological sample is established. Names applied to this amount of artificial radiation include paleodose, equivalent dose, post-dose, total dose, accrued dose, accumulated dose, and archaeological dose (Aitken 1997). Following Aitken (1989, 1997), the term paleodose, P, shall refer to the total amount of artificial radiation throughout this text. The term equivalent dose, Q, shall refer to the paleodose less the supralinearity correction (I) for initial non-linear growth of the TL and OSL signal (Fig. 10.1). This simplifies the age determination to

$$\text{Age} = \text{Paleodose}/\text{annual dose}$$

To determine annual dose, the current intensity of radiation falling on the sample each year is measured and assumed to be constant throughout the entire period of radiation accumulation.

All types of trapped charge dating techniques follow the same general process:

Step 1: Measure the cumulative effect of exposure to natural radiation. This is the magnitude of signal that arises from the trapped charges.

Step 2: Artificially irradiate the sample with varying doses of radiation to assess its sensitivity. The equivalent dose is determined by extrapolation; this value is used to calculate ESR dates. TL and OSL dates use the paleodose, which includes a correction for initial supralinear signal growth.

Step 3: Divide the paleodose (or equivalent dose in the case of ESR) by the natural dose rate to yield an age or, more precisely, the time when the accumulation of radiation effects began. The "age" represents the last time the mineralogical clock was set, or reset, to zero.

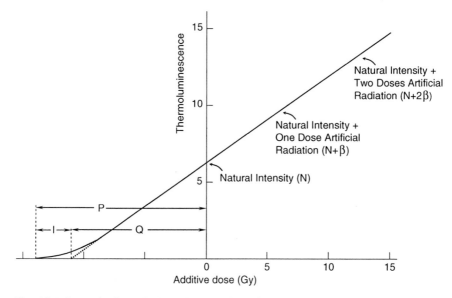

Fig. 10.1 Determination of the paleodose (P) using the additive method showing the extrapolated equivalent dose (Q) and correction for supralinear growth (I) (Reprinted from "Thermoluminescence Dating", M. J. Aitken, Figure 2.1, page 19, Copyright 1985, Academic Press, with the permission of M. J. Aitken and Elsevier.)

A date provided by TL, OSL, or ESR corresponds to the last time all, or certain shallow, traps within the sample were vacant. Electrons are freed from crystalline material through exposure to heat or sunlight; the energy from a campfire is sufficient to release all previously trapped electrons. For this reason, mineral temper in fired pottery and heat-treated lithics can be dated using TL; burnt lithics that were exposed to high temperatures can be dated using TL or ESR. Exposure to sunlight or heat, a process called bleaching, is sufficient to remove electrons from certain shallow traps. Sediments representing the former living floor at a buried site can be dated using OSL and occasionally TL. New minerals, such as calcite deposits formed in caves or springs, glass cooled from a molten state, and tooth enamel, contain no trapped electrons at the time of their formation; dates for these materials are often obtained by ESR.

Measuring the Population of Trapped Charges

The minerals suitable for trapped charge dating are semiconductors (Chapter 1). If a semi-conducting material is either newly formed or has undergone a zeroing process, all electrons are in the valence band, which is the ground state and the lowest energy level possible. When the mineral absorbs natural radiation, electrons detach from atoms; the loss of a negatively charged electron creates a positively charged

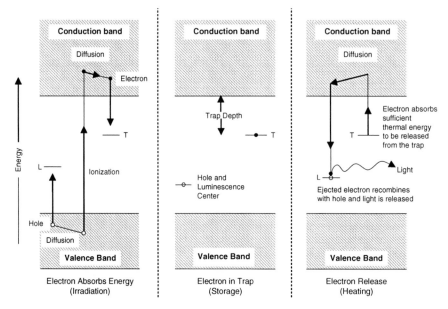

Fig. 10.2 The trapping, storage, and release of electrons at intermediate energy levels (Reprinted from "Thermoluminescence Dating", M. J. Aitken, Figure 3.2, page 44, Copyright 1985, Academic Press, with the permission of M. J. Aitken and Elsevier.)

vacancy called a "hole" (Fig. 10.2). This ionization process moves the higher energy electron into the conduction band. After a brief period of energy diffusion, the electron can drop back to the low-energy valence band and recombine with the hole. When this occurs, the mineral returns to its starting ground state condition.

Inorganic crystals consist of a metal ion with a positive charge and a negatively charged non-metal (Chapter 1). When a negatively charged ion is absent from its proper place in the crystal lattice, a trap carrying a positive charge forms between the conduction and valence bands. Ionized electrons can be attracted to and held in the trap at an intermediate energy level. The positively charged hole exists until it recombines with a released electron; if the hole is a luminescence center, light will be emitted when this occurs.

In order for an electron to escape from an intermediate energy level trap, it must absorb enough energy to return to the conduction band, diffuse the excess energy, and then drop down to the valence band (Fig. 10.2). Traps hold electrons at discrete energy states between the low-energy valence band and the high-energy conduction band and are described in terms of depth. Shallow traps are located much closer to the conduction band than the valence band. Electrons situated in these traps need only absorb a relatively small amount of energy to be ejected from the trap and move to the conduction band. Deep traps are located closer to the valence band so electrons must absorb much more energy in order to move to the conduction band.

Electrons are less likely to remain in shallow traps for an extended period of time because the addition of a relatively small amount of energy, even exposure to sunlight, can release them. Electrons generally occupy shallow traps for only a few years. Electrons in deep traps are more likely to remain there until the mineral is subjected to enough energy to release all trapped electrons.

The number of traps in crystalline material is constant over time. Immediately after the clock is reset to zero, all traps are vacant. Over time, the absorption of radiation causes a gradual increase in the population of trapped electrons. While the population of electrons in deep traps is stable, leakage of shallowly trapped electrons can occur. Over very long periods of time, a saturation point may be reached where all existing traps in the mineral are full.

TL, OSL, and ESR are similar in that all manipulate the electrons trapped in the gap between conduction and valence bands in the sample to determine its age. With OSL, electrons in shallow traps are released and the emitted light is measured. TL focuses on more deeply trapped electrons, which are released and the emitted light is measured. ESR does not release trapped electrons; instead, they are made to resonate in a strong magnetic field and the signal from them is measured.

Procedures for measuring the signals in TL, OSL, and ESR dating are discussed below. Techniques for assessing material sensitivity for the estimation of the paleodose and assessing signal reliability are then described.

Thermoluminescence Dating

Thermoluminescence dating is most often applied to crystalline minerals exposed to heat at the time of the archaeological event being dated. In order to date pottery, the minerals added as temper or naturally occurring in the clay, such as quartz, feldspar, and zircon, are targeted. Burnt flint is suitable for dating, whether it was intentionally exposed to heat to improve flaking or simply discarded in a hearth. Newly formed minerals, including stalagmitic calcite, and volcanic glass have also been subjected to TL dating. When TL is used to date unburned sediments, special procedures are used to measure only light emitted by electrons released from shallow "bleachable" traps.

Samples submitted for TL dating are processed under dim red light to avoid releasing trapped electrons. If the sample is large enough, the surface, or a portion thereof, is removed with a diamond wheel and discarded. As described in detail later in this chapter, surface removal simplifies determination of annual dose by eliminating the effects from external sources of radiation from alpha particles. The sample is then gently crushed and the minerals are sorted according to size fractions. The smallest minerals (2–8 μm fraction) are collected through suspension in acetone and analyzed together as the polymineral fine fraction. Larger quartz and potassium feldspar grains (upward of 100 μm) are processed separately because radioactive ^{40}K causes the internal dose of feldspar grains to be much higher than that of quartz. Acid is used to remove the surface of these large grains to further simplify the determination of annual dose by eliminating areas exposed to alpha particles. Each mineral-size grouping is further subdivided into several aliquots, each of which is placed on a small metal disc.

The thermoluminescence signal from each aliquot is obtained by rapidly heating it and measuring the light emitted (Fig. 10.3). The metal disc with a portion of the sample is set on a heating element in an oven flushed with nitrogen or an inert gas. The temperature is rapidly increased to 500°C at a rate of 10–20°C per minute. Thermal energy releases electrons from traps; if they recombine with holes that are also luminescence centers, a weak emission of light occurs. A sensitive photomultiplier detector (see Chapter 35) amplifies the light and measures the strength of the TL signal. Colored filters placed in front of the detector reduce light interference from the heating element and non-TL emissions from the mineral.

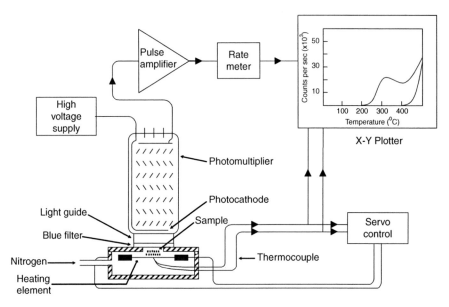

Fig. 10.3 Thermoluminescence apparatus (Reprinted from "Thermoluminescence Dating", M. J. Aitken, Figure 1.2, page 5, Copyright 1985, Academic Press, with the permission of M. J. Aitken and Elsevier.)

A glow curve is a plot of the TL emission (N) detected at each temperature (Fig. 10.4). Electrons are released from progressively deeper traps as temperature increases. The glow curve of the first heating includes both thermoluminescence and the red-hot glow of the sample and heating element (curve a). In order to measure the strength of the TL signal, the sample is heated again. The glow curve from the second heating represents only the red-hot glow (curve b); there is no thermoluminescence because all electrons were ejected from their traps during the first heating. The natural TL is the difference between the glow curve from the first heating (curve a) and the glow curve of the second heating (curve b).

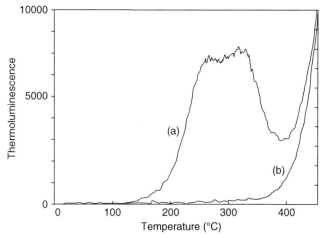

a) Curve depicts the light emission during the first heating.
b) Curve depicts the light emission during the second heating.

Fig. 10.4 Thermoluminescence glow curve (Reprinted from "Thermoluminescence Dating", M. J. Aitken, Figure 1.1, page 2, Copyright 1985, Academic Press, with the permission of M. J. Aitken and Elsevier.)

Optically Stimulated Luminescence Dating

Optically stimulated luminescence dating is typically used to date unburned sediments and involves only electrons in relatively shallow, or bleachable, traps. Exposure to sunlight at the time of the archaeological event prevents electrons from occupying shallow traps; these traps remain vacant as long as the material is on the surface. Once depositional processes bury the sample and cut off the sunlight, electrons begin to accumulate. Electrons in shallow traps require relatively little energy to release them. Special care is required to protect samples from light during their collection and processing; otherwise, procedures are quite similar to those for TL dating.

Samples are processed under dim red light and unconsolidated sediment is sorted by grain size. As with TL, the fine-grained fraction is polymineral; coarse grains are sorted by mineral type. Although OSL targets electrons in shallow traps, electrons in the shallowest of these are highly transient and their presence does not relate to the event being dated. Samples are gently preheated to remove electrons from these unstable traps prior to luminescence signal measurement.

Instead of heat, light with relatively low photon energy is used to evict electrons from shallow traps. This light does not affect electrons in deep traps so determining the paleodose is more straightforward and material from poorly bleached contexts can be examined. Ultraviolet, visible, or infrared light (Chapter 2) can be used, but the range of wavelengths must be restricted to make it monochromatic. The color of the excitation light depends on the material being dated. Green light from a laser is typically used to release luminescence from quartz crystals; red light is used for feldspar. Less expensive light sources, including xenon and quartz halogen lamps and diodes, can be used but filters are needed to restrict the spectral range. Filters are also placed in front of the photomultiplier to block the excitation light so that only the blue violet light of the luminescence signal is measured (Fig. 10.5).

OSL differs from TL in that the energy of the light used to release electrons is constant. The luminescence emitted is plotted against the duration of light exposure, or shine time, producing a shine-down curve. In general, the initial signal is strongest followed by a rapid decline over time. The signal produced by samples from poorly bleached contexts may increase over time; this indicates that not all electrons were removed from shallow traps at the time of site occupation. The single aliquot regeneration (SAR) technique is used to establish the equivalent dose in OSL dating (see below).

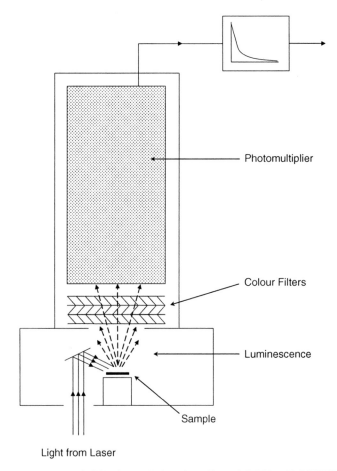

Fig. 10.5 Measurement of OSL for optical dating (From LONGMAN SCIENCE BASED DATING IN ARCHAEOLOGY © 1990 Pearson Education, Inc. or its affiliates. Used by permission. All rights reserved.)

Electron Spin Resonance Dating

Electron spin resonance (ESR) or electron paramagnetic resonance (EPR) dating is less precise than other forms of trapped charge dating but the array of potentially dateable materials and broad time range of applicability enhance its value. While ESR uses the same pool of trapped electrons, it differs from TL and OSL in that electrons are not released during the measurement of natural intensity. Instead, trapped electrons in the sample are detected by the absorption of energy.

ESR dating is often applied to materials that are hundreds of thousands years old. The process is complicated by the fact that some dateable materials are initially free of uranium, but absorb it over time. The challenges of modeling uranium uptake in order to produce an accurate date are identical to those of uranium series dating. As described in Chapter 9, by applying both uranium series and ESR dating, the uranium uptake model and the date of a sample can be more precisely established.

The criteria for identifying samples suitable for ESR dating are similar to those for uranium series dating. Geological minerals including calcite speleothems and travertine are often dated; it is also possible to date burnt flint. Dateable biological samples include tooth enamel, eggshell, and mollusk shell; bone and tooth dentine are not suitable for ESR dating.

Typically, 5 g or less of material is required to date a sample. As with TL samples, calculation of the annual radiation dose can be simplified by removing the surface. Surface removal is not possible when dating thin material, such as eggshell, mollusk shell, and tooth enamel. The sample is ground up and sieved to ensure homogeneity and then divided into about ten portions. Portions being used to calculate equivalent dose are subjected to artificial gamma irradiation prior to signal measurement.

The ESR signal is measured with an electron spin resonance spectrometer, which consists of a powerful magnet with a microwave cavity (see Chapter 36). When a sample is placed inside the cavity, the trapped electrons behave like small spinning bar magnets and the magnetization of most is aligned in the direction of the strong magnetic field. Microwave radiation is applied and, at resonance, electrons absorb the energy and flip so that their magnetization is aligned in the opposite direction. The amount of energy absorbed is related to the number of trapped electrons in the sample, which, in turn, is related to the age of the sample. Older samples of a given material will contain more trapped electrons than younger samples of the same material, so the higher energy absorbance will result in a stronger ESR signal (Fig. 10.6).

To enable comparisons between data collected from ESR spectrometers with different magnetic field strengths, results are reported with reference to g values. The g value is a ratio of the absorption frequency in gigahertz, f, to the magnetic field strength at resonance in Tesla, H:

$$g \text{ value} = f/14H$$

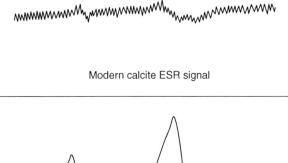

Modern calcite ESR signal

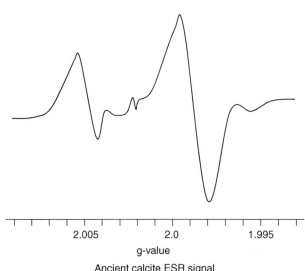

2.005   2.0   1.995

g-value

Ancient calcite ESR signal

Fig. 10.6 Electron spin resonance signals from modern and ancient material (From Chronometric Dating in Archaeology, Advances in Archaeological and Museum Science Vol. 2, 1997, Electron Spin Resonance Dating by R. Grün, Figure 8.2, with kind permission of Springer Science and Business Media)

The g value for a given trap does not vary with magnetic field strength; consequently, the first derivative of the ESR signal (Chapter 36) depicted as a function of the g value is independent of the operating conditions under which it was obtained.

Determining the Equivalent and Paleodose Dose

The signals obtained by TL, OSL, or ESR indicate the size of the trapped electron population. The signal is a measure of the natural intensity of the sample, that is, the dose of natural radiation from the environment to which the sample was exposed. While buried, samples are exposed to multiple types of radiation (α, β, γ, and cosmic) from both internal and external sources. With trapped charge dating, the size of the artificial dose required to produce a trapped charge signal equal to the natural intensity of the sample is determined. In trapped charge dating, units of radiation are now given in Grays (Gy); 1 Gy is equivalent to 100 rad, the unit used previously.

The amount of artificial radiation required to reproduce the signal can be determined through simple extrapolation (giving the equivalent dose, Q) and may include a correction for initial supralinear growth for calculations of TL and OSL ages (giving the paleodose, P). Since the signal from young material with many traps may be identical to that from much older material with fewer traps, the sensitivity of material to radiation is established with a dose-response curve. This usually involves irradiating portions of the sample with differing amounts of artificial β (or γ) radiation prior to or after measuring its natural intensity.

Applications of the additive method and regeneration technique are suitable for samples completely zeroed at the time of the targeted archaeological event. These crystalline samples were exposed to sufficiently high temperatures for a period of time long enough to evict all trapped electrons. By contrast, samples such as heat-treated lithics and unburned sediments likely contain electrons trapped prior to the archaeological event. These are called residual electrons. When dating incompletely bleached samples, it is necessary to correct the paleodose for the signal produced by residual electrons. This is accomplished using approaches such as the partial bleach technique and the regeneration method. Single aliquot regeneration can also be used to detect partial bleaching at the time of deposition.

Additive (Dose) Method

The additive dose method involves irradiating certain sample aliquots prior to measuring its signal or natural intensity. The signal from these samples arises from electrons trapped in the archaeological context plus those newly trapped due to exposure to artificial radiation. Separate portions are exposed to different levels of radiation and the resulting signal is measured. The strengths of the natural signal and natural + artificial signals are then plotted against the size of the additive doses (Fig. 10.1).

In samples that can be reliably dated, the intensity of the signal increases in proportion to the amount of radiation received. The mathematical equation that "best fits" the dose–response curve is determined and the equivalent dose, Q, is extrapolated. The advantage of the additive method is that the resulting signals reflect both the unknown natural intensity and the known artificial dose. If the dose–response curve is fitted with an appropriate mathematical equation, the sensitivity of the material should be accurately depicted. The equivalent dose is given by the intersection of the dose–response curve with the X-axis.

The additive method is commonly used for TL and OSL and is the only method used to determine the equivalent dose in ESR dating. The only disadvantage is that the initial growth of the signal is not exactly the same as that depicted in the dose–response curve. The dose–response curve generated by the additive method is quite linear; in reality, the initial growth of TL and OSL signals is supralinear. As shown in Fig. 10.1, the material becomes increasingly sensitive to radiation after the zeroing event. The initial signal generated by a given amount of radiation is smaller than expected; over time, it gradually increases and then stabilizes. Ages calculated

using the equivalent dose without considering the effects of supralinearity would
be too young. The regeneration technique is used to determine the correction for
supralinearity.

Regeneration Technique

As noted above, trapped electrons are released in the process of measuring TL and
OSL signals. The regeneration technique involves irradiating portions of the sample that have been reset to zero. Exposure to different, known amounts of artificial
radiation causes electrons to re-occupy some of the vacated traps. Signals from the
newly irradiated samples are measured and their intensities plotted against the artificial dose applied. By subjecting several samples to doses less than the natural
intensity, the initial supralinear growth of the signal is simulated. The technique
is generally not used to calculate paleodose because the process of releasing the
electrons, or bleaching, can change the sensitivity of the material to radiation. The
technique establishes the supralinearity correction factor, I, for the equivalent dose,
Q, measured using the additive method. The paleodose, P, is the sum of Q and I.

An alternative approach is the slide technique, which more fully combines the
data from both the additive method and the regeneration technique. The data from
the additive method are plotted on a dose–response curve as signal vs. dose; the data
from the regeneration technique are similarly plotted on a second dose–response
curve. The horizontal distance between the two curves is the paleodose (Fig. 10.7).

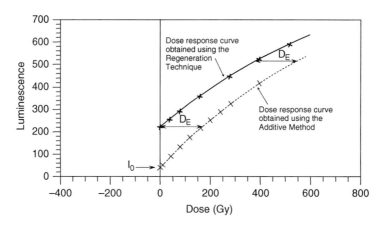

Fig. 10.7 Slide method for dose determination (Reprinted from R. Grün, Figure 4.4c., "Handbook
of Archaeological Sciences", copyright 2001, with permission of Wiley-Blackwell)

Partial Bleach(ing) Technique

This technique is primarily used to establish the actual equivalent dose when a portion of the trapped charge signal is due to the release of residual electrons. An

example would be when unburned sediments are dated using TL. Sunlight produces enough energy to remove only electrons from shallow, bleachable traps, yet measurement of the TL signal removes electrons from both shallow and deep traps.

In order to quantify the signal from only bleachable traps, two dose curves are established, one of which is generated by the additive method as normally applied and the other consists of artificially irradiated samples that have been exposed to light. For the latter, re-zeroed samples are irradiated with doses equal to levels of the first curve: N, N+β, N+2β, etc., but prior to measurement, the irradiated samples are subject to natural or artificial light for a short period of time. The wavelength of light used for the partial laboratory bleaching depends upon the material type and its depositional context.

When properly applied, partial bleaching only removes electrons from the shallowest traps, that is, those that would be lost even in poorly bleached contexts. As long as the laboratory bleaching is less than the bleaching the sample received at the time of deposition, the intersection point of two curves will indicate the equivalent dose, Q (Fig. 10.8). If the laboratory bleaching exceeds the degree of bleaching at the time of deposition, the age of the sample will be overestimated.

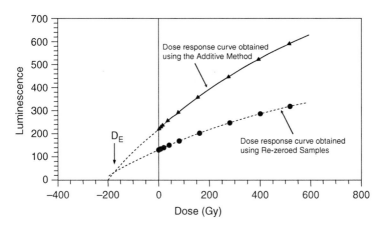

Fig. 10.8 Partial bleaching technique (Adapted from Aitken (1990) and Grün (1990) with permission. From Longman Science Based Dating in Archaeology © 1990 Pearson Education, Inc. or its affiliates. Used by permission. All rights reserved. Reprinted from R. Grün, Figure 4.4e., "Handbook of Archaeological Sciences", copyright 2001, with permission of Wiley-Blackwell)

Single Aliquot Regeneration (SAR)

The single aliquot regeneration is a variation of the regeneration technique that can be used for samples that may not be completely homogeneous. Instead of dividing it into portions, all measurements are made on the whole sample. First, the natural intensity of the sample is measured, then it is irradiated with an artificial dose and the regenerated signal is measured. The process is repeated with artificial radiation doses of different magnitudes until enough points are generated to produce a reliable dose–response curve. The process is then applied to different portions of sample to examine its homogeneity. This technique is used to establish the equivalent dose in OSL dating.

A variation of this technique, called single grain dating, is used if there is the possibility that bleaching of the sample was uneven. The paleodose is evaluated from single grains so that grain-to-grain variation in residual electrons can be detected. After differences in bleaching are established, the estimated age from all grains should be the same.

The Plateau Test

The plateau test is used to assess the depth of traps from which the signal arises. As mentioned above, shallow traps are unstable; electrons occupy them for only a short time. Electrons in the deepest traps are likely to remain in them for millions of years. The plateau test is performed to ensure that only signals arising from electrons ejected out of stable traps are used to calculate an age estimate. In the case of TL, a portion of the natural sample is radiated with a fixed amount of artificial β-radiation. The glow curve of the natural sample (N) is compared to that of the sample with natural and additional β-radiation (N+β). The N to N+β ratio (N/N+β) is plotted over the entire range of temperatures. In samples that can be reliably dated, N/N+β is constant from about 300°C to about 400°C; in the plot, this region appears as flat plateau (Fig. 10.9). The most reliable dates have plateaus extending over a temperature range of 125°C. If the sample fails the plateau test, it should be discarded because it cannot provide an accurate age estimate. When

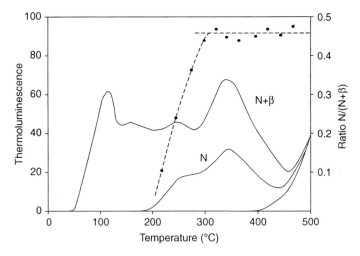

Fig. 10.9 The plateau test (Reprinted from "Thermoluminescence Dating", M. J. Aitken, Figure 1.3, page 8, Copyright 1985, Academic Press.)

using the regeneration method to measure the paleodose for an OSL or TL date in samples from poorly bleached contexts, the plateau test is passed when the degree of laboratory bleaching closely corresponds to that of the depositional environment.

An analogous test in ESR dating involves plotting the equivalent dose, Q, obtained by the additive method against magnetic field strength (Grün 1998). If the signal was generated by electrons exhibiting the same dose–response and thermal stability, Q should be constant at magnetic field strengths ranging from 3455 to 3465×10^{-4} T.

Sources of the Annual Dose

The annual dose is the amount of natural radiation to which the sample was exposed each year since its clock was reset. The largest component of the annual dose is due to the decay of radioactive elements and their unstable daughter products. Small amounts of uranium (^{238}U and ^{235}U), thorium (^{232}Th), potassium (^{40}K), and sometimes rubidium (^{87}Rb) exist within the sample and its surrounding environment. Their decay leads to the production of α-, β-, and γ-radiation (see Chapter 3); cosmic radiation may also affect the sample. Accurate determination of the annual dose is critical for the calculation of the sample date. The contributions from external and internal sources are considered separately.

External Contribution

External sources of radiation include radioactive elements in the surrounding sediment and cosmic rays from outer space. In general, if the sediment is completely homogeneous for a distance of 30 cm in all directions around the sample for the majority of time since its clock was re-set, elemental analysis of the sediment will provide an accurate estimate of its radioactivity. For this reason, a sample from a paleosol at a stratified site likely will not comply; a sample from the fill of a large, deep pit feature may comply (Fig. 10.10). It is best to avoid selecting samples from near large boulders, feature boundaries, or any other discontinuity.

Direct measurement of radiation is recommended for samples from inhomogeneous contexts. A portable γ-spectrometer can be positioned to measure the radiation in the context from which the sample was recovered. The instrument must remain in place for about 1 h. Alternatively, a capsule containing thermoluminescence phosphor can be buried in the sample context for an extended period, from several months to one full year.

In practice, external alpha radiation is not usually considered because the particles are too large to deeply penetrate the surface. An alpha particle only travels about 0.03 mm so the effects of external alpha radiation are restricted to the outermost 50 μm of the sample. By discarding the outer 2 mm of the sample, effects of external α- and β-radiation are eliminated (Fig. 10.11). Portions of the sample potentially

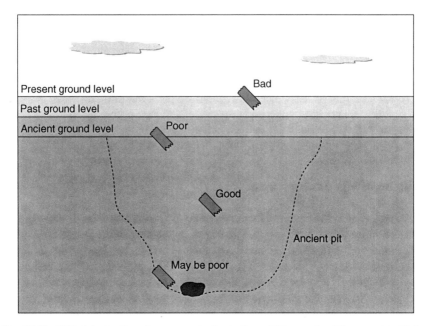

Fig. 10.10 Differing situations for success in the testing of TL samples (Chronometric Dating in Archaeology, Advances in Archaeological and Museum Science Vol. 2, 1997, Luminescence Dating by M. J. Aitken, Figure 7.2, with kind permission of Springer Science and Business Media)

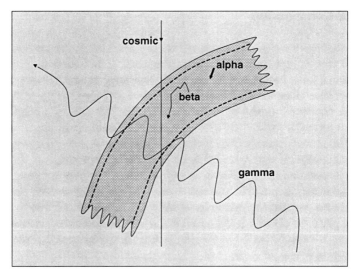

Fig. 10.11 Sources of annual dose when the outer surfaces of the sample are removed (From LONGMAN SCIENCE BASED DATING IN ARCHAEOLOGY © 1990 Pearson Education, Inc. or its affiliates. Used by permission. All rights reserved.)

affected by exposure to light during excavation and handling are also removed; this procedure is particularly important when processing translucent minerals such as flint.

Gamma rays have a range of 30 cm and the annual dose can be established through the analysis of sediment samples or through the use of dosimeter capsules and portable gamma ray spectrometers. The amount of cosmic radiation to which the sample was exposed depends upon its burial depth, latitude, and sea level. At sea level, the annual cosmic dose rate is about 300 μGy and this amount decreases with depth of burial (Grün 1997).

Another factor that affects external radiation is the water content of the soil over the entire burial period. Water absorbs radiation so contexts that have been continuously dry are ideal. A common procedure is to determine the water content of the sample at the time of processing and use this "as found" measurement as an estimate for the entire burial period. For this reason, potential trapped charge dating samples should be placed in a plastic bag with any adhering soil as soon as possible after excavation.

Assessing Sources of the External Dose in a CRM Context

Given these conditions, trapped charge dating requirements seem onerous, especially in a cultural resource management context. It is noteworthy that Dunnell and Feathers (1994) report they were able to obtain reasonable dates from surface samples and suggest that TL dating should be the method of choice for material, such as opaque pottery, recovered from tillage zone sites. They argue that the calculation of the external radiation contribution is simpler for material from surface contexts because the atmospheric contribution is essentially constant. They suggest that portions of the matrix collected from strata within 30 cm of the artifact can be used to assess external sources of radiation. States of disequilibrium in U and Th decay chains due to soil leaching presented calculation challenges, but Dunnell and Feathers (1994) consider them correctable. They propose that soil moisture levels be used to estimate water content for samples collected from surface contexts.

Internal Contribution

The contribution from internal sources varies with the amounts of radioactive elements in the sample. Once elemental analysis is conducted to determine the concentrations of U, Th, and K in the sample, the annual contribution of radiation can be calculated from published sources. The α-efficiency or k-value of the sample must also be determined because the large size of the α-particle limits its effect; only 3–30% of the radiation from α-decay actually contributes to the annual internal dose. Because it is often free of radioactive elements, the internal contribution for quartz is minor and only considered if the contribution from external sources is very low. By contrast, concentrations of U and Th are so high in zircon crystals that only the internal contribution needs to be determined; the external contribution is considered negligible. Potassium is a major source of the internal contribution in feldspars.

Calculations of internal radiation from the concentrations of radioactive elements must include radiation from radioactive daughters in the decay series. The annual dose increases as radioactive daughter products form or "grow in" (Chapter 3); the internal component of annual dose doubles for the first 100,000 years due to uranium series daughter isotopes. Mathematical corrections for this assume that there has been no change in the amount of parent uranium due to leaching or uptake.

Dunnell and Feathers (1994) found that the gaseous daughter radon was lost from highly porous pot sherds. The resulting disequilibrium caused highly porous sherds to date younger than expected. Once corrected for the loss of radioactive daughters, the TL dates conformed to expectations.

Uranium Uptake

ESR dating of biogenetic materials poses special challenges because the amount of uranium in the sample changes over time. When initially formed, teeth, shell, and bone do not contain uranium; the level increases as these materials absorb uranium from the environment. There are two models for uranium absorption, early uptake and linear uptake. Under the early uptake (EU) model, it is assumed that uranium was introduced shortly after the death or deposition of the sample. Consequently, the effects of the radiation from the decay of absorbed uranium would have the greatest impact on the sample by increasing the number of trapped electrons. Sample ages calculated using the EU model represent its minimum age. Under the linear uptake model (LU), the radiation effects gradually increase over time as more uranium accumulates in the sample. In samples with low levels of uranium, the difference between LU and EU age calculations is less than 10%; under extreme conditions, the LU age can be twice the EU age estimate. Uncertainties associated with age estimates on older teeth containing significant amounts of uranium are upward of 25% (Grün 1997).

As described in Chapter 9, one method of solving the problem of modeling uranium uptake is to obtain both an ESR and a uranium series date on the same sample. When the correct model is determined, the ESR and U-series age estimates should agree. This approach cannot account for uranium lost through leaching.

Dating Limitations and Sources of Error

All trapped charge dating techniques share similar limitations that affect their applicability and accuracy. It may be difficult to accurately estimate the annual external dose, especially if preferential leaching of radioactive parent or daughter isotopes has occurred. Problems have been encountered with samples collected from certain geological contexts, such as caves and near boundaries between different types of rock/geological zones. The uncertainty associated with TL and OSL dates is usually between ±5 and ±10%. The reliability of ESR dates increases if the uranium uptake model can be established.

Incomplete zeroing of the clock can also produce problems, especially in the case of incomplete bleaching of sediments dated by OSL. It is correctable with additional measurements to establish the size of the residual electron population.

The trap saturation may be a factor in very old samples. Saturation occurs when all or almost all of the traps are full so that accumulation slows or stops. When this occurs, the relationship between the size of the trapped electron population and the annual radiation dose to which the sample is exposed changes from linear to sublinear (see Fig. 9.1 for an example of sublinear growth). The number of traps in a sample varies with material type; the number should be constant or change in a predictable manner. Processes, such as recrystallization, new crystal growth, and transitions from solid to liquid phases, alter the number of traps and make the material unsuitable for dating.

As noted previously, signals arising from electrons in stable traps enable reliable age estimates. The purpose of the plateau test in TL and preheating the sample in OSL is to ensure that this condition is met. With ESR, it is necessary to individually evaluate the nature of the trapped electrons producing the signal in each sample. As a rule, the trapped charge signal should be obtained from electrons residing in traps for a period of time ten times greater than the archaeological event. For example, the natural intensity of a sample zeroed during an archaeological event 1000 years ago should be measured with electrons from traps that are stable for at least 10,000 years.

The length of time an electron resides in a trap is related to its thermal stability. Trap stability is highest in polar regions and lowest in equatorial regions. At the thermal mean life, 63% of the original population of electrons will have left their traps and recombined with holes. If the thermal mean life is at least ten times greater than the age of the sample, the loss of original electrons can be ignored.

Potential Benefits of Trapped Charge Dating

Trapped charge dating has not been extensively applied in North America but these techniques have the potential to resolve, at least partially, some dating problems. Feathers (2000) notes that erroneous assumptions about external dose and sample homogeneity prevented American labs from producing viable dates in the 1960s and 1970s. Despite technical advances and widespread acceptance in the United Kingdom, luminescence dating was considered the less precise "poor cousin" of radiocarbon dating, applied only when organic material is not well preserved or poorly associated with cultural materials (Feathers 2000:159). Feathers (2000) argues that the limitations of radiocarbon dating are not fully appreciated by

North American archaeologists. The age of material less than 300 years old is especially difficult to determine with radiocarbon dating due to recent and pronounced fluctuations in the amount of ^{14}C in the atmosphere (Taylor 1997, 2001).

Feathers (2000) suggests that TL dating of pottery is well-suited for detecting changes in material culture, such as the introduction and spread of shell-tempering, clarifying chronologies, and tracing settlement patterns and OSL dating of sediments could provide information about the construction of earthen mounds. Dykeman et al. (2002) applied TL dating to pottery and fire-cracked rock from protohistoric sites with available tree ring dates and found that TL outperformed radiocarbon dating. Their argument that thermoluminescence dating should become part of a comprehensive site dating strategy is especially strong in regions where dendrochronology cannot be applied.

The varieties of material that can potentially be dated make trapped charge dating especially valuable. Thermoluminescence dating of fired pottery and fire-cracked rock is useful; both OSL and TL dating can be applied to ancient living floors. It may be possible to obtain TL dates from burnt flint; however, the TL signal may be weak unless the sample is many thousands of years old. Minimally, radiocarbon and trapped charge dates on different materials from a site could provide multiple independent lines of evidence to refine chronologies.

Chapter 11
Other Dating Techniques

Amino Acid Racemization

Amino acid racemization (AAR) uses systematic, natural changes in the structure of amino acids in a substance to measure relative and absolute time. The process of racemization, where the configuration of amino acids is converted from their initial L-form to the D-form, is described in Chapter 5; the principles of stereochemistry are presented in Chapter 6. Racemization occurs in amino acids with one chiral center; when amino acids with two chiral centers undergo interconversion, the process is referred to as epimerization. As AAR utilizes amino acids that undergo both racemization and epimerization, the technique is more precisely referred to as amino acid diagenesis dating. The history and development of the technique are outlined in Hare et al. (1997); Bada (1985) discusses theoretical and analytical aspects and archaeological applications; Johnson and Miller (1997) also describe various archaeological applications but emphasize approaches that can improve the reliability of dates.

The potential of using structural changes in amino acids as a clock was introduced in the 1950s by Abelson. It reached its heyday in the 1970s, prior to the advent of accelerator mass spectrometry (AMS) radiocarbon dating. At the time, it was regarded as a means of obtaining an age on materials too old to be reliably dated by conventional radiocarbon dating. As outlined by Pollard and Heron (1996), the optimism quickly faded with the publication of several improbable dates on New World Paleoindian skeletal material. It is now known that diagenetic processes leading to collagen loss affect the reliability of AAR dates. Because racemization and epimerization reactions are temperature dependent, it is also important to accurately estimate the average burial temperature. While its application has been controversial, Johnson and Miller (1997) suggest that with careful selection of samples from certain depositional contexts, the technique can make valuable contributions to archaeology.

Diagenetic Changes That Affect Proteins and Amino Acids

As described in Chapter 5, amino acids are compounds that have an amino group, a carboxyl group, and a hydrogen atom bonded to the same carbon atom, the

M.E. Malainey, *A Consumer's Guide to Archaeological Science*, Manuals in Archaeological Method, Theory and Technique, DOI 10.1007/978-1-4419-5704-7_11, © Springer Science+Business Media, LLC 2011

α-carbon. The fourth group on the α-carbon is one of 20 side chains, called the R group, resulting in 20 different amino acids. Most amino acids produced by living organisms have the same configuration and are referred to as L-amino acids because they rotate plane polarized light to the left (see Chapter 6). Proteins are long chains of amino acids formed by condensation reactions, where a water molecule is lost and the acid group of one amino acid links to the amino group of another amino acid.

Because only L-amino acids occur in living animals, the initial ratio of D-amino acids to L-amino acids is 0.0; when an organism dies, the proteins begin to degrade. Degradation processes include hydrolysis, decarboxylation, deamination, and racemization. Hydrolysis is the reverse of condensation reaction (Fig. 5.18); the addition of water breaks the peptide bond between amino acids. Decarboxylation is loss of the carboxylic acid group leading to the formation of an amine. The loss of the amino group, called deamination, results in the formation of carboxylic acids. Both amino acids freed by hydrolysis and those bound together in polypeptide chains also undergo structural alterations leading to the appearance of D-amino acids. The term racemization is used when the α-carbon is the only chiral carbon in the amino acid and only two forms are possible, the L-amino acid and the D-amino acid. When the first carbon of the R-group is also a chiral carbon that undergoes structural alteration, the process is referred to as epimerization and results in the production of four different amino acid forms. In addition, the total concentration of amino acids in the sample decreases as they are removed by ground water action.

Racemization is called the interconversion of L- and D-amino acids because the process is reversible (Fig. 11.1). It is believed that the relatively loosely bonded hydrogen on the α-carbon can become temporally disconnected from the L-amino acid. If it reattaches on the same side from which it originated, the L-amino acid reforms. If the hydrogen reattaches on the opposite side, the D-amino acid is produced. The rate at which the process occurs is fastest initially when the concentration of L-amino acids is highest and then it slows. At equilibrium, the mixture is racemic, meaning that the concentration of L-amino acids equals the concentration of D-amino acids, so the D/L ratio is 1.0. Because four different amino acids occur when L-isoleucine undergoes epimerization, the ratio of D-alloisoleucine to L-isoleucine is 1.3 at equilibrium. The racemization rate is largely dependent upon the

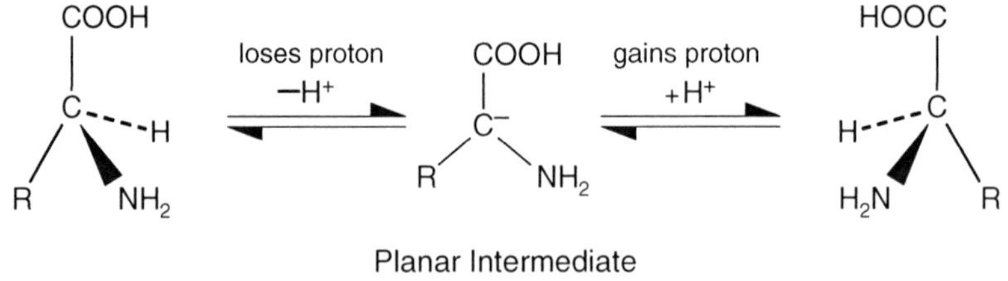

Fig. 11.1 Amino acid racemization (From A. Mark Pollard and Carl Heron, "Archaeological Chemistry" Figure 8.2, page 280, © 1996 - Reproduced by permission of The Royal Society of Chemistry)

ease with which the hydrogen can be detached from and reattached to the α-carbon. While the temperature dependence of the racemization rate was acknowledged from the outset, over time it became clear that a number of other factors are also involved.

The amino acid racemization half-life is the time required for a particular amino acid to reach a D/L ratio of 0.33. Bada (1985) provides examples of racemization half-lives of different amino acids extracted from different tissues in various parts of the world. The half-life of aspartic acid from bones and teeth can vary between 5000 and 200,000 years in East Africa. The racemization half-life of aspartic acid from bones, teeth, and certain types of plant material from California is between 20,000 and 30,000 years. The epimerization half-life of isoleucine in bivalve mollusk shell ranges from 60,000 years in Southern Florida to 300,000 years in the Canadian Arctic.

Analytical Procedures

The procedure described by Bada (1985) requires only about 5 g of clean, dense, compact bone for analysis. The mineral component is dissolved (hydrolyzed) using hydrochloric acid; when working with young samples, care must be taken to reduce acid-catalyzed racemization. Liquid chromatography employing different columns is used to first isolate amino acids from other sample components then to isolate aspartic acid from other amino acids. The aspartic acid is then treated with a chemical to produce L-dipeptides and D-dipeptides that can be separated using an automatic amino acid analyzer. Alloisoleucine/isoleucine ratios and amino acid compositions of the sample can be directly determined using the same instrument. Gas chromatography is used to determine the D/L ratios of alanine, glutamic acid, and leucine from amino acid methyl ester derivatives. Johnson and Miller (1997) describe the use of high-pressure liquid chromatography and gas chromatography to measure concentrations of amino acids in samples (see Chapter 33 for instrument descriptions).

AAR Dating of Archaeological Bone

Bada (1985) cautions that, while different analytical techniques give comparable D/L ratios, the conversion of the value to a reliable age estimate is only possible if a rate constant for the racemization reaction (k_i) has been determined for the location where the bone was found. The technique is stigmatized by applications of AAR dating by Bada and his coworkers in the mid-1970s when this condition did not hold. Using a technique known as the "calibration" method, the D/L ratio was determined in one bone from a site then it was subjected to radiocarbon dating. The calculated k_i of the "known age" bone was used to estimate the exposure temperature (also known as the effective diagenetic temperature or EDT) for the site. This temperature value was then employed in calculations of AAR ages for all other bones.

Aspartic acid racemization dating was performed on skeletal material from California using a human bone from the Laguna site dated by conventional radiocarbon to 17,150 ± 1450 years for calibration. The resulting AAR-inferred (or amino) ages on anatomically modern human skeletal material from Del Mar and Sunnyvale ranged from 48,000 to 70,000 years ago. Subsequent uranium series and accelerator mass spectrometer radiocarbon dates on the material indicated it was less than 12,000 years old. Bada (1985) and Johnson and Miller (1997) point out that the discrepancy was due to inaccurate radiocarbon age estimates on the calibration material and poor preservation of the dated bone.

As noted by Pollard and Heron (1996), at the time about 400 g of bone was required to obtain a radiocarbon age compared to only about 10 g for AAR dating. The prospect that one expensive, in terms of cost and sample size requirement, radiocarbon age estimate could enable the determination of many inexpensive AAR dates on other samples was attractive. Johnson and Miller (1997) suggest that cost-effectiveness and small sample requirements of AAR are still highly advantageous.

A number of factors affect the rate at which racemization and epimerization occur and can affect the reliability of AAR age inferences (Hare et al. 1997; Johnson and Miller 1997). The technique is best applied to materials from contexts with moderately low temperature and moisture levels. Under very cold conditions, racemization rates are reduced; some water is required for the process to proceed. The racemization rate of free amino acids is not identical to those bound in peptides and is also affected by the position of the amino acid in a peptide chain. Experimental data obtained by simulating diagenetic processes through oven storages or from well-dated samples with constant thermal histories are used to develop kinetic models. These models provide a mathematical relationship between D/L values, absolute age, and temperature; the model employed depends upon the material analyzed and the D/L value. Johnson and Miller (1997) suggest the current mean annual temperature can be used as the average Holocene temperature. Paleotemperature records are required to date older materials because racemization rates during interglacials were different than those during glacial periods.

Depth of burial must also be considered, as materials closer to the surface are more likely to be affected by temperature changes that occur throughout the day and on a seasonal basis. In general, samples from the upper 30 cm are considered to be in the "kinetically active" soil zone; however, soil loss due to erosional processes may cause more deeply buried materials to enter this zone.

Several other factors can affect the rate at which racemization occurs in bone. The presence of chemicals with alkaline properties increases racemization rates; the precipitation of secondary minerals can also affect the racemization process. D-Amino acids may be present in the sample at time zero and, perhaps most importantly, the degree to which intact collagen is preserved profoundly affects the reliability of the date. Variability in AAR dates on bone samples of similar age and temperature histories has been related to the retention of intact collagen. Contamination by microorganisms and amino acids from the environment can occur. In addition, Child et al. (1993) state that certain bacteria will preferentially attack either the

L-form or the D-form of aspartic acid and could potentially alter D/L ratios. Since the racemization rate of certain amino acids is faster than others, the extent of racemization follows a particular pattern (Bada 1985; Poiner et al. 1996). Since aspartic acid has the fastest racemization rate, its D/L ratio (D/L_{asp}) should always be highest. The D/L ratios of alanine should be lower and approximately the same as glutamic acid. The D/L ratio of leucine should be lowest and approximately equal to the ratio of alloisoleucine and isoleucine. A deviation from the expected pattern is an indication of serious contamination (Poiner et al. 1996). In general, isoleucine is less susceptible to contamination than acidic amino acids, such as aspartic acid (Bada 1985).

Given the problems associated with AAR age inferences on bone, Hare et al. (1997) suggest that they should be accepted only if they are consistent with radiocarbon and uranium series dates on material from the same context. Johnson and Miller (1997) contend that AAR gives reliable age estimates in well-preserved bone, especially if collagen or non-collagenous proteins isolated from bone are analyzed. Age estimates of material with D/L values between those of the calibration samples can often be interpolated with a precision of 10% or better.

AAR Dating of Mollusk and Eggshell

Unlike the amino acids in bone which racemize upon the death of the animal, amino acid racemization in both mollusk shell and eggshell begins immediately after formation. While AAR dating of these materials is possible, careful sample selection is required. The technique has been successfully applied to marine, freshwater, and terrestrial mollusks from a variety of contexts. While calibration samples are still required, reliable AAR dates have been obtained on material from geological contexts. Some species, such as the cherrystone clam (*Mercenaria mercenaria*), are known to produce reliable dates; the marine oyster (*Ostrea angasi*) does not and should be avoided (Johnson and Miller 1997).

The thick eggshells of large flightless (ratite) birds can be dated; according to Johnson and Miller (1997) ostrich eggshell more closely approximates a closed system than either bone or mollusk shell. Eggshells from archaeological contexts are more problematic as cooking or proximity to fire accelerates the racemization process. Burning can be detected by comparing the rates of leucine hydrolysis and isoleucine epimerization in a known age sample. Hare et al. (1997) suggest that as a general rule, in the absence of burning or other types of anomalous heating, amino acid ratios of samples within a level should differ by 10% or less. The presence of decarboxylation and deamination products, low concentrations of amino acids, and variable ratios of isoleucine epimeration products, as indicated by the ratio of alloisoleucine to isoleucine, within a sample are also considered evidence of burning. Johnson and Miller (1997) consider an increased abundance of decomposition products, such as methyl and ethyl amines, to be the most important indicator of exposure to fire; heat can also cause levels of ammonia to increase from less than 5% to over 15% in burnt samples.

Other Archaeological Applications

While the use of AAR as an absolute dating technique are most frequently described, Johnson and Miller (1997) note that AAR studies can provide other information. As with using AAR for dating, the D/L values are only comparable if samples have experienced similar thermal histories, i.e., their effective diagenetic temperatures are ±1°C. If this condition is met AAR can provide relative-age determinations of stratigraphic units since older material will have higher D/L values than younger material. The use of AAR enables the correlation of strata, detection of sediment mixing, or recognition of intrusive features. If a reliable kinetic model is employed and calibration samples are available, the D/L value can be used to determine EDT or paleotemperature of the sample.

The racemization of aspartic acid can be used to determine the age at which an individual died under certain conditions. Aspartic acid racemization in human teeth starts at the time of tooth formation and can virtually cease at the time of death in cold environments or if individuals were rapidly buried. The rate of D-aspartic acid formation in human teeth is about 0.1% each year. Care must be taken to ensure that the teeth analyzed have not been affected by postmortem bacterial contamination.

Summary

Amino acid racemization dating is best applied in conjunction with other chronometric dating techniques. Diagenetic processes that affect the reliability of AAR dates, such as collagen loss and leaching, make bone unsuitable for radiocarbon dating, uranium series dating, and electron spin resonance dating as well. While some suggest that whole bone should be avoided altogether, it may be possible to obtain reliable dates from isolated proteins extracted from bone, such as collagen and non-collagenous protein. Tooth dentine are more resilient and better sample material for AAR-inferred dates. Reliable dates have been obtained on shell from marine environments but unburnt ostrich and emu eggshells are the best material for AAR dating. AAR can also be used as a relative dating technique and for determination of paleotemperature conditions and age at death.

Because racemization and epimerization reactions are temperature dependent, the average burial temperature must be accurately estimated. Consequently, the same uncertainties associated with determining the hydration rate for obsidian hydration dating affect amino acid racemization dating. Ideally, samples should come from contexts minimally affected by diurnal and seasonal temperature fluctuations. The technique is most effectively applied to terrestrial samples from caves and rockshelters because the material was shaded from the sunlight; material from open-air sites should be recovered from depths greater than 1 m.

Obsidian Hydration

Obsidian is volcanic glass that was highly valued for the production of flaked stone tools; because of its superior flaking properties it was extensively used and widely traded (Chapter 18). A fresh surface, exposed by flintknapping or other processes, slowly absorbs water from its environment and a hydration rind forms. Friedman and Smith (1960) recognized that the thickness of the hydration rind and its rate of growth provided a means of estimating the time of exposure for archaeological applications. Friedman and Smith (1960) described the theoretical basis of the technique and outlined in detail the process of sampling an artifact, preparing a thin section, and measuring the thickness using optical microscopy in *American Antiquity*. The primary advantages were that the technique was inexpensive, highly accessible, and enabled the dating of an artifact, rather than material that may or may not be strongly associated with it.

Friedman and Smith (1960) suggested the following relationship between the thickness of the resulting hydration rind in microns, x, and time in years, t:

$$x^2 = kt \text{ or } x = kt^{1/2}$$

where k is a diffusion constant for a particular temperature. In order to enhance an understanding of the process, obsidian hydration was induced at elevated temperatures in the laboratory. Over time, an increasing number of variables have been shown to affect the rate at which hydration occurs.

Beck and Jones (2000) identify decadal trends in the development and refinement of the technique in response to the recognition of complicating factors, including the effect of composition and relative humidity. Accurately estimating the temperature the obsidian experienced in the archaeological context over time has proven to be particularly challenging as it is affected by a host of variables (Jones et al. 1997; Ridings 1991, 1996; Stevenson et al. 1989). Friedman et al. (1997) recognize certain aspects of the technique which require improvement and acknowledge interobserver variability in measurements made using optical microscopy, but exhibit steadfast confidence in its ability to provide absolute dates without the use of expensive equipment.

This position is at odds with Anovitz et al. (1999) and Stevenson et al. (2001; 2004) who recommend the use of instrumental techniques, such as infrared photoacoustic spectroscopy and secondary ion mass spectrometry, to improve the precision and accuracy of hydration rind measurements. Recently, the need for a greater understanding of the effects of surface dissolution and changes in the diffusion rate due to differences in water concentration and obsidian composition has been expressed (Ambrose 2001; Anovitz et al. 1999; Rogers 2008; Stevenson et al.1989).

Archaeological applications of obsidian hydration dating are provided in Chapter 18. Analytical instruments used for precise measuring of the hydration rind, secondary ion mass spectrometry (SIMS) and combined infrared and photoacoustic spectroscopy (IR-PAS), are described in Chapters 31 and 35, respectively.

The Effect of Temperature on Hydration Rind Formation

Friedman and Smith (1960) recognized that a freshly exposed surface of obsidian has a strong affinity for water and slowly absorbs it from the environment at a certain rate. Obsidian forms from magma with very low water content; when no longer under high temperature and high pressure, it is under-saturated (Ambrose 2001). Hydration increases the water content of unhydrated obsidian from about 0.1–0.2% to as much as 9% in the hydrated obsidian known as perlite (Ambrose 2001). The process of absorption continues until the rind reaches a thickness of 50–100 μm, at which time it spalls off and a fresh surface is exposed (Friedman et al. 1997).

The importance of temperature on the rate at which water diffuses, k, was acknowledged immediately. On the basis of obsidian artifacts from contexts dated by radiocarbon, the hydration rate in northern Alaska was determined to be 0.36 μm^2/1000 years while the rate in the tropics was more than 30 times faster, 11 μm^2/1000 years (Friedman and Smith 1960). The relationship between temperature and the rate at which many chemical reactions occur is named after Arrhenius, who first described it (Levine 1978):

$$k = Ae^{-E_a/RT}$$

where A is the pre-exponential factor (or Arrhenius A factor), e is the base of natural logarithms, E_a is the Arrhenius activation energy, R is the universal gas constant, and T is the absolute temperature in degrees Kelvin. If the Arrhenius equation is obeyed, a plot of log k vs. $1/T$ is a straight line whose slope is $-E_a/2.303R$ and the Y-intercept is log A. Induced hydration experiments by Friedman and Long (1976) indicate the Arrhenius equation holds for the obsidian hydration reaction; however, Anovitz et al. (1999) disagree with this assertion. The need to obtain an accurate assessment of temperature is critical for obsidian hydration studies as a $\pm 1°C$ change in temperature results in $\pm 10\%$ difference in hydration rate (Ambrose 2001).

Estimating Site Temperature

Friedman and Smith (1960) noted that mean annual air temperature could only give an approximate indication of the actual temperature an artifact would experience over time, called its effective hydration temperature (EHT). In the absence of actual data for a particular site or depositional context, meteorological averages and mathematical equations were used to estimate the temperature. Some employed Lee's (1969) temperature integration equation:

$$T_e = \frac{(T_a + 1.2316) + (0.1607R_i)}{1.0645}$$

where T_e is the effective hydration temperature, T_a is the mean annual air temperature, and R_i is the annual range in air temperature based on monthly means. Stevenson et al. (1989) consider this to be an incorrect application of Lee's equation because the magnitude of temperature fluctuations is reduced as soil depth increases. Instead, both Stevenson et al. (1989) and Ridings (1996) calculate EHT with complex heat flow models, which take additional factors into account, such as depth of burial and thermal diffusivity of the soil.

While these approaches may produce EHTs and resulting obsidian hydration dates that are consistent with expectations under some circumstances, they are not ideal. Temperature differences due to the environmental conditions at a microregional scale can result in significant and predictable variations in EHT (Jones et al. 1997). Jones et al. (1997) report approximately 21% of the total variation in EHT is due to factors such as the direction of exposure (aspect), degree and type of vegetation cover, depth, exposure, and variation in soil properties. Daily and annual fluctuations in temperature have the most pronounced effect on the uppermost 100 cm of soil, with the largest variation occurring from depths of 0–60 cm (Ridings 1991; Stevenson et al. 1989).

Methods of Measuring Site Temperature

Techniques for direct monitoring of temperature changes that occur at a site, or at specific locations within a site, have been developed or proposed to attain more reliable EHT estimations for obsidian hydration date calculations. Friedman et al. (1997) suggest the EHT could be estimated from daily maximum–minimum air temperatures measured at, or close to, the site or from ground temperatures measured at the depth of the artifacts at 6-month intervals. Alternatively, if A and E_a have been determined (see below), the EHT of archaeological obsidian securely dated by another technique (such as radiocarbon) can be obtained by re-arranging the Arrhenius equation and solving for T.

Thermal cells are devices designed by Ambrose (1976) and later modified by Trembour et al. (1986, 1988) specifically for obsidian hydration dating studies. A cell or multiple cells are buried at depths corresponding to the location of obsidian finds at a site for a 1-year period. During this time, water diffuses into a polycarbonate barrier (i.e., plastic cell) in a manner similar to the hydration of obsidian; the increase in weight corresponds to the amount of water absorbed over the burial period, which is related to burial temperature. While the relative humidity of the burial environment is often close to 100%, the hydration rate is reduced if it is not fully saturated. In order to obtain a hydration rate adjusted for relative humidity, paired "wet," i.e., the cell immersed in water, and "dry," i.e., the cell not immersed in water, thermal cells are buried at the site (Ridings 1991; Stevenson et al. 2004).

The activation energy of the plastic cell is not identical to obsidian so errors occur in contexts with daily and annual temperature fluctuations (Ambrose 2001; Ridings 1996). In addition, Ridings (1996) reported inexplicable differences in the values of effective hydration temperatures generated from two thermal cell types buried in identical contexts. Data collected by Stevenson et al. (2004) from paired thermal cells buried at depths of 20, 40, and 80 cm below the surface also deviated from expected results. Ridings (1996) further noted that these devices do not replicate the hydration history of the obsidian because artifacts are not instantly buried to the depth from which they were recovered.

Inducing Hydration Rind Formation

As noted previously, the hydration rate can be calculated on the basis of rind development on obsidian from a context dated by another technique, such as radiocarbon. Attempts to experimentally and independently establish the hydration rate of a particular variety of obsidian involve oven storage. Obsidian fragments are heated at a constant temperature in the presence of liquid water or water vapor for a known period of time and the development of the hydration rind is monitored. If powdered obsidian is used, water diffusion is monitored through weight gain. As the rate constant is determined for a specific temperature, other variables in the Arrhenius equation, such as the pre-exponential factor (A) and activation energy (E_a) can be determined.

Over time, procedures have evolved to address problems as they were recognized. For example, induced hydration experiments are usually conducted at elevated temperatures because a measurable rind can form in just a few days. Stevenson et al. (1989) found that distilled and deionized water actually dissolves induced hydration rinds at temperatures greater that 200°C. To prevent this from occurring, Stevenson et al. (1989) suggest maintaining a low-pH environment and either saturating the aqueous solution with silica or conducting reactions in vapor environments at 100% relative humidity. Friedman et al. (1997) recommended that induced experiments be conducted in a vapor environment to prevent the leaching of sodium, potassium, and other trace elements as this may alter the hydration rate. Friedman et al. (1997) consider the use of whole pieces preferable to measuring weight gain of powdered samples, as hydration will depend upon the surface area of the powder. Ambrose (2001) suggests that rates determined through high temperature-induced hydration should be verified through experimentation at low temperatures rather than simply extrapolated.

Factors Affecting Hydration Rate Constants

Practitioners of obsidian hydration dating are also criticized for assumptions about the integrity of the rind and the compositional uniformity of the sample (Ambrose 2001; Anovitz et al. 1999; Rogers 2008). The original water content is an important factor with higher density lower water content obsidian hydrating at a slower rate than water-rich varieties (Ambrose 2001). While possible differences in hydration rate due to variations in the concentrations of major elements in the obsidian were acknowledged, Friedman et al. (1997) reported that one obsidian source usually has only one hydration rate. To the contrary, Rogers (2008) found significant differences in intrinsic water content, based on the concentration of hydroxyl ions, within obsidian sources and even within individual specimens from the Coso field in eastern California. Rogers (2008:2013) states, "even if the [OH$^-$] of a given specimen can be measured, point-to-point variation within the specimen will still yield a variation of approximately $\pm$ 21% in rim thickness unless [OH$^-$] can be measured at exactly the same spot as the hydration rim."

As noted previously, Stevenson et al. (1989) reported surface dissolution occurs if hydration is induced with deionized water at high temperatures. Anovitz et al. (1999) and Ambrose (2001) argue that dissolution also occurs naturally at ambient temperatures and under soil conditions seemingly favorable for preservation. Factors contributing to surface loss in the burial environment include clay content, soil microorganisms, and humic acids, so Ambrose (2001) suggests measuring hydration rinds on internal fracture surfaces rather than exposed outer surfaces. The loss of the hydrated surface through dissolution can be the source of significant error in an age calculation because it affects the hydration rate. The hydration rate of obsidian slows as the hydration rind thickens; removal of the rind will increase the hydration rate (Ambrose 2001).

Measurement of the Hydration Rind

Optical Microscopy

Rind thickness can be measured using optical microscopy because the hydrated layer has a higher density and refractive index than the original glass. The hydration front appears as a bright line that often occurs with a dark edge, called a Becke line. In theory, precision of 0.1 μm is attainable yet Friedman et al. (1997) discuss examples of measurements on the same sample differing by up to 1.4 μm when made by different researchers. Anovitz et al. (1999) suggest that inconsistency occurs because Becke line position is partially an artifact of the optical properties of visible light and moves with focus position. Consequently, the precision attainable with optical measurements is limited by sample-to-sample and inter-observer variability (Anovitz et al. 1999).

Instrumental Techniques

Anovitz et al. (1999) demonstrated the advantages of making obsidian hydration rind measurements using secondary ion mass spectrometry (SIMS). A tightly focused beam of primary ions removes (sputters off) and converts a portion of the sample into secondary ions, which are then analyzed using mass spectrometry. The technique is capable of detecting changes in the composition of the sample with depth at the nanometer level (1 nm = 0.001 μm); for obsidian hydration studies, the concentration of water in the sample is monitored. Anovitz et al. (1999) and Stevenson et al. (2004) used SIMS to verify that the drop in water concentration occurs abruptly between the hydrated and the non-hydrated regions, producing a depth/concentration profile that resembles a backward S curve (Fig. 11.2). Friedman and Smith (1960) originally suggested the diffusion front measured about 0.1 μm, but Anovitz et al. (1999) and Stevenson et al. (2004) both found the region of rapidly decreasing water content is actually up to 1 μm wide. Anovitz et al. (1999) reported

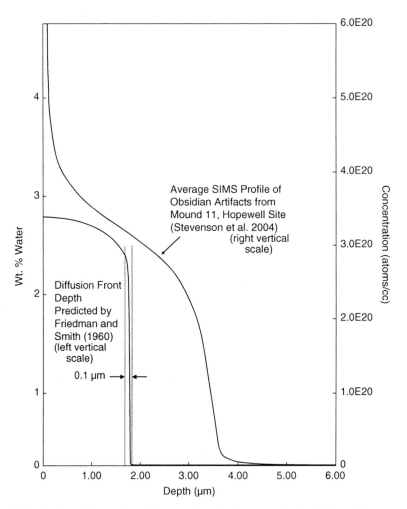

Fig. 11.2 Obsidian diffusion front (Adapted from Friedman and Smith (1960) and Stevenson et al. (2004). Reproduced by permission of the Society for American Archaeology from American Antiquity, volume 25(4), 1960 and American Antiquity, volume 69(3), 2004.")

the optically measured hydration boundary falls within the region of decreasing water content but varies within it.

Stevenson et al. (2001) argued for the use of either SIMS or IR-PAS as an alternative to optical measurements. They favored a combined technique where quantitative measurements of absorbance from IR-PAS are calibrated with depth measurements using SIMS. Stevenson et al. (2004) compared SIMS to optical measurements of hydration rind thickness and noted that optical measurements were consistently overestimated and the data sets generated using the two techniques are statistically different. Their data support the findings of Anovitz et al. (1999) that high levels of precision are not attainable with optical microscopy.

Summary

After nearly 50 years, obsidian hydration dating is still a technique under development. While SIMS can greatly enhance the precision with which measurements of hydration rind are made, variation in water content and surface dissolution reduce the certainty of obsidian hydration dates. Friedman and Smith (1960) originally envisioned that obsidian hydration dating would be performed in a central laboratory or a few closely cooperating laboratories. Limiting the process to a small group of professionals would control inter-observer variability in measurements made using optical microscopy; however, obsidian hydration rates are affected by many variables. According to Ambrose (2001), the optimal conditions for obsidian hydration dating involve younger age specimens, warmer climates, low temperature variation, high moisture levels, and under least aggressive soil conditions. As Ridings (1996) points out, it may be difficult to accurately model changes in EHT an artifact experiences from the time it is deposited on the ground, gradually buried, then recovered at some depth below the surface even when the obsidian is recovered from ideal conditions.

Friedman et al. (1997) regard obsidian hydration dating as a means of providing absolute dates, while Beck and Jones (1994, 2000) stress its utility as a relative dating technique, which can be applied to surface assemblages. Anovitz et al. (1999) see the technique as fixable if highly precise measurements can be combined with a high-precision diffusion model that addresses the effect of water content, surface dissolution, and the sharpness of the diffusion front. Given the findings of Rogers (2008) with respect to water content variability, the challenges associated with attaining reliable dates may remain for some time.

Recently, Ericson et al. (2004) proposed that the hydration of quartz could be used as a dating technique. The hydration rate of quartz is much slower than that of obsidian for a given temperature. The error associated with dates is estimated to be 35%, but may be improved through careful sample selection and the use of SIMS to measure the hydrated layer. Ericson et al. (2004) suggest its use may be appropriate for relative dating or in situations where no other techniques are available.

Cation Ratio Dating and Rock Varnish

Cation ratio (C-R) dating was developed by a geographer, Ronald I. Dorn (1983), and used to infer the age of archaeological materials, primarily ancient art on exposed bedrock surfaces, which are not easily dated by other chronometric techniques. Rock (or desert) varnish is a coating that naturally develops on exposed surfaces in arid or semi-arid regions. Human activities, such as making petroglyphs and flaking stone tools, remove old rock varnish and expose fresh surfaces. The chemical composition of rock varnish that developed after the new surface was exposed is used to estimate its age. A date for the onset of rock varnish formation represents a minimum date for rock art panels, surface artifacts, and certain archaeological features. This technique garnered considerable interest because rock varnish forms in regions with stable landforms where sedimentation rates are negligible. Archaeological materials deposited on the surface can remain there for millennia and no associated organic material is preserved.

Rock or Desert Varnish

In general, rock varnish is a dark, often shiny coating that appears on surfaces of rocks exposed to the air (subaerial) in arid environments. It is primarily derived from wind-blown dust from the external environment rather than the substrate on which it forms. Rock varnish that is darker in color, thicker, and more expansive in coverage is considered older than lighter, thinner, and patchier deposits.

The chemical composition of rock varnish has been assessed using a variety of techniques including PIXE, ICP, electron microprobe, and XRF (see Part IV). Up to about 70% of rock varnish consists of clay minerals (oxides of aluminum and silica); manganese oxides and iron oxides typically form between 25 and 33% of its composition. Minor and important trace elements include magnesium (Mg), titanium (Ti), calcium (Ca), potassium (K), phosphorus (P), barium (Ba), and strontium (Sr). These metallic elements form ionic bonds as positively charged ions or cations.

A fundamental hypothesis of cation ratio dating is that the concentration of these cations in rock varnish changes as a function of time. Both Ca and K are considered "mobile" cations that are prone to leaching. As mobile cations are lost over time, the relative level of "stable" ions, such as Ti, increases. C-R dating involves comparing concentrations of Ca and K to Ti in the rock varnish as follows: $(Ca + K)/Ti$. A low cation ratio is considered evidence that the rock varnish, and the surface on which it formed, was exposed to the environment for a long period of time.

In order to determine an absolute age, cation ratios must be calibrated using another dating technique. AMS radiocarbon age estimations on several samples of rock varnish with known cation ratios are used to produce a calibration curve for the particular region. The calibration curve is then employed to correlate cation ratios of other rock varnish samples in the region with dates in radiocarbon years.

Examples of archaeological applications of cation ratio dating and concerns regarding its utility for dating archaeological materials are considered in this section. Detailed reviews of rock varnish, its composition, hypotheses about its formation as well as other characteristics and methods of assessment appear in Schnieder and Bierman (1997) and Dorn (2007).

Archaeological Applications

Quite simply, archaeological applications of cation ratio dating have been controversial. The technique has most frequently been used to establish, test, or refine petroglyph chronologies, although it has also been applied to surface artifacts and features. In this section, reviews of an apparent successful application of C-R dating to a petroglyph site and an apparent unsuccessful application to a quarry site with surface artifacts, as well as criticisms leveled against each of these approaches, are presented.

Petroglyph Chronology

The Pinon Canyon Maneuver Site in southeastern Colorado contains more than 175 rock art sites. Rock varnish on petroglyphs from a sample was collected and analyzed by Dorn (Loendorf 1991). Loendorf (1991) reported the C-R inferred ages on petroglyphs corresponded well to the previously established relative chronology. By correlating subject matter and manner of execution, totally pecked figures were deemed to be relatively older than pecked outline figures; incised figures were the youngest. Seriation of forms was used to further refine the relative chronology. Curvilinear abstract forms were considered to be the oldest; quadrupeds appear early and increase in frequency over time. Anthromorphs (human figures) appear late and also increase in popularity through time. Loendorf (1991) noted the C-R age estimates matched the relative sequence and were internally consistent with respect to superimposed petroglyphs. Representations of bows and arrows and the mountain spirit are associated with Apache, who arrived in the area after A.D. 1500. The C-R date on a bow and arrow figure was 350 ± 100 B.P.; those on mountain spirits range between 400 and 300 B.P.

Loendorf (1991) also provided examples of the independent confirmation of the C-R ages by radiocarbon dating associated materials. Dates on a petroglyph were older than charcoal in cultural deposits overlaying it (one assumes the deposits were not directly on the figure as rock varnish only forms when the surface is exposed to air). The radiocarbon age of soil that covered a buried petroglyph corresponded well to the C-R age on a petroglyph with a similar rectangular grid form located at another site. Radiocarbon dates on carbon, from the lowest level of a pit at one site, and charcoal, from one of seven house rooms at a different site, overlapped the C-R ages on nearby petroglyphs. Loendorf (1991) concluded that the C-R ages support the established rock art chronology in all but two cases where human visitors or animals may have disturbed the rock varnish.

A general appraisal of C-R dating by Lanteigne (1991) appeared in the same issue of *Antiquity*. The first of five concerns addressed the irregular appearance and growth of rock varnish on surfaces and the number of variables affecting its development; the second involved the potential effect of incomplete removal of old varnish on this process. A third concern involved the assumption of climatic stability, as an increase in humidity and associated increase in acidic chemical erosion could cause differential removal of varnish. Lanteigne (1991) also noted potential problems related to field sampling of rock varnish, sample processing, and laboratory analysis due to the complexity and degree of subjectivity involved. Finally, Lanteigne (1991) rather astutely pointed out that the accuracy of C-R dates ultimately rests upon the calibration leaching curve produced from AMS radiocarbon dates.

While Loendorf (1991) reported C-R ages on petroglyphs in Colorado were in-line with expectations, sometimes there is considerable divergence. Dorn et al. (1988) examined rock engravings at the Karolta site in Olary province in South Australia. Cation ratio dates on 24 randomly selected figures ranged from about 1,400 to 31,700 radiocarbon years. Other research suggested arid regions of Australia were not occupied by humans until around 22,000 radiocarbon years. The rock art dates pushed back the presence of humans 8,000–9,000 radiocarbon years. Dorn (1998) later admitted that heterogeneities in varnish organics made the radiocarbon dates on material from Olary, and those from several other regions, ambiguous. Issues regarding the dating of rock varnish are presented in more detail below.

Surface Artifacts

Harry (1995) examined rock varnish on chert artifacts found on the surface of site CA-KER-140, near an outcrop in the western Mojave Desert in California. The bedrock was well varnished but varnish on artifacts was thin, patchy, and highly variable in composition. Of the five types of chert in the area, proportionately more varnish developed on artifacts fashioned from coarser grained white chert and appeared least often on fine-grained, caramel-colored chert. Harry (1995) expressed concerns that rock varnish development was influenced more by substrate texture than the duration of exposure and the delay of onset could greatly exceed 100 years on finer grained materials. Rock varnish was also more likely to grow on artifacts resting on desert pavements and bedrock outcrops than those lying on cobbles and sands. Variability in the measured cation ratio of rock varnish did not appear to correlate with the duration of surface exposure. Rock varnishes on some artifacts had lower cation ratios than varnishes that had formed on bedrock. Since varnish development on surface artifacts could be interrupted by burial or overturning, Harry (1995) also suggested cation ratio dating was better applied to stable surfaces of similar composition and texture.

In his comment to the paper, Bamforth (1997) suggested that many problems Harry (1995) reported would have been avoided if she had adopted the methodology followed at Intermountain Power Project (IPP). Consistency of cation ratios on artifacts should have been verified by analysis of varnish on refitted pieces. Artifacts that may have moved should have been excluded from C-R dating analysis. Bamforth (1997) reported that the IPP calibration leaching curve was based on 16 absolute dates acquired using radiocarbon and K–Ar dating and strongly correlated to cation ratios. No absolute dates were obtained on rock varnish from site CA-KER-140. His suggestion that all rock varnish should have been scraped off and examined in the lab, instead of being analyzed in situ, appears to be a misunderstanding of the procedures. Instead of removing the varnish in the field, a sample of rock was collected and the adhering varnish was analyzed in a laboratory (Karen Harry, personal communication 2008).

In other papers about Harry's study area, Dorn (1995) and Harrington and Whitney (1995) both concluded varnishes at site CA-KER-140 were unsuitable for

C-R dating. Dorn (1995) suggested varnish from bedrock outcrops gave problematic cation ratios because the varnish did not originate on freshly exposed surfaces but spread from rock joints. Harrington and Whitney (1995) suggested cation ratio dating should not be attempted on chert because its inherent properties made it unsuitable for varnish development, in general. Bierman and Gillespie (1995) countered that rock varnish was well developed on bedrock surfaces and cortical surfaces. The site was originally selected for the test because chert was a non-contaminating substrate, whereas other researchers identified time-dependent trends on potentially contaminating substrates.

Evaluation of Technique for Dating

An accurate and precise age estimate can be obtained when a number of conditions are met:

1. the present composition of the sample must be accurately measured with a high degree of precision;
2. its original composition must be known or conform to assumptions made about it;
3. changes in sample composition must be a function of the passage of time; if any other changes occurred, they must be detectable and correctable.

If any of these conditions are not fully met, the accuracy and precision of age estimate are reduced. The process of inferring the age of desert varnish using cation ratios will be considered with respect to each of these points.

Assessment of Chemical Composition

Concerns have been raised with respect to the measurement of Ba in rock varnish using XRF and PIXE in that it may "masquerade" as Ti. Any interference by Ba would result in an overestimation of the amount of Ti. Due to its position in the denominator of the formula, an inflated level of Ti lowers the C-R value so the age calculated for a desert varnish sample would be too old. Bierman and Gillespie (1995) reported that cation ratios at CA-KER-140 did not reflect relative substrate age whether Ba was excluded using Dorn's formula $(K+Ca)/Ti$ or included as $(Ca + K)/(Ti + Ba/3)$, a formula suggested by Harrington and Whitney (1995).

The Formation and Original Chemical Composition of Desert Varnish

Other than that it is absent from freshly exposed surfaces, little is known about the development of desert varnish. Dorn et al. (1988) estimated that the time lag of varnish formation in arid western United States is about 60–100 years and, based on examination of historic engravings, the time lag between engraving and the onset of varnish formation was about 100 years in Australia.

Variables affecting the development of rock varnish are not fully understood. There appears to be some evidence that it does not develop as a uniform layer and that the micro-topography of the sample is a significant factor. Desert varnish appears first in depressions, perhaps because these areas trap water and dust particles. From these "nucleation centers," the desert varnish spreads in all directions to cover the surface. Rocks with relatively smooth exteriors do not promote the growth of desert varnish. Consequently, the onset of development may be more closely related to surface roughness than the duration of surface exposure.

Varnish growth may also be affected by several variables that determine the specific microenvironment of the surface upon which it forms. According to Schnieder and Bierman (1997), these may include amount of precipitation, rock (clast) size, exposure to sunlight, daily and seasonal temperature variation, wind direction and velocity, amount of atmospheric dust, geographic location, distance above the ground, and direction of exposure.

Changes in Sample Composition Over Time

As previously mentioned, Dorn (1983) suggested that the chemical composition of desert varnish systematically changes with the passage of time. He proposed that the positively charged ions (cations) of certain elements were "mobile" or more likely to be lost through leaching; "stable" elements were more likely to be retained in the desert varnish. The cation ratio, comparing Ca and K to Ti, is sensitive to these changes. Schneider and Bierman (1997) expressed concerns that a number of fundamental processes, the rate at which they occur and their impact on the development and composition of rock varnish have not been well documented. These processes include cation leaching, weathering, and varnish accretion. The effects of the substrate, the microenvironment, long-term climate change, and the addition of organics from the environment on varnish development and composition are not well understood.

The process of preferential leaching of K and Ca has been disputed. Dorn (2007) suggested that attempts to replicate experiments supporting the occurrence of leaching led to problematic results because of differences in field and laboratory sampling procedures. Schneider and Bierman (1997) support suggestions by others that changes in varnish composition may be due to incorporation of the underlying substrate into the varnish rather than the leaching of cations out of the varnish. Schneider and Bierman (1997) stressed that Harry (1995) found no relation between cation ratios and relative age when desert varnish formed on chert, a non-contaminating silica substrate. Bednarik (1996) noted that even if preferential leaching of K and Ca occurred as predicted, cation ratios on a surface would vary laterally; rock varnish closest to nucleation centers would be leached to a greater degree. Cation ratios would depend on how the varnish spread out from its points of origins rather than the duration of surface exposure.

Another concern regards the need for calibration. Cation ratios cannot directly date rock varnish; instead, chronometric ages and/or historical dates are required to develop a calibration curve. Potassium–argon dating is suitable for geological applications because it provides an age estimate for the volcanic rock substrate upon which the desert varnish develops. Even so, this age is valid only if rock varnish development on the surface has been continuous and completely unaffected by erosional processes since the time of its formation.

Calibration curves of archaeological applications are based on radiocarbon dates of organic material trapped within the deepest, and presumably oldest, layer of rock varnish. Schneider and Bierman (1997) caution that these organics may not have been derived from a single source but may include microorganisms, pollen, or other types of plant particles, lichen, fungi, algae, or other materials. Multiple sources of carbon in the rock varnish would undermine the validity of both the radiocarbon age estimations and the calibration curves used to provide C-R ages.

The debate about AMS dating of rock varnish escalated when Beck et al. (1998) reported two sources of carbon in rock varnish samples processed by Dorn. The two types of carbon were each subjected to AMS dating: Type I carbon, which resembled coal, produced ages between 27,500 B.P. and 36,600 B.P. Type II carbon, which resembled pyrolized wood charcoal, produced dates between about 800 B.P. and 4,180 B.P. Upon examining available remains of other samples Dorn had submitted to the lab, Type I, Type II, or both types of carbonaceous materials were identified in all but one sample. Neither type of carbonaceous material appeared in comparable samples that Beck et al. (1998) independently prepared. Serious concerns were raised about the possible origin of the carbonaceous material in Dorn's samples.

Dorn (1998) countered that Beck et al. (1998) failed to replicate his results because different sample processing techniques were used. Overlapping AMS radiocarbon age estimates on rock varnish from a site in Portugal independently analyzed by Dorn and another researcher were offered as proof of the validity of his processing procedures. Furthermore, Dorn (1998) argued that potential problems with AMS age estimations on rock varnish due to the mixing of carbon of different ages had been identified before Beck et al. (1998) raised similar concerns.

Dragovich (2000) examined rock varnish in the arid to semi-arid tropical Pilbara region in Western Australia. Upper, lower, and lowest layers were collected separately and submitted for AMS radiocarbon dating. Although an age of 17,000 years had been suggested for varnish in the region, all samples produced ages of less than 2,800 B.P. In some cases, dates on samples from the lowest layer were considerably younger than those above. Dragovich (2000) concluded that the rock varnish in the area did not persist for long periods of time. Weathering appeared to have completely or partially removed old varnish deposits, which were then replaced by younger material. In some cases, microorganisms may have created pits in the varnish, allowing contaminating young carbon to reach the lowest layer adjacent to the substrate.

Summary

Schneider and Bierman (1997) expressed a variety of concerns about rock art dating. First, rock varnish collection is a destructive process and it is not possible to obtain enough samples for statistically significant analyses. Second, given the variables surrounding the onset of varnish development, C-R dates are only comparable if samples are collected from substrates with similar microtopography and perhaps lithology as well. Finally, the rock varnish must have formed under identical regional environmental and microenvironmental conditions.

The presence of carbonaceous material from multiple sources in rock varnish renders it unsuitable for radiocarbon dating. Dorn (2007) recognized that carbon that pre-dates and post-dates the exposure of the rock surface can become trapped in rock varnish. The implication is that calibration curves cannot be developed on the basis of radiocarbon ages. While geological applications are still feasible using K–Ar age estimates on the substrate, archaeological applications are not. Even if assumptions about rock varnish accumulation and cation leaching are shown to be valid, in the absence of calibration, cation ratios only provide relative ages. Given the ability of archaeologists to develop solid relative chronologies in the absence of absolute age estimations, there is little reason to analyze cation ratios.

Archaeomagnetism

Archaeomagnetism is the study of the magnetization retained in certain cultural materials in order to gain insight into past human activities. The magnetization induced in a substance is a product of the Earth's magnetic field, called the geomagnetic field, and the inherent properties of the substance, such as iron content. It is of archaeological significance because the direction and intensity of the geomagnetic field changes over time and human activities can affect magnetization. Iron-rich particles in rock, clay, or soil that were heated to high temperature at an archaeological site acquired a magnetization reflecting the geomagnetic field at the time of occupation. While primarily used as a dating technique (Eighmy 2000), the integrity of features as well as the heating and cooling patterns of artifacts can be examined through archaeomagnetic (or paleomagnetic) analyses (Gose 2000; Takac 2000).

Eighmy (2000) and Sternberg (2008) present historical reviews of archaeomagnetism. It is a subdiscipline of paleomagnetism, which is the study of the magnetic field over geological time as evidenced through geological specimens in the pursuit of answers to geophysical questions. Archaeomagnetism was developed primarily by Emile Thellier (Aitken 1970; Eighmy 2000) and involves the study of the Earth's magnetic field during archaeological time and the direction and intensity of magnetization of artifacts and other historically dated materials. Aitken (1970), Tarling (1975), and Sternberg (1997, 2001) present summaries of the technique for archaeologists; the volume edited by Eighmy and Sternberg (1990) provides detailed considerations of techniques, issues, and research. Another excellent resource is the

Archaeomagnetic Applications for the Rescue of Cultural Heritage (AARCH) web site (http://dourbes.meteo.be/aarch.net/frameset_en.html) and articles by Hus et al. (2003) and Linford (2006) which can be downloaded from the site.

Magnetism and Magnetization

Magnetism causes an object to produce a magnetic field and have an attraction for iron. A magnetic field will deflect a charged particle and attract or repulse another magnetic object. A magnetic field with north and south poles is called a dipole; the force between opposite poles is attractive whereas the force between like poles is repulsive. Tiny grains of certain iron-rich minerals in sediments will naturally align parallel to the ambient magnetic field at the time of deposition. The ambient magnetic field acting on minerals is usually the Earth's magnetic field, but exceptions occur when the sample itself is highly magnetic or when it is near highly magnetic materials.

Iron sulfides and iron oxides have high-magnetic susceptibilities, which means they can acquire an induced magnetization in the presence of a permanent magnetic field. Materials that are also ferromagnetic, such as magnetite (Fe_3O_4), or antiferromagnetic, such as hematite (Fe_2O_3), are able to retain a previously acquired, or remanent, magnetization. The orientation of iron-rich minerals is fixed within a solid, but if heated sufficiently, these minerals in artifacts or features lose their original magnetization. The particles act like the magnetized needles of compasses and re-align in the direction of the ambient magnetic field. Upon cooling, this new alignment is "frozen" into the material and becomes a permanent record of the magnetic field that existed at the time of the heating and cooling event.

The Earth's Magnetic Field

An essential component of archaeomagnetism is the documentation of changes in the Earth's magnetic field that occurred in the past. The position of the main geomagnetic dipole, commonly referred to as the magnetic North and South Poles, varies over time with respect to the geographic North Pole. In addition, localized magnetic disturbances affect the geomagnetic field at a regional scale. The magnetic North and South Poles can also flip, remain reversed for a period of time, then flip back. The term secular variation is used to describe the changes in the geomagnetic field over time.

The Earth is believed to consist of a central solid iron core surrounded by a liquid outer core, which is encompassed by a solid mantle and the outer crust. Electric currents within the liquid outer core produce a magnetic field similar to that of a bar magnet. The geomagnetic field is dipolar with the magnetic North Pole of the Earth corresponding to the south pole of the bar magnet and the magnetic South Pole corresponding to the North Pole of the magnet (Fig. 11.3). The angle between

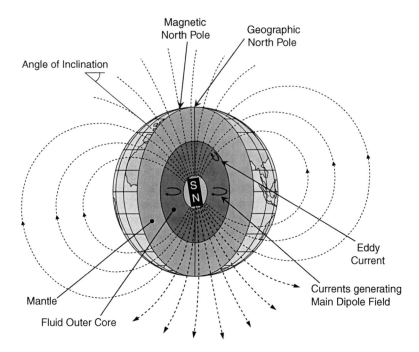

Fig. 11.3 The geomagnetic field (© English Heritage)

the Earth's magnetic dipole and its rotational axis (i.e., the line connecting the geo-graphic North and South Poles) is about 11°. In addition, eddies form in the current pattern at the boundary between the outer core and the mantle and produce short-lived localized magnetic disturbances. Disturbances measure about 1000 km across and generally drift westward at a rate of 0.2° of longitude per year (Aitken 1970). Sternberg (1997) reports the typical rate of change in the direction of the magnetic field at the Earth's surface is about 1° every 10–20 years, but changes of up to 20° are possible. Variations in the strength of the field range from a few percent to as much as 50% over a 100-year period.

The geomagnetic field at any point on the Earth's surface has both intensity and direction. Intensity is measured in nanoTesla (nT) and is highest at the magnetic poles, over 60,000 nT, and lowest near the equator, about 24,000 nT (Sternberg 1990). The direction of the geomagnetic field is described in terms of declination, which is the horizontal component, and inclination, which is the vertical compo-nent. Declination is the angle between magnetic north and true north (the geographic North Pole) at the point of measurement. It can be given as degrees clockwise from the geographic North Pole (Gose 2000) or in degrees east or west of the geographic North Pole (Aitken 1970). The magnetic North Pole is currently situated in the Queen Elizabeth Islands in northern Canada, so along a line running through east-central North America the declination is 0° (Sternberg 1990). From the perspective of an observer along this line, the magnetic North Pole is aligned with the geographic North Pole. Inclination of the magnetic field is measured with respect to the local horizontal plane. In most parts of the northern hemisphere, measurements of inclination are positive, meaning they dip below the horizontal plane. This is because a magnetized needle points to the source of the magnetic field at the Earth's core, specifically, to the south pole of the dipole field (Fig. 11.3). Consequently, inclination is usually negative in the southern hemisphere and 0° near the equator. When standing at the magnetic North Pole, inclination is +90°; at the magnetic South Pole, it is −90°.

Changes in the direction of the magnetic field over time for a region can be depicted as a curve on a stereonet or area projection (Fig. 11.4). Declination is indicated along the circumference of the full circle (or portion thereof), with north at 0°. Inclination is depicted as the distance from the middle of the circle with 0° at the outer edge and 90° at the center; different symbols are used for positive and negative values.

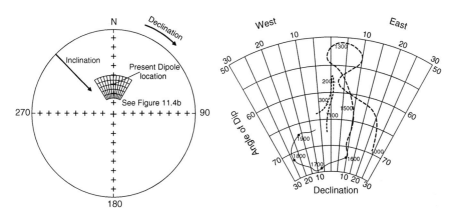

a) The declination is measured along the circumference and the inclination increases from 0° at the perimeter to 90° at the centre.

b) Inset shows the secular variation for London based on recorded observations (AD1580 to present) and archaeomagnetic data as degrees east and west of geographic north.

Fig. 11.4 Explanation and Examples of Stereonet Projections showing Direction of Field by Declination and Inclination. (Adapted from Gose (2000) and Aitken (1970) with permission. (**a**) Modified from "Journal of Archaeological Science" 27(5), Wulf A. Gose, Paleomagnetic Studies of Burned Rocks, page 410 Copyright 2000, with permission of Elsevier. (**b**) Reprinted from "Philosophical Transactions of the Royal Society, Series A, Mathematical and Physical Sciences" 269:78, M. J. Aitken, Fig. 1, copyright 1970, with permission.)

When variation in the geomagnetic field is small, secular variation is shown as a fairly straight or gently curving line; however, major changes are also possible. On the basis of archaeomagnetic data from Western Europe and the Mediterranean spanning a 3000-year period, Gallet et al. (2003) identify sharp changes in magnetic field direction and intensity, to which they apply the term "archaeomagnetic jerks." The two most dramatic examples date to approximately A.D. 200 and approximately A.D. 1400 when hairpin-like changes, measuring 170° and 140°, respectively, occurred over a 100-year period (Gallet et al. 2003).

Properties of Suitable Samples

Materials containing small amounts of magnetite and hematite are most suitable for analysis. These minerals have very high magnetic susceptibilities (i.e., an external magnetic field readily induces magnetization) and the magnetization is retained. Iron-rich minerals in sediments and sedimentary rocks align themselves parallel to the geomagnetic field at the time of their deposition; this is referred to as depositional or detrital remanence (DRM) or post-depositional remanence (PDRM) (Eighmy 2000; Sternberg 1997). Depositional remanence is typically the subject of paleomagnetic studies but is of relevance to paleoanthropologists as magnetic reversals provide chronological data for the remains of early hominids.

Archaeomagnetism involves the study of materials whose magnetization was affected by human activity. Igneous rocks and fired clay associated with thermal features, such as hearths, ovens, and kilns, often contain iron-rich minerals. When heated to their Curie point, 580°C for magnetite and 680°C for hematite, the minerals lose their previously acquired magnetization and acquire a new magnetization as the material cools down. If the position of the sample has not changed, the heat-induced magnetization, or thermoremanent magnetization (TRM), of the minerals corresponds to the direction and intensity of the geomagnetic field at the time of cooling. A partial thermoremanent magnetization (pTRM) is produced when the maximum temperature is below the Curie point and the mineral retains a portion of the original magnetization.

The temperature at which the magnetization of a cooling material becomes "frozen" is called its blocking temperature. This temperature is less than the Curie point of a given mineral and is affected by the size and nature of the iron-rich mineral grains. Grains measuring from 10^{-5} to 10^{-3} cm in diameter have the most stable magnetizations; the magnetizations of smaller and larger grains may change over time (Tarling 1975). For this reason, a sample will generally have a continuum of blocking temperatures. The stable component of the total magnetization arises from minerals with blocking temperatures in excess of 150°C. The magnetization from minerals with blocking temperatures less than 150°C can be altered by changes in the ambient geomagnetic field. This unstable portion, called viscous (remanent) magnetization (VRM) or the viscous component (Aitken 1970; Takac 2000), usually forms less than 5% of the total magnetization. Prior to measurement of TRM, the viscous component is removed by heating the sample to 150°C, then cooling it in a zero magnetic environment.

Archaeomagnetic Dating

Archaeomagnetic dating involves matching the TRM orientation of a sample to the secular variation of the magnetic field at, or close to, a specific location. For this reason, application of this dating technique is restricted to regions where changes in the direction of the geomagnetic field over time are known. In North America, the master curves of secular variation are primarily reconstructed through the use of magnetized archaeological features of known age as control samples. The depiction of secular variation at London, England, is based on more than 400 years of actual observations as well as archaeomagnetic data (Aitken 1970; Fig. 11.4b). Establishing master curves is a cooperative and mutually beneficial endeavor. Well-dated samples from archaeological sites are the baseline data geophysicists require to establish the secular variation curve needed to determine archaeomagnetic dates for archaeologists.

An alternative method of illustrating secular variation uses the locations of the virtual, or equivalent, geomagnetic poles (VGPs) (Fig. 11.5). A VGP is a mathematical transformation that uses archaeomagnetic data from a particular location

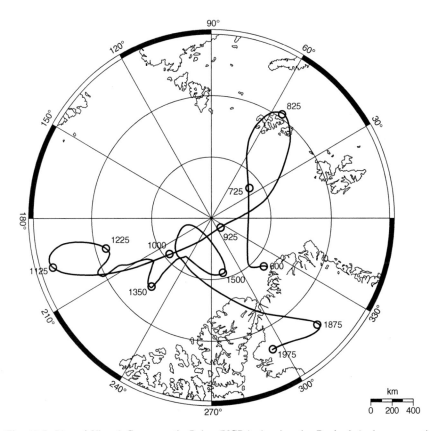

Fig. 11.5 Plot of Virtual Geomagnetic Poles (VGPs) showing the Revised Archaeomagnetic Secular Variation Curve for the U.S. Southwest as Proposed by Lengyl and Eighmy (2002). (Reprinted from Journal of Archaeological Science 29(12), Stacey N. Lengyel and Jeffrey L. Eighmy, A Revision to the U.S. Southwest Archaeomagnetic Master Curve, page 1431 Copyright 2002, with permission of Elsevier.)

to calculate the former position of the magnetic North Pole with respect to the geographic North Pole (Sternberg 1997). If the calculation includes an additional correction for variation in the non-dipole field, the mean angular error associated with VGPs is 2° per 1000 km (Sternberg 1997).

Because measurements have both a horizontal (declination) and a vertical (inclination) component, the error associated with sample directions is two dimensional. Results from multiple samples for a particular artifact, feature, or structure are used to calculate a mean value and the precision of the archaeomagnetic direction, α_{95}. The α_{95} for a sample mean is the angular radius of a cone of confidence and there is a 95% probability that the true direction lies within the cone. According to Aitken (1970), good archaeomagnetic results for a structure, such as a pottery kiln, vary less than $\pm 1°$ in inclination and less than $\pm 2°$ in declination. Poor results vary by more than $\pm 3°$ in inclination and more than $\pm 6°$ in declination. Sternberg (1997) suggests the minimum age range achievable for an archaeomagnetic date is ± 25 years, which is only slightly more than the ± 20 years originally suggested by Thellier (Aitken 1970).

While it is possible to use DRMs measured on sediments from a site for archaeomagnetic dating, Sternberg (2001) regards them as generally less reliable directional indicators than TRMs. While the declination may be accurate, gravitational forces can increase the error associated with the measurement of inclination by flattening the orientation of ferromagnetic grains.

Archaeointensity Studies

The intensity of the geomagnetic field in the past, or paleointensity, is linearly proportional to the strength of the TRM for a sample (Sternberg 1997). When used in combination with directional information, the precision of an archaeomagnetic date can sometimes be improved because matching paleointensities can reduce the number of dating options. If the orientation of the samples is not known, the archaeointensity of a sample alone can provide chronological information.

Measuring paleointensity involves comparing the strength of the original TRM of a sample to TRM induced in a laboratory magnetic field in demagnetization studies. As described by Sternberg (1997), the Thellier method involves separating the original TRM from the laboratory TRM by heating and cooling a sample twice at each step through a series of progressively higher temperatures until the Curie point is reached. Another option is to determine intensity through alternating field demagnetization. The sample is placed inside a coil that produces a magnetic field, which then decays to zero. The strength of the laboratory magnetic field is progressively increased over a series of 10 or more demagnetization steps (Gose 2000). These studies are conducted with the assumption that the magnetic susceptibility of the sample remains constant throughout the demagnetization process (Sternberg 1997).

It is also possible to measure changes in both direction and intensity of the TRM through demagnetization studies (Gose 2000; Takac 2000). After each step in the thermal or alternating field demagnetization process, the sample is placed in a zero magnetic field then the direction and intensity of the remaining magnetization are measured. The advantage of this procedure is that it can be used to discern different components of the TRM. The high-temperature component of the TRM reflects the magnetization acquired prior to the move and a low-temperature component is due to magnetization acquired after the move. This procedure can be used to examine the integrity of thermal features and test hypotheses that certain rocks were used as boiling stones (Gose 2000; Takac 2000).

Sample Selection and Collection Procedures

Samples best suited for archaeomagnetic analysis were heated to temperatures ranging from 400°C to 500°C at the time of site occupation. In the case of soil, increased hardness can be used as an indicator of higher firing temperatures but clay content, mineral content, sample homogeneity, and texture are also important factors (Eighmy 1990).

The orientation of samples for archaeomagnetic analysis must be precisely recorded at the time of collection. The position of geographic north can be determined using a compass or transit and recorded on a plastic disc, which is then attached to the sample. If the sample is enclosed in a plaster of Paris mold or jacket, the orientation can be inscribed directly into the plaster before it dries. Gose (2000) cautions that plaster of Paris can be as magnetic as an archaeological sample so it is important to measure the magnetic remanence of a cured plaster sample before using for sample collection.

Careful planning is required to ensure the material best suited for archaeomagnetic dating is collected in the best manner. The procedures outlined by Eighmy (1990) require the use of a pocket transit, cross-test levels, non-magnetic plaster, as well as hand tools and molds made of aluminum or brass. Typically, individual samples are isolated as small pedestals then a clay collar is placed around the base. An oiled mold is positioned over the pedestal, onto the clay collar and leveled. Plaster is poured into the mold and allowed to set for a short time. The magnetic declination reading is made with the pocket transit while it is against one side of the mold and level. Excess plaster is removed, the edge from which the reading was made is marked, and the cube is extracted. The base is then capped with plaster and appropriately marked with a sample identification number.

Prior to making a reading with a pocket transit, the area should be cleared of all metal; however, large fired archaeological features and igneous rocks that are strongly magnetized will distort compass readings (Eighmy 1990; Hathaway and Krause 1990). To avoid these problems and eliminate the need to correct for local, present-day declination, some prefer to take readings with a sun (or solar) compass positioned atop the leveled mold (Tarling 1975). A sun compass consists of a circular platform with 360° protractor and central vertical pin. The angle of the Sun's shadow cast by the vertical pin is recorded along with the precise time, location, and date of collection.

Eighmy (1990) and Sternberg (1997) suggest collecting 8–12 oriented samples, each measuring several cubic centimeters in volume, for each *in situ* feature. Aitken (1970) recommends that a minimum of 12 well-distributed, oriented samples be collected for structures, such as kilns. Sternberg (1997) notes that samples collected by inexperienced personnel can yield poor results. Eighmy (2000) regards only about 50–60 individuals in North America as "competent collectors."

Measuring Magnetism

Instruments used to measure the magnetization of an archaeomagnetic sample are called magnetometers. A spinner magnetometer, used to determine the direction of magnetization, involves rotating the sample fairly rapidly (300–420 revolutions per minute) inside an electric coil (Aitken 1970; Tarling 1975). The magnetization of the rotating sample induces a weak alternating current in the surrounding electric coil. The phase of the induced current is related to the direction of the magnetization. A disadvantage is that a sample experiences considerable centrifugal force during rotation; however, the device can accommodate samples measuring up to 20 cm across (Aitken 1970).

A cryogenic magnetometer is used to measure the magnetization of a sample and then incrementally remove a portion of the contribution through thermal demagnetization. The technique used by Takac (2000) involves measuring the total magnetization, then demagnetizing the sample at 50°C intervals from 100°C to 650°C, followed by a final heating at 685°C. The sample is monitored for possible changes in magnetic susceptibility by measuring its response to a small, applied electromagnetic field between heating intervals.

Summary

Although the potential for growth in the field is high, Eighmy (2000) suggests that three issues contribute to a crisis in archaeomagnetism in North America. First, there are few specialists in the field; second, even fewer are associated with Ph.D. granting universities. Third, archaeology students in North America generally lack the requisite strong background in geophysics, physics, and chemistry. In many ways, archaeomagnetism is still under development in North America. Although the technique is most widely applied in the American Southwest, the secular variation curve for the region is still under revision (Lengyel and Eighmy 2002). Outside of chronological applications, the work of Gose (2000) and Takac (2000) demonstrates that archaeomagnetism studies can make valuable contributions by providing data on feature integrity, artifact function, and site activities.

Chapter 12
Provenance Studies

Provenance studies involve the use of particular artifact traits to establish where the piece was manufactured or the source of the raw materials from which it was made. Distinct physical characteristics, such as style, shape, and decoration, may provide sufficient evidence to assign an artifact to a particular geographic location. In other cases, it may be necessary to rely on data that cannot be obtained through visual inspection to test a hypothesis or verify a conclusion.

Sayre (2000) notes that structural analysis is important for provenance studies as different techniques were employed at different times and in different places. Imaging techniques, such as X-radiography (Chapter 37), can reveal how an artifact was manufactured but it is important to match the energy of the X-rays (Chapter 2) used to the composition of the artifact under investigation. For example, radiographs obtained with very high-energy photons or thermal neutrons (Chapter 3) are appropriate for examining a bronze statue (Sayre 2000); low energy or "soft" X-rays can reveal the structure of woven fiber sandals (see Yoder 2008 in Chapter 24).

The composition of lithic and ceramic artifacts can be described on the basis of their mineralogy, using petrographic techniques through microscopic examination of thin sections. This approach can yield valuable data alone (Stoltman 1989, 1998) or when used in conjunction with chemical analysis techniques (Baxter et al. 2008). Thin section analysis is widely used in geology and the procedures involved are well documented (Moorhouse 1959; Williams et al. 1982) and will not be discussed here.

While petrographic and structural analysis can provide valuable provenance information, the primary focus of this chapter will be issues related to the use of instrumental analysis to determine the chemical composition of archaeological materials. Wilson and Pollard (2001), Sayre (2000), and Neff (2000, 2001) can be consulted for more detailed discussions of the issues related to the use of elemental composition for archaeological provenance studies.

Sourcing on the Basis of Composition

Provenance studies represent a special application of composition analyses involving comparisons between different artifacts, known raw material sources, and/or

M.E. Malainey, *A Consumer's Guide to Archaeological Science*, Manuals in Archaeological Method, Theory and Technique, DOI 10.1007/978-1-4419-5704-7_12, © Springer Science+Business Media, LLC 2011

known manufacturing sites. Similarity in composition is regarded as evidence of common origin. Interpretations are strengthened by clearly demonstrating that (1) the composition of the artifact is homogeneous; (2) the composition of the artifact has not been altered by processing or mixing of raw materials or the effect of the alteration is known; (3) the composition of each potential raw material source is homogeneous; and (4) the composition of all potential raw material sources is distinct. Comparisons typically involve some type of multivariate statistical analysis or bivariate plots of element pairs whose presence or concentration is source specific.

The process may appear relatively straightforward but complications may arise at a number of levels when dealing with archaeological materials. The criterion of homogeneity is often demonstrated by making a number of measurements on one artifact and then assuming all others exhibit the same degree of variability. The desire to minimize the impact on artifacts may limit an analyst's ability to thoroughly investigate sample homogeneity. Non-destructive techniques are desirable but care must be taken as the composition of the exterior surface of an artifact may differ from its bulk. The surface may have been intentionally coated with a different substance, such as paint, glaze, or metal plating, or the chemical composition of the outermost surface may have been altered by oxidation, corrosion, or contamination. These issues are discussed further in Chapter 30.

The second criterion typically holds for artifacts manufactured from stone, animal tissue, or plant material, but the elemental composition of refined or synthetic substances, such as metals, ceramics, and glass, may differ significantly from their constituent raw materials. Even if only one source was exploited, the removal of impurities or the addition of colorants, fluxes, and temper can result in significant differences between the composition of an artifact and the raw materials. Instances involving the mixing of raw materials from different sources or recycling are more complicated. In these cases, it may be necessary to make comparisons between the elemental compositions of finished artifacts from different sites and/or material recovered from possible manufacturing sites (see Sempowski et al. 2001 in Chapter 26).

Possible problems associated with the third and fourth criterion may arise because the location of raw material sites exploited during ancient times may be unknown, the source may be exhausted or destroyed by later natural processes or human activities. As in the case of turquoise (Hull et al. 2008), considerable effort may be required in order to identify a unique geochemical signature for raw material deposits. If sources cannot be distinguished on the basis of major, minor, or trace elements, discrimination may be possible using ultra trace [also called rare earth elements (REE)] or the stable isotope ratios of elements such as carbon, oxygen, lead, strontium, hydrogen, and copper (see Chapter 13). In some cases, regional differences in the composition of additives, such as colorants, may be used to identify manufacturing sites (see Shortland et al. 2007 in Chapter 26).

The need to have a representative sample from each potential source has long been recognized. To demonstrate that the deposit is homogeneous, however, it is important to collect several reference samples from all accessible parts of the deposits. In general, the recommended number of reference samples for a deposit has increased over the last two decades. Provenance studies can only be conducted if the compositional variability within a single deposit is less than the variability that exists between different deposits.

Wilson and Pollard (2001) caution that provenance studies are only capable of systematic elimination of possible sources because other deposits with identical geochemical fingerprints may exist. It is possible to rule out a source on the basis of elemental composition but not unequivocally prove a source designation.

Sample Considerations

Analysis of chemical composition using the techniques described in Part IV usually involves removal of a small, representative portion of the artifact. While it is possible to analyze small samples non-destructively, solid samples are typically finely ground for instrumental neutron activation analysis (INAA); for particle-induced X-ray emission analysis (PIXE), a fused bead is formed from ground material. Prior to analysis with atomic absorption (AAS) or inductively coupled plasma (ICP) techniques, solid samples are usually dissolved in strong acids. Special care is required to ensure the concentration of components with very low solubilities in the acid solution matches that in the solid (Young and Pollard 2000). These sample constituents may precipitate out of solution, causing their concentration in the acid solution to change during the course of analysis.

The ability to determine the chemical composition of whole samples or miniscule portions can be advantageous, especially when analyzing valuable, rare, or sacred objects. It is now possible to perform compositional analysis with little or no impact on the artifact using X-ray fluorescence (XRF), PIXE, Raman spectroscopy, secondary ion mass spectrometry (SIMS), or by coupling mass spectroscopy or ICP with laser ablation. However, one must bear in mind that the chemical composition of a solid is not necessarily homogeneous, so very small samples, especially those taken from or near the surface, may not be representative of the whole. In some cases, the shape of the artifact must also be considered. Kuhn and Sempowski (2001) report non-destructive analysis of the matrix of ceramic pipes is possible using XRF and PIXE as long as measurements are made on a flat, smooth area of matrix that is free of temper (see Chapter 17).

As discussed in Chapter 30, the surface is a boundary layer between a solid and its environment. Reactions with water and oxygen may lead to the development of a patina or corrosion. Contaminants from the burial environment or traces of previously processed materials may be trapped in voids at the surface. Slips or paints may have been applied to the surface as decoration. If a ceramic item was glazed or a metallic object electroplated, the chemical composition of a thin layer at the surface will be completely different from the underlying material. Differences between the composition of the surface of a sample and its bulk composition generally diminish with depth.

Wilson and Pollard (2001) advise that the selection of instrument should be based on the specific aims of the project and the nature of the material involved. While most instrumental techniques measure multiple elements during an analysis, if potential sources can be successfully discriminated on the basis of a single element, any technique that can accurately assess that particular element can be selected. It is also important to recognize that different techniques measure different elements with varying degrees of precision, which can make it difficult to compare data sets. Technological improvements may even complicate comparisons between measurements obtained today and data obtained decades ago using the same type of instrument.

Raw Material Processing and Mixing

Raw materials collected directly from the source can require processing before they are suitable for use. One or more refining techniques may be needed to remove impurities prior to artifact manufacture. Substances may be added to improve the workability of raw material or impart desirable qualities to the finished product. In some cases, raw material from different sources may be mixed together. While this is not a concern for lithics, the chemical compositions of metal and ceramic artifacts are more likely to differ from their raw material sources because of processing and mixing.

The composition of metal precisely matches its source if it has not been refined or purified in any way; examples of this include artifacts fashioned from native copper and meteoric iron. In many cases, however, refining processes may have altered the elemental composition of metals. Because metal was so valuable, it was often recycled and reused; different metals can be mixed together to form alloys. The effects of raw material processing and mixing with respect to the utility of lead isotopes for provenance studies are described in Chapter 13.

Typically, ceramic items are not fashioned directly from raw clay and the processes involved in pottery manufacture may affect its chemical composition. In addition to removal of plant roots and pebbles, naturally occurring sand and silt may be separated from the clay. Tempering materials are often added; traces of parting agents used to separate formed objects from molds may adhere to the surface. Paints, slips, and glazes may be applied before, after, or between firings.

Neff et al. (1988) developed computer models to simulate changes in chemical composition of pottery caused by the addition of temper to clay. The models showed the effect of tempering on paste composition was minimized if the added material was chemically homogeneous. The addition of heterogeneous temper was most pronounced if the clays in the hypothetical paste mixtures were initially similar in composition. However, a large amount (75–80%) of heterogeneous temper would have to be added before the separation of two chemically distinct clay sources was lost. If the chemical characterizations of the clays were based on many elements and a moderate proportion of temper was added (about 40%), most types of temper would not obscure the separation between the clay groups (Neff et al. 1988). Despite the results of these mathematical models, Wilson and Pollard (2001) still suggest that ceramic provenance studies are impeded by the dilution effect of added temper.

The boiling points of several elements routinely analyzed in ceramic studies are below 1100°C. In particular, high firing temperatures could affect concentrations of arsenic, potassium, rubidium, cesium, sodium, and zinc. Cogswell et al. (1996) investigated this possibility and their results demonstrated that no loss of these elements through volatilization occurred when clay is fired at 1100°C.

The practice of using debris from pottery manufacturing sites, in particular kiln wasters, as reference material is well established. The benefits of this approach are that chemical changes due to clay purification, temper addition, and firing are reflected in the fabric of the vessels that failed during firing. Wilson and Pollard (2001) caution, however, that differences in composition may have existed between the vessels that survived and those that did not.

Neff (1993) contends that the compositional analysis of ceramic artifacts contributes to our understanding of past cultural practices beyond provenance. Due to the relationship between paste preparation and chemical composition of ceramics, similarities may signify shared resource procurement patterns. Continuity of composition over time may indicate the intergenerational transmission of knowledge required to make pottery with the desired mechanical performance characteristics.

The Composition of Raw Material Sources

There are two general approaches to provenance determination: one requires the recognition of localized sources of raw materials and the other does not (Neff 2000, 2001). The first approach is most relevant to the study of artifacts fashioned from either metal or rock. If locations of quarries, mines, outcrops, or deposits of flakable rock and metal ores are known and accessible, it may be possible to establish the chemical composition of potential raw material sources. The number of samples required to establish the range of compositional variation of the source depends upon its degree of homogeneity. Neff (2000) suggests that between 5 and 10 samples from each source is adequate for obsidian, because its chemical composition is known to have a high degree of uniformity.

A large number of samples are required to characterize a source that may have a heterogeneous chemical composition. In their study of chert, Lyons et al. (2003:1156) analyzed between 10 and 20 hand samples for each source, but admitted that quantity was well below the number some consider necessary for robust principal component analysis, specifically, "as many source samples per source group as there are elements in the analysis."

While some have suggested that at least 20 samples are required to characterize an ore body on the basis of lead isotope ratios, Baxter et al. (2000) note that this minimum would only be sufficient if the data were normally distributed. They further suggest between 40 and 50 samples are needed to test an ore deposit for normality. If the lead isotope field for a deposit was bimodal or multimodal, then between 60 and 100 samples would be required for adequate characterization.

The second approach typically applies to the study of pottery. Wilson and Pollard (2001) note that the chemical composition of clay depends upon the mineralogy of its parent material, the weathering and transport processes that acted on the material and the chemical environment in which the clay was deposited and matured. Deposits of clay may be widespread and boundaries between them may be indistinct; the same may apply to certain tempering materials. In situations such as this, it is impractical or impossible to sample and characterize all possible raw material sources. It is possible, however, to recognize patterns in the chemical compositions of samples that relate to the exploitation of clay from geographically restricted locations, such as drainage basins (Bollong et al. 1997). Once the range of chemical compositions of ceramics produced in a particular region is established, ceramics manufactured outside the area can be recognized. This approach is also appropriate in areas where clay deposits were purified through levigation, or slaking, to remove naturally occurring silt or sand.

Statistical Analysis of Compositional Data

Detailed descriptions of statistical techniques used to examine compositional data are beyond the scope of this work. Recommended books on the topic include *Quantifying Archaeology* (second edition) by Stephan Shennan (1997) and *Exploratory Multivariate Analysis in Archaeology* by M.J. Baxter (1994). Chapters summarizing the topic include Baxter and Buck (2000) and Baxter (2001).

Bivariate plots are used to examine how the concentration of one element varies with respect to another in the samples. Although instrumental techniques can generate data for more than 30 elements, groupings based on one or more element pairings can often be clearly illustrated using this approach (Bellot-Gurlet et al. 1999; Hull et al. 2008; Lyons et al. 2003; Neff 2000; Yacobaccio et al. 2004). Multivariate techniques are required to describe data with respect to many variables simultaneously. Techniques commonly used to identify patterns in compositional data sets include principal component analysis and cluster analysis. The use of correspondence analysis for the examination of compositional data has been recently advocated.

Discriminant analysis can be used to confirm the existence of discrete groupings and assess the likelihood that a particular sample belongs to a previously established group. Discriminant equations incorporate the variables that most significantly account for differences between the groups weighted to maximize the separation between them. Once established, these equations can be used to determine the group to which an unknown sample belongs. A new sample can be evaluated by inputting applicable values into the discriminant equation for each group; the equation generating the highest score signifies the most likely group. Mahalanobis distance calculations are used to determine the probability that a specimen is a member of any of the identified groups. Lyons et al. (2003) apply this approach to assign chert artifact compositions to recognized raw material sources (see Chapter 18).

Bishop and Neff (1989) note that problems can arise when multivariate techniques are applied without complete awareness of their requirements and inherent assumptions. It is important to be cognizant that each statistical approach evaluates data with respect to a particular mathematical model. These models may reveal structures in the data, such as discrete groupings; however, they can also impose a structure on a data set. Bishop and Neff (1989) illustrate the possible outcomes associated with the inappropriate application of statistical methods to compositional data. For example, the use of cluster analysis to examine a data set containing correlated variables can lead to spurious aggregations. Principal component analysis (PCA) removes these correlations, so cluster analysis of PCA scores provides the correct groupings (Bishop and Neff 1989; Sayre 2000).

When comparing major, minor, and trace elements, it is necessary to apply some form of scaling or data transformation to prevent variables with higher concentrations from having excess weight in calculations of many coefficients of similarity. In this regard, base-10 logarithms of element concentrations are often used in data analysis. Under some circumstances, k-means cluster analysis is more appropriate than hierarchical clustering approaches. Q-mode factor analysis can be applied to data representing the mixing of three "pure" substances or end members. A familiar example may be the ternary diagram used for soil classification, with which every soil type is defined as a variable mixture of sand, silt, and clay. The example provided by Bishop and Neff (1989) involved two chemically distinct clay sources and one widespread temper source of relatively homogeneous composition.

Many methods of multivariate analysis that have been applied to compositional data assume normal distributions of variables; consequently, recognition of non-normality is important (Baxter et al. 2000). Baxter et al. (2000) caution that multivariate statistical methods that assume normality should be avoided if this condition has not been demonstrated through extensive sampling of the deposit.

The key lesson is that statistical approaches should not be applied blindly. Commercially available statistical software packages, such as Statistica®, enable a user to analyze a data set with a wide assortment of multivariate analysis techniques both rapidly and easily. Basic knowledge of the statistical approach is required to ensure the appropriate settings for the data analysis are selected. Analysts who furnish compositional data should be able to recommend suitable multivariate statistics. Universities have statisticians available for consultation on a fee-for-service basis.

Conclusions

The strongest case for attributing an artifact to a particular location rests on multiple lines of evidence derived from analysis of style, structure, and chemical composition. Even under ideal circumstances, it is only possible to rule out potential sources and report high statistical probabilities of location assignment. Consultation with the analyst or a statistician may ensure that the appropriate statistical tool is selected and applied properly to the data set. It may not be possible to identify the source of materials whose composition is altered by refining, mixing, or recycling. In these cases, provenance studies must rely on comparisons between finished artifacts and/or debris from manufacturing sites. As Neff (1993) pointed out, the compositional analysis of artifacts contributes to our understanding of past cultural practices even if the source of raw materials cannot be identified.

Chapter 13
Isotope Analysis

Isotope analysis is a highly versatile technique that uses the ratio of two non-radioactive isotopes of an element within a sample to provide insights about the conditions under which it formed. This is possible because certain biochemical and geochemical processes favor one isotope, usually the lighter one, over the other. This results in measurable differences between the isotopic composition of the initial reactants (e.g., the food an animal consumes) and the resulting products (e.g., animal tissue formed from food components). The technique can be applied to a wide variety of materials including the remains of plants and animals, organic residues, metals, rocks, and minerals. Depending upon the sample and isotopes measured, information can be gained about diverse issues ranging from the diet and mobility patterns of an animal, global climate change, and the source of building material. Interpretations may be based on the value of one or more ratios of isotopes of the same element or the isotope ratios measured on more than one element. General descriptions of elements commonly targeted in archaeological applications, namely carbon, nitrogen, strontium, lead, oxygen, and hydrogen, appear in Chapter 4; isotopic ratios are measured using mass spectrometry, which is described in Chapter 31. Examples of specific applications of stable isotope analysis to archaeological materials are presented in Part III.

Stable isotope analysis of human remains is a well-established technique within the realm of biological anthropology. Vogel and van der Merwe (1977) were the first to demonstrate that differences in the ratios of the stable isotopes of carbon in human bone could mark the introduction and assess the importance of maize to Eastern North American populations. Over time, it became clear that the relationship between diet and stable isotope composition was not necessarily straightforward due to environmental factors, differential fractionation, and diagenetic processes (Sillen et al. 1989). Studies by physical anthropologists of the effects of preservation, tissue selection, and sampling techniques on stable isotope ratios are equally relevant to the study of archaeological faunal remains. Several excellent reviews and overviews exist including van der Merwe (1982, 1992), Schwarcz and Schoeninger (1991), Pate (1994), Katzenberg and Harrison (1997), Sealy (2001), Larsen (1997), Bentley (2006), Tykot (2006), Schwarcz (2006), Katzenberg and Saunders (2008), and Lee-Thorp (2008). The principles of stable isotope analysis for geological applications are presented in Sharp (2007) and Hoefs (2009).

M.E. Malainey, *A Consumer's Guide to Archaeological Science*, Manuals in Archaeological Method, Theory and Technique, DOI 10.1007/978-1-4419-5704-7_13, © Springer Science+Business Media, LLC 2011

Stable Carbon Isotope Ratio Analysis

Stable carbon isotope analysis is most frequently used to address questions related to foods eaten and prepared; however, it can also be applied to provenance studies of carbonate rocks and minerals. The factors that affect stable carbon isotope values in carbonates are the same as those that affect oxygen isotopes and are described later in this chapter. The stable carbon isotope ratio of a sample is in some way related to the material from which it was formed, but the relationship is complicated by isotopic fractionation. Isotopic fractionation occurs because certain chemical reactions are biased against the heavier ^{13}C isotope. Stable carbon ratios are expressed as δ (delta)^{13}C values and represent the ratio of ^{13}C to ^{12}C in a sample relative to the international standard, Pee Dee belemnite (PDB), a limestone from South Carolina:

$$\delta^{13}C = \left[\frac{^{13}C/^{12}C \text{ sample}}{^{13}C/^{12}C \text{ standard}} - 1 \right] \times 1000$$

which is identical to the formula

$$\delta^{13}C = \left[\frac{^{13}C/^{12}C \text{ sample} - \, ^{13}C/^{12}C \text{ standard}}{^{13}C/^{12}C \text{ standard}} \right] \times 1000$$

The standard contains less ^{12}C and more ^{13}C than most natural materials, so $\delta^{13}C$ values of samples are usually negative, ranging between -37 and $-8‰$. Differences between samples are very small so values are given as per mil (‰), rather than percent (%). The $\delta^{13}C$ value of atmospheric CO_2 is approximately $-8‰$. During photosynthesis (Chapter 4), the ratio between ^{13}C and ^{12}C changes because plants use carbon dioxide containing the lighter isotope, $^{12}CO_2$, more readily than molecules with the heavier isotope, $^{13}CO_2$. Isotopic fractionation is more pronounced in C_3 plants than in C_4 plants. Carbon fixation with ribulose diphosphate, the first step of the Calvin cycle for C_3 plants, occurs much faster with $^{12}CO_2$ than $^{13}CO_2$. Since more ^{12}C is incorporated into the plant, the average $\delta^{13}C$ value for C_3 plants is about $-26‰$ with a range between -22 and $-34‰$. Photosynthetic processes are also affected by temperature so plants growing at higher latitudes have different $\delta^{13}C$ values than those growing closer to the Equator. By contrast, the range of $\delta^{13}C$ values for C_4 plants is from -8 to $-16‰$, with an average of $-13‰$. Crassulacean acid metabolism (CAM) photosynthesis of succulent plants produces intermediate $\delta^{13}C$ values, ranging from -10 to $-20‰$. CAM plants may follow C_3 photosynthetic processes during the day and C_4 photosynthetic processes at night. For this reason, CAM plants that fix a greater amount of CO_2 during the day have more negative values than those that primarily undergo CO_2 fixation at night.

   The isotopic composition of an animal depends upon the food it eats (carbon input) and the carbon lost through respiration (i.e., exhaling carbon dioxide), and excretion (feces output). When herbivores eat plants, the $\delta^{13}C$ value of the tissue of the animal is usually less negative than the plants they consume. With the exception of nematodes, DeNiro and Epstein (1978) found this process, called enrichment, consistently occurs regardless of either diet or species. Analysis of the whole animal provides the most accurate estimate of the $\delta^{13}C$ value of its diet, but this is usually not practical. Controlled feeding experiments are used to show that the degree of enrichment varies with the type of animal and type of tissue examined (DeNiro and Epstein 1978; Hare et al. 1991). Schwarcz (2000, 2006) describes the biochemical processes that affect the distribution of carbon isotopes in human tissue and considers possible implications for paleodiet studies.

The $\delta^{13}C$ value of the muscle tissue of an herbivore is 3‰ higher (less negative) compared to the plants they consume but its adipose tissue is 3‰ lower than (more negative) its diet. Human hair and nail keratin produce similar $\delta^{13}C$ values, which are 1.41‰ lower than bone collagen from the same living individual (O'Connell et al. 2001). The difference between human collagen and keratin is smaller in archaeological samples, about 0.5‰; the difference between collagen and keratin in animals is between 2 and 4‰. For large animals, the $\delta^{13}C$ values of bone apatite ($\delta^{13}C_{ap}$) are 12–14‰ more positive than their plant-based diets; a difference of about 10–11‰ is observed between human bone apatite and diet (Schwarcz 2000, 2006).

The $\delta^{13}C$ values of the bone collagen ($\delta^{13}C_{co}$ or $\delta^{13}C_{col}$) of herbivores are 5‰ higher than the plants they consume; a 5‰ fractionation between diet and bone collagen appears to occur in most animals and humans. Despite this consistency, the amino acid constituents of bone collagen exhibit different degrees of enrichment with respect to diet. In a controlled feeding experiment involving pigs, the amino acid glycine from bone collagen was 8‰ more positive than its total C_3 diet whereas the amino acid aspartate was only 3‰ more positive (Hare et al. 1991). This difference occurs because certain "non-essential" amino acids can be either obtained from the diet or synthesized by the body; "essential" amino acids are only obtained from the protein component of the diet. For this reason, it is important to compare stable isotope values made on either the same amino acids or the tissue samples that contain a number of different amino acids (Schoeller 1999).

Schwarcz (2000, 2006) suggests that, for human bone collagen, the level of protein intake may also affect the degree of bias toward protein sources. Individuals consuming significant quantities of meat would derive both non-essential and essential amino acids constituents of collagen from consumed protein. As a result, the difference between bone apatite and collagen ($\delta^{13}C_{ap} - \delta^{13}C_{co}$ or Δ_{ap-co}) would be very large. Conversely, individuals consuming little meat would derive only essential amino acids from protein but would manufacture non-essential amino acids from all dietary components (protein, carbohydrates, and fats), so the Δ_{ap-co} would be smaller.

When carnivores eat herbivores, another 1‰ enrichment in $\delta^{13}C$ values occurs. If a carnivore eats the flesh of a carnivore, there is a further 1‰ enrichment. A similar enrichment occurs when an infant feeds on its mother's breast milk. Examples of how the process of enrichment affects the $\delta^{13}C$ value of tissues at different trophic levels are shown in Fig. 13.1. If a consumer has a mixed diet, rather than purely C_3, C_4, or marine plants and animals, the $\delta^{13}C$ value of its bone collagen will lie

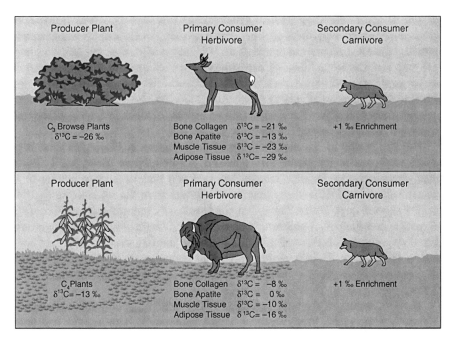

Fig. 13.1 Average isotopic fractionation between plants, herbivore tissues, and carnivores (Original Figure by M. E. Malainey and T. Figol)

between two extremes. Simple linear interpolations have been used to estimate the relative amounts of the different groups in the diet.

The $\delta^{13}C$ values of freshwater fish are similar to terrestrial plants and herbivores but values of marine species are quite different. Carbon exists in marine environments as dissolved carbon dioxide and carbonic acid, both of which are less negative than atmospheric CO_2. The average $\delta^{13}C$ value for marine plants and phytoplankton is $-19.5‰$, halfway between the average C_3 and C_4 values for terrestrial plants. Likewise, $\delta^{13}C$ values of marine fishes and mammals consuming marine plants or animals would fall between animals at the same trophic level eating either C_3-based or C_4-based terrestrial diets. The average $\delta^{13}C$ value for marine animals living in temperate oceans is about $-16‰$ (Sealy 2001). The $\delta^{13}C$ values of organisms in near-shore environments are affected by organic material, primarily the remains of terrestrial plants, washed in from coastal environments.

Sample Selection

The $\delta^{13}C$ values of both plants and animals depend upon the tissue sampled. Stable isotope analyses involving animal remains may target apatite in order to gain information about all dietary components. Apatite is found in tooth enamel and forms the inorganic component of bone, but tooth enamel is the preferred sample material because bone is much more susceptible to diagenetic processes (Chapter 19). Over time, carbon in bone apatite can be exchanged with carbon from the environment, primarily from atmospheric CO_2 and CO_2 dissolved in ground water. The degree of diagenetic alteration, through processes such as recrystallization, is usually assessed in bone apatite samples prior to measurement of stable isotope ratios.

Bone collagen is often used for stable carbon isotope analysis but $\delta^{13}C$ values largely reflect the protein component of an animal's diet. As mentioned earlier, this protein consists of essential amino acids that an animal consumed and non-essential amino acids from dietary protein or synthesized from dietary carbohydrates. Hare et al. (1991) caution that only intact collagen provides reasonable dietary signals because selective preservation of the different amino acids alters the isotope ratios of partially hydrolyzed collagen. Balzer et al. (1997) studied the effect of soil bacteria on bone collagen *in vitro* (in a petri dish). Bacteria penetrated the bones through holes and canals and attacked the collagen; this bacterial action resulted in $\delta^{13}C$ value shifts of up to $-2.9‰$.

For studies of human diet, bone collagen is often extracted from a spongy, cancellous (trabecular) bone, such as a rib. Because bones are constantly remodeled only a portion of the dietary record is retained. Analysis of human rib bone collagen provides an average of the individual's diet over the last 10 years of life. Sealy et al. (1995) suggest that stable isotope values of tissues formed at different periods, or reformed at different rates, provide information about specific phases in the life of an individual. Tooth enamel and primary dentine forms during childhood, while secondary (and tertiary, if present) dentine is added later. A more detailed picture of diet has been obtained by comparing teeth formed early in childhood with those formed later (Fuller et al. 2003).

Bone is continuously resorbed and reformed throughout the life of an animal. The turnover rate depends upon the type of bone, the age of the individual, and nutrition. Dense, compact bone from a long bone shaft turns over slowly; spongy, cancellous bone in mammalian ribs is believed to turnover at a faster rate. In order to detect dietary changes that occurred throughout life, Sealy et al. (1995) compared stable isotope values from the third molar, a long bone shaft, and a rib of each individual. The cholesterol and noncollageneous proteins, such as osteocalcin, preserved in bone (see Chapter 19) probably turnover at a faster rate than bone collagen and may provide information about diet close to the time of death (Schwarcz 2006).

Preserved soft tissue, including skin and muscle, is suitable for stable carbon isotope analysis. Nail and hair samples provide information about the 6-month period just prior to death (O'Connell et al. 2001).

Factors Affecting $\delta^{13}C$ Values

As outlined by Tieszen (1991), the $\delta^{13}C$ value of a plant is modified by genetic and environmental components. Genetic factors determine a plant's response to differing environmental conditions. The rate of photosynthesis is affected by factors such as water and nutrient availability, temperature, and altitude; the isotopic composition of the CO_2 processed during photosynthesis can be altered by the canopy effect.

The canopy effect occurs where tall, thick vegetation restricts the movement of air and can cause very low $\delta^{13}C$ values. Soil respiration produces ^{13}C-depleted CO_2; in dense forests, this carbon dioxide is then recycled during photosynthesis under low-light conditions. Van der Merwe and Medina (1991) show that a $\delta^{13}C$ gradient exists in the Amazon forest. The undergrowth at the forest floor has the most negative $\delta^{13}C$ value, $-37‰$; the leaves from the upper canopy have the least negative $\delta^{13}C$ value, $-31‰$. Catfish with a diet of ^{13}C-depleted forest detritus produce $\delta^{13}C$ values as low as $-29.3‰$, while terrestrial herbivores that ate the fruit of ^{13}C-depleted plants have $\delta^{13}C$ values only slightly (2‰) more negative than expected. The canopy effect is not restricted to tropical rainforests and can occur in temperate forests as well.

Stable carbon isotope studies of archaeological animals provide information about the environment that they inhabited; inventories of modern plant foods growing naturally in a region are used to estimate the isotopic value for herbivores that inhabited similar environments in the past. Diet selectivity and seasonal variation can affect $\delta^{13}C$ values as well as the paleoecological and paleodietary studies upon which they are based (Tieszen 1991). In cases where both C_3 and C_4 plants are available in the environment, animals may preferentially eat or avoid plants following one or the other photosynthetic pathway. Consequently, the average $\delta^{13}C$ value of local vegetation based on a random diet will be different from the actual $\delta^{13}C$ values of herbivores. The $\delta^{13}C$ values of plants can vary throughout a growing season so the timing of consumption must be considered.

The proportions of C_4 and C_3 plants in the diet can be calculated using the following equation (Schwarcz et al. 1985):

$$\text{Percent } C_4 = \frac{(\delta_c - \delta_3 + \Delta_{d-c}) \times 100}{(\delta_4 - \delta_3)}$$

where δ_c is the $\delta^{13}C$ of the sample bone collagen, Δ_{d-c} is the difference between diet and tissue due to fractionation (e.g., for bone collagen it is $-5‰$), and δ_3 and δ_4 are the average values of C_3 and C_4 plants prior to the Industrial Revolution, respectively. The $\delta^{13}C$ of plants that grew prior to the Industrial Revolution are 1.5‰ more positive than modern plants. It should be noted, however, that Hart et al. (2007) reported a similar formula used by Morton and Schwarcz (2004) to calculate proportions of C_3 and C_4 foods in carbonized pottery residues could significantly over- or underestimate the percentage of C_4 foods. Hart et al. (2007) suggest erroneous estimations occur because the carbon content of the foods is not equal and each contribute different amounts of carbon to the residue.

Stable Nitrogen Isotope Ratio Analysis

Stable nitrogen isotope analysis involves the most abundant isotope of nitrogen, ^{14}N, and ^{15}N, which occurs in nature at levels less than 0.4%. The $\delta^{15}N$ value is the ratio of ^{15}N to ^{14}N in the sample compared to the ratio in the international standard, atmospheric nitrogen (sometimes called the Ambiant Inhalable Reservoir, AIR):

$$\delta^{15}N = \left[\frac{^{15}N/^{14}N \text{ sample}}{^{15}N/^{14}N \text{ standard}} - 1 \right] \times 1000$$

The $\delta^{15}N$ of air is 0‰; atmospheric nitrogen is depleted in ^{15}N so most materials have positive $\delta^{15}N$ values. The $\delta^{15}N$ values measured in plants range from -5 ‰ for nitrogen fixers to $+20$‰ for those growing in hot, dry, saline environments. The $\delta^{15}N$ values of animals fall between about $+1$‰ for animals that must drink water regularly (obligate drinkers) and $+20$‰ for those that obtain most or all water from their food and retain water under stress (water-conserving species) (Ambrose 1991; Sealy 2001).

The $\delta^{15}N$ values of an organism can be affected by a number of variables related to diet and the environment. Stable nitrogen isotope analysis is particularly useful for discriminating consumers with a dietary component of marine plants and animals, which are enriched in ^{15}N, from those consuming only terrestrial plants and animals. Schoeninger et al. (1983) demonstrated that the $\delta^{15}N$ values of bone collagen from individuals with primarily marine-based diets are very high, $+17$ to $+20$‰, compared to individuals eating terrestrial plants and animals, $+6$ to $+12$‰. An exception is marine fish living around coral reefs, which have $\delta^{15}N$ values comparable to terrestrial mammals due to the higher level of nitrogen fixation that occurs in this environment. The $\delta^{15}N$ values of freshwater fish fall between those measured for marine and terrestrial organisms. Stable nitrogen isotope studies can also be used to track weaning practices because a $\delta^{15}N$ enrichment of $2-4$‰ occurs between a breastfeeding infant and its mother (or any other non-nursing member of a population) (Katzenberg and Harrison 1997). The degree of enrichment between consumers and their foods has prompted the use of $\delta^{15}N$ values to address questions of trophic level (Lee-Thorp 2008).

The technique is also useful for detecting the presence of nitrogen-fixing plants, such as legumes, in the diet of an animal. The difference in $\delta^{15}N$ values between individuals eating varying amounts of legumes and non-legumes ranges from $+6$ to $+10$‰ (DeNiro and Epstein 1981). As described below, the degree of enrichment can vary with environmental factors. Typically, the $\delta^{15}N$ values obtained on bone collagen from consumers are about $2-3$‰ higher than the foods they eat (Hare et al. 1991; Schoeller 1999; Schoeninger et al. 1983). Ambrose (1991) reported the difference between $\delta^{15}N$ values of plants and the herbivores that consume them is significantly higher in Africa: 4.73‰ in central Rift Valley forest and 5.36‰ in open habitats.

Sample Selection

The $\delta^{15}N$ values of plants can vary with location and time of year they were harvested (see below). In their controlled feeding experiments, DeNiro and Epstein

(1981) found the distribution of stable nitrogen isotopes in animals depends upon the tissue examined and their diet. O'Connell et al. (2001) report an average difference of 0.65‰ between hair and nail keratin from the same individual. Nitrogen occurs in proteins so $\delta^{15}N$ values are often obtained from bone collagen, although care must be taken to ensure it is well preserved. Balzer et al. (1997) studied the decomposition of bone collagen by soil bacteria *in vitro* (in a petri dish) and found bacterial action caused shifts in $\delta^{15}N$ values up to +5.8‰. If present, muscle tissue can also be used but the nitrogen content of bone apatite is too low for this type of analysis (Sealy 2001).

Factors Affecting $\delta^{15}N$ Values

Ambrose (1991) outlined the influence of a myriad of factors on $\delta^{15}N$ values, which are summarized in Table 13.1. In addition to its diet, the environment in which it lives and its physiology greatly affect the $\delta^{15}N$ value of an organism but some generalizations are possible.

Within a single ecosystem, the $\delta^{15}N$ value of nitrogen-fixing plants, primarily legumes, is lower than plants that do not fix nitrogen. Animals consuming nitrogen-fixing plants, either directly (herbivore) or indirectly through their prey (carnivore), have lower $\delta^{15}N$ values than animals that do not. The $\delta^{15}N$ of a carnivore will be

Table 13.1 Environmental factors that affect $\delta^{15}N$ values based on Ambrose (1991)

Environmental factors that increase $\delta^{15}N$ values in plants and soils	Environmental factors that decrease $\delta^{15}N$ values in plants and soils
Marine	Terrestrial
Aridity	High rainfall amounts
Savanna	Forest
Higher temperatures	Low temperature
Lower altitudes	Higher altitudes
Non-fixation	Nitrogen fixation
Fertilization with Guano	No fertilization
Clay	Sand and silt
High soil salinity	Low soil salinity

Environmental factors that increase $\delta^{15}N$ values in animal tissue	Environmental factors that decrease $\delta^{15}N$ values in animal tissue
Increased excretion of urea	Decreased excretion of urea
Water stress	Water abundance
Higher trophic level	Lower trophic level
Carnivore	Herbivore
Grazing	Browsing and mixed feeding
Water independence/conservation	Obligate drinking of water
Long distance to water	Short distance to water
Marine diet	Terrestrial diet (especially legumes)

higher than its herbivore prey. Within the same species, animals living close to water will usually have lower $\delta^{15}N$ values than animals living long distances from permanent sources of water. When comparing animals occupying the same trophic level (i.e., herbivores to herbivores, carnivores that eat herbivores to carnivores that eat herbivores), the $\delta^{15}N$ values of those physiologically adapted to retain water under stress (water conservers) will be higher than those without this adaptation (obligate drinkers). Water-independent animals retain water by excreting urine with high concentrations of urea. Urea is depleted in ^{15}N, so the lighter nitrogen isotope is lost and the $\delta^{15}N$ value of the animal tissue is elevated (Ambrose 1991; Schoeller 1999).

Generalizations can also be made about organisms living in different ecosystems. The $\delta^{15}N$ value of marine organisms is higher than terrestrial organisms at the same trophic level. Organisms living at lower elevations usually have higher $\delta^{15}N$ values than organisms of the same trophic level living at higher elevations. Ambrose (1991) reported that in East Africa, organisms living in arid, open habitats tend to have higher $\delta^{15}N$ values than similar organisms living in well-watered, forested areas. Open grassland areas have low levels of rainfall, fewer water sources, and soil that is more saline and alkaline (higher pH). Forested areas have high levels of rainfall, more water sources, and more acidic (lower pH) soils with low levels of salinity.

Organisms living in hotter environments will have higher $\delta^{15}N$ values than similar organisms living in cooler environments. For these reasons, the $\delta^{15}N$ value of animals in arid terrestrial conditions may appear similar to those living in marine environments. In temperate, well-watered situations, the $\delta^{15}N$ value of terrestrial animals is usually less than 10‰.

Reconstructing Paleodiets

Paleodietary studies often present the results of stable carbon and nitrogen isotope analyses plotted on a graph with $\delta^{13}C$ values on the X-axis and $\delta^{15}N$ values on the Y-axis (Fig. 13.2). The dietary contribution of different foods can be estimated if $\delta^{13}C$ and $\delta^{15}N$ values of the consumer and the foods they ate are combined with information about isotopic fractionation for the particular tissue analyzed (Morton et al. 1991; Schwarcz 1991). If an animal eats only one type of food, the difference between the isotope value of the food and the consumer will be due to fractionation. If foods with different isotope values were exploited, isotopic fractionation and the proportional contribution of each food to the animal's diet affect the $\delta^{13}C$ and $\delta^{15}N$ values of the consumer.

The dietary contribution of different foods to the paleodiet can be reconstructed through the use of a linear mixing model (Morton et al. 1991; Schwarcz 1991). As with any mathematical model, certain conditions must be met and limitations exist. First, if data were collected for N number of isotopes, the dietary contribution of $N+1$ foods can be explored. In cases where both $\delta^{13}C$ and $\delta^{15}N$ values were determined,

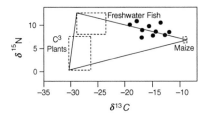

a) Other isotopically distinct food groups were eaten.

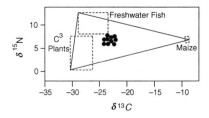

b) Diet of entire population was homogenous.

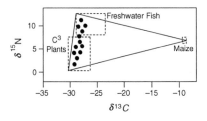

c) Limited range of plants and fish were consumed.

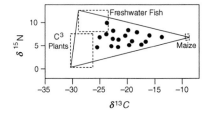

d) Differing proportions of the three foods were eaten.

Fig. 13.2 Interpretations of data point distributions in linear isotope models (Adapted from Morton et al. (1990) and Schwarcz (1991) with permission. From Archaeometry '90, 1991, Estimation of Palaeodiet: A Model from Stable Isotope Analysis by J. D. Morton, R. B. Lammers, and H. P. Schwarcz, Figure 2, with kind permission of Springer Science and Business Media. Reprinted from Journal of Archaeological Science 18(3), Henry P. Schwarcz, Some Theoretical Aspects of Isotope Paleodiet Studies, page 264, Copyright 1991, with permission from Elsevier)

N equals 2 so the dietary contribution of three foods can be examined. If data were collected for four different stable isotopes, the dietary contribution of five foods can be modeled. Second, the foods selected must have different $\delta^{13}C$ and $\delta^{15}N$ values (i.e., they cannot be co-linear on the $\delta^{13}C$ and $\delta^{15}N$ plane). When this condition is met, changing the dietary proportions of the foods shifts the position of the animal tissue on both the X-axis and the Y-axis.

The foods appearing in a linear mixing model must actually account for the vast majority of the animal's diet. Selections should be based on the information obtained from site materials (artifacts, faunal remains, and plant remains) and any other relevant sources (historical and ethnographic accounts, oral histories) that may be available. A single point in space is used to represent each of the foods in the linear mixing model; however, the $\delta^{13}C$ and $\delta^{15}N$ values of foods can vary throughout the year. The model is improved if selected foods were consumed at a site at specific times of the year, as the possible range of $\delta^{13}C$ and $\delta^{15}N$ values would be quite narrow. The precision of paleodietary models is further increased when the foods selected have markedly different $\delta^{13}C$ and $\delta^{15}N$ values. The degree of uncertainty associated with the measurement of isotopic ratios and estimation of isotopic fractionation also impacts model precision.

When $\delta^{13}C$ and $\delta^{15}N$ values are obtained, the dietary contribution of three foods (f_1, f_2, and f_3) can be investigated if the sum of the proportional dietary contributions (X_1, X_2, and X_3) equals the whole diet (Morton et al. 1991; Schwarcz 1991):

$$X_1 + X_2 + X_3 = 1$$

A theoretical $\delta^{13}C$ value for the consumer, $\delta^{13}C_{theoretical}$, can be calculated by multiplying the $\delta^{13}C$ of each food by its proportional contribution to the diet:

$$\delta^{13}C_{theoretical} = X_1\delta^{13}C(f_1) + X_2\delta^{13}C(f_2) + X_3\delta^{13}C(f_3)$$

Likewise, a theoretical $\delta^{15}N$ value of the consumer, $\delta^{15}N_{theoretical}$, can be calculated as follows:

$$\delta^{15}N_{theoretical} = X_1\delta^{15}N(f_1) + X_2\delta^{15}N(f_2) + X_3\delta^{15}N(f_3)$$

Theoretical values are compared to the actual stable isotope values of the samples, with appropriate corrections for isotopic fractionation. If samples fall outside of the boundaries of the triangle formed by the $\delta^{13}C$ and $\delta^{15}N$ values of the three foods, the foods selected do not represent the full range of isotopic variation of the animal's diet and different dietary components must be considered (Fig. 13.2a).

The distribution of sample points on a plot of $\delta^{13}C$ and $\delta^{15}N$ values can reveal certain aspects of the consumption patterns of a population (Schwarcz 1991). A tight cluster indicates the diet of the population was homogeneous (Fig. 13.2b). A linear distribution between two foods indicates different individuals consumed them in varying amounts and the third food was not eaten (Fig. 13.2c). If the points are fairly evenly spread over the region defined by the three foods, all were consumed in substantial proportions (Fig. 13.2d). Depending on the nature of the sample set analyzed, linear models can be used to test a variety of different hypotheses related to diet. Differences in the distribution of samples examined can be considered with respect to the status of an individual, changes in consumption patterns over time or variations due to the availability of resources in different geographic regions.

When using linear mixing models, there is an inherent assumption that changes in diet proportion will be reflected in $\delta^{13}C$ and $\delta^{15}N$ values of the animal sample tissues corrected for fractionation. The relationship between diet and stable isotope composition is actually very complex, which limits the utility of models. While stable isotope values obtained on bone apatite analyses are generally considered to reflect total diet, fractionation is somewhat variable (Schwarcz 2000). Isotopic routing, which is the inhomogeneous internal distribution of dietary constituents, complicates the relationship between tissue composition and diet (Katzenberg and Harrison 1997; Schwarcz 1991, 2000; Schwarcz and Schoeninger 1991). As noted earlier, bone collagen is primarily formed from dietary protein but there is some indication that the level of protein intake also affects the stable carbon composition of collagen (Lee-Thorp 2008; Schwarcz 2000). Schwarcz (2000) suggests the effects of isotopic routing are diminished among populations with low levels of protein intake.

The extent to which dietary models reflect the paleodiet of the individuals sampled largely depends upon the extent to which the foods selected represent the total intake. Their utility is diminished if the stable isotope values of the modeled foods are similar. For example, it may be difficult to accurately model the dietary proportions of individuals that consumed only non-nitrogen fixing C_3 plants and herbivores that ate C_3 grasses or browse. While linear mixing models primarily involve stable isotope values obtained on human tissue, Morton et al. (1991) attempted to interpret $\delta^{13}C$ and $\delta^{15}N$ values obtained on human bone collagen and carbonized cooking residues with a model that included maize, freshwater fish, and C_3 plants. Problems encountered by Hart et al. (2007) due to differential carbon content of foods suggests that this approach may not be valid for carbonized residues.

Strontium Isotope Analyses

Strontium naturally occurs in igneous bedrock and becomes incorporated into the local ecosystem through erosional processes, such as water dissolution and soil formation. Strontium in water and soil is passed up the food chain to plants growing in the area, herbivores eating the plants and carnivores preying upon the herbivores. Strontium isotope analyses are particularly valuable to the study of the movement of animals and people because the ratio of ^{87}Sr to ^{86}Sr serves as a geochemical signature for a region. Archaeological applications of strontium analysis are outlined by Sealy (2001); Bentley (2006) provides an extensive review of the technique.

There are four strontium isotopes. The heaviest, ^{88}Sr, is the most common with an abundance of 82.5%; the lightest, ^{84}Sr, is the least common. About 10% of strontium occurs as ^{86}Sr, whereas ^{87}Sr has an abundance of approximately 7%. Although strontium-87 is the β-decay product of radioactive rubidium-87, $^{87}Sr/^{86}Sr$ values are stable for a given mineral because the half-life of ^{87}Rb is 49 billion years. The strontium isotope ratio of a mineral depends on (1) the $^{87}Sr/^{86}Sr$ at the time the mineral formed, (2) the time since the mineral formed and (3) its initial content of ^{86}Sr and rubidium. The range of $^{87}Sr/^{86}Sr$ values is between 0.702 and 0.750; the lowest values are associated with young basaltic rocks with low initial Rb/Sr ratios and the highest values from very old granitic rocks with high initial Rb/Sr ratios. The $^{87}Sr/^{86}Sr$ value of river water depends upon the geology of the area through which it flows, the rate at which the underlying bedrock erodes and amounts of precipitation. The $^{87}Sr/^{86}Sr$ value for the ocean, 0.7092, arises from the weathering of landmasses from around the world. While this figure has been stable for the last 10,000 years, it previously ranged between 0.707 and 0.709. The strontium isotope ratios of marine shells and carbonates depend upon the $^{87}Sr/^{86}Sr$ of the ocean at the time of their formation.

Strontium atoms are about the same size and carry the same 2+ charge as calcium. During the formation of minerals such as calcite, dolomite, gypsum, plagioclase feldspar, and apatite, strontium sometimes appears in place of calcium. The ratio between ^{87}Sr and ^{86}Sr provides information about the environment in which these calcium-containing minerals formed. This geochemical signature can be obtained from both inorganic materials and organic substances with a mineral component. Strontium is such a large atom that the effects of isotopic fractionation are negligible; [87]Sr is only 1.1% heavier than [86]Sr.

Strontium isotope ratios are not affected by the extreme pressure and heat associated with metamorphic processes or the magmatic processes associated with volcanic activity so they are useful in provenance studies of sedimentary, metamorphic, and igneous rocks (Herz 1990; Herz and Garrison 1998). Limestone is a sedimentary rock composed of the biogenic carbonate shells of marine organisms and inorganic carbonates that precipitate from seawater. The [87]Sr/[86]Sr variation of seawater over time is known and limestone reflects the composition of seawater at the time of its formation. Marble is limestone exposed to high temperature and pressure that converts the sedimentary rock without altering its [87]Sr/[86]Sr ratio (Herz 1990; Herz and Garrison 1998). Studies indicate that particular marble deposits are internally uniform with respect to strontium, oxygen, and carbon isotope ratio composition, enabling their use to differentiate quarries (Chapter 20). The source of gypsum (alabaster), which is also deposited in marine environments, has been established on the basis of [87]Sr/[86]Sr and sulfur isotopes. Obsidian provenance studies typically employ trace element composition (Chapter 18); however, Gale (1981:49) demonstrated that precise separations of Mediterranean obsidian was possible by combining strontium isotope ratios with rubidium content or by simply determining rubidium and strontium content in the samples.

The [87]Sr/[86]Sr ratio of terrestrial organisms depends upon what was biologically available at the time of tissue formation (Bentley 2006; Sealy et al. 1991). The mineral component of tooth enamel and animal bone provides information about the specific environment in which it formed. Much of the strontium that could become incorporated into an organism is the product of bedrock weathering; in areas with multiple bedrock sources, the [87]Sr/[86]Sr ratio may be affected by differential weathering. The contribution from the atmosphere is important in mountainous regions; in regions where soils support spruce trees, strontium recycled from the litter on the forest floor may be a significant factor. The ocean water can be a significant source of strontium in coastal areas because it reaches land as spray or evaporates then falls as rain. When multiple sources of strontium are present, it is possible to use mixing equations to estimate relative contributions of each identified source within a specific area (Bentley 2006). Today, fertilizers and air pollution contribute strontium to the modern environment.

Sealy et al. (1991) described how [87]Sr/[86]Sr ratios could be used as a dietary indicator. The [87]Sr/[86]Sr is directly related to the levels in the environment. Plants that grow in a region reflect the [87]Sr/[86]Sr values of the soil and underlying bedrock, which is then passed on to consumers at different trophic levels without fractionation or enrichment. All tissues from an individual who resides within an area lacking variation in bedrock will have identical [87]Sr/[86]Sr values. If an area includes more than one rock type with differing [87]Sr/[86]Sr ratios or the individual resides in a coastal region with terrestrial [87]Sr/[86]Sr values different from that of the sea, then bone strontium isotope ratio measurements provide a measure of the relative importance of foods from each isotopic zone. If tissues that form at different stages of life are analyzed, the movement of the individual between the isotopic zones can be traced.

For archaeological applications, the average strontium isotope ratios of small wild animals that lived in the vicinity of a site at the time of occupation can be used to estimate biologically available strontium. Small wild animals do not migrate and have $^{87}Sr/^{86}Sr$ ratios that reflect local vegetation. The $^{87}Sr/^{86}Sr$ values of domestic pigs would more closely reflect the ancient human diet than domestic cattle, sheep, or goats.

Strontium to Calcium Ratios

The ratio of strontium to calcium can be useful for assessing the trophic level of an individual for paleodiet studies. Strontium is not as readily absorbed from food as calcium. Only about 10–40% of strontium is passed to the consumers, whereas 40–80% of calcium in food is absorbed. The Sr/Ca ratio of a plant will be about one-fifth or 20% of the Sr/Ca ratio of the soil in which it grew. The Sr/Ca of the herbivore will be one-fifth that of the plants that it eats; a carnivore that consumes herbivores will have a Sr/Ca ratio about 20% of the level of its diet. This process, termed biopurification, reduces the variation in both Sr/Ca and $^{87}Sr/^{86}Sr$ ratios.

Sillen et al. (1989) note that while the reduction in Sr/Ca holds between specific prey and their predators, it may not necessarily exist for all herbivores and carnivores. In particular, Sillen et al. (1989) caution against the use of human skeletal Sr/Ca as an indicator of the meat-to-plant ratio in the diet. Burton and Price (2000) report that Sr/Ca ratios only relate to dietary intake when either a pure plant or a pure meat diet is followed. If the diet is mixed, the mineral content of the plant food greatly affects Sr/Ca ratios. The amount of meat in the diet has little impact on Sr/Ca ratios if leafy vegetables with high mineral contents are eaten but has significant impact if grains with low plant mineral contents are consumed (Burton and Price 2000). Among pure herbivores, differences arise due to variable levels of strontium in plants, which are higher in shrubs and trees and lower in grasses.

Sample Selection and Processing

Measurements of strontium isotopes can be made on the inorganic, crystalline component of bone and teeth; however, tooth enamel is preferred as it has a higher mineral content. In the burial context, secondary minerals containing strontium can fill pore spaces and cracks and coat original hydroxyapatite crystals. Tooth enamel is much less susceptible to contamination because it is harder, denser, and less porous than bone apatite. Secondary minerals containing diagenetic strontium can usually be successfully removed by leaching tooth enamel with weak acid, such as 5% acetic acid.

Attempts to remove diagenetic Sr from well-preserved bone with weak acid leaching have been made however, if the apatite has undergone recrystallization, the original biogenic strontium in the mineral may have been replaced. Bone affected by microbial degradation is particularly vulnerable to high degrees of contamination. In these cases, acid leaching may begin to remove biogenetic strontium before all diagenetic material is gone. The microbial degradation of bone tends to be most severe in hot and humid environments and it may be impossible to analyze bone apatite from these areas. If at all possible, tooth enamel should be used instead of bone.

Using Strontium Isotope Ratios to Track Movements

Certain body tissues form at a particular time and provide information about the organism's diet at that phase of life (Sealy et al. 1991). While the crown enamel of human teeth forms in childhood, different teeth, and even different areas of a single tooth, mineralize at different ages. For large mammals, it may be possible to track seasonal changes by sequentially sampling dentine layers, which are deposited incrementally, and enamel, which forms along the cervical margin of the molar.

Lead Isotope Analysis

The use of lead isotope analysis for provenance studies of metal began after Barnes et al. (1978) reported ratios of ore bodies were unaffected by metal processing procedures and were retained in the finished artifact. The technique was subsequently used to examine the movement of Bronze Age metal artifacts in the Mediterranean. Gale and Stos-Gale (1992) presented a comprehensive summary of work conducted in the Aegean; ironically, this was followed shortly thereafter by a series of papers questioning the fundamental assumptions upon which the research was based (Budd et al. 1996; Pollard and Heron 1996). The principles of lead isotope analysis are presented in this section as well as potential limitations of the technique with respect to metal sourcing and potential promise to address other questions. Applications involving the study of lead in human tissues are also described. More detailed accounts of lead isotope analysis, metal provenance studies, and ancient metal processing techniques can be found in Gale and Stos-Gale (1992, 2000), Pollard and Heron (1996), and Henderson (2000).

The Occurrence of Lead in Metal

The abundances of four stable isotopes of lead, ^{204}Pb, ^{206}Pb, ^{207}Pb, and ^{208}Pb, are measured during the analysis of an artifact. The occurrence of ^{204}Pb is solely due to its synthesis at the time of the formation of the solar system some 4.5 billion years ago. While other lead isotopes have a primeval (or primordial) component, they are also products of different radioactive decay series. Radiogenic ^{206}Pb is formed from the decay of the most common isotope of uranium, ^{238}U, which has a half-life of about 4.5 billion years (Fig. 3.3). Radiogenic ^{207}Pb results from the decay of ^{235}U with a half-life of 0.7 billion years and radiogenic ^{208}Pb is the product of the decay of thorium, ^{232}Th, with a half-life of 14 billion years. In archaeological studies, isotope abundances are usually measured as ratios between pairs of isotopes, ^{206}Pb/^{204}Pb, ^{208}Pb/^{207}Pb, and ^{206}Pb/^{207}Pb.

While few artifacts are fashioned from lead, the element naturally occurs in commonly used metals. The mineral galena is lead sulfide (PbS); argentiferous galena contains silver, but sometimes only a small amount. Using ancient metal-refining processes it was possible to extract relatively pure silver with a lead content of a few percent from this ore. Copper metal extracted from chalcopyrite ($CuFeS_2$) typically contains small amounts of lead.

As outlined by Gale and Stos-Gale (1992) and Pollard and Heron (1996), the isotope composition of an ore body depends upon its age, the relative abundance of the radioactive parent isotopes, and the retention of decay products. While the ratio of ^{238}U to ^{235}U is constant, the ratio between thorium and uranium varies with rock type. Gale and Stos-Gale (1992:66) report that most ore deposits contain "vanishingly small amounts" of these elements so the isotopic composition of lead does not change after formation. Amounts of radiogenic lead in a deposit depend upon the half-life of its radioactive parent. The stable lead end product only forms if none of the intermediate daughter products in the decay chains are removed by weathering, leaching, or another process.

The geological history of a lead deposit affects its isotopic composition. Lead formed by magma directly from the mantle and lower crust contains only primeval isotopes; these are referred to as "common" or "ordinary" deposits. Anomalous lead deposits are formed by a combination of material from the mantle and subducted continental crust; these contain both primeval and radiogenic lead. Hydrothermal lead deposits are most variable in composition and contain high levels of radiogenic material. These deposits form when previously dissolved metal precipitates out of the warm water. In theory, ore bodies that formed at about the same time by similar processes in similar geological environments should have similar isotope ratios.

Changes in trace element composition that occur during the refining process make it impossible to match a metal artifact to its source (Chapter 26). Experimental results of Barnes et al. (1978) indicate that the lead isotope ratio of metal is unaltered by the metal-refining techniques employed in antiquity or any subsequent corrosion; this finding opened a new avenue of inquiry (Gale and Stos-Gale 1992; Pollard and Heron 1996). Using a combination of two-dimensional plots of ^{207}Pb/^{206}Pb vs. ^{206}Pb/^{204}Pb and ^{207}Pb/^{206}Pb vs. ^{208}Pb/^{206}Pb and trivariate stepwise discriminant function analysis, Gale and Stos-Gale (1992) were able to separate Eastern Mediterranean copper ore deposits and identify the source of metal artifacts, primarily oxhide ingots. These large (25–30 kg) copper ingots shaped like a flayed skin of an ox are found throughout the Mediterranean. Gale and Stos-Gale (1992) suggest any recycling and mixing of metals from a variety of sources is detectable and report that "smeared out" lead isotope compositions indicative of mixing were not observed in Bronze Age Mediterranean artifacts. Gale and Stos-Gale (1992) suggest that at least 20 geologically well-selected ore samples are required to characterize a deposit and report that, except in rare circumstances, the isotope composition of lead in an ore deposit varies less than about ± 3%. Only ore bodies for which there is archaeological evidence of exploitation and/or minimum metal content should be considered as potential sources. The isotopic composition of copper artifacts should be compared to those of copper deposits, not lead/silver deposits.

While garnering support from Tite (1996), the validity of provenance studies of metal using lead isotopes has been challenged. Budd et al. (1996) suggest that the limitations of trace element analysis of metals apply equally to lead isotope studies. Budd et al. (1996) questioned assumptions regarding the homogeneity of ore bodies, the validity of groups defined on the basis of statistical analyses and the assumption that isotope values of ore samples will have a normal distribution and called for publication of archaeological isotope data. Pollard (a co-author on the Budd et al. paper) and Heron (1996) expanded on these points and expressed additional concerns about the effect of high-temperature metal processing techniques. Any loss of vapor (containing primarily the lighter lead isotopes) during the refining process would result in the enrichment of the liquid phase; as a result, the refined metal would be isotopically heavier than the ore. Assumptions regarding the homogeneity of ore fields were also questioned because samples that did not conform to expectations were dismissed as outliers. Pollard and Heron (1996) also questioned the processes employed to detect mixing or recycling and suggested the isotopic homogeneity observed in oxhide ingots may indicate that extensive mixing had occurred in the past. Gale and Stos-Gale (2000) have responded to each of these concerns.

Gale et al. (1999) suggest that copper isotope analysis may be a valuable adjunct to lead isotope analysis for provenance studies. They reported variable levels of two copper isotopes, ^{63}Cu and ^{65}Cu, in geographically separate ore deposits. Smelting processes produced no statistically significant difference between the original copper ore (malachite from Zaire), the resulting metal or slag; no change was observed when metal was subjected to an additional fire-refining process. Compositional differences at the per mil level were observed in the ^{63}Cu/^{65}Cu ratio of Late Bronze Age oxhide ingots using the National Bureau of Statistics/National Institute of Standards and Technology copper isotope standards. The observed variations in isotope composition were consistent with the typological and temporal differences between the ingots.

The Study of Lead in Body Tissue

The presence of high levels of lead in human body tissue is generally assumed to be due to inadvertent ingestion. Studies may involve linking the lead in the body tissue to lead-containing artifacts; those with the same isotope ratios are considered potential sources. This approach has been used to match the stable isotope composition of lead in soft body tissue samples from members of the Franklin expedition to the lead–tin solder used to join the seams of food cans (Kowal et al. 1991).

Keenleyside et al. (1996) investigated the timing of lead exposure using the remains of other Franklin expedition members by comparing levels in the bones with different turnover rates. Keenleyside et al. (1996) demonstrated that the lead content was highest in trabecular bones with the fastest turnover rates and lowest in long bones with the slowest turnover rates, which is consistent with short-term exposure prior to death.

The findings of Budd et al. (1998) suggest that common processing procedures cause the loss of lead from tooth samples prior to measurement. Budd et al. (1998) report that the enamel within about 30 μm of the surface tends to be enriched in lead relative to levels present in core enamel and dentine. This enrichment was found in modern teeth that were never buried as well as archaeological teeth, which suggests its occurrence is due to biogenic, rather than diagenetic, processes. Budd et al. (1998) propose that the presence of lead relates to the original process of tissue formation. Cleaning processes employing acid or solvent washing remove the metal, resulting in a systematic underestimation of lead.

Oxygen Isotope Analysis

Oxygen isotope analysis involves a comparison between the heaviest, ^{18}O, and the most common, ^{16}O, isotopes; the ratio between them provides information about the environment in which a mineral formed. Minerals that contain oxygen can be examined, including carbonates, silicates, phosphates, and crystalline hydrates. Oxygen isotope analysis can be applied to both geological and biological mineral precipitates, making it a highly versatile technique. Depending upon the sample, oxygen isotope analysis can provide information about changes in the global climate over time, the source of raw material or where an animal was born. Readers should consult Aitken (1990), Herz (1990), Herz and Garrison (1998), Hoefs (2009), Kohn (1996), Kohn et al. (1996), Lowe (2001), Sharp (2007), Sponheimer and Lee-Thorp (1999), Stephan (2000), and the references therein for more detailed discussions of oxygen isotope analysis. Examples of archaeological applications are presented in Part III.

Oxygen isotope ratios are determined by comparing the ratio in the sample to that measured in an international standard, either standard mean ocean water (SMOW), Vienna SMOW, or Pee Dee belemnite (PDB):

$$\delta^{18}O = \left[\frac{^{18}O/^{16}O \text{ sample}}{^{18}O/^{16}O \text{ standard}} - 1 \right] \times 1000$$

Oxygen Isotope Fractionation

The ratio between ^{18}O and ^{16}O is affected by factors operating on vastly different scales. The relevancy of each depends on the goals of the analysis and the nature of the sample. In general, if a phase change occurs where liquid water becomes water vapor, the lighter isotope will preferentially enter the vapor phase, which causes the liquid phase to be enriched in the heavier isotope. Specific factors influencing the ratio of oxygen isotopes for the interpretation of global climate, mineral sourcing, and animal behavior are discussed separately.

Climate Studies

The amount of water on Earth is constant but the portion existing as solid ice, liquid water, or gaseous water vapor at any time depends on the climate. Oxygen isotopic fractionation occurs because the water vapor in the atmosphere and the precipitation that falls as rain and snow are overly represented by $H_2{}^{16}O$, water molecules containing the lightest oxygen isotope. When the amount of water that evaporates is equal to the amount that runs off the land and returns to the sea, the system is in equilibrium and the $\delta^{18}O$ value of the world's oceans is constant. If the climate warms, more ice melts and releases isotopically light water; this dilution causes a relative decrease in the $\delta^{18}O$ value of the ocean. When the global climate is colder, more of the isotopically light water becomes trapped on land as glaciers. The loss of the light ^{16}O causes a relative increase in the amount, or enrichment, of heavy isotopes in the ocean.

During the last glacial period, seawater levels are estimated to have dropped by 130 m and massive ^{18}O-depleted ice sheets formed on land. The difference in $^{18}O/^{16}O$ or $\delta^{18}O$ between the seawater at the coldest part of the last glacial period and the warmest interglacial period was about 0.1% or 1‰. Consequently, every 10 m change in sea level corresponds to a 0.1‰ change in $\delta^{18}O$. The $\delta^{18}O$ values of marine organisms reflect the $\delta^{18}O$ value of the ocean during their lifetime so microfossils recovered from deep marine cores provide a record of changes over time. The most significant changes in $\delta^{18}O$ values recorded in these cores are related to astronomical cycles (Lowe 2001). Prolonged periods of elevated $\delta^{18}O$ values correspond to glacial episodes; periods of low $\delta^{18}O$ values are interglacials. Interglacials are marked by abrupt decreases in $\delta^{18}O$ values while the onset of glacial episodes is characterized by more gradual decreases. Oxygen isotope stage (OIS) numbers have been assigned to warm (odd numbered) and cold (even numbered) climatic episodes starting from the Holocene, OIS-1. Oxygen isotope stages 3, 5a, and 5c are actually substages representing relatively warmer periods during the last glaciation; OIS-5e is a true interglacial. It is estimated that for each 1°C change in seawater temperature, the $\delta^{18}O$ value changes by 0.2‰.

Annual variations in $\delta^{18}O$ values corresponding to seasonal changes in sea surface temperatures and salinity are also recorded in the shells of marine organisms. Higher levels of salinity are associated with $\delta^{18}O$ enrichment (Herz 1990; Herz and Garrison 1998). Jones et al. (2002) found a seasonally related difference in $\delta^{18}O$ values of about 0.9‰ in both modern and archaeological specimens of Washington clam (*Saxidomus nuttalli*). The oxygen isotope composition, and hence the utility, of mollusk shell as a paleothermometer will be compromised if a mollusk shell contains aragonite, dolomite, or organic protein rather than pure calcite (Herz 1990; Herz and Garrison 1998).

Sourcing of Carbonate and Other Minerals

Oxygen isotope analyses provide information about the environment in which oxygen-containing minerals formed. The isotopic composition of the ocean at the time of carbonate formation affects the $\delta^{18}O$ value of marine shells, carbonate rocks, and chert that forms in association with limestone deposits (Herz 1990; Herz and Garrison 1998). When used together, $\delta^{18}O$ and $\delta^{13}C$ values contribute to provenance studies when isotopic variation exists between different sources of carbonate materials exploited by humans. Carbonate rock can form from inorganic carbonates that precipitate from the ocean, the organic carbonate shells of marine organisms, or a combination of both. De Vito et al. (2004) found two different limestone lithotypes, calcareous dolomitic and calcarentic, were distinguishable on the basis of $\delta^{18}O$ values.

Exposure to high temperature and pressure is required to convert sedimentary limestone into metamorphic marble. Unlike strontium, metamorphic processes cause the fractionation of oxygen and carbon isotopes so the temperature of metamorphism is an important variable (Herz 1990; Herz and Garrison 1998). Thick and pure marble deposits can have fairly uniform isotopic composition if (1) both the marble and its parent limestone were in isotopic equilibrium with the environment in which they formed and (2) the temperature increase during metamorphism was not too abrupt (Herz 1990; Herz and Garrison 1998). Weathering processes can alter the isotopic composition of marble surfaces because oxygen in the rock exchanges with oxygen in meteoric water. Herz (1990) reported a 0.6‰ decrease in the $\delta^{18}O$ value of marble exposed to 50 years of weathering.

Terrestrial Organisms

The oxygen isotope values of plants and animals depend upon the $\delta^{18}O$ value of water incorporated into and lost from their tissues. The number of factors influencing the $\delta^{18}O$ value of plant tissues is small compared to those affecting animals. The $\delta^{18}O$ values of plant roots and stems closely correspond to local meteoric water, but leaf water is enriched in $H_2^{18}O$ because the lighter isotope is preferentially lost during evapotranspiration. The $\delta^{18}O$ values of plants change throughout the day, peaking at midday. Plants that follow the C_4 photosynthetic pathway tend to be enriched in the heavier oxygen isotope because many C_3 plants stop photosynthesis during drought conditions. Exceptions are deeply rooted C_3 plants that can continue evapotranspiration unhindered while it is reduced in shallowly rooted C_4 plants (Sponheimer and Lee-Thorp 1999).

Kohn (1996), Kohn et al. (1996), and Sponheimer and Lee-Thorp (1999) outline the relationship between oxygen isotope values obtained on mammalian tooth enamel, the environment, as well as animal physiology and diet. Mammals are warm-blooded with constant body temperature, so the $\delta^{18}O$ value of an animal is affected by the ways oxygen enters and exits its body. Oxygen enters the body through the intake of air during breathing and the consumption of food and water; oxygen exits when the animal exhales, sweats, and urinates. Consequently, both environmental and physiological factors affect the $\delta^{18}O$ value of an animal. In this respect, the factors that affect the $\delta^{18}O$ value of an animal are quite similar to those that affect its $\delta^{15}N$ value.

Physiological factors include body size, its degree of water dependency, how it conserves water, its metabolic rate, and whether the animal sweats or pants. Water dependency refers to whether it is an obligate drinker, an occasional drinker, or drought tolerant. Drought tolerant animals obtain all necessary water from the plants they consume, so their $\delta^{18}O$ values reflect the relative humidity of the environment, whereas the $\delta^{18}O$ values of obligate drinkers depend upon the water available for them to drink (meteoric water). Diet of herbivores is also an important factor because C_4 plants are enriched in the heavier oxygen isotope. While there are some exceptions, browsers and mixed feeders within the same region tend to have higher $\delta^{18}O$ values (isotopically enriched) compared to grazers. Grazers usually drink more meteoric water and browsers may be able to obtain more of their water from leaves enriched in ^{18}O. Carnivores tend to have lower $\delta^{18}O$ values than herbivores. In general, the $\delta^{18}O$ values of animals that pant are higher than those that sweat because the water vapor exhaled when an animal pants contains a higher proportion of $H_2^{16}O$ (i.e., isotopically depleted). Animals with high metabolic rates have lower $\delta^{18}O$ values than animals with low metabolic rates. Diurnal herbivores eat plants that are isotopically enriched and should have higher $\delta^{18}O$ values than a nocturnal herbivore in the same region consuming similar foods. Animals active during the hottest part of the day lose water when they cool themselves through sweating or panting and should be enriched in ^{18}O compared to similar animals that are active at night.

The climatic factors that affect the $\delta^{18}O$ value of water were outlined above. The source of meteoric water is usually the precipitation that falls as either rain or snow and feeds local water bodies and river systems so it has lower $\delta^{18}O$ values than the ocean. The $\delta^{18}O$ value of precipitation at a specific geographic location depends upon a number of variables including temperature/climate, altitude, and distance from the ocean. Exceptions occur when the water consumed by an animal is a distantly sourced river or glacial meltwater streams (White et al. 2004). The relationship between the phosphate component of human skeletons, $\delta^{18}O_P$, and the $\delta^{18}O$ value of the meteoric water it consumed is given by the equation:

$$\delta^{18}O_P = 0.78\ \delta^{18}O + 22.7(\%o)$$

Due to the relationship between the environment and the $\delta^{18}O$ values of its drinking water, the $\delta^{18}O$ values of larger animals may provide evidence of migration and transhumance. For sedentary animals, such as domesticated animals and livestock, seasonal variation in $\delta^{18}O$ values is about 2.4‰. The value observed among sedentary human populations, such as the Maya, is slightly lower at 2.0‰. White et al. (2004) found that the variation in $\delta^{18}O$ values in tooth enamel due to nursing and weaning observed in black bears did not occur in human populations, indicating animal models may not match humans.

The ratio between ^{18}O and ^{16}O in the hydroxyapatite of teeth can provide good information about migration and transhumance. Seasonal variation in oxygen isotope ratios enables the movement of people and animals to be tracked. The information from cortical bone represents mean values over several years depending on the age of the individual and the species under investigation. Species-specific regression equations have been developed to relate $\delta^{18}O$ values obtained from animal tissue to the $\delta^{18}O$ of the water they drank.

Sample Selection and Processing

The oxygen isotope composition of mollusk shell, limestone, and marble is measured on carbonates. In studies involving human and animal bone apatite, both the phosphate and carbonate component can be targeted; some believe the phosphate component is more resistant to diagenesis. Prior to processing, the extent to which recrystallization has affected the sample should be established using a technique such as Fourier transform-infrared spectrometry (Chapter 35) in conjunction with the crystallinity index proposed by Shemesh (1990) and described in Chapter 19. A correlation between the $\delta^{18}O$ values and the crystallinity index indicates diagenetic processes have affected the sample.

When oxygen isotope analysis was first developed, fluorinating agents were used to extract and isolate phosphate from bone (Tudge 1960). O'Neil et al. (1994) introduced a three-step procedure involving (1) treatment with HF to remove calcium and dissolve phosphate; (2) addition of silver ammine and precipitation of Ag_3PO_4 crystals from the solution; and (3) production of carbon dioxide for isotopic analysis by the thermal decomposition of Ag_3PO_4 crystals at 1200°C in the presence of graphite. Kohn et al. (1996) suggest that, after bleaching, it is possible to rapidly obtain a precise measurement of oxygen isotopes in tooth enamel using laser fluorination with bromine trifluoride, BrF_3.

Stephan (2000) proposed a two-step pre-treatment process to remove organic substances and humic materials from bone. The first step involved treatment of clean, powdered bone with 2.5% sodium hypochlorite (NaOCl) to remove organic substances. The second step involved treatment with 0.125 M sodium hydroxide (NaOH) to remove humic acid.

The first step in the pre-treatment is capable of removing lipids, microorganisms, or stabilizing agents (epoxy resins and glues). Some of the animal bones collected for a study of Holocene climate had been treated with organic consolidants such as polyvinyl acetate, shellac, cellulose nitrate, and wood glue made from polyvinyl acetate (Stephan 2000). After pre-treatment, oxygen isotope values obtained from experimentally stabilized bone closely matched those on untreated controls; however, the standard deviation of values from poorly preserved archaeological bone was greater than values obtained from well-preserved material.

Hydrogen Isotope Analysis

Hydrogen isotope analysis has only recently been applied to archaeological materials, including hair, bone collagen, and turquoise (Hull et al. 2008; Reynard and Hedges 2008; Sharp et al. 2003). The ratios between the common isotope, ^1H, and deuterium, ^2H or D, are measured. Hydrogen isotope ratios are determined by comparing the ratio in the sample to that measured in the international standard, standard mean ocean water (SMOW) or Vienna SMOW:

$$\delta D = \left[\frac{D/^1H \, \text{sample}}{D/^1H \, \text{standard}} - 1 \right] \times 1000$$

The environmental factors that affect hydrogen isotope values are identical to those affecting oxygen isotopes and the two values are related (see Chapter 4). Due to temperature-related effects, δD values tend to be lower in the winter than in the summer. Plants and hydrogen containing inorganic substances, such as mineral hydrates, derive their hydrogen isotopes from local meteoric water. The isotopic value of meteoric water in a particular region depends on several factors including latitude, temperature, aridity, altitude, distance from the coast, and rate of precipitation. For this reason, δD values are suitable for provenance studies (Hull et al. 2008; see Chapter 18). In the United Kingdom, the difference between baseline δD values in marine environments is about 45‰ higher than terrestrial environments (Reynard and Hedges 2008).

Animals derive hydrogen isotopes from both food and water; Sharp et al. (2003) showed that in humans about two-thirds of the value reflects food intake and one-third is from drinking water. If an organism ingests local water and food sources, the δD of its body water is similar to that of local meteoric water. Sharp et al. (2003) determined that the average fractionation between body water and human hair is 17‰. Sharp et al. (2003) attributed the cyclical fluctuations of δD values along the length of hair from an Inca mummy to seasonal variations. The small range of variation in δD values observed in mammoth hair suggests the animal may have avoided extreme seasonal temperature fluctuations by migrating (Sharp et al. 2003).

Reynard and Hedges (2008) report an increase of 30–50‰ in δD values between herbivores and omnivores and a further 10–20‰ increase from omnivores to humans. Their value as trophic level indicators is enhanced because, unlike $\delta^{15}N$, δD values are not affected by aridity, soil chemistry, and manuring.

The hydrogen atoms in a sample that are bound to oxygen and nitrogen are easily exchanged with hydrogen atoms in the local environment, such as a laboratory, but hydrogen atoms bound to carbon are not exchangeable. It is necessary to determine and correct for the exchangeable fraction of hydrogen atoms as they have no relevance to the research question. This involves equilibrating the sample with water (Sharp et al. 2003) or water vapor (Reynard and Hedges 2008) of known isotopic composition to measure the exchangeable population. This solution is not ideal because temperature-dependent fractionation occurs between exchangeable hydrogen in the sample collagen and environmental water vapor and its effect cannot be determined (Reynard and Hedges 2008).

Conclusions

The stable isotopes of several different elements found in a wide variety of materials provide information that is of interest to archaeologists. Differences in stable carbon isotope values in most organisms are largely due to (1) the photosynthetic pathways of plants, (2) their diet, as primary, secondary or higher trophic level consumers, (3) isotopic fractionation, and (4) isotopic routing. A wide variety of environmental factors influence the value of stable nitrogen, oxygen, hydrogen, and, to a lesser degree, strontium isotopes in a sample. While concerns about the effects of refining, mixing, and recycling have been raised, lead isotopes appear to make a valid contribution to provenance studies of metal artifacts. Lead isotopes also provide information about the source of lead inadvertently ingested by humans.

The utility of the information obtained depends upon the sample, the isotopes measured, and our understanding of the processes that affect isotopic fractionation. Stable isotope analysis undoubtedly will continue to enhance our understanding of diet, mobility patterns, climate, and material source.

Chapter 14
Lipid Residue Analysis

Lipids are a broad category of compounds that are insoluble in water (Chapter 5); those of archaeological interest include fatty acids, triacylglycerols, sterols, waxes, and terpenes. Rottländer (1990) noted that lipid analysis is suitable for the study of vessel contents because they are present in virtually all human food, they have a relatively high stability with increased temperature (up to 400°C), and their decomposition from cooking temperatures is minimal, compared to carbohydrates and proteins. Over the last four decades, different instrumental techniques have been used to obtain information about archaeological lipid residues. The most commonly employed involve component separation with gas chromatography: gas chromatography with a flame ionization detector (GC), gas chromatography with mass spectrometry (GC/MS), and, recently, gas chromatography-combustion-isotope ratio analysis (GC-C-IRMS). Researchers in the United Kingdom have made extraordinary advances in the recognition of biomarkers able to provide precise identifications of archaeological residues over the last 20 years (Evershed 1993a, 2000, 2008a, b; Heron and Evershed 1993; Evershed et al. 1992, 2001). The compilation edited by Barnard and Eerkens (2007) includes examples of the use of both biomarkers and criteria based on fatty acid composition to identify lipid residues.

Residues can be characterized on the basis of fatty acid composition, the presence of general or specific biomarkers, and/or by determining the stable carbon isotope values of specific lipid components. Fatty acid compositions provide a general characterization of an archaeological residue; while the origin of a residue cannot be proven, potential sources can be eliminated. General biomarkers can clarify the source of the residue as either animal or plant. Specific biomarkers can provide solid evidence or unequivocally prove the residue identification. Stable carbon isotope ratio analysis of components separated by gas chromatography can be used to discriminate different animal sources or provide additional confirmation of the identity of a residue.

Residues preserved in the walls of pottery vessels are commonly targeted, but lipids extracted from a wide variety of contexts have been examined. Sampling strategies and the relationship between lipid accumulation and vessel function will be presented. The introduction of lipids into thermally altered rock and other materials will also be discussed.

M.E. Malainey, *A Consumer's Guide to Archaeological Science*, Manuals in Archaeological Method, Theory and Technique, DOI 10.1007/978-1-4419-5704-7_14, © Springer Science+Business Media, LLC 2011

The Nature and Occurrence of Lipid Residues

Many different types of artifacts or other site materials previously in contact with plant or animal tissue can retain traces of lipids. Lipids occur in the walls of pottery; on tool surfaces; in rocks used for stone boiling; in the soils of features, mixed with pigments as binders; in resins, pitches, tars, and waxes; and in countless other places. Lipids from some contexts are better than others for the purposes of archaeological analysis. Accumulation, preservation, and contamination are important considerations. Certain artifacts, or certain portions thereof, are more likely to accumulate lipids than others. This is sometimes related to the density of lipids and their hydrophobic nature. For example, the density of oils is less than water so these lipids form a separate layer on the surface. This greatly affects the distribution of lipids in boiling vessels. While some lipids are very stable, others readily decompose. Lipid preservation is enhanced by protection from, or minimizing exposure to, oxygen, heat, water, and sunlight (Frankel 1991). For this reason, lipids absorbed into the matrix tend to be better preserved than lipids occurring on the surface of an artifact. Lipids absorbed into pores are also less vulnerable to post-depositional contamination than lipids on the surface. Each of these factors is considered below.

Simulation experiments are an important part of the analysis of archaeological lipid residues. In addition to verifying the composition of residues from specific sources, methods of lipid introduction, accumulation, and preservation can be tested. Areas of lipid accumulation can also be correlated with the functions of archaeological materials.

Accumulation

With respect to pottery, lipids tend to accumulate in specific areas on different vessel forms and can suggest variations in food processing techniques and cooking temperatures. For this reason, patterns of lipid distribution that vary according to vessel form may be useful for functional determinations. Much of our understanding of the distribution of lipids in archaeological pottery is due to the work of Charters and associates. Charters et al. (1993) showed the lipid content of a sherd depends on the whole vessel form and the original location of the sherd on the vessel. The lipid content of 105 sherds from the rim, base, and, for larger vessels, the body of 62 vessels of various shapes from sites in the Northamptonshire, United Kingdom, was assessed.

Comparisons of lipid content were made between sherds from (1) different parts of the same vessel, (2) vessels of the same form, and (3) vessels of different form. Lipid content varied significantly between and within some vessel forms. The average lipid content for jugs and bowls fell into a narrow range, but the range observed for jars was very wide. Spouted bowls had the highest values for mean lipid content, while the bowls without spouts had the lowest. The lipid content was consistently low in sherds taken from the base of cooking vessels (Charters et al. 1993). Vessel forms with a wide range of lipid contents were interpreted as having been used for a variety of functions; those with narrow ranges of lipid contents were interpreted as serving similar functions. Charters et al. (1997) examined the distribution of plant waxes in vessel walls that occurs when cabbage is boiled. The concentration of lipids in rimsherds could be 10 times higher than body sherds and 30 times higher than the concentration in sherds from the base of the vessel.

These studies suggest rimsherds should be targeted for residue analysis. It is important to note, however, that the area of fat accumulation can extend over the entire upper third of boiling pots. Vessels are not filled to capacity and evaporative loss reduces water levels during boiling, so sherds from the neck or shoulder area are also suitable and may be more easily obtained for analysis than rimsherds. Vessel morphology, paste characteristics, decoration, and use alterations, such as the location of soot and carbonized residues, provide clues as to the function of a vessel if it is not known (Hally 1983; Henrickson and McDonald 1983; Skibo 1992; Smith 1983).

Burned rock features and cooking pits are widely distributed in the archaeological record in several parts of the United States (Ellis 1997). Thermally altered rock used in hot rock cooking may carry absorbed food residues. Depending upon the function of features, it may be possible to identify surfaces most likely to have incorporated residues and predict residue preservation (Quigg et al. 2001). Ethnographic research indicates a variety of foods were likely prepared in pits using hot rocks as heat reservoirs. Wandsnider (1997) reported that differences in food preparation relate to fat, protein, and carbohydrate content. Foods prepared in the ovens may have been dry-roasted or water may have been added so that the food was steamed. The construction of earth ovens in terms of the physical arrangement of hot rocks, insulating material, and the food within ovens is also known to have varied (Ellis 1997). On the Northern Plains, there is ethnographic evidence that some pits were lined with hide and served as receptacles for stone boiling; some of the pits in the Southern Plains may have served similar functions (Quigg et al. 2001). Features resembling platforms or beds of burned rock may have been heated and then used to broil, sear, or parch foods (Ellis 1997). If rocks were exposed to secondary heating after the introduction of food residues, thermal degradation will further change the lipid composition.

In the case of earth ovens, there is ethnographic evidence that insulating material served to absorb fat and other cooking juices lost by roasted animals (Wandsnider 1997). Thick layers of insulation between the food and the heat source may have prevented residues from reaching the rocks. In this case, soil from the pit wall at the apparent food layer should be targeted and a natural control from the same depth below surface should also be collected so that the composition of natural soil lipids can be assessed (Malainey 2007).

Lipid residues are most likely to accumulate on the working surfaces of tools.

Preservation

Oven storage is a technique widely used by food scientists to accelerate oxidative decomposition of fatty acids and estimate shelf life. Pieces of vessels containing experimental residues can be stored in an oven to simulate the effects of decomposition due to the passage of time. A temperature of 75°C maximizes the effects of oxidative degradation that would occur over time without causing thermal degradation (Malainey et al. 1999a).

A study examining changes in the relative fatty acid composition of absorbed residues over a 68-day period, supported the findings of Charters and colleagues (1993, 1997) by clearly showing the upper portion of the vessels in which foods were boiled should be targeted for residue analysis (Malainey 2007). The relative percentages of the C18:1 isomers, which are monounsaturated fatty acids, and, where relevant, C18:2, a polyunsaturated fatty acid, in the residues were plotted and logarithmic decay curves were fitted to the data.

Results showed initial levels of C18:1 isomers are highest and rate of decomposition is slowest in residues recovered from the upper portion of vessels. Extrapolations from fitted curves predict that the relative percentage of C18:1 isomers in smoked trout residues from the upper portion would remain above 15% for 25,600 days oven storage, compared to 1167 days for the average values for the cylinder. Similarly, C18:1 isomer levels in the decomposed cooking residues of pickerel, another freshwater fish, would exceed 15% for more than 19,000 days oven storage if selected from the upper portion.

In some cases, C18:2 was also preserved in residues from the upper portion of clay cylinders after extended periods of oven storage. With large herbivore meat (deer) and Spanish dagger (medium–low fat content plant) residue, extrapolations suggest C18:2 would disappear after 700 days oven storage and high levels of C18:1 isomers would be maintained throughout this period. In the case of the high fat content seed mescal, the decomposition of C18:2 caused the relative amount of C18:1 isomers to increase from about 46 to 56% over the 68-day period of oven storage. Logarithmic decay curves predict C18:1 isomer levels would peak after 190 days oven storage when the levels of C18:2 reach zero. This extraordinary preservation was attributed to the presence of antioxidants (Malainey 2007).

In order to monitor the effects of microbial decomposition, pieces of experimental boiling vessels were buried in moist soils at room temperature and residue compositions were analyzed at 2-month intervals over a period of 300 days. Tiles buried for 10 months were also stored in an oven at 75°C and the residues were compared to those from unburied controls. While the precise placement of the tile on the clay cylinder may account for some of the variation observed, the microbial activity in parkland, prairie grassland, and forest soil did not adversely affect C18:1 isomer preservation in large herbivore (bison), fish (pike), and moderate–high fat content plant (Texas ebony seed) cooking residues. After the additional period of oven storage, the degree of preservation in residues extracted from tiles buried for 10 months was often equal to or better than that in unburied controls. In most residues, 10-month burial in any type of soil produced little or no difference in the preservation of C18:1 isomers. It appears that in some cases, the burial environment was protective in that residues were less susceptible to oxidative degradation.

Contamination

The effects of two potential sources of vessel residue contamination have been examined: the soil in the burial environment and archaeological processing. Several experiments have shown that contamination of archaeological residues by soil lipids is negligible. Condamin et al. (1976) examined the extent of possible external contamination of absorbed residues in their pioneering work with Roman amphorae. Fatty acids from the interior vessel wall were compared with those from the exterior vessel wall and those in the soil outside of the amphora.

The migration of oil into the amphora walls was demonstrated by the existence of a concentration gradient. The concentration of fatty acid esters extracted from the vessel interior was eight times higher than from the vessel exterior; the concentration of fatty acids detected from the soil in contact with the vessel wall was very low compared to those found inside the vessel wall (Condamin et al. 1976:199). The researchers concluded that the fatty acids in the vessel fabric originated from its former contents and the degree of contamination of the vessel with fatty acids from the surroundings was negligible.

Heron et al. (1991) compared the lipids present in ten potsherds with lipids in soil adhering to the sherds and three control soil samples from other locations within the West Cotton site, England. The three control samples contained a complex mixture of soil lipids "characterized by a series of long-chain both odd- and even-carbon number saturated fatty acids ($C_{16:0}$—$C_{33:0}$), principally pentacosanoic acid ($C_{25:0}$), odd-carbon number *n*-alkanes (C_{21}—C_{33}) and wax esters (C_{42}—C_{56})" (Heron et al. 1991:646). Numerous other compounds were also present in very small amounts.

The soil adhering to the sherd typically contained small amounts of lipids (30–510 μg/g) whereas the amount of lipids from the sherds varied from 60 to 4800 μg/g. In one instance, the yield of lipid from the sherd was 4800 μg/g whereas the yield from adhering soil was only 80 μg/g. Even when the lipid yield from sherds was similar to that of the soil, there was marked variation in the composition of the two extracts. In general, the lipid content of soil adhering to the potsherds closely matched the lipid "fingerprints" of the control soil samples. By contrast, the sherd extracts were composed of acyl lipids, such as intact triacylglycerides and free fatty acids (Heron et al. 1991:648). Heron et al. (1991) argued this clearly demonstrated that negligible migration of soil lipids occurs during burial and attributed the absence of contamination to the hydrophobic nature of the lipid molecules. Similarly, Oudemans and Boon (1991:223) found that the total pyrolysis product profile of the peat adhering to sherds was very different from the residue samples.

The possibility of contamination or modification of the residue through archaeological processing is of wider concern. Improper handling of sherds could result in the introduction of fingerprint oils (Evershed 1993a). Ideally archaeological samples for lipid analysis should be selected in the field. To prevent the introduction of contaminants, samples should only be handled with clean tools and gloved hands. It is preferable to examine unwashed artifacts, but it is possible to extract lipid residues from samples washed only in clean water. Washing the sherd with water may lead to the loss of soluble residue components and can alter stable isotope values of residues (Morton 1989; Oudemans and Boon 1991); the addition of detergents, including baking soda, will cause some lipid components to dissolve. Visible residues may be lost if a brush is used to clean soil from the sherd.

Phthalates, which are industrial plasticizers, are commonly detected in archaeological residues (Deal and Silk 1988; Oudemans and Boon 1991:223). Possible sources of these chemicals include the plastic bags in which artifacts are stored after excavation and plastic pipette tips used to handle the residue extract in the laboratory.

The Extraction and Analysis of Lipids

Lipids are extracted from the sample with non-polar solvents, usually a mixture of chloroform and methanol. In the past, a Soxhlet apparatus was used to repeatedly wash the sample with solvent continuously recycled by evaporation and condensation (Harris and Kratochvil 1981). Following Evershed et al. (1990), most samples are now agitated with the solvent for several minutes in an ultrasonic bath. At this point the extract solution can be centrifuged or washed to remove water-soluble contaminants, both processes result in the separation of the solvent layers. Lipids in the chloroform layer are taken and the solvent is removed under vacuum and gentle heat with a rotary evaporator. This process yields the total lipid extract (TLE) because it contains fatty acids, sterols, waxes, terpenoids, and other lipids. In order to make the lipids more suitable for analysis by gas chromatography, derivatives are produced of all or specific classes in a portion of the extract.

Fatty Acid Methyl Esters (FAMES)

Fatty acids can be converted into fatty acid methyl esters in either an acidic or an alkali environment. Fatty acid methyl esters can be prepared by treating the TLE with methanol that has been combined with an acid, such as hydrochloric acid or sulfuric acid (Malainey 2008; Malainey et al. 1999), or boron trifluoride and methanol combined with a strong base, such as NaOH (Condamin et al. 1976; Deal 1990; Deal and Silk 1988; Heron 1989; Hill and Evans 1988; Patrick et al. 1985). After a period of heating in the oven at 68°C, the FAMES are recovered, dried, and then dissolved in a volatile solvent, such as iso-octane. The methylation process releases fatty acids still attached to the glycerol backbone and converts them and other free fatty acids in the sample into methyl ester derivatives by replacing the hydrogen (H–) of the carboxyl group with a methyl (CH_3–) group. FAMES are more suitable for analysis with gas chromatography because they are more volatile (have a lower boiling point) and more stable (less likely to decompose during analysis) than fatty acids.

Trimethylsilyl (TMS) Derivatives

Evershed et al. (1990) outline procedures for preparing of ester and ether derivatives by adding an excess of N,O-bis (trimethylsilyl)-trifluoroacetamide (BSTFA) to the TLE of archaeological material. The mixture is heated in the oven at 70°C for about 30 min. The resulting trimethylsilyl (TMS) derivatives can be injected directly onto the GC column or dried and re-dissolved in hexane. Derivatization involves replacing a hydrogen atom (H–) on the lipid with a TMS group, consisting of a silicon atom to which three methyl groups are attached, $-Si(CH_3)_3$.

Identifications Based on Fatty Acid Composition

Fatty acids are the major constituents of fats and oils (lipids) and occur in nature as triacylglycerols, consisting of three fatty acids attached to a glycerol molecule through ester linkages (Chapter 5). They have been targeted for residue analysis because of their insolubility in water and relative abundance compared to other classes of lipids, such as sterols and waxes. Gas chromatography (Chapter 33) is an effective and efficient method of examining fatty acids in the form of methyl ester derivatives and instruments are widely available and relatively inexpensive to operate. When using a gas chromatograph with a flame ionization detector, the identity of individual fatty acids within a sample is based on comparisons with known standards.

The degree of certainty with which the identity of the separated components is determined is enhanced through the use of gas chromatography combined with mass spectrometry (GC/MS) (Chapters 31 and 33). Molecules of the separated component typically undergo electron impact ionization and the identifications are based on the fragmentation pattern. While routine GC/MS can be used to analyze either FAMES or TMS derivatives, only TMS derivatives are suitable for high temperature GC (HT-GC) and GC/MS (HT-GC/MS) (Evershed et al. 1990).

Criteria Based on the Uncooked Reference Materials

For many years, research has focused on the fatty acid component of the lipid extracts. Attempts to identify archaeological vessel residues involved direct comparisons between the fatty acid composition of the residue and fresh foods considered potential sources. Condamin et al. (1976) compared amphorae believed to have once held oil but lacked visible residues to samples of olive oil and (unused) amphorae broken during manufacture. Fatty acids characteristic of olive oil were detected in the walls of used amphorae but not in unused amphorae.

Similarly, Deal and Silk (1988) reported that elevated levels of stearic acid (C18:0) indicated the presence of animal materials, likely moose, while behenic acid (C22:0) indicated the presence of plant oils in archaeological residues. Hill and Evans (1988) suggested palmitic (C16:0), palmitoleic (C16:1), and oleic (C18:1) acid suggested a fish or fish product origin; the presence of C18:1 and the polyun-saturated fatty acids, linoleic acid (18:2) and linolenic (18:3), suggested the presence of seed oils.

Later, attempts were made to identify archaeological residues using fatty acid ratios capable of discriminating uncooked foods (Loy 1994; Marchbanks 1989; Skibo 1992). Marchbanks (1989) defined the relative percentage of saturated fats in a sample, %S, as a ratio between two saturated fatty acids, lauric (12:0) and myristic (14:0), and two polyunsaturated fatty acids, linoleic (18:2) and linolenic (18:3):

$$\%S = \frac{C12:0 + C14:0}{C12:0 + C14:0 + C18:2 + C18:3}$$

The %S values of modern reference materials clearly separated different foods; the %S value for modern plants was less than 18%, modern fish ranged between 22 and 39% and modern land animals exceeded 47% (Marchbanks 1989:97). Marchbanks (1989) recognized that vessel residues with very different fatty acid distributions could produce similar %S values. Marchbanks' (1989) percent of saturated fatty acids (%S) criteria has been applied to residues from a variety of materials includ-ing pottery, stone tools, and burned rocks (Collins et al. 1990; Marchbanks 1989; Marchbanks and Quigg 1990).

Skibo (1992) found that linoleic acid (18:2) and linolenic acid (18:3) present in raw foods were usually not detected in the vessel walls after cooking. For this reason, Skibo (1992) characterized residues from ethnographic Kalinga rice and meat/vegetable cooking vessels using two ratios of fatty acids, C18:0/C16:0 and C18:1/C16:0, which were unaltered by cooking. It was possible to link the uncooked foods with residues extracted from modern cooking pots used to prepare one type of food; however, the ratios could not identify food mixtures. These ratios could not be applied to residues extracted from archaeological potsherds because the ratios of major fatty acids had changed with decomposition (Skibo 1992).

Loy (1994) proposed the use of a Saturation Index (SI), determined by the ratio: SI = 1 − [(C18:1 + C18:2)/(C12:0 + C14:0 + C16:0 + C18:0)]. Loy (1994) admitted, however, that poorly understood decompositional changes to the original suite of fatty acids make it difficult to develop criteria for distinguishing animal and plant fatty acid profiles in archaeological residues.

The major drawback of the distinguishing ratios proposed by Marchbanks (1989), Skibo (1992), and Loy (1994) is they have never been empirically tested on degraded material. Ratios able to discriminate food classes on the basis of their original fatty acid composition may not apply to decomposed cooking residues.

Criteria Based on Decomposed Reference Material

The composition of uncooked plants and animals provides important baseline infor-mation, but it is not possible to directly compare modern uncooked plants and animals with highly degraded archaeological residues. Unsaturated fatty acids, which are found widely in fish and plants, decompose more readily than satu-rated fatty acids, sterols, or waxes. In the course of decomposition, simple addition reactions at points of unsaturation or peroxidation might lead to the formation of a variety of volatile and non-volatile products that continue to degrade (Frankel 1991). Peroxidation occurs most readily in fatty acids with more than one point of unsaturation.

The study of fatty acid decomposition has lead to the development of criteria for residue identification. General criteria based on the relative fatty acid composition of a residue, specific fatty acid ratios, and specific degradation products which can be used to indicate the source material of archaeological residues are described below.

General Criteria Based on Fatty Acid Ratios

Although it slowly decomposes over time, monosaturated C18:1 fatty acids have proven valuable for characterizing ancient residues. Many archaeological vessel residues were likely formed by boiling, i.e., cooking foods in water at a temper-ature of 100°C. Patrick et al. (1985) was the first to demonstrate that it may be possible to characterize these types of archaeological residues on the basis of fatty acid composition if the effects of cooking and decomposition over long periods of time are understood. Patrick et al. (1985) examined brown flaky residue found on the inside of potsherds from South Africa and reported, although no polyunsaturated fatty acids were present in the residue, its composition compared favorably to Grey Atlantic and Cape Fur seals, the remains of which were recovered from the site. An experimental residue was prepared by heating boiled seal tissue until it became a thick, yellow, crusty mass. Analysis of the residue after oven storage at 55°C for 72 days and again after storage at 120°C for 17 days indicated the ratio of the 18:1 fatty acid isomers, oleic and vaccenic acid, was not altered by decomposition. The oleic acid to vaccenic acid ratio supported the conclusion that residue on pottery from South Africa was related to the preparation of seal (Patrick et al. 1985).

Evershed et al. (1997) noted several differences in the fatty acid compositions of non-ruminant and ruminant animals. Residues from pigs, which are monogastric, non-ruminants, contained only one C18:1 isomer, Z-9-octadecanoic acid. The fat of ruminants, such as sheep and cattle, contained a mixture of C18:1 isomers with double bonds located at the 9, 11, 13, 14, 15, and 16 positions. The formation of different positional isomers was attributed to the biohydration of dairy fats in the rumen. In addition, a higher abundance of C15:0, C17:0, and C19:0 and branched-chain fatty acids was observed in the residues of ruminants.

Malainey (1997) proposed criteria for characterizing cooking residues in pre-contact vessels from Western Canada, likely formed by boiling, using gas chro-matography. The fatty acid compositions of more than 130 uncooked food plants and animals from Western Canada formed groupings that generally corresponded to divisions that exist in nature when subjected to hierarchical cluster and princi-pal component analyses (Malainey 1997; Malainey et al. 1999b). Changes in fatty acid composition of experimental residues were determined after 4 days and 80 days at room temperature and again after 30 days in an oven at 75°C. Relative percentages were calculated on the basis of the ten fatty acids (C12:0, C14:0, C15:0, C16:0, C16:1, C17:0, C18:0, C18:1ω9, C18:1ω11, C18:2) that regularly appeared in Precontact period vessel residues from Western Canada. Degraded experimental cooking residues were distinguishable on the basis of the relative levels of medium chain fatty acids (the sum of C12:0, C14:0, and C15:0), C18:0 and C18:1 isomers in the sample (Malainey 1997; Malainey et al. 1999b). The identifications made under the eight criteria strongly coincided with groups generated by principal component analysis of the total relative fatty acid composition (Malainey et al. 1999c).

The criteria were expanded, refined, and generalized to accommodate foodstuffs from other regions. Categories included *large herbivore*, which includes residues from flesh of animals such as deer, bison, cow, and moose. Levels of C18:1 isomers in large herbivore residues were found to be a reliable indicator of the degree of meat fattiness. The category *low fat content plant* includes residues of roots/bulbs/tubers, greens/leaves, and berries; but decomposed cooked camel's milk produces residues of similar composition (Barnard et al. 2007). Residues with elevated levels of medium chain fatty acids and C18:0 are called *plant with large herbivore*, but include decomposed cooking residues of yucca root. The category *medium–low fat plants* includes foods with slightly higher levels of the C18:1 isomers, such as fleshy fruit of prickly pear and Spanish dagger. The category *medium fat content foods* includes fish, corn, and other animal and plant foods that produce similar residues. Foods with levels of C18:1 isomers in excess of 25% were subdivided into *moderate–high*, *high*, *very high*, and *extremely high fat content foods*. These divisions reflect differences in C18:1 isomer levels in the fatty meat of medium-sized mammals, rendered fat of mammals such as bear, and the different seeds and nuts exploited by humans in the past.

In general, an elevated level of C18:0 is associated with the presence of large herbivores, but javalina and tropical seed oils must be considered as possible sources if they were locally available. The relative amount of C18:1 isomers in the residue indicates the fat content of the material of origin. Medium and very long chain saturated fatty acids facilitate the discrimination between foods of plant origin and those of animal origin. It must be understood that the identifications given do not necessarily mean that those particular foods were actually prepared because different foods of similar fatty acid composition and lipid content would produce similar residues. It is possible only to say that the material of origin for the residue was similar in composition to the food(s) indicated.

Specific Criteria Based on Fatty Acid Ratios

Copley et al. (2001) report date palm and doum palm fruit and ancient vessel residues are characterized by unusually high abundances of C12:0 and C14:0 and low levels of C16:0 and C18:0. Graphic depictions show C12:0 levels approach 50% and C14:0 levels occur between 25 and 30% while levels of C16:0 and C18:0 are in the range of 20 and 5% or less, respectively. Relatively higher levels of C18:0 in the residue of a second vessel indicated a combination of date palm and animal fat (Copley et al. 2001).

Summary

Relative fatty acid compositions of cooking residues change over time; but experiments show these variations can be modeled. In particular, decreases in the relative amounts of C18:1 isomers in partially decomposed experimental residues strongly correlate with logarithmic curves which can be used to extrapolate further change. Better than expected preservation of monounsaturated and, occasionally, polyunsaturated fatty acids is observed in the cooking residues of certain plant and plant/meat combinations, likely due to the presence of antioxidants. With careful selection of archaeological samples for analysis, good preservation of residues is possible. The effects of microbes found in parkland, prairie, and forest soils appear to be mediated by the reduced availability of oxygen in a burial environment.

Residue Identification with Biomarkers

General Biomarkers

The presence of certain molecules or the presence and distribution of certain types of lipid provide information about the source of a residue. For example, HT-GC separates a very broad range of lipids in a single analytical run; the distribution of these components provides a general categorization of the residue. The HT-GC profile of typical degraded animal fats are characterized by a range of odd and even numbered saturated and unsaturated fatty acids, with C16:0 and C18:0 forming the major peaks, monoacylglycerols with 16 and 18 acyl carbon atoms, diacylglycerols with 32, 34, and 36 acyl carbons, and triacylglycerols with 44–54 acyl carbon atoms; a wider distribution of triaclyglycerols, with 40–54 acyl carbon atoms, occurs in dairy fats (Dudd et al. 1999). Beeswax can be identified on the basis of the distribution of *n*-alkanes, ranging from 23 to 33 carbon atoms in length, and long-chain palmitic acid wax esters (Evershed et al. 2001).

Evershed et al. (1992:199) suggest "sterols are amongst the most important of the minor components of fats and oils used diagnostically in classifying the lipid extracts of potsherds." Although they are present in only small amounts, these molecules could be used to distinguish animal-derived residues, which contain cholesterol, from plant-derived residues, indicated by sitosterol and campesterol (Evershed 1993a).

Specific Biomarkers

Biomarkers are molecules that are associated with a narrow range of substances so their presence enables a residue to be identified with a high degree of precision. Gerhardt et al. (1990) demonstrated the potential utility of biomarkers in their examination of absorbed residues from intact Corinthian figure vases. Residues were extracted by washing the interior with non-polar solvents then analyzed without saponification or conversion to methyl or TMS esters. Gerhardt et al. (1990) reported that one type of vase contained a series of straight-chain hydrocarbons in the C_{14-27} range, characteristic of flowers and other waxy plant parts. As well, an oleoresin of pine, cypress, or juniper was found, indicating the presence of perfumed oil in these vases. The major components of another vessel were cedrol and cedrene, found in cedarwood oil.

More recently, R. P. Evershed has championed residue identification through the application of the "archaeological biomarker concept." Evershed (2008a) notes that structures or distributions of molecules that appear in fresh substances and retained in decomposed residues can serve as "chemical fingerprints." Altered structures are also suitable as biomarkers if the chemical pathway connecting it to the source material has been established. In other words, a molecule formed by the thermal alteration or oxidative degradation of a specific biomarker can also serve as a biomarker (Evershed 2008a).

Altered structures able to serve as biomarkers include stable structures that form when saturated and unsaturated fatty acids are heated to very high temperatures. Raven et al. (1997) reported that fatty acids (or triacylglycerols) found in animal fats could undergo condensation reactions at temperatures in excess of 300°C. Free radicals can cause two saturated fatty acids (or other fatty acyl lipids) to lose one water molecule (dehydration) and a carboxyl group (decarboxylation) and form long mid-chain ketones, a process called ketonic decarboxylation. The resulting ketones are similar to those found in higher plant waxes, but instead of one dominant ketone, a mixture of ketones comprised of 31, 33, and 35 carbons in a 1:2:1 ratio are observed. Experiments show that particularly high levels of ketones were produced when fatty acids were pyrolyzed in the presence of a metal salt, such as CaO, and clay matrix material. Exposure to temperatures of about 800°C resulted in the formation of a variety of shorter chain secondary pyrolysis products, but these were not detected in archaeological residues. The origins of ketonic decarboxylation products can be confirmed by compound-specific stable carbon isotope analysis (Raven et al. 1997).

Different stable structures occur when plant oils and animal-based fuels are burned in ceramic lamps for illumination. Two hydroxyl groups (–OH) are added at the position of the double bond in monounsaturated fatty acids to form dihydroxy fatty acids, some of which can serve as highly diagnostic biomarkers for the original oils in the lamps (Copley et al. 2005). In particular, it is possible to recognize the use of castor oil, radish oil, and animal fat as illuminants.

If heated to temperatures in excess of 250°C, polyunsaturated fatty acids typically found in marine animals that contain conjugated double bonds can form cyclic structures and transform into ω-(o-alkylphenyl) alkanoic acids with 16–20 carbon atoms (Hansel et al. 2004). Evershed et al. (2008) experimentally established the conditions required for the formation of these cyclic compounds from unsaturated fatty acids. A variety of pure fatty acids as well as horse fat, rape seed oil, and cod liver oil were applied to powdered pottery sherds and then heated to 270°C either in the presence of oxygen or in sealed, evacuated tubes for 17 h.

No ω-(o-alkylphenyl) alkanoic acids formed when a polyunsaturated fatty acid with three double bonds (i.e., a C18:3 isomer) was heated at 120°C for 17 h, demonstrating the need for exposure to very high temperatures. Heating saturated fatty acids at high temperatures did not produce these compounds, demonstrating that ω-(o-alkylphenyl) alkanoic acids are only produced by the thermal degradation of unsaturated fatty acids. Results showed that different ω-(o-alkylphenyl) alkanoic acids form from thermal degradation of unsaturated fatty acids with one, two, and three double bonds. Any ω-(o-alkylphenyl) alkanoic acids formed at high temperature are more likely to preserve in low-oxygen environments; in the presence of oxygen, they appear to degrade into di- and tri-benzoic acids (Evershed et al. 2008). Metal ions in the silicate structure of clays appear to catalyze the reaction; consequently, the formation of cyclic structures could depend upon clay composition.

Evershed et al. (2008) note that in order to establish the marine origin for a residue, ω-(o-alkylphenyl) alkanoic acids with 18, 20, and possibly 22 carbon atoms as well as isoprenoid fatty acids should be present. The value of these cyclic structures as archaeological biomarkers stems from their stability because polyunsaturated fatty acids rarely preserve in ancient residues.

A compilation of established lipid biomarkers appears in Table 14.1. More detailed descriptions about the recognition and application of several of these

Table 14.1 Examples of lipid biomarkers

Biomarker	Source material	Chapters
Cholesterol	Animal products	14 and 23
β-Sitosterol, stigmasterol, and campesterol	Plant material	14
ω-(o-Alkylphenyl) alkanoic acids with 16–20 carbons together with 2 isoprenoid fatty acids	Marine mammal fats	14 and 21
High levels of C12:0 and C14:0 and low levels of C16:0 and C18:0	Palm fruit lipids	14
n-Dotriacontanol,	Panicoid grasses, such as maize	14 and 21
nonacosane, nonacosan-15-one, and nonacosan-15-ol	Genus *Brassica*, such as turnip and cabbage	21
Cedrol and cedrene	Cedarwood oil	21
n-Alkanes with chains of 23–33 carbon atoms and palmitic acid wax esters with chains of 40–52 or 54 carbon atoms	Beeswax	23
Moronic acid and 28-norolean-17-en-3-one	*Pistacia* resin	23
Two unidentified components with a mass spectral base peak at *m/z* 453	Heated *pistacia* resin	23
Dehydroabietic acid and 7-oxo-dehydroabietic acid	Conifer products	23
Betulin	Birch products	23
5β stanols and secondary bile acids	Animal manure	23
Steranes and terpenes	Petroleum bitumen embalming products	23
Boswellic acids and their *O*-acetyl derivatives	Frankincense	23

biomarkers appear in Part III. Examples of lipid biomarkers for foods can be found in Chapter 21; lipid biomarkers for pitches, resins, tars, and beeswax appear in Chapter 23; lipid biomarkers used to detect manuring are also presented in Chapter 23.

Compound-Specific Stable Isotope Analysis

Compound-specific stable isotope analysis using gas chromatography-combustion-isotope ratio mass spectrometry, GC-C-IRMS (Chapters 31 and 33), has proven valuable for the identification of archaeological animal and plant residues. After separation by gas chromatography, an isolated component undergoes combustion (e.g., at 850°C with CuO) and the resulting CO_2 passes into the isotope ratio mass spectrometer. Stable carbon isotope ratio analysis (see Chapter 13) can be performed on individual fatty acids in an archaeological residue, such as C16:0 and C18:0, or a biomarker. The $\delta^{13}C$ of lipids in C_3 plants native to temperate zones is significantly more negative than those from C_4 plants, which are tropical grasses such as maize, millet, sugarcane, and sorghum (Chapter 4). Smaller differences in stable carbon isotope values are due to variations in how animals metabolize food.

The potential of GC-C-IRMS to resolve the origins of animal fats was first recognized by Evershed et al. (1997). The $\delta^{13}C$ values of methyl esters of C16:0 and C18:0, the major n-alkanoic acids, in the residues were found to correlate to vessel form in Neolithic Grooved Ware dating to about 4500 B.P. The $\delta^{13}C$ values on C16:0 and C18:0 in residues from vessels interpreted as "dripping dishes" were about -27 and $-25.5‰$, respectively. This differed markedly from the $\delta^{13}C$ values on C16:0 and C18:0 in residues from vessels interpreted as lamps, which were about -29.5 and $-31‰$, respectively. In addition to differences in the magnitude of the values, the fatty acid displaying the higher degree of ^{13}C enrichment also differs, i.e., C16:0 is enriched relative to C18:0 in the dripping dishes and the reverse is observed in the lamp (Evershed et al. 1997). Differences in stable carbon isotope values observed in archaeological residues correspond to those in reference fats from non-ruminant and ruminant animals. Pigs were fed diets with a bulk $\delta^{13}C$ value of $-24.4‰$; the bulk $\delta^{13}C$ value of the cattle diet was $-29.3‰$.

Evershed et al. (1997) reported similar isotopic differences in pottery residues from the Radnor Valley in Wales. The isotopic values on absorbed and carbonized residues of Groove Ware pottery corresponded to values for reference pig adipose fats while the values on Peterbough Ware vessels were similar to reference ruminant fats.

Mukherjee et al. (2008) conducted a larger study of 222 Grooved Ware sherds from 11 sites in the British Isles. Although some residues contained evidence of beeswax and plant waxes, all were identified as degraded animal fat. Preserved tri-acylgylcerols were detected in only 19% of the residues and indicated the presence of dairy products, ruminant adipose tissue, or degraded dairy products and porcine (pig) fats. Stable carbon isotope values of C16:0 and C18:0 were obtained on 126 absorbed and 5 surface residues; 20 of these were predominantly porcine fats. Other vessels contained ruminant dairy and carcass products (alone or combined) or mixtures of ruminant and porcine products. The number of vessels containing primarily pig fat was estimated using a mixing model; vessels with residues identified as more than 75% porcine fat were considered as predominantly porcine. When compared to site function, the incidence of porcine fats was found to be much higher in vessel residues from ceremonial sites (40%) than in vessel residues from domestic sites (4%) (Mukherjee et al. 2008).

Compound-specific stable isotope analysis of C16:0 and C18:0 was able to discriminate animal sources of residues extracted from the Late Saxon/Medieval site of West Cotton in the United Kingdom (Evershed et al. 2001). A plot of $\delta^{13}C$ of C16:0 against $\delta^{13}C$ of C18:0 clearly separated the fat of sheep and cows from the fat of pigs and horses; the $\delta^{13}C$ values of the milk from sheep and cows overlapped as did the $\delta^{13}C$ values of chicken and goose fat.

The identification of maize in lipid residues has been examined by Reber and Evershed (2004a, b) and Reber et al. (2004). Maize contains the long-chain alcohol, n-dotricontanol, but it also occurs in waxes formed by panicoid grasses and insects. Consequently, an unequivocal identification of maize is not possible solely on the presence of this biomarker or on the basis of fatty acid composition. Because maize is a C_4 plant, more precise identifications are possible through the stable carbon isotope analysis of n-dotricontanol. Stable carbon isotope values between -19 and $-21‰$ suggest the n-dotricontanol was entirely or almost entirely derived from C_4 panicoid grasses, of which maize is the most probable source; $\delta^{13}C$ values on residues between -26 and $-33‰$ would indicate a C_3 plant source or mixture of C_3 and C_4 plants. Additional confirmation of maize can be obtained by determining the $\delta^{13}C$ values on C16:0 and C18:0, which would also have a C_4 signature.

Spectroscopic Techniques

Researchers have had success with the application of infrared spectroscopy (IR), Raman spectroscopy, and nuclear magnetic resonance spectroscopy (NMR) to archaeological residues. IR and Raman spectroscopy are used to detect the presence of specific groupings of atoms in the sample through the absorbance of IR light of a specific wavelength (Chapter 35). With NMR, the absorbance of radiofrequency energy by specific nuclei, usually hydrogen and/or carbon, provides information about neighboring atoms and bonds in a molecule (Chapter 36). Both techniques can be used to make unambiguous identifications of pure and pristine substances. The size and position of peaks of the sample form a chemical signature that can be matched to the reference material. In mixed samples, the techniques provide a means of verifying the presence of certain functional groups or types of bond.

Infrared and Raman Spectroscopy

Robinson et al. (1987) demonstrated general similarities between pitch and tar samples from the Mary Rose, an Etruscan shipwreck and Stockholm tar with IR spectroscopy. Hadži and Orel (1978) examined archaeological pitches derived from birch using IR spectroscopy. Regert (2007) suggests IR can be used to detect a variety of substances including pistacia and pine resins, birch bark tar, beeswax, copal, and frankincense. In the past, researchers also used infrared spectroscopy to simply screen sample extracts for the presence of organic molecules (Davies and Pollard 1988; Heron 1989; Hill and Evans 1987). Edwards et al. (1997) non-destructively examined the composition of organic resins adhering to the lids and lips of burial jars from a site in Vietnam and compared the composition to North American resin samples with FT-Raman spectroscopy; however, results from the destructive analysis of residues with GC/MS facilitated the establishment of the non-destructive FT-Raman database.

Nuclear Magnetic Resonance Spectroscopy

Nuclear magnetic resonance spectroscopy was proven useful in the characterization of pitches and tars found in archaeological contexts. Hadži and Orel (1978) used NMR to show that three archaeological pitches were derived from birch, supporting results obtained using IR spectroscopy. Sauter et al. (1987) used both ^{1}H and ^{13}C NMR to identify betulin in archaeological tar from an Austrian site, demonstrating the Iron Age tar was produced from birch. Robinson et al. (1987) used ^{1}H NMR spectroscopy to demonstrate a common *Pinus* origin for tars and pitches obtained from Tudor and Etruscan shipwrecks.

Solution-state NMR has been applied to the study of lipids derived from archaeological contexts. Cassar et al. (1983) established the composition of wax and resin seals using solution-state ^{13}C Fourier transform NMR, enabling museum conservators to repair them with the appropriate materials. They concluded that the non-metallic royal seals of King Stephen (A.D. 1135–1154) and King John (A.D. 1199–1216) were made of beeswax and the royal seal of King William IV (A.D. 1830–1837) consisted of a mixture of beeswax and resin. The NMR spectra showed that the chemical compositions of the beeswax seals were essentially unaltered over time. The spectra arising from King William's seal was more complex, containing peaks associated with beeswax as well as others similar to those found in the resin, colophony.

Davies and Pollard (1988) subjected lipid extracts of soil from an Anglo-Saxon grave to ^{1}H NMR analysis. The resulting spectra provided little information "because the amounts of residues obtained were too small and the mixture of substances making up the residues gave rise to complex multiplets" (Davies and Pollard 1988:397). It was possible to identify a fragment of a cholesterol molecule in only 1 of the 21 soil samples taken.

Although nuclear magnetic resonance spectroscopy was first applied to the study of vessel contents 20 years ago, until recently only residues in solution were studied. Beck et al. (1974) used ^{1}H NMR spectroscopy to analyze a dark brown liquid in a flask dating from the fourth to the sixth century B.C. and found that it consisted mainly of oleic acid or a soluble metal oleate. When the technique was applied to a residue dissolved in carbon tetrachloride, myristic and palmitic acids were identified (Beck et al. 1974).

Addeo et al. (1979) applied ^{1}H NMR and a variety of other analytical techniques to the methyl esters of extracts from wine amphorae. They identified cyclic diterpenes, primarily dehydroabietic and abietic acids, which suggested the amphorae formerly contained resinous wines.

Sherriff et al. (1995) applied ^{13}C CP/MAS NMR spectroscopy directly to carbonized residues scraped from globular vessels and oval plates from Northern Manitoba. By making comparisons with modern residues, they were able to demonstrate the absence of plant materials from the vessels. Most of the archaeological samples contained peaks between 65 and 70 ppm, from oxygen–carbon bonds, and strong signals at 30 ppm, due to methylene chains. The presence of these peaks was attributed to a high fat content in the residue (Sherriff et al. 1995). Sherriff et al. (1995) produced modern charred encrustations in replica earthenware vessels over open fires as well as under laboratory conditions to provide comparative standards for ^{13}C CP/MAS NMR spectroscopy and stable isotope analysis. They found that the peak positions in the NMR spectra of the experimental residues produced by charring meat and fish were most similar to the spectra of the archaeological residues (Sherriff et al. 1995). A recent application of this technique by Oudemans et al. (2007) is described in Chapter 21.

Summary

Spectroscopic techniques can be used to identify pure substances and screen for the presence of organic molecules. Evershed et al. (1992) suggest the small amount of lipid available is rarely sufficient for analysis by ^{1}H and ^{13}C nuclear magnetic resonance spectroscopy. Heron and Evershed (1993:263), however, admit that ^{13}C NMR was useful for the analysis of the carbohydrate content of carbonized visible residues.

Conclusions

The potential for valuable information to be acquired through the study of archaeological vessel residues depends upon several factors including (1) the nature and selection of the residue, (2) the degree and nature of sample processing prior to analysis, (3) the analytical technique chosen and its implementation, and (4) interpretation of the data. Almost any analytical technique will provide some information about the vessel residue, however, some types of analyses are more appropriate than others under certain conditions.

One can maximize the potential information by selecting the best technique(s) for the analysis of a particular residue in light of the research question being addressed. If an archaeologist wishes to test for the presence of specific well-preserved compounds in the residue, then infrared spectroscopy or high-pressure liquid chromatography may be sufficient. If detailed analysis of the overall lipid or molecular composition is desired then gas chromatographic techniques may be the most useful. It may be necessary to subject a residue or different components of a residue to a variety of analyses in order to characterize it fully.

Fatty acid analysis, especially when combined with general biomarkers, is a relatively rapid, inexpensive, and highly accessible method for characterizing archaeological residues. It is especially valuable when dealing with the material remains of cultures in periods for which there is no written history, such as those in North America prior to European contact. It provides an independent line of evidence that can be used with other types of information, from artifact recoveries, faunal and archaeobotanical remains, and other types of residue analyses, to paint a more complete picture of the activities of the former site inhabitants.

Specific biomarkers, including degradation compounds, enable more precise identifications and can provide information about food preparation techniques or the decompositional processes that have acted on the residue. Stable carbon isotope analysis of components separated by GC-C-IRMS can be used to distinguish animal sources of residues. These approaches are required to prove that a specific food or food combination was the source of a particular residue.

Chapter 15
Blood and Protein Residue Analysis

Protein and Blood

Proteins are giant molecules made up of subunits, called amino acids (Chapter 5). There are two major categories, fibrous and globular. Collagen is a fibrous protein and when extracted from archaeological bone is the material of choice for both radiocarbon dating (Chapter 8) and stable isotope analysis (Chapter 13). The rate at which amino acids transform from the L-form to a 50:50 mixture of both L-form and D-form enantiomers is the basis of amino acid racemization dating (Chapter 11). The techniques discussed in this section were originally only applied to residues believed to represent ancient blood. Protein residues from other sources are now examined, so the generic term "protein analysis" is used.

Archaeological protein analysis generally refers to the study of globular proteins found in blood, muscle tissue, and milk, but plant proteins have also been targeted. In the 1980s, chemical and biological techniques developed for the analysis of fresh blood were introduced as a means to detect and identify ancient proteins in archaeological residues (Gurfinkel and Franklin 1988; Lowenstein 1985; Loy 1983; Newman and Julig 1989). Screening tests involved microscopy and the use of chemicals that react in the presence of protein. Animal source determinations were made on the basis of the shape of hemoglobin crystals and techniques utilizing the immune response between a foreign protein (antigen) and antibodies. Concerns about the preservation and recovery of ancient blood proteins and the reliability of immune-based techniques grew in the mid- to late 1990s.

The various approaches employed to study protein residues associated with archaeological materials are described in this chapter. Concerns expressed about their application and recent efforts to adjust the methodologies in order to enhance the veracity of protein analysis techniques are also discussed.

Screening Techniques

Microscopy

Selection of artifacts for protein analysis often begins with examination under a light microscope. Under low magnification, the presence of residue can be established;

M.E. Malainey, *A Consumer's Guide to Archaeological Science*, Manuals in Archaeological Method, Theory and Technique, DOI 10.1007/978-1-4419-5704-7_15, © Springer Science+Business Media, LLC 2011

under high magnification, the presence of red blood cells, which form a major fraction of blood, has been reported (Loy 1983; Loy and Dixon 1998; Loy and Hardy 1992). Recent studies by Hortolà (2002) using SEM analysis verify that the integrity of red blood cells can be preserved on the surface of rock for an extended period of time. Blood was dripped or smeared on obsidian, limestone, and chert and stored from 7.5 to 10.17 years at temperatures ranging from about 11–34°C and relative humidity ranging from 38 to 84%. One sample was given limited (2 h) exposure to natural sunlight. The majority of red blood cells maintained "normal" shapes from a hematological perspective, although the effects of drying resulted in the occurrence of cells with distorted shapes (Hortolà 2002).

While microscopy cannot be used to identify the origin of archaeological residues, it can establish the presence of material suitable for analysis. Although Fiedel (1996) was critical of researchers for not submitting a random sample of artifacts for analysis, the analysis of archaeological proteins is time-consuming and a service offered by a limited number of laboratories. The use of microscopy to determine which artifacts are most likely to retain protein residues is recommended.

Colorimetric Methods

Colorimetric methods use reagents that change color in the presence of protein. One well-established test involves the compound, ninhydrin, which reacts with most amino acids to produce a purple product but in the presence of the amino acid proline, a yellow-colored product forms (Lehninger et al. 1993). Other reagents that detect the presence of proteins include 1-fluoro-2,4-dinitrobenzene, fluorescamine, dansyl chloride, and dabsyl chloride.

Loy (1983) cited a positive test to the ninhydrin-Schiff reaction as verification that an archaeological residue contained amino acids. Gurfinkel and Franklin (1988) tested seven methods, including the ninhydrin-Schiff reaction and found all methods lack specificity because they react with clay, soil, and humic acids. Gurfinkel and Franklin (1988) suggest that the false positives occur because most methods merely test the binding of the colorimetric reagent to protein by electrostatic attraction and/or hydrogen bonding (Chapter 1). Clay, soil, and humic acids all contain sites where these same types of bonds can occur.

Urinalysis Test Strips

Urinalysis test strips, or dip-sticks, sold under the names Labstix®, Hemastix®, and Chemstrip®, are a specific type of colorimetric test developed to detect blood in urine. The reaction actually involves portions of the hemoglobin molecule called heme groups. Heme contains a ring structure, called protoporphyrin IX (Fig. 15.1). This structure contains unshared (lone) pairs of electrons that bind Fe^{2+}, forming

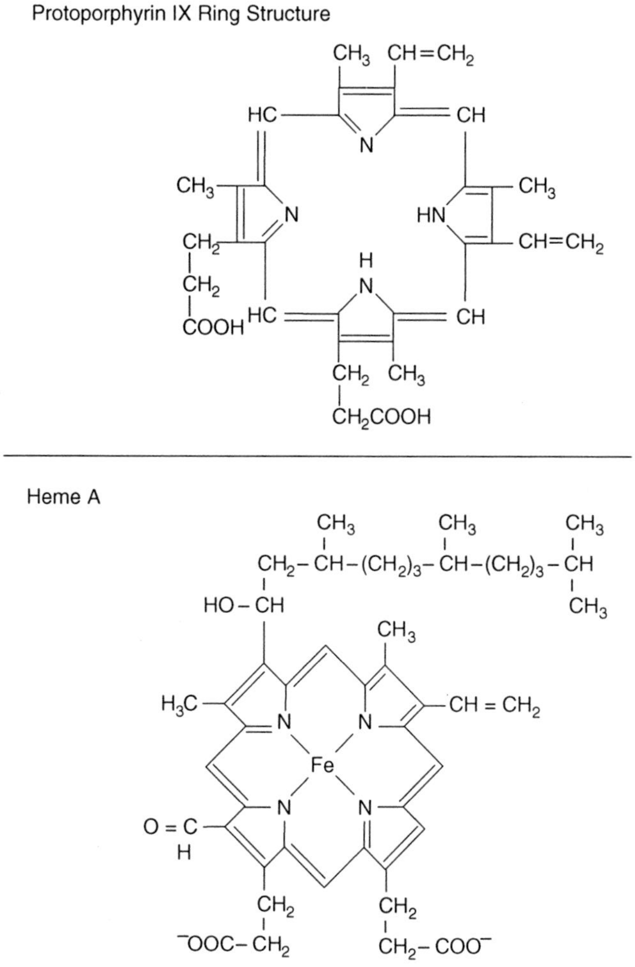

Protoporphyrin IX Ring Structure

Heme A

Fig. 15.1 Protoporphyrin ring structure and the heme A chelate complex

a Fe(II) chelate complex, which in turn binds oxygen atoms. Protoporphyrin IX is found in both hemoglobin and myoglobin.

The colorimetric dip-stick test is based on the behavior of porphyrin ring structure of heme in the presence of peroxide, which is called its peroxidase activity. Mixtures of peroxide and certain colorless compounds oxidize rapidly and change color in the presence of heme. Heme serves as a catalyst; it does not participate directly in the reaction but speeds up the reaction rate. The colorless compounds most often utilized in urinalysis test strips are tetramethylbenzidine (TMB) and diaminobenzidine; phenolphthalein is also used but considered to be less sensitive.

$$\text{colorless compund} + H_2O_2 \text{ (peroxide)} \xrightarrow{\text{Heme (catalyst)}} \text{oxidized colored compound}$$

Urinalysis test strips have been employed to screen archaeological residues for the presence of hemoglobin (Loy 1983, 1993; Loy and Dixon 1998; Loy and Hardy 1992; Loy and Wood 1989; Loy et al. 1990); the basic procedure was outlined by Loy (1983). A site on an unwashed artifact with a thick accumulation of residue but free of algae or lichen growth is selected. The residue is dissolved with a small amount (50 µL) of distilled and deionized water, then at least 20 µL is applied to the pad or dip-stick. If the color change to green occurs after about 1 min, the reaction is positive and indicates the residue should be subjected to further analysis. If there is no color change after 2 min, the reaction is negative. After 4–6 min, the reagent reacts with atmospheric oxygen, a process known as auto-oxidation, and a slight color change will occur.

The potential benefits of using urinalysis test strips for archaeological residues seemed clear; the method was fast, easy, quite cheap and testing could be performed in the field. The problem is that any compound containing porphyrin rings can catalyze the color change reaction. This includes myoglobin, which accepts the oxygen delivered by hemoglobin, and chlorophyll, which is a plant pigment involved in photosynthesis. The structure of chlorophyll is similar to hemoglobin but the porphyrin ring forms magnesium (Mg^{2+}) chelates, instead of iron chelates. All heme-containing enzymes, called peroxidases, will also react in a similar manner.

A number of substances that naturally occur in soil may also produce a color change. Gurfinkel and Franklin (1988) found that the reagent TMB reacts with clay, humic acid, compost, and lawn fertilizer. They suggest that testing residues on unwashed artifacts, as recommended, may lead to false-positive reactions. Custer et al. (1988) found that manganese oxide, which commonly occurs in the soils of eastern North America, produces a positive reaction. Copper and magnesium can also produce positive reactions. On the basis of blind comparative tests, Downs and Lowenstein (1995) reported that urinalysis test strips failed to predict subsequent responses from more definite tests to identify protein and consider them to be unreliable in archaeological settings.

Loy and Dixon (1998) suggested modifications to testing procedures to reduce the possibility of false-positive reactions. They recommend heating the sample to reduce reactions to bacterial and vegetable peroxidases and treatment of the residue with ethylenediaminetetraacetic acid (EDTA) to reduce reactions with chlorophyll, manganese, and copper. Williamson (2000) followed this procedure to directly test South African rock painting pigments from Rose Cottage Cave for hemoglobin. Two of 15 residues applied to urinalysis test strips gave positive results and were submitted for further analysis.

Techniques Used for Protein Identification

Techniques used to identify archaeological proteins fall into two categories. The first involves crystallization of hemoglobin molecules preserved in the residue. As outlined below, the technique was developed in order to make species-specific identifications of fresh blood; archaeological applications have met with skepticism. The second category includes techniques that exploit the interaction between proteins and antibodies to provide family-level identifications.

Hemoglobin (Hb) Crystallization

In addition to the four heme groups, hemoglobin consists of four amino acid chains (or protomers), which are all linked together by hydrogen bonds. The shape of hemoglobin is referred to as its closest packing arrangement or minimum free energy conformation (MFEC). The structure of hemoglobin molecules vary from species to species due to differences in the primary structure of the amino acid chains, which result in unique MFECs and higher level configurations.

Thomas Loy (1983) introduced the hemoglobin crystallization technique to the study of archaeological residues. Biologist developed the technique to identify fresh blood on the basis of species-specific hemoglobin crystal shapes. Closely related animal species do not necessarily have similar or even related hemoglobin crystal shapes. As long as the primary structure is retained, the hemoglobin crystal can be recovered. Hemoglobin, like other globular proteins, is soluble in water and the addition of salt causes globular proteins to precipitate out of water-based (aqueous) solutions. Precipitation of hemoglobin crystals is determined by several interrelated variables: (1) temperature, (2) solution pH, (3) concentration of salt solution, (4) type of salts, and (5) concentration and solubility of the protein itself. Different phosphate salts can be used for this purpose including tri-ammonium phosphate, monosodium phosphate, and dipotassium phosphate.

Loy (1983) applied this procedure to archaeological residues believed to represent ancient blood. The residue was dissolved in 80 μL of deionized and distilled water, then 50 μL was placed on a microscope slide. Small amounts of concentrated (3 M) sodium phosphate buffer (20 μL) and ammonium oxalate solution (10 μL), both at pH 6.4, were added. Small amounts of peroxide and 50% ethyl alcohol were sometimes added prior to the salt solutions. The slide was warmed to 35°C, then cooled to 22°C; a cover slip was placed over the solution and the slide was set aside. Crystals believed to be hemoglobin could develop in as little as 20 min or require up to 24 h and were identified on the basis of appearance. Examples of the use of hemoglobin crystallization to identify archaeological residues believed to be blood include Loy (1983), Loy and Wood (1989), and Loy and Dixon (1998). Although there is little evidence of independent attempts to replicate Loy's procedures, the technique has been widely criticized on a number of levels.

Smith and Wilson (1992) criticized several aspects of the species-specific identifications of ancient blood on the basis of hemoglobin crystallization. Their analyses of degraded blood using enzyme-linked immunosorbent assay (ELISA) and spectroscopic and chromatographic techniques showed that hemoglobin is not preserved intact. Smith and Wilson (1992) reported that the molecule survives without the heme group and that the four polypeptide chains were degraded. Since the unmodified molecule does not exist, it cannot be recrystallized. Smith and Wilson (1992) deem identification on the basis of crystal appearance to be unreliable and suggest that, at a minimum, axis lengths and angles of crystals are required for identification. They continue that the formation of perfect crystals is unlikely due to hemoglobin degradation and impurities in reagents. Finally, Smith and Wilson (1992) suggest that species-specific hemoglobin crystals cannot be recovered from mixtures of blood because the hemoglobin of each animal has a different solubility and crystal hybrids may develop. They challenge Loy to prove that the crystals he examined were actually hemoglobin.

Hyland et al. (1990) suggest that the crystals may actually be environmental contaminants, such as salt. Downs and Lowenstein (1995) suggest the technique is difficult to replicate and consider it to be neither objective nor sensitive. Tuross et al. (1996) argue that hemoglobin crystallization was discredited in the 1950s and simple, less expensive tests should not be applied to the study of ancient blood residues on lithic material.

Rebuttals to these criticisms of hemoglobin crystallization appear in Loy and Hardy (1992) and Loy and Dixon (1998). Loy and Hardy (1992) used high-pressure liquid chromatography to examine archaeological blood, 6-year-old caribou blood, and fresh blood. They reported that after 90,000 years, 13% of the serum albumin and 4% of hemoglobin molecules were preserved in a stone tool residue. These intact hemoglobin molecules in the degraded residue were available for recrystallization. Loy and Dixon (1998) argue that hemoglobin crystals are very different in form from salt crystals, so there is no chance of confusion. They continue that most crystals that precipitate out of solution are very small, ranging from $100 \ \mu m^3$ or less to $750 \ \mu m^3$, so the imperfections that occur in large crystals are avoided. Loy and Dixon (1998) suggest it was possible to identify extinct species using reference hemoglobin crystals grown from frozen or desiccated tissue or bone.

Using Immunological Methods to Characterize Blood Proteins

Immune response techniques harness the action of certain blood proteins, called antibodies, to bind potentially harmful foreign substances, called antigens. Specific antibodies differ with respect to their hypervariable regions; specific antigens differ with respect to structures called epitopes. The hypervariable region ensures that an antibody only binds to an antigen carrying a particular epitope. This interaction is so specific that the source of the antigen can be determined by testing antigens against a variety of antibodies. Proteins in the archaeological residue serve as the antigen, which is exposed to antibodies from known animals. In theory, a positive immune response reaction indicates the residue represents the blood of that particular species or one closely related. The most common applications examine residues from stone tools believed to be the ancient blood of prey animals. The technique has also been used to identify human tissue proteins from pottery vessel residues. Crossover (or counter) immuno-electrophoresis (CIEP), enzyme-linked immunosorbent assay (ELISA), radioimmunoassay (RIA), dot-blot tests, and related techniques use the reaction between antibodies and antigens.

Immunology methods target the other major fraction of mammalian blood, plasma proteins. Plasma proteins of vertebrates include serum albumin and alpha (α), beta (β), and gamma (γ) globulins. Gamma-globulins, also known as type G immunoglobulins (IgG) or antibodies, are produced by plasma cells in the lymph system as the body's first line of defense against an attack by a foreign substance. The foreign substances, called antigens, are usually proteins or polysaccharides in the form of viruses and bacteria. The presence of antigens stimulates the production of specific antibodies to neutralize those particular attackers; this is called an immune response. Studies of plasma proteins in archaeological residues often exploit the specific nature of the immune response process.

The antibodies defend an organism by binding the potentially harmful antigens that stimulated their formation. Antibodies usually possess two or more binding sites believed to complement structural features of the antigen. A three-dimensional lattice, called precipitin, is produced when antibodies attach to antigens (Fig. 15.2). After a certain amount of growth, it precipitates from the serum. Precipitin

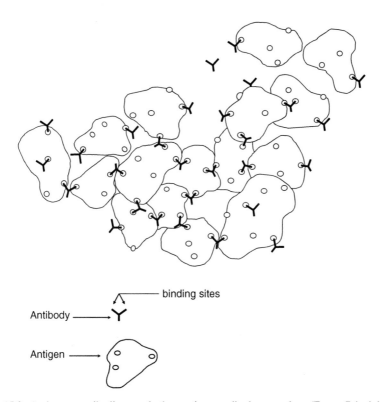

Fig. 15.2 Antigens, antibodies, and the antigen–antibody complex (From Principles of Biochemistry, 2e by A.L. Lehninger, D.L. Nelson, and M.M. Cox. © 1993 by Worth Publishers. Used with the permission of W.H. Freeman and Company.)

formation is an indicator that specific antibodies recognized and reacted to specific antigens.

Because each organism produces specific proteins, the injection of a small amount of protein from one animal into a different animal species stimulates the formation of protective antibodies. Once produced, the newly formed antibodies provide the animal with immunity to that particular antigen; this is the theory behind vaccination. A vaccine contains the antigen required to stimulate the production of antibodies needed to make the recipient immune to a particular disease.

In theory, it is possible to produce antibodies to any protein and then use them to test for its presence. For example, the immune system of a rabbit injected with whole bison blood produces antibodies for all proteins present, including hemoglobin, serum albumin, and globulins. Blood from the inoculated rabbit is now anti-bison antiserum; if mixed with bison blood, an immune response is triggered. Because antibodies for a variety of proteins are present, the serum is called polyclonal. In published papers, these types of antisera are typically identified by family only (e.g., anti-deer antiserum). In general, antisera that contain polyclonal antibodies are raised in living animals. Monoclonal antisera contain antibodies for only one specific type of protein and are grown in cell cultures. These types of antisera are identified by both animal family (or other taxonomic designation) and protein (e.g., anti-bovine albumin).

Antibodies are highly specific to the foreign proteins that evoked their formation. Antigens altered by denaturation or otherwise modified will not react with the antibody. Functionally different proteins from any single species lead to the formation of different antibodies. Anti-bison hemoglobin does not react with either bison albumin or bison immunoglobulin; however, functionally similar or homologous proteins may cause a reaction. Anti-bison hemoglobin antibodies react maximally with bison hemoglobin and are less reactive to hemoglobin of other animal species. The strength of the reaction with antibodies for functionally similar proteins depends upon the degree of similarity between epitopes from different animal sources. Homologous proteins of closely related animals will cause stronger reactions; reactions from proteins of more distantly related animals will be progressively weaker. Anti-bison hemoglobin will react to hemoglobin from cow, elk, and deer more strongly than to bird, rodent, or fish hemoglobin.

Detecting the Antibody–Antigen Complex

All methods utilizing the immune response identify an archaeological residue by exposing it to a variety of antisera containing either monoclonal or polyclonal antibodies for different animals. If the antibody–antigen complex forms, the reaction is considered positive for that animal. Several different techniques can be employed to detect positive reactions. The simplest techniques use direct visual inspection for evidence of precipitin while others use indirect methods to amplify and quantify the strength of the reaction.

Gel Diffusion–Ouchterlony (OCH)

Gel diffusion or the Ouchterlony method is the oldest and simplest method of detecting immune response reactions. The archaeological residue and antisera are placed in opposing wells in an agar gel plate and left at room temperature for several hours. The residue solution and antisera gradually diffuse through the gel and come into contact. Precipitin appears as a milky line if the reaction is positive and the archaeological residue is judged to contain the proteins targeted by the antisera. The reaction must be very strong in order to be detected with the unaided eye. Downs and Lowenstein (1995) used the Ouchterlony method to examine archaeological materials.

Crossover (or Counter) Immuno-electrophoresis (CIEP)

The first application of CIEP to characterize archaeological residues was by Newman and Julig (1989). The technique is similar to OCH but uses gel electrophoresis (Chapter 33) to drive the antigen and antisera together. Rows of paired wells are cut into an agarose gel. Residue solutions (i.e., the antigens) are placed in wells closest to the negative cathode and antisera are placed in wells closest to the positive anode. The gel is placed in an electrophoresis tank buffered at pH 8.6 and a 130 V electrical current is applied. Antigens are drawn toward the anode and the antisera move to the cathode (Fig. 15.3). Observation of precipitin formation is evidence of a positive reaction. Color enhancement by gel staining and magnification is used to detect and assess the strength of the reaction.

Positive reactions can be assessed with a test for non-specific proteins, which checks for interactions between the sample and the blood of an animal that was not immunized (pre-immune serum). For example, if the anti-deer antiserum was raised in rabbit, the residue sample is tested with the blood of a rabbit that was not inoculated with deer protein. If precipitin forms, the original reaction could be a false positive and may not be based on the immunological specificity of the antibody. This procedure has been extensively applied to archaeological residues (Downs and Lowenstein 1995; Kooyman et al. 1992, 2001; Newman and Julig 1989; Newman et al. 1996; Yohe et al. 1991) but was extensively criticized in the 1990s (see below). Recently proposed revisions to CIEP protocols are described at the end of this chapter.

Absorption Methods

With absorption methods, the test for an immune response is performed on a solid medium, rather than a gel. Typically, the residue is placed in a small nitrocellulose or plastic micro-titer (titration) cup (Fig. 15.4). Antigens in the residue adhere to the solid membrane, which is then washed to remove any unbound antigens. A

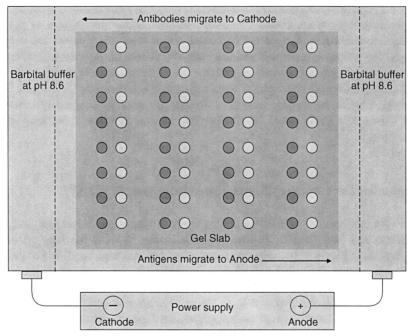

Fig. 15.3 Crossover immunoelectrophoresis (Adapted with permission from M. Newman and P. Julig, 1989, Figure 4, "Canadian Journal of Archaeology" 13:123.)

milk blocking solution is applied to prevent antibodies from binding to the solid, then antiserum is added. In the case of a positive reaction, the antibodies bind to the antigen; if the reaction is negative, the antibodies will remain unattached. The membrane is washed again to remove any unbound antibodies. Another approach is to sandwich the antigen between one antibody bound to the membrane and one or more antibodies.

Antibodies can be joined (conjugated) to materials that enhance their detection, such as color-producing enzymes, colloidal gold, or radioactive isotopes. Most antigens and antibodies have multiple binding sites so more than one antibody can bind to a single antigen. A single positive reaction may involve two or more antibodies with conjugated material, making their detection easier. If desired, a second or even a third antibody can be used to detect the presence of the first bound antibody; only the final antibody applied has the conjugated material. By using this approach, a single positive reaction is indicated by several antibodies that carry the conjugated material. Absorption methods are regarded as more specific and more sensitive than gel electrophoresis methods. The reaction is assessed on the basis of the presence and amount of conjugated material.

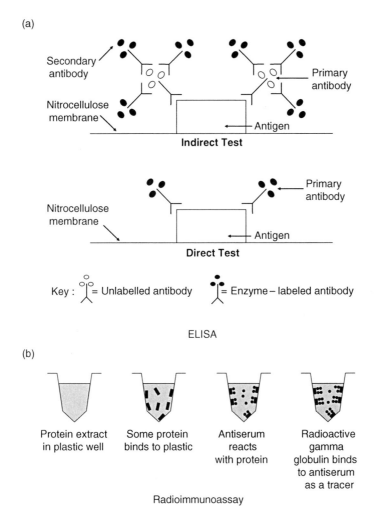

(a)

Secondary antibody

Primary antibody

Nitrocellulose membrane

Antigen

Indirect Test

Nitrocellulose membrane

Primary antibody

Antigen

Direct Test

Key : = Unlabelled antibody   = Enzyme – labeled antibody

ELISA

(b)

Protein extract in plastic well | Some protein binds to plastic | Antiserum reacts with protein | Radioactive gamma globulin binds to antiserum as a tracer

Radioimmunoassay

Fig. 15.4 Absorption methods for detecting the immune response (a) Reproduced by permission of the Society for American Archaeology from American Antiquity, volume 55(1), 1990. b) Reproduced from La génétique des fossiles, Jerold M. Lowenstein, p. 1268, n° 148, vol. 14, octobre 1983 with the permission of Sophia Publications/La Recherche.)

The specific type of conjugated material linked to the antibody gives rise to the name of the test. When the antibody is conjugated with a color-producing enzyme, such as horseradish peroxidase, the technique is called ELISA. The membrane is either examined (read) under ultraviolet light with a wavelength of 400–450 nm or a color-developing solution is added. The presence of a specific color, deep purple in the case of horseradish peroxidase, indicates the assay is positive. The intensity of the color provides a qualitative measure of the strength of the reaction. Gold immunoassay (GIA) is the term applied to tests involving antibodies linked to colloidal gold. The particles of gold are enhanced with silver to improve their visibility. Each particle of gold serves as a nucleation site and binds several atoms of silver, so positive reactions appear as dark spots. Again, the color intensity provides a qualitative indication of reaction strength. Radioimmunoassay (RIA) or protein radioimmunoassay (pRIA) is similar, except that a radioactive label, such as the radioactive isotope of iodine (^{125}I), is conjugated to the antibody. Radioactivity is directly measured with a scintillation counter (see Chapter 8). The level of the radioactivity provides a quantitative measure of the strength of the reaction. Single and multiple antibody tests are described below. After each step involving the addition of an antibody or antigen, unbound material is washed away and a blocking solution is applied.

Single Antibody Tests

During single antibody tests, a drop of the sample solution is immobilized on a solid membrane and tested against one antibody. These dot-blot tests have been used to demonstrate the mammalian origins of residues believed to be blood (Loy 1993; Loy and Hardy 1992; Loy and Dixon 1998; Loy et al. 1990). The test uses the broad-spectrum antibody, staphylococcal protein A (SpA), to detect the presence of mammal IgG in the sample; however, the blood of domestic chickens is known to produce false-positive results. The SpA antibody can be conjugated with either horseradish peroxidase or colloidal gold (Loy 1993; Loy and Dixon 1998; Loy and Hardy 1992).

Multiple Antibody Methods

With multiple antibody methods, the formation of complexes between the first antibody and the antigen is detected using a second or third antibody conjugated to the enhancing material. These tests can detect antigens at the nanogram or picogram level, making them 100–10,000 times more sensitive than CIEP. The simplest multiple antibody method involves immobilizing sample proteins onto a solid medium, then adding the first antibody to detect the targeted protein. The second antibody typically detects the animal in which the first antibody was raised. For example, if the first antibody was raised in rabbit, the second antibody may be anti-rabbit antisera raised in goat. If desired, a third antibody is used to detect the second. Immunological analysis of archaeological and/or experimental samples using two antibodies (double antibody tests) has been employed by Hyland et al. (1990), Cattaneo et al. (1990, 1993), and Tuross et al. (1996) using ELISA, Downs and Lowenstein (1995) using RIA, and Eisele et al. (1995) using GIA with silver enhancement.

Sandwich ELISA Tests

A variation on the standard multiple antibody test involves "sandwiching" the sample between antibodies. Instead of binding the antigen to the membrane, the first antibody is bound to the solid medium. The sample solution is added and if any complementary antigens are present, the first antibody "captures" them. A second antibody is added to detect the antigen, then a third antibody is added to detect the second antibody.

Marlar et al. (2000) used a sandwich ELISA test to examine archaeological residues and control samples from pottery and stool samples. First, the captured antibody, immuno-purified rabbit anti-human myoglobin, was applied to the solid medium. This particular antibody was selected because it did not cross-react with other potential food source animals. A second antibody, mouse monoclonal anti-human myoglobin, was added to detect the antigen. Finally, a third antibody, sheep anti-mouse IgG conjugated to horse radish peroxidase, was added to detect the second antibody. Each sample was tested three times by two different people, for a total of six repetitions. Positive reactions were used to support the identification of Anasazi cannibalism in the American southwest.

The Application of Immunological Methods to Archaeological Residues

While the specificity of the immune response provides a means of identifying an antigen, there is some debate as to whether epitopes preserve in archaeological proteins. The specificity of reactions has been questioned due to the apparent occurrence of cross-reactions. Inaccurate identifications have been reported when experimental residues were analyzed in blind tests. In some cases, identifications based on archaeological residues differ markedly from the composition of faunal assemblages recovered from the site. These issues are discussed in this section.

Survival of Proteins

In order for the antibody–antigen complex to form, the antibody must be able to recognize the protein as the targeted antigen. Different opinions about the preservation of proteins have been presented. Lowenstein (1985) argued that identification of ancient proteins is not governed by actual age, but the ability of the protein to retain source-identifying traits. He reported the detection of proteins in fossil hominid specimens, including an *Australopithecus* specimen dated to about 1.9 million years ago. Later, Lowenstein and Scheuenstuhl (1991) suggested that immunological methods, like RIA, could readily detect degraded protein fragments as long as they retain some source-specific conformations. Newman and Julig (1989) and Newman et al. (1996) contend that antibodies bind to antigens even if protein denaturation has occurred. While the protein may not be in tertiary form, the epitope (antigenic determinant) is preserved.

On the other hand, Downs and Lowenstein (1995) suggest that particular amino acid chains may break up; these modified (partially degraded) proteins may react to tests in unpredictable ways. In a highly critical review of the literature, Eisele et al. (1995) contend that antibodies raised against native protein do not react with the denatured form. Immunological methods may give misleading results when used with poorly preserved material. At best, immunological methods should be considered a screening procedure. Amino acid sequencing of preserved proteins is necessary for verification of results.

Survival of Blood Proteins

Other researchers have turned to experimental archaeology approaches where modern lithic tools are coated with animal residues and subjected to various treatments. Experimentally generated artifacts have been buried in soil, exposed to ultraviolet light, and/or simply stored in laboratory environments for a period of time.

Cattaneo et al. (1993) examined the effect of burial environment on (1) bovine and human bone, (2) lithic tools used to scrape or cut fresh bovine material, and (3) lithic flakes coated with blood. The materials were either buried in garden soil, buried in chalk soil, or stored in the laboratory. The samples from the buried caches were periodically exhumed and tested for bovine albumin or human albumin and immunoglobulin using ELISA with monoclonal antibodies; the longest storage time was 28 months. Proteins were only recovered and identified from bone and tool residues in a small fraction of cases. Overall, the success rate for identifying proteins from material buried less than 2.5 years ranged from 15 to 33%. Proteins were well preserved in blood-coated flakes stored in the laboratory.

Eisele et al. (1995) examined blood stored in a clean, closed bottle in a laboratory for 8 years. Immediately after the bottle was open, the immunological and enzymic activity was determined to be 30–40% that of fresh blood; after a few months, the activities dropped to 10%. In their tests using replica stone tools coated with blood, Eisele et al. (1995) found that the blood experienced considerable degradation during storage. Immunological activity of blood on tools buried in damp soil lasted less than 1 month but lasted 10 months in dry soil. Exposure to sunlight did not seem to affect the immunological activity of the blood.

Tuross et al. (1996) examined residues extracted from four flakes used to dismember a goat; each flake was used for a different butchering step. Two of flakes were also exposed to ultraviolet light to simulate sunlight exposure equivalent to several weeks or a few months prior to burial. They found that serum albumin is preserved in the residue but IgG is not. Tuross et al. (1996) suggested that the blood residues on flakes exposed to sunlight did not react because ultraviolet light destroyed the epitopes, their explanation was not verified through the use of controls.

Contrary to these reports, Craig and Collins (2002) demonstrated that epitopes could withstand conditions that cause protein denaturation and degradation. Bovine serum albumin heated for 7 days at 85°C retained the same immunoreactivity as the fresh protein.

The Question of Reliability

Another criticism concerns the applicability of tests designed to examine fresh blood in a clinical environment to archaeological materials assumed to be ancient blood residues. Rigorous tests to ensure the material is actually blood or verify that the result represents a true positive, rather than a false positive due to reactions with other substances, are generally not performed.

Loy et al. (1990) reported an accelerator mass spectrometry date of rock art site DMP6 from Laurie Creek, Northern Territory, Australia. The material was described as a brown-red "pigment" that had naturally exfoliated from a rock shelter. Under microscopic examination, hematite, silicate, and carbonate were observed. The material was dissolved in water and applied to an Ames Hemastix® urinalysis test strip and gave a positive reaction for the presence of blood. The solution was then subjected to a dot-blot test using SP-A and produced a positive identification for IgG. The substance was then examined with a sandwich ELISA test. The first antibody was supposed to bind primate serum albumin in the residue; the second antibody was monoclonal for human serum albumin. The test was positive for human blood. The AMS radiocarbon date on the material was 20, 300 +3100/−2300 B.P.

Nelson (1993), the radiocarbon scientist who analyzed the Laurie Creek sample, was concerned that the carbon and nitrogen concentrations (Chapter 38) were atypical for protein, so he collected and re-examined a second sample from the same sandstone rock face. The red mineral "skin" was 17–18% carbon and nitrogen levels were less than 0.5%, giving it a carbon to nitrogen (C/N) ratio greater than 36. Authentic proteins are 45–55% carbon and 15–20% nitrogen, resulting in C/N ratios of 3. The ninhydrin reaction result showed the total protein concentration in the sample was 1–2%. The stable carbon isotope value ($\delta^{13}C$) of the sample was −22‰; the same value obtained on wallaby bone was −9.3‰. He (1993:894) concluded that the substance was not human blood, or even proteinaceous, suggesting instead the colored layer "may be nothing more than dissolved cement re-precipitating on the surface." Without a connection to human activity, the date lacked archaeological meaning (Nelson 1993).

Cross-Reactions and Antibody Selection

Unexpected results to immunological tests of archaeological materials, are often attributed to cross-reactions. Newman and Julig (1989) indicated that antisera used for CIEP were diluted to the lowest point at which a positive reaction was obtained with control blood, usually 1:10 or 1:20. Even after dilution, they found that cross-reactions could still occur, that is, antisera for members of the family Bovidae (bison) could react with the blood of a member of the family Cervidae (deer). Cross-reactions were used to explain positive reactions of stone tool residues from Head-Smashed-in-Buffalo Jump to anti-elk antisera (Kooyman et al. 1992). Tuross et al. (1996) tested the ability of antibodies to discriminate between animals of different species, family, or even order. Antisera for goat, pig, horse, and cow diluted at 1:1600 were all equally reactive to goat blood; only anti-goat antisera was maximally reactive after a 1:2000 dilution. Had the researchers not known the animal of origin and not tested the blood residues with goat antisera, samples could have easily been misidentified as cow, pig, or horse (Tuross et al. 1996).

Some have suggested that problems with immunological methods of blood residue identification may be reduced or eliminated through the use of monoclonal antibodies. In support of this proposal, Cattaneo et al. (1990) obtained positive results on archaeological human bone using monoclonal antibodies for human serum albumin. Positive results were obtained on bone from 9 of 12 early Saxon (A.D. 450–600) individuals and two medieval adults (A.D. 1100–1400). The promise of monoclonal antibodies to resolve blood identification problems is not universally accepted. Loy et al. (1990) used monoclonal antibodies to identify the controversial "pigment" in Australia. Eisele et al. (1995) argued that even monoclonal antibodies exhibit cross-reactions. Tuross et al. (1996) questioned the application of monoclonal antibodies for the analysis of archaeological materials. Tuross et al. (1996) contend that monoclonal antibodies usually recognize higher order (conformational) determinants in proteins, not primary structure (sequential determinants), and that native protein is unlikely to be preserved in archaeological samples.

Blind Tests

Downs and Lowenstein (1995) conducted blind tests of three archaeological blood analysis techniques: OCH, CIEP, and RIA. After confirming the basic reliability and accuracy of all techniques using modern blood controls, 30 residues from tools, soils, mummy skin, and shell from New World sites dating from about 40,000 B.P. to A.D. 1300 were examined. Some material was gathered from museum collections while other samples were recently excavated; modern controls were also tested. About 80% of the ancient samples tested negative for blood; however, there was no consensus with respect to the identity of those that tested positive. No clear positive reactions were observed using the OCH. CIEP produced twice as many identifications as RIA, including some obvious mistakes; the skin from a human mummy tested positive as rabbit. None of the identifications made using RIA matched those made using CIEP. A substance that gave a weak reaction to human antibodies was later identified as copal incense; the true identifications of other archaeological materials were not established. The researchers concluded that archaeologists should be wary of blood residue identifications.

Leach and Mauldin (1995) prepared 54 freshly knapped experimental stone tool residues by cutting or scraping the flesh of one of ten animal species for several minutes. After drying, visible hair was removed by rubbing the tool with cloth or sand; some samples were also exposed to heat. After 1 to 2 months of refrigerator storage in plastic bags, the tools along with "blanks" without residues and fire-cracked rock from a hearth used to cook rabbit were submitted for analysis. CIEP was conducted using polyclonal antibodies for a variety of mammals, two birds, and six plants. Only 20 identifications were deemed correct or partially correct; another nine were judged to be partially wrong or wrong. No blood was detected on 25 tools that carried residues. Some misidentifications could be attributed to cross-reactions between closely related animals, but others were well beyond the level of family. The success rate of CIEP was judged to be 37%, but approached 50% if samples exposed to heat were excluded.

Blood Residues vs. Faunal Assemblages

Fiedel (1996, 1997) presented a highly critical case-by-case assessment of all methods used to analyze archaeological blood residues. The techniques were criticized for producing unexpected identifications, such as mouse blood on a large spear point and elk blood at a large bison jump site; archaeologists were criticized for their attempts to reconcile bizarre results. A number of revisions to protocols were suggested including the use of antisera from both probable and improbable sources, such as elephant or kangaroo. In order to test the methodology, Fiedel (1997) recommended that a random sample of tools should be submitted for analysis instead of only archaeological materials believed to carry blood residues.

Newman et al. (1997) responded that the residue identifications were credible. The results of the blind tests may have been affected by decomposition while the samples were in transit to the lab. They recommended the use of pre-immune serum to test for non-specific protein reactions.

Recent Studies

Shanks et al. (1999) proposed and employed new protocols to improve the confidence in identifications made with CIEP. They used only forensic quality antisera obtained from professional laboratories. The sensitivity of each was established by serial dilution to determine the concentrations at which the antiserum accurately identified a known amount of protein. Cross-reactivity analysis was conducted to identify inconsistencies in antisera specificity so that cross-reactions could be distinguished from true positives. If the antiserum reacted with the whole blood of animals in the same family taxon, the range at which cross-reactions occurred was established through serial dilution. Both soil samples taken from the site and artifact residues were tested with pre-immune serum.

Procedures used to recover blood residues from mineral surfaces have recently been considered. Shanks et al. (2001) examined the recovery of blood preserved in micro-cracks in obsidian micro-blades after surface washing. Shanks et al. (2001) found that between 60 and 80% of fluorescently labeled blood residues could be recovered by either agitating the specimen in 4 M guanidine hydrochloride for 18 h or a combination of sonication (3 min) and agitation (30 min) in 5% ammonium hydroxide. Shanks et al. (2001) found that the extraction procedures employed by Hyland et al. (1990) and Eisele et al. (1995) were ineffective. Shanks et al. (2004) report that even after extensive and rigorous surface washing, sonication with 5% ammonium hydroxide recovered sufficient amounts of both protein and DNA for analysis. Craig and Collins (2002) report that extended baking (85°C for 168 h) causes blood serum albumin to become so tightly bound to mineral surfaces (finely ground ceramic) that it cannot be effectively removed with any of the seven techniques previously used to extract blood from stone tools.

Recently, Reuther et al. (2006) reported that the reliability of RIA was markedly improved simply by using a different iodine 125-labeled second antibody. Blood residues on experimental obsidian tools were identified in blind tests conducted in 2000 and 2004. In 2000, polyclonal antisera raised in rabbits were detected with radioactive goat anti-rabbit gamma-globulin; in 2004, radioactive donkey anti-rabbit gamma-globulin was used. The 20 specimens examined in 2000 produced 40 significant reactions. Of these, 75% were due to cross-reactions; only 25% produced positive identifications to the taxonomic level of class, order, or family. By comparison, 39 of the 42 (93%) significant reactions were correct in 2004; 31% of these identifications were made to the taxonomic level of genus. One of the 43 residues reacted with both anti-sheep and anti-musk ox antisera; the blood of both animals was actually present.

Conclusions

Since the 1980s, a number of techniques have been used to screen for, or identify, proteins in archaeological residues believed to represent ancient blood. While successful applications were reported, all techniques have encountered controversy. Colorimetric tests for protein determination and urinalysis test strips for hemoglobin have been largely abandoned due to reactions with substances found in soil. Although hemoglobin crystallization of archaeological residues has never been independently tested, it is unlikely to be resurrected. The adoption of new extraction protocols, soil controls, and tests for cross-reactivity may restore confidence in identifications using immunological techniques.

The findings of Shanks et al. (2001) support the position of Newman et al. (1996) that the poor success rate reported by Eisele et al. (1995) stems from the extraction process they employed. The choice of extraction technique by Hyland et al. (1990) may have also affected the analysis of Shoop site material. Although Hyland et al. (1990) regarded their analysis of blood residues as successful; only 1 of the 45 artifacts examined produced a positive test. Suggestions made by Fiedel (1997) with respect to the use of antisera from improbable sources and analysis of tools lacking obvious residues may be useful as well. In addition, Shanks et al. (2001) found that blood is preserved in micro-cracks formed during percussion and pressure flaking, which may explain reports of the poor residue preservation on simple flakes. While the success rate of immunological techniques of protein analysis is generally low, it exceeds that of DNA analysis and likely has a place in the analysis of archaeological materials.

Chapter 16
Ancient DNA and the Polymerase Chain Reaction

The development of the polymerase chain reaction (PCR) in the mid-1980s by K. B. Mullis, an American biochemist, stimulated the rapid growth of a new avenue of anthropological inquiry. Using this process, miniscule amounts of genetic material from ancient materials can be studied. Targeted segments of preserved deoxyribonucleic acid (DNA) are copied and re-copied until there is enough present for further analysis. The area selected for amplification depends upon the specific research question. The majority of anthropological applications involving DNA analysis fall under the realm of physical anthropology. Both modern and ancient human DNA are used to determine the status of genetic traits with restricted distributions in order to address questions of kinship, population movement, the occurrence of disease, gender distributions, and evolutionary lineage. Reviews focusing primarily on the analysis of ancient DNA (or aDNA) from human remains recovered from archaeological contexts appear in Kaestle and Horsburgh (2002) and Jones (2003); Mulligan (2006) provides a critical review of aDNA extracted from the remains of humans, other animals and plants, as well as aDNA preserved in archaeological residues. Descriptions of the procedures and issues surrounding the analysis of aDNA are provided in Brown (2001), Cano (2000), Kaestle and Horsburgh (2002), Mulligan (2006), and O'Rourke et al. (2000). Further information about the amplification and analysis of DNA using PCR, DNA cloning, and DNA sequencing can be found in biochemistry textbooks such as Lehninger et al. (1993) and Nelson and Cox (2008).

While procedures for amplifying aDNA are similar regardless of its source, sample handling considerations are especially important when dealing with human bone. Modern DNA from excavators, laboratory personnel, or even those who prepared the DNA reagents can contaminate the sample and compromise the analysis results. Recent studies have demonstrated the modern DNA of domesticated animals, including house pets and livestock, can appear when archaeological faunal remains are analyzed (Leonard et al. 2007).

The topics considered in this section include: the preservation of aDNA, the polymerase chain reaction, issues surrounding sample selection, handling, and processing, the factors that limit the success rate of aDNA analysis and typical research questions.

M.E. Malainey, *A Consumer's Guide to Archaeological Science*, Manuals in Archaeological Method, Theory and Technique, DOI 10.1007/978-1-4419-5704-7_16, © Springer Science+Business Media, LLC 2011

The Preservation of Ancient DNA

Ancient DNA does not preserve well except under very cold and very dry conditions. Although one would expect the hydrogen bonds between complementary nucleotides to break as DNA degrades (see Chapter 5), single strands of DNA generally do not occur. Decomposition processes attack the covalent bond between the phosphate group and sugar, causing the DNA molecule to fragment into shorter sections. Basically, the twisted DNA ladder breaks on the sides but the rungs stay intact, so several shorter ladders are produced. In addition, nitrogen bases may be lost from particular sites, a process called deamination. The number of pairs of complementary nucleotides, or base pairs (bp), in the fragments is used to describe them. Ancient DNA fragments are usually quite short, generally not more than a few hundred base pairs in length.

Most of the DNA present in cells is found in the nucleus. In humans, the nucleus contains extremely long double helices representing 2 copies of each of the 22 autosomes and 1 pair of sex chromosomes. The chances that single or double copies of nuclear DNA will survive intact are quite low. For this reason, the majority of research has targeted mitochondrial DNA (mtDNA) and the ribonucleic acids of ribosomes (rRNA). Mitochondrial DNA is frequently studied because many identical copies occur in each cell; even so, many extracts from ancient samples do not contain preserved mtDNA that can be amplified. An extensive computerized database exists for mtDNA and a number of software programs have been developed that enable comparisons between the nucleotide sequence of the unknown sample and reference material. For example, the basic local alignment search tool (BLAST), provided by the National Center for Biotechnical Information, can be used to compare the nucleotide sequence in an unknown to sequence databases and calculate the statistical significance of the match. Commonly targeted mtDNA sequences include cytochrome *b* and the D-loop region.

Ancient DNA can only be amplified if it is well preserved. As with all organic materials, the principal factors that affect the preservation of DNA are the presence of water and oxygen as well as the temperature and pH of the environment. Reactions with water, or hydrolysis, attack the bond between the sugar and the base causing depurination, which is loss of either adenine (A) or guanine (G), and depyrimidination, which is loss of either cytosine (C) or thymine (T). The resulting baseless site is a minor lesion, but the DNA molecule is more likely to break at that location. Oxygen in the environment can cause the loss or modification of bases and distortions of the helix. The rate at which hydrolysis and oxidation reactions occur depends upon the pH and temperature. Dental samples are less likely to be affected by the environment, so DNA extracted from teeth tends to be better preserved than DNA extracted from bone. Exposure to sunlight (UV radiation) and X-rays can also damage DNA.

The same factors that affect DNA preservation, especially depurination, affect the rate at which L-amino acids transform from their original state into the D-amino acid enantiomers. Poinar et al. (1996) suggest that amino acid racemization (see Chapter 11) can be used to predict aDNA preservation and detect contamination. The degree to which a particular amino acid has undergone racemization is measured by the ratio of D-enantiomers to L-enantiomers, $D/L_{amino\ acid}$. Poinar et al. (1996) examined three amino acids, aspartic acid (Asp), alanine (Ala), and leucine (Leu), and were unable to retrieve DNA sequences from samples with D/L_{Asp} that exceeded 0.08. They also note that since Asp racemization occurs the fastest, the D/L_{Asp} should be higher than either D/L_{Ala} or D/L_{Leu}. If this is not the case, the sample has probably been contaminated by modern (or at least younger) amino acids. Poinar et al. (1996) suggest that DNA could be preserved for a few thousand years in a hot climate but upwards of 100,000 years in cold regions.

Haynes et al. (2002) tested whether gross morphology and tissue appearance under $70\times$ magnification could be used to predict DNA survival. Bones were scored on the basis of their outward appearance and histological preservation. Amplifiable DNA was recovered from only 55 (17%) of the 323 archaeological goose humeri examined. Haynes et al. (2002) report that 42% of the best-preserved bone contained amplifiable DNA compared to only 13% of the worst preserved specimens. They emphasized, however, that the majority of successful amplifications (61%) were actually from the latter group and would have been missed if only very well-preserved bones had been analyzed. Their results were consistent with those of Hagelberg et al. (1991), who found that microscopic preservation was related to DNA recovery and histological screening could be used to identify regions of unmodified tissue in poorly preserved bone.

While screening methods can increase the chances of recovering amplifiable DNA, Mulligan (2006) considers none sufficiently reliable to predict the outcome prior to actual DNA analysis. Götherström et al. (2002) found that both hydroxyapatite and collagen have roles in the preservation of DNA in bone and suggest that levels of collagen, determined by radiocarbon laboratories when a bone is dated, can be used as a guide for sample selection.

The Polymerase Chain Reaction

As described in Chapter 5, enzyme-catalyzed DNA replication requires that a pre-existing strand serves as the template. First, two complementary DNA strands separate and then primers complementary to one or the other end of the region to be replicated attach to the templates. Nucleotides present in the environment as deoxynucleoside-5′-triphosphates (dNTPs) attach to the 3′-ends of the growing strands marked by the primers. The polymerase chain reaction (PCR) mimics natural enzyme-catalyzed DNA replication and results in a doubling of the targeted section of DNA after each cycle (Fig. 16.1).

The first step in the amplification process involves using heat (94°C) to separate the two strands of DNA. Typically, the initial heating is of longer duration, lasting between 3 and 7 min; only 45 s may be required for strand separation in subsequent cycles. The second "annealing" step is a cooling phase where the temperature is reduced to about 55°C for about 30 s. At the lower temperature, synthetic

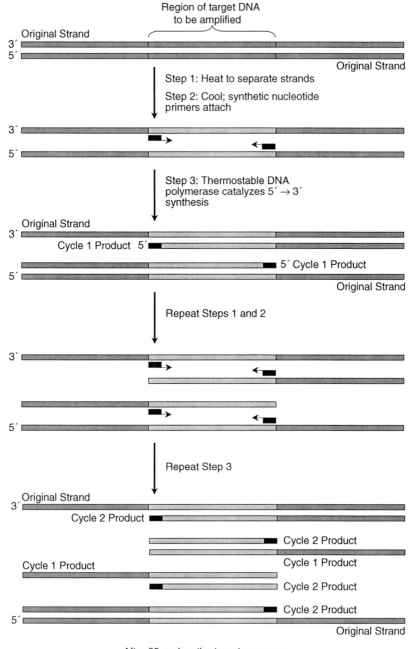

After 25 cycles, the target sequence
has been amplified about 10^6 - fold.

Fig. 16.1 Amplification of a targeted DNA segment using the polymerase chain reaction (From *Principles of Biochemistry*, 2e by A. L. Lehninger, D. L. Nelson, and M. M. Cox. © 1993 by Worth Publishers. Used with the permission of W. H. Freeman and Company.)

nucleotide primers in the solution have the opportunity to attach to the individual strands. DNA amplification can proceed only if the primers attach; usually, there are many more primers present than could possibly be used in the reaction. In the third step, the temperature is increased to about 72°C. In early applications of PCR, heat (or thermo)-stable DNA polymerase and the four deoxynucleoside triphosphates (dNTPs) were added at this time; now they are present in the starting solution.

The DNA polymerase used is called *Taq* (I), which is isolated from *Thermus aquaticus* or *Thermus thermophilus* bacteria. These bacteria grow in hot springs where water temperatures hover around the boiling point. Unlike most other naturally occurring enzymes, *Taq* (I) DNA polymerase can tolerate the temperatures required to separate the DNA. Commercially produced dNTPs can be synthesized by adding phosphate groups to deoxynucleoside monophosphates obtained from animal DNA by hydrolysis (Leonard et al. 2007). The dNTPs are added to the 3'-end of the primer and the targeted segment of DNA is selectively replicated. The duration of the replication step is about 90 s.

Upon successful completion of the first cycle, the concentration of the targeted DNA region has been doubled. Since only a tiny amount of genetic material is extracted from archaeological samples, it is necessary to repeat the three steps many times before there is sufficient DNA for further analysis; samples are routinely subjected to 25 or 30 cycles. The popularity of PCR led to the development of automated systems using a thermocycler programmed to perform each step in a cycle a specified number of times. The polymerase chain reaction takes place in 0.1–0.5 ml PCR tubes, to which the DNA, primers, polymerase, and dNTPs have been added. After 25 cycles of doubling, the original amount of DNA has increased a million-fold. The technique is so sensitive that it may be capable of detecting a single molecule of DNA. This is important in the context of archaeological studies since, at best, only a small amount of amplifiable DNA is recovered. More than one sequence can be targeted for amplification if they are of different base pair lengths.

Analysis After Amplification

Attempts to replicate aDNA are often unsuccessful, so analysis involves determining: (1) whether or not amplification products formed and (2) whether these products, called amplicons, represent the targeted region of aDNA or modern contaminants. Agarose gel electrophoresis is used to separate the DNA according to size; electrophoresis is described in Chapter 33. The PCR product is placed in wells cut into the gel slab and an electric current is applied. The amplified DNA fragments migrate through the gel according to their sizes; the presence of the targeted sequence can be established on the basis of its base pair length.

DNA cloning and sequencing can also be used to examine amplified products; these processes are described below in two separate "Special Topic" sections, based on Lehninger et al. (1993) and Nelson and Cox (2008). Cloning is a process that uses bacteria to make multiple identical copies (clones) of the amplified products. It is performed to determine the ratio of authentic DNA arising from the amplification of the targeted region (endogenous sequences) to contaminants (exogenous sequences). The presence of more than one PCR product indicates multiple sources of DNA in the sample. Cloning can also be used to detect damage-induced errors because authentic sequences produce overlapping fragments but damaged templates do not. Contamination by nuclear mitochondrial DNA insertions (NUMTs) can also be detected. NUMTs are mitochondrial DNA sequences that appear in nuclear DNA (Mishmar et al. 2004). If the nucleotide sequence in the NUMTs matches the targeted mtDNA segment, they may be accidentally amplified along with the mtDNA target. This occurs most often when very little mtDNA is present in the sample (Goios et al. 2008).

DNA sequencing is performed to determine the order of nucleotides in the DNA after cloning. It is required to compare the nucleotide sequence in the amplified products to reference material. When a nucleotide sequence is associated with a specific group, it can be used as a genetic marker. For example, a simple tandem repeat (STR), or micro-satellite repeat, is a short sequence of nucleotides that reoccurs several times. The number of times the repeated nucleotide unit appears can vary between different populations. In other cases, the occurrence of a particular nucleotide at a specific location may be a genetic marker. The term, haplotypes, refers to nucleotide changes that occur on the same sequence or chromosome. Haplogroups are groups of related DNA haplotypes that share one or more key markers.

In order to make comparisons between populations, regions of DNA most likely to exhibit differences are examined. In the study of ancient human DNA, this often involves the first and second hypervariable regions (HVRI and HVRII). This region of mtDNA accumulates nucleotide substitutions at a particularly high rate because there are no mechanisms in place to correct these types of errors.

Sample Selection

Previous research has shown that it is possible to extract DNA from a variety of materials including bone, teeth, soft tissues, and even residues preserved on stone tools. As noted above, it can be difficult to predict which samples contain well-preserved DNA. Sometimes it is impossible to extract and amplify DNA from samples that at the outset showed great promise; other times samples that seem poorly preserved contain amplifiable DNA.

The composition of bone makes it suitable for DNA analysis because hydroxyapatite and collagen seem to bind DNA and slow its degradation (Götherström et al. 2002; O'Rourke et al. 2000). Cortical bone is often selected but cancellous or spongy bone has the potential to provide significantly higher DNA yields. A cortical bone sample can be drilled from the midshaft of a long bone or a small fragment can be used; cancellous bone is typically obtained from rib fragments. DNA can be preserved in teeth that are intact and free of dental caries. Extraction from the whole tooth generally yields the most DNA; a less destructive approach involves sectioning the tooth, collecting the pulp, then gluing the empty enamel and dentine shell back together.

Soft tissue from desiccated remains can yield well-preserved DNA due to the absence of water. Ancient DNA was recovered from skin and hair samples of extinct animals, including the quagga and marsupial wolf, housed in museum collections (Higuchi et al. 1984; Pääbo 1989; Thomas et al. 1989). Coprolites can yield lots of DNA; however, special processing may be required to remove DNA inhibitors (see below). In general, subsurface samples tend to be better protected from environmental contaminants. In the case of stone tools, ancient DNA has been recovered from micro-cracks that formed during percussion and pressure flaking, even if the surface has been washed and no visible surface residues are present (Shanks et al. 2001). Plant macrofossils and seeds preserved in cold and dry environments or by charring can be examined. O'Rourke et al. (2000) suggest that pollen is also a promising source; however, Mulligan (2006) stresses that these male gametophytes are not suitable for chloroplast DNA analysis. Chloroplast-specific primers are useful as they are able to specifically target plant DNA in complex samples, such as soil, that may also contain bacterial or vertebrate DNA. Examples of studies involving the analysis of ancient DNA are provided in Chapters 19, 21, 22, 24 and, 27.

Extraction

A variety of methods can be used to recover aDNA for PCR analysis (Cano 2000; Hagelberg et al. 1991; Kaestle and Horsburgh 2002; Kemp et al. 2006; O'Rourke et al. 2000; Pääbo 1989; Shanks et al. 2001; Yang et al. 1998). The first step is removal of possible surface contaminants from the sample. This can be achieved mechanically by cutting or flaking off the surfaces that have been exposed to the environment. Another option is to soak the sample in a bleach solution or irradiate it under UV light for a period of time.

Prior to DNA extraction, it is common to break apart or crush the sample to increase the available surface area. The chemical ethylenediaminetetraacetic acid (EDTA) is often added to decalcify bone samples prior to extraction. Substances that break apart protein, such as proteinase K (PK) and guanidinium thiocyanate (GuSCN), are typically used for DNA extraction; collagenase, which targets collagen, can also be employed. Proteinase K will also digest hard tissue so bone samples do not require prior crushing. When using GuSCN, any undigested, solid cellular debris is removed from the extraction solution by centrifugation.

Phenol/chloroform and/or silica can be used to purify the DNA extract. When using the phenol/chloroform method, DNA separation is achieved on the basis of solubility and miscibility. DNA is a water-soluble acid that will occur in aqueous solutions; many other substances found in tissues, such as lipids, are not water soluble but will dissolve in the organic solvents phenol and chloroform. In each of the three separate extractions, the sample is vigorously combined with the solvents using a vortex mixer and then centrifuged. The density of these particular organic solvents is greater than water, so after centrifugation, the aqueous layer containing DNA is the supernatant, separate from and above the organic solvents (if EDTA is not present). The first DNA extraction involves phenol alone; the second is performed with a mixture of phenol, chloroform, and alcohol; and the third is performed with chloroform and alcohol.

The aqueous layers containing the DNA are filtered to separate the largest molecules, or macromolecules (which are more likely to be DNA fragments), from the smaller ones (which are likely to be chemical impurities). Solid DNA is precipitated out of the solution using an alcohol, such as ethanol, isopropanol, or both, then redissolved in TE buffer. TE buffer is a solution of *Tris* [tris(hydroxymethyl) aminomethane] treated with hydrochloric acid and EDTA. When isopropanol is used, the final solution is less likely to be colored; color is considered an indication that PCR inhibitors are present.

The silica method involves digesting the protein, with either GuSCN or proteinase K, for an extended period of time. After centrifuging, the DNA in the supernatant can be mixed with a solution in which tiny silica particles are suspended. The DNA binds to silica, forming a silica–DNA pellet, but other substances remain in solution. The DNA is released from the silica when it is redissolved in TE buffer. Alternatively, DNA can be collected by passing the supernatant through a silica-based spin column (Yang et al. 1998). The DNA is released from the silica by pouring TE buffer through the column.

Greater DNA yields are associated with proteinase K extractions; but successful amplification is more likely when the silica GuSCN approach is used. This may relate to the enhanced selectivity of the silica GuSCN process so that fewer PCR inhibitors are co-extracted.

Selection of DNA Primers

DNA consists of two complementary strands, called Watson and Crick, in honor of the scientists that first described its structure. One primer complementary to the Watson strand and another complementary to the Crick strand are used to mark the ends of the region to be amplified. Synthetic oligonucleotide primers, usually 20–30 bp in length, are designed to ensure they are only complementary to the targeted region of the nucleotide sequence. The specific primers used for the analysis of DNA depend upon the goals of the research project. Leonard et al. (2007) compiled a list of primer sets they and others used to amplify regions of mitochondria. While some primers enabled identifications to the level of species (pig) or subfamily (Phasianinae or pheasant), most simply targeted mammalian and avian DNA. The term "universal primers" is applied to those capable of amplifying the DNA of a variety of animal species.

Challenges Associated with Analyzing Ancient DNA

The amplification of aDNA is difficult due to poor preservation, the presence of PCR inhibitors, and the potential for contamination. The implications of each of these factors and procedures recommended to mitigate their effects are described in this section.

Amplification of Poorly Preserved DNA

As noted earlier, a number of degradation processes can damage DNA. Pääbo et al. (1989) regard PCR to be ideal for amplifying aDNA because a damaged molecule will either not amplify or replicate more slowly than intact ancient molecules. Minor lesions, such as baseless sites, can cause replication errors but the overall effect will be minor if the number of undamaged templates greatly exceeds those with damage. Low amplification efficiency and short amplification products are other indications of DNA damage. By contrast, when living bacteria are used to make DNA clones the organisms attempt to repair damaged areas; mis-repairs result in the appearance of cloning artifacts.

Special protocols must be followed when amplifying aDNA to reduce the chances of mispriming and amplification of molecules other than those targeted. For example, the concentration of primers is considerably lower than levels typically used for modern samples, so two primers are less likely to bind to each other, forming a primer dimer. Other unwanted reactions, such as large numbers of primers binding to form primer oligomers or binding to areas other than the target sequence, can occur if reagents begin reacting before DNA strands are fully separated. The use of a "hot start" procedure may reduce these types of premature reagent reactions; it involves keeping the sample extract and reagents refrigerated or stored on ice until they are placed in the thermocycler preheated to about 94°C. In addition, one can use polymerase that remains inactive until temperatures capable of fully separating DNA strands are reached. The duration of the initial DNA strand separation step can also be extended from 3 min to between 5 and 7 min.

Another option is to keep the temperature of the second annealing step very high and then reduce it gradually over the course of the first ten cycles. Using this approach, called "touchdown" PCR, high temperatures prevent primers from binding to anything for the first several cycles. When the temperature is low enough for annealing, primers are more likely to bind to the target sequences.

Alternatively, "booster" PCR can be used to increase the number of DNA templates in the sample. Amplicons produced in the first set of 12 cycles of low annealing temperature PCR form the DNA templates used to "seed" a second set and then a third set of low annealing temperature PCR. Up to three cycles of booster PCR are performed, then the template-enriched sample undergoes PCR under normal operating conditions. O'Rourke et al. (2000) caution that the increased sample handling required for booster PCR presents an additional opportunity for the introduction of modern contaminants.

Strategies for Managing PCR Inhibitors

The possible outcomes of aDNA analysis using the polymerase chain reaction are (1) amplification of authentic endogenous DNA, (2) amplification of contaminating DNA, or (3) amplification fails (Mulligan 2006). In some cases, PCR amplification fails because no DNA is preserved in the sample; at other times failure is due to the presence of substances that inhibit the reaction even though aDNA is present. PCR inhibitors include substances that are commonly found in soil, naturally occur in the sample, or are used during sample processing (see Kemp et al. 2006 for a recent review). Inhibitors that can be co-extracted with DNA include humic acids, tannins, fulvic acids, porphyrin products, phenolic compounds, hematin, and collagen type I. Silica and organic solvents used in the extraction process will also inhibit PCR if they are not completely removed. Maillard products are condensation products of sugar that cross-link macromolecules, such as DNA and protein, so the entangled DNA cannot be amplified by PCR. The presence of color in the DNA extract, from faint yellow to reddish brown, typically indicates the presence of PCR inhibitors. When subjected to gel electrophoresis and examined under UV light, inhibitors cause the DNA extract to appear as a blurred blue-green fluorescent cloud.

The effects of inhibitors can be subdued by diluting the extract to the point where the concentration of inhibitors is too low to block the reaction but the concentration of DNA is still high enough to be amplified. Another approach is the use of booster PCR to increase the initial concentration of DNA templates. Increasing the amount of *Taq* DNA polymerase or adding bovine serum albumin to prevent inhibitors from interfering with the reaction can enhance polymerase activity.

Several other strategies are used to remove PCR inhibitors, including the addition of cetyltrimethylammonium bromide (CTAB) alone or with polyvinyl pyrrolidone (PVP). Small molecules, such as humic acid or polysaccharides, can be filtered out while DNA is retained as either intact double-stranded or denatured single-stranded segments. Precipitating DNA with isopropanol rather than ethanol appears to reduce levels of co-extracted inhibitors. The addition of *N*-phenacylthiazolium bromide (PTB) can break apart Maillard products in coprolites so that the DNA is released. Silica-based extractions, using gel-packed columns, affinity beads, or spin columns, appear to remove more inhibitors than extractions with phenol/chloroform. Collagen type I can be removed by digesting the sample with collagenase, instead of proteinase K.

Kemp et al. (2006) propose a protocol that combines phenol/chloroform and silica DNA extraction with a test for the presence of PCR inhibitors; gel electrophoresis is used to confirm DNA amplification. After digestion with proteinase K and the triple extraction with phenol, chloroform, and/or alcohol, DNA precipitation is performed with ethanol followed by isopropanol. The DNA extract is then tested for the presence of inhibitors by running three reactions in parallel: (1) the sample alone, (2) the sample "spiked" with an aDNA positive control, and (3) the aDNA positive control alone. The presence of inhibitors will prevent the amplification of the sample alone as well as the sample spiked with the positive aDNA control, but the reaction of the positive control alone should proceed. Inhibited DNA extracts are subjected to consecutive silica extractions, which they call "repeat silica extraction," until either amplification occurs (all three reactions are successful) or the uninhibited sample fails to amplify because it contained no DNA (only positive controls are amplified). Kemp et al. (2006) believe that this protocol can effectively remove all types of PCR inhibitors from the DNA extract.

Avoiding Contamination

Modern DNA contaminants invalidate or reduce the certainty of results; they undergo PCR faster and more efficiently than old and damaged templates and can even prevent the amplification of aDNA. Contamination can occur from any one of a variety of sources. Pääbo et al. (1989), Handt et al. (1994), and Cooper and Poinar (2000) encouraged researchers to adhere to certain guidelines with respect to the laboratory setting, use of controls, and verification procedures to ensure accurate and reproducible results. Ancient DNA analysis should be performed in a dedicated laboratory equipped with HEPA-filtered positive pressure airflow so that contaminants are not introduced by entering or exiting the lab or through the heating/air-conditioning systems. Technicians need to be properly clothed with masks, gloves, and body suits. All equipment and the workbenches must be cleansed between samples with UV light, bleach, or DNase. Extraction procedures should be spatially separated from amplification; ideally, the procedures should be conducted in completely separate rooms. Replicate samples should be run together with closed and open controls to confirm the initial results; preferably, different personnel should perform the process in a different laboratory. Closed controls are run to monitor the quality of reagents; the reagents are placed in reaction vials and immediately closed. Open controls are used to monitor the laboratory environment; reagents are placed in reaction tubes prior to the start of DNA extraction and remain open until all other sample vials are sealed and are ready for amplification.

The amplified product should also be consistent with expectations for damaged aDNA. Pääbo et al. (1989) suggest that DNA extracted from archaeological samples should not exceed 150 bp in length. Better-preserved museum samples may produce longer segments, but amplification products that exceed 500 bp in length probably arise from contaminants. The amplification efficiency is lower for damaged aDNA because the sample may lack template molecules that span the full length of the targeted segment. If regions complementary to the primers exist, amplification may proceed as a series of extension steps. With each cycle, the lengths of the primer products increase until they span the full length of the target segment, at which time exponential amplification can proceed.

The presence of modern human DNA in PCR reagents is "expected" because they are manufactured by humans and humans set up PCR experiments (Leonard et al. 2007:1364). On the basis of results from four independent labs, Leonard et al. (2007) report that commercially produced dNTPs can be a source of animal DNA contaminants. Researchers conducted hundreds of no-template reactions, i.e., no DNA was intentionally added. At one lab, animal DNA contamination, from cattle (87.5%), pigs (6.25%), and chickens (6.25%), was found in 2% (16 of 763) of no-template reactions. Previous studies indicate that contamination from mouse, goat, and guinea pig DNA could also occur. Another lab reported that 7 of 175 (4%) no-template reactions could be amplified with pig-specific primers; pig-specific primers also amplified 7% of DNA extracted from the feces of bonobos. The most likely source of contamination was determined to be dNTPs, which are synthesized from deoxynucleic monophosphates extracted from animal tissue. Leonard et al. (2007)

regard this to be a serious problem because often only about 10% of aDNA samples are amplified; a 5% contamination rate implies that up to 50% of successful amplifications may be false positives due to contaminants. The implications are most significant for archaeological studies attempting to establish the presence of domesticated animals, in particular cattle, pigs, and chickens.

Conclusions

The processes involved in the analysis of aDNA are complex. A number of precautions at all stages of the project are necessary to ensure the validity of results, beginning with sample collection. Even under the most ideal conditions, chances of success are low. If the study focused on the remains of humans or domestic animals, the presence of contaminants in DNA reagents can confound positive results. Mulligan (2006) produced a checklist for archaeologists to assess the feasibility of a proposed aDNA project. In addition to factors related to sample availability, preservation, and contamination, researchers are encouraged to consider the appropriateness of genetic analysis to address their particular research question. The possibility of success is enhanced through close collaboration with other experts, including an experienced molecular genetic researcher.

Special Topic: DNA Cloning

Amplified products can be cloned by living bacteria, such as *Escherichia coli* (*E. coli*), using recombinant DNA technology. Pieces of foreign DNA are inserted into a small DNA molecule, called a cloning vector, which is capable of replicating itself. When introduced into an *E. coli* host cell, identical copies are made of the recombinant DNA, i.e., cloning vector DNA into which foreign DNA has been inserted. Plasmids are the most commonly used cloning vectors used in aDNA studies. The use of bacterial vectors to examine DNA predates the polymerase chain reaction.

Insertion of Foreign DNA into Cloning Vectors

In archaeological applications, the amplified aDNA is the foreign DNA being inserted into the cloning vector. The first step in the cloning process is the precise cutting of the PCR amplicons with special enzymes at sequence-specific locations. These enzymes, called restriction endonucleases, have names based on the bacteria from which they were derived; *Bam* is from *Bacillus amyloliquefaciens* and *Eco* is from *E. coli*. Roman numerals are used to distinguish different endonucleases from the same bacterial source. The specific location on the DNA molecule targeted by a particular enzyme is called its recognition sequence.

Restriction endonucleases can sever the DNA between two complementary base pairs, producing blunt ends, or the cut can be staggered so that unpaired nucleotides are left exposed on one strand, producing sticky, or cohesive, ends. The PCR amplicons can then be inserted into a cloning vector, usually a plasmid, cleaved with the same restriction endonuclease. This process is called ligation; afterward the molecules are sealed together with DNA ligase. Ligation is not 100% efficient; many cloning vectors do not receive foreign DNA and are resealed unmodified. A stronger joint is produced when the molecules have sticky ends because the nucleotide bases of the cloning vector are complementary to the amplified aDNA. The different types of cloning vectors and the process of introducing recombinant DNA into bacteria are described below.

Cloning Vectors

Plasmids are circular DNA molecules found in many types of bacteria that are usually only a few thousand base pairs long. The ones used for molecular cloning are often modified to include genes, called selectable markers, to facilitate their identification and provide different recognition sequences for restriction endonucleases (Fig. 16.2). The process of introducing plasmids into *E. coli* is called transformation. The plasmids and host cells are incubated in a calcium chloride solution at 0°C, then the temperature is rapidly increased to between 37 and 43°C. The heat shock causes a few of the cells to take up the plasmids. The *E. coli* cells are grown in agar and those carrying recombinant DNAs are identified through their selectable markers. For example, if the selectable marker provides a cell with antibiotic resistance, that antibiotic is added to the agar so that only cells with recombinant DNA survive (Fig. 16.3). If the plasmid contains two selectable markers providing resistance to different antibiotics, the foreign DNA can be inserted into one of them (which destroys it). This facilitates the separation of cells that contain unmodified plasmids, with resistance to two antibiotics, from those only resistant to one due to the foreign DNA insert (Fig. 16.3).

Plasmids are used to clone short DNA fragments, less than 15,000 bp in length. Bacteriophages and cosmids are used as cloning vectors for fragments up to about 45,000 bp in length. Bacteriophages are viruses that deliver recombinant DNA by infecting the *E. coli*. Cosmids are plasmids that deliver viable recombinant DNA in a manner similar to bacteriophages.

Special Topic: DNA Sequencing

DNA sequencing is used to determine the order in which nucleotides are connected in a strand of DNA. The Sanger (dideoxy) method of DNA sequencing was developed in the 1970s and is commonly used today. It is similar to the polymerase chain reaction in that the unknown is used as a template from which complementary

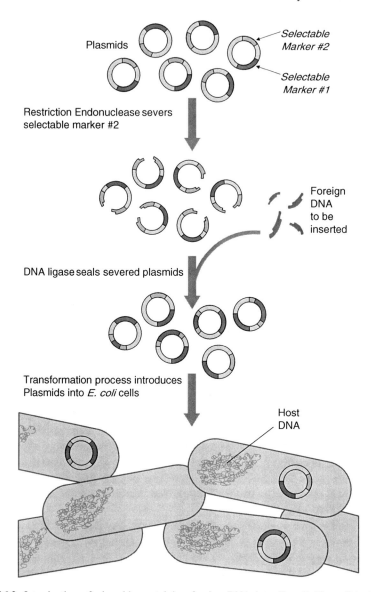

Plasmids

Selectable Marker #2

Selectable Marker #1

Restriction Endonuclease severs
selectable marker #2

Foreign
DNA
to be
inserted

DNA ligase seals severed plasmids

Transformation process introduces
Plasmids into *E. coli* cells

Host
DNA

Fig. 16.2 Introduction of plasmids containing foreign DNA into *E. coli* (From Principles of
Biochemistry, 2e by A. L. Lehninger, D. L. Nelson, and M. M. Cox. © 1993 by Worth Publishers.
Used with the permission of W. H. Freeman and Company.)

strands of DNA are produced. The difference is that dideoxynucleoside triphosphates (ddNTPs) are used to stop the replication at certain points. The process
involves a primer strand, DNA polymerase, dNTPs along with one of the four different types of ddNTPs, either ddATPs, ddTTPs, ddCTPs, or ddGTPs. The structure

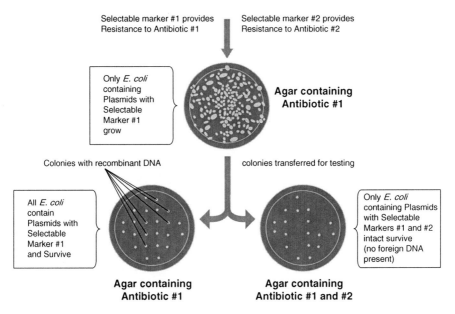

Fig. 16.3 Selectable markers enable identification of *E. coli* containing recombinant DNA (From Principles of Biochemistry, 2e by A. L. Lehninger, D. L. Nelson, and M. M. Cox. © 1993 by Worth Publishers. Used with the permission of W. H. Freeman and Company.)

of ddNTPs is slightly different from dNTPs in that it lacks the 3′-hydroxyl (-OH) group; when ddNTP is added instead of dNTP, elongation of the complementary strand is terminated. In order for the reaction to occur to at least some extent, the concentration of ddNTPs in the reaction solution is low in comparison to dNTPs. As a result, the lengths of the fragments in each set vary according to their distance from the 3′-end of the primer.

For example, in order to determine the position of every T in the DNA template, ddATPs are added to the reaction solution along with primers, DNA polymerase, and all four types of dNTPs. Primers attach and DNA polymerase directs the elongation of the complementary strands. Whenever T appears in the template strand, there is a chance that ddATP may be added instead of dATP. Since the number of templates in the reaction solution is high, the reaction will almost certainly terminate at least once at every single position of T in the DNA template (if it is not too long). The reaction is repeated using ddTTPs to determine the positions of every A, using ddCTPs to determine the positions of every G, and using ddGTPs to determine the positions of every C.

The technique was originally developed for use with gel electrophoresis and primers often carried a radioactive label. The nucleotide-specific reaction products were placed in individual wells cut into one end of the gel slab and separated on the basis of their size when the current was applied. The radiation emitted from each spot was recorded on a photographic plate; the nucleotide sequence of the complementary strand was then read directly from the autoradiogram.

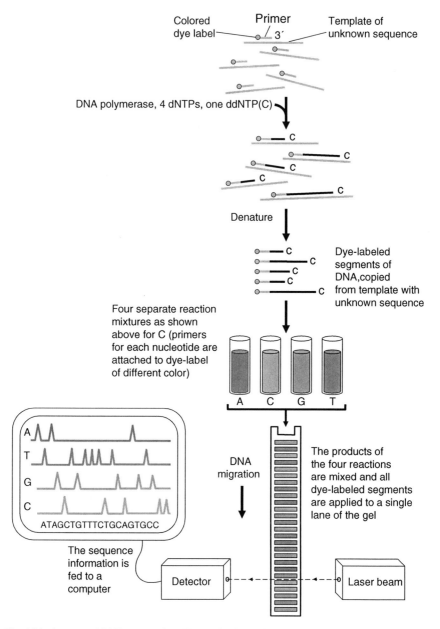

Fig. 16.4 Automated DNA sequencing (From Principles of Biochemistry, 2e by A. L. Lehninger, D. L. Nelson, and M. M. Cox. © 1993 by Worth Publishers. Used with the permission of W. H. Freeman and Company.)

Instead of radioactive substances, primers can now be labeled with fluorescent molecules that give each nucleotide its own color when viewed under UV light. This permitted the development of automated DNA sequencing. Because the nucleotides are distinguishable on the basis of color, the four sets of reaction products are combined and then separated on the basis of size using electrophoresis. Instead of a slab, the DNA fragments are separated in a gel contained in a long, small diameter, capillary tube. Individual nucleotides are identified on the basis of color as they pass by a detector equipped with a laser that produces light in the UV range (Fig. 16.4).

Part III
Materials

Part III includes case studies illustrating the applications described in Part II using the analytical techniques described in Part IV. A wide variety of materials are discussed including pottery (Chapter 17), flaked and groundstone tools (Chapter 18), bone and teeth (Chapter 19), cultural rock (Chapter 20), organic artifacts (Chapter 24), paint, pigment and ink (Chapter 25), metal and glass (Chapter 26), plant remains (Chapter 27), matrix and other environmental deposits (Chapter 28) and other materials (Chapter 29). The analysis of food residues from pottery (Chapter 21) and other artifacts (Chapter 22) are discussed separately from non-food residues (Chapter 23). Compositional analyses are presented at the beginning of the each chapter and applications of dating techniques appear at the end.

The majority of studies involve archaeological remains recovered from New World contexts. Accessibility to original papers was a major consideration in the selection process. Most case studies presented were published in widely distributed peer-reviewed journals. It is assumed that readers are familiar with the basic science concepts presented in Part I. Brief descriptions of statistical techniques appear in Chapter 12.

Chapter 17
Pottery and Other Ceramics

A wide variety of analytical techniques have been applied to archaeological pottery to address a diverse array of questions (Tite 1999); this chapter provides an overview of the applications. In addition to the composition of pastes, glazes, and slips, methods used to assess manufacturing techniques and firing conditions are presented. While instrumental neutron activation analysis (INAA or NAA) is extensively used for the analysis of the bulk composition of pottery (Speakman and Glascock 2007), several alternative approaches are presented. Particular attention is paid to methods capable of analyzing the composition of the paste or substances applied to the surfaces with little or no sample destruction.

Paste Composition

Pottery is not usually formed from raw clay; prior to vessel manufacture, the clay is subjected to purification, tempering, and/or mixing processes. These procedures may alter the chemical composition of clay and make it difficult to match the paste of a ceramic vessel to the raw clays from which it was formed (Neff 1993). In these cases, it is often more useful to compare the paste compositions of vessels recovered from different archaeological sites.

Raw clay is purified through the process known as slaking or levigation, which involves adding excess water to create a thin slurry and allowing coarser grained sand and silt to settle out while the fine-grained clay particles remain suspended (Rice 1987). Tempering agents are mineral or organic materials intentionally added to clay. As outlined in Chapter 12, Neff et al. (1988) used computer models to demonstrate that the magnitude of the effect on composition depends on the elemental concentrations in the clay and temper and the degree of similarity of the clay groups. In addition, the compositions of pastes prepared for vessel manufacture are not necessarily homogeneous. Garrigos et al. (2003) employed several different analytical techniques in a recent "ethnoarchaeometric" study of modern potters in Spain and found their practices resulted in pastes that were highly variable in composition. Cogswell et al. (1996) found that firing temperature has little effect on the elemental characterization of pottery.

M.E. Malainey, *A Consumer's Guide to Archaeological Science*, Manuals in Archaeological Method, Theory and Technique, DOI 10.1007/978-1-4419-5704-7_17, © Springer Science+Business Media, LLC 2011

Bulk Composition Analysis

Instrumental neutron activation analysis (INAA) (Chapter 32) has been applied to the study of archaeological materials, in particular the bulk chemical characterization of ceramics, for 50 years (Speakman and Glascock 2007). Studies illustrating how statistical analyses of compositional data generated by the INAA enables the recognition of significant groupings of archaeological vessel pastes and/or raw materials regularly appear in the literature (Baxter et al. 2008; Fowles et al. 2007; Jamieson and Hancock 2004; Mommsen et al. 2002; Schwedt and Mommsen 2004; Steponaitis et al. 1996; Vaughn and Neff 2004; Vaughn et al. 2006).

While a large number of elements can be measured using INAA, the technique is expensive, the analysis is performed over several weeks and requires a source of neutrons. In addition, the availability of nuclear reactors for sample irradiation is decreasing as more facilities are decommissioned (Kennett et al. 2002). X-ray diffraction (XRD) can be used to gain some compositional information about archaeological ceramics (Tang et al. 2001; Zhu et al. 2004); but it is more often employed for studies of firing analysis. Analyses of the bulk composition of paste using a variety of other techniques are also considered.

Bulk Composition Analysis with INAA

Herrera et al. (1999) used a combination of decorative analysis and INAA to investigate interactions between different sites in the Oaxaca Highlands and detect possible influences from the Olmec community of San Lorenzo on the Gulf Coast of Mexico. The composition of Early Formative Pottery from four sites in the Valley of Oaxaca was compared to pottery from the Etlatongo site in the Nochixtlán Valley. A total of 92 sherds were analyzed, primarily samples of Leandro Gray and Delfina Fine Gray decorated with pan-Mesoamerican motifs (also known as Olmec motifs); a small number of Xochiltepec White and other fine wares were examined. In addition, three samples from the Classic period were analyzed but were not included in the statistical analyses. The compositions of archaeological samples were also compared to previously analyzed modern ethnographic pottery and raw materials.

Raw data obtained were converted to log base-10 values prior to the application of statistical techniques. Groups initially identified with principal component analysis were refined through the use of Mahalanobis distances. Of the 89 Early Formative samples, 59 sherds were encompassed by the 5 compositional groupings identified; 30 sherds, including 8 of the 12 Xochiltepec White samples, could not be assigned to a compositional group. Four Xochiltepec White and unclassified fine paste ceramics occurred in one compositional group. The wide distribution and narrow compositional range of fine white paste ceramic suggest the vessels may have been produced at a central location near high-quality clay resources.

By contrast, Leandro Gray pottery was produced from diverse geological materials throughout the Highlands, which is consistent with decentralized production without exchange. Delfina Fine Gray was also manufactured at several places, but the range of composition was less diverse. More than one-half of the Delfina Fine Gray sherds were associated with a single compositional group.

Decorative motifs were not associated with any one compositional group, meaning their use was not restricted. Herrera et al. (1999) report that potters throughout the Highlands were familiar with the motifs and pottery decorated with these symbols was manufactured throughout the study area. Many motifs on pottery from Etlatongo in the Nochixtlán Valley also appeared on Oaxaca Valley pottery; however, some Etlatongo motifs were only shared with Gulf Coast Olmec sites. Conversely, the were-jaguar motif, which is common in the Oaxaca Valley, did not appear on Etlatongo pottery.

Bulk Composition Analysis with ICP-AES and AA

While INAA is used to analyze powdered sherds, atomic absorption spectroscopy (AA or AAS) (Chapter 34) and inductively coupled plasma-atomic emission spectroscopy (ICP-AES or ICP) (Chapter 34) require that the samples be dissolved in acids. Compared to AA, ICP-AES is able to identify more elements with lower detection limits and higher degrees of precision. Applications involving the use of ICP-AES for analysis of the bulk composition of archaeological ceramics, alone or in conjunction with another technique, appear in the literature more often than analysis with AA (Monette et al. 2007; Pérez-Arantegui et al. 1996; Zhu et al. 2004).

Hatcher et al. (1995) analyzed 22 samples of pottery and silicate materials with both AA and ICP-AES. The samples were selected to reflect a range of compositions and expose potential problems that relate to sample dissolution. For analysis with AA, 25 mg samples were treated with four different acids: hydrochloric acid (HCl), nitric acid (HNO_3), hydrofluoric acid (HF), and perchloric acid ($HClO_4$); all except nitric acid were used to dissolve 100 mg samples for ICP-AES.

In general, AA was able to detect major and minor elements even when they appeared in a sample at low levels; ICP-AES could identify major, minor, and a wide range of trace elements. For many elements, values obtained by AA were higher than certificate values of standards; those obtained by ICP-AES were lower than the certificate values. Data from the two techniques agreed well for calcium (Ca), magnesium (Mg), iron (Fe), sodium (Na), titanium (Ti), and manganese (Mg). Variation in data obtained for aluminum (Al), potassium (K), nickel (Ni), and chromium (Cr) was greater and may relate to a systematic shift (i.e., the values generated consistently differ by a specific amount). Consequently, multivariate statistical analyses of data obtained by both AA and ICP-AES would likely produce groups relating to the analytical technique employed rather than the actual differences in paste composition.

Burton and Simon (1993) suggested that compositional analysis of ceramics with AA and ICP-AES not require total sample dissolution. The technique they described, called weak acid extraction, involves soaking an intact sherd in 1 M HCl acid solution for 2 weeks at room temperature with periodic agitation. Burton and Simon (1993) reported the data obtained on 12 elements were both reproducible and highly precise. Although the data are different from and cannot be compared with bulk compositional analysis using techniques, such as INAA, Burton and Simon (1993) indicated it was sufficient for interregional comparisons of pottery and recognition of compositional groups. Triadan et al. (1997) countered this assertion by showing that the compositional groups identified by data from weak acid extraction-ICP-AES do not match those generated by INAA.

Using a combined approach, Monette et al. (2007) analyzed 298 kiln wasters from a site in 16 potter's workshops in Québec with ICP-AES and then a subsample of 40 sherds was analyzed with ICP-MS. Instead of relying solely on acid dissolution, 200 mg of the powdered samples was mixed with lithium metaborate flux, heated to a high temperature, and then the molten mixture was dissolved in nitric acid. After the removal of three outliers, five main groupings, which correspond to pottery production centers, were identified through hierarchical cluster and principal component analysis. The five main groups were further subdivided into at least 20 reference chemical groups that could be used to source local coarse wares. Monette et al. (2007) report that, although data on rare earth elements (REE) provided by ICP-MS were useful, the same groups could be recognized on the basis of major, minor, and trace elements measurable by ICP-AES alone.

Bulk Composition Analysis with ICP-MS

As with ICP-AES, ICP-MS (Chapters 31 and 34) involves the analysis of samples in solution; the difference is that the mass of ions is analyzed rather than emissions from excited atoms. Mallory-Greenough et al. (1998) used ICP-MS to measure trace element compositions of pottery from two sites in ancient Egypt. With respect to the 30 elements reported, the precision and accuracy of the data generated by ICP-MS were found to be as high or higher than values generally obtained by INAA. More recently, Li et al. (2005) used ICP-MS to obtain highly precise data for 40 elements in white porcelains. Compared to INAA, sample size requirements and costs of ICP-MS are lower, sensitivity is higher with elemental detection limits in the parts per billion range (ppb), and more species (up to about 70) can be measured, including isotope ratios of some elements (Kennett et al. 2002; Li et al. 2005; Mallory-Greenough et al. 1998).

Kennett et al. (2002) suggested microwave digestion ICP-MS (MD-ICP-MS) is a faster and safer alternative to typical, open-vessel multi-acid digestion of samples. After removal of surface contaminants and grinding in an agate mortar, 100 mg of the powdered samples was placed in a reaction vial with a combination of HF, HCl, and HNO_3 and heated to 200°C for 30 min in a specially designed microwave oven. After digestion, 25 mL of 4% boric acid was added to prevent loss of REE and HF was removed through evaporation. The sample was then rehydrated in HNO_3 and diluted ten-fold prior to analysis. The protocol was applied to standard reference materials and archaeological pottery samples previously analyzed with INAA.

Standard deviations for most elements analyzed by MD-ICP-MS were in the range of 5–10% but much higher for arsenic and thorium; whereas the standard deviation for most elements analyzed by INAA is 5%. It was difficult to measure calcium with ICP-MS because of interference with the argon carrier gas and the signal from uranium was very small. The data obtained by MD-ICP-MS overall compared very well to those from INAA and demonstrated the technique was effective for discriminating chemically distinct groups of pottery (Kennett et al. 2002:449). In addition, MD-ICP-MS is able to detect several elements that cannot be easily measured by INAA, such as nickel, copper, and lead.

Bulk Composition Analysis with PIXE

The protons or alpha particles used to induce X-ray for PIXE (Chapter 37) cannot penetrate deeply into solid materials. In order to analyze bulk composition, a flat cross section of the sample can be analyzed or it can be ground into a powder and formed into a pellet. Bollong et al. (1997) used PIXE (with protons) to analyze thick pellets formed from 56 sherds of sand-tempered Khoi and fiber-tempered pottery recovered from sites located on the central plateau of South Africa. Data were obtained for 13 elements: K, Ca, Ti, vanadium (V), Cr, Mn, Fe, Ni, copper (Cu), zinc (Zn), rubidium (Rb), strontium (Sr), and zirconium (Zr). Statistical analyses were performed on the data presented as proportional values normalized against iron. Differences in composition were mainly due to levels of Ca, K, and Cu, which correspond to differences in the compositions of the clays (shale-based vs. doleritic with caliche fragments) and temper (sand vs. grass).

Bulk Composition Analysis with X-Ray Fluorescence Techniques

García-Heras et al. (1997) compared results obtainable from total reflection X-ray fluorescence (see Chapter 37) with INAA. Whereas INAA involves the analysis of finely ground material in a capsule, TXRF analysis is performed on a thin film of sample particles deposited on a flat carrier. The advantage of TXRF over ED-XRF is that the small angle with which the X-rays strike the sample reduces background (noise) from the sample carrier and improves the peak–background ratio (i.e., signal to noise ratio). In the procedure outlined by García-Heras et al. (1997), TXRF analyses were conducted using X-rays that struck the sample with an angle of incidence of 4 min, or 1/15th of 1°.

Three requirements for proper thin film geometry outlined by García-Heras et al. (1997:1005) were "(1) chemical homogeneity in order to avoid the selective sedimentation of micro-particles on the basis of their density; (2) average particle sizes of less than 10 μm; (3) homogeneous distribution of particles deposited as a mono-particle layer." Meeting the first and second criteria involved more than 1 h of sample grinding/pulverization, ultrasonic disaggregation of the particles suspended in water, and verification of their size distribution using quasi-elastic light scattering spectroscopy (QELS). After 5 μl of the suspension was evaporated onto the carrier in a vacuum, the resulting layer was examined with scanning electron microscopy (see Chapter 37). Due to the formation of concentric-shaped rings of particles as the water evaporated, the third requirement could not be completely met.

Semi-quantitative results were obtained for 19 elements with TXRF, while routine INAA provides quantitative data on 33 elements. Comparisons with reference clays demonstrated excellent agreement for 11 elements, but poor agreement for 5 others; the concentrations of 3 elements were not obtained by INAA. Analysis of archaeological pottery from Celtiberian sites in Spain indicated that elemental data from TXRF could be used to establish different compositional groupings.

García-Heras et al. (1997) present a sample-processing protocol that yielded good results using TXRF; however, they did not indicate whether all procedures would be required for routine analysis. It appears that analysis of the particle suspension with QELS would be a necessary step but it is unclear whether examination of the thin film with SEM was also needed. While it is clear that TXRF is capable of providing good compositional data, García-Heras et al. (1997) did not elaborate on the consequences of not fully meeting the requirements for proper thin film geometry. García-Heras et al. (1997:1012) describe sample preparation for TXRF as "somewhat more laborious," than INAA. Considering that INAA sherd preparation typically requires only grinding and cleaning prior to sealing the powder in a capsule, this may be a significant understatement.

Compositional Analysis Using Electron Microprobe Analysis

Freestone (1982) promoted the analysis of pottery with electron microprobes (Chapter 37) as an adjunct to thin section analysis for the compositional analysis of pastes, slips, and glazes and to estimate firing temperature. Electron microprobes are still employed to analyze ceramics but the need to prepare polished cross sections is a liability as the availability of minimally destructive techniques increases. A recent study by Abbott et al. (2008) is presented below.

Virtually all earthenware pottery from the upland zone adjacent to the Phoenix Basin, known as the Hohokam Northern Periphery, was tempered with phyllite. Phyllite, a dense, fine-grained, platy metamorphic rock, was easily accessible at numerous bedrock exposures that occur over a broad area. Unlike the village inhabitants of the Phoenix Basin who had extensive irrigated fields, these Hohokam relied on a combination of wild game, wild plants, and agricultural products planted in the narrow floodplains of tributaries of the Salt River. In the absence of compositional analysis of the full range of raw materials from the region, Abbott et al. (2008) hypothesized that the most common ceramic variety recovered from a site is most likely to have been locally manufactured. Similarities between the composition of temper from these vessels and the phyllite from nearby bedrock exposures were considered support for this assumption. Additional support was garnered when the clay composition of the most common pottery variety, i.e., most likely to be locally manufactured, differed from types that occurred with less frequency, i.e., most likely to be imported.

A total of 224 potsherds from 14 early Classic period sites situated in the eastern half of the upland zone were examined. Some of the sherds were selected randomly; others were selected because they were distinctive with respect to phyllite color and texture. The use of electron microprobe analysis was considered advantageous because it allowed for very small areas (0.109 mm^2) of the cut and polished cross section of a sherd to be analyzed. In this way, data from both the clay fraction and the temper fraction could be obtained independently. A total of ten different spots were analyzed on each potsherd, five for clay composition and five for phyllite temper composition. Because electron microprobe analysis requires that the surface conduct electricity, samples had to be coated with a thin layer of carbon.

Samples of the clay fraction altered by post-depostional chemical leaching and cation exchange, 32 of 1120, were excluded from the study. Statistical analyses of the clay fraction were performed on the average percent compositions of eight elements measured at the ten spots on each sherd. The composition of each spot of temper was analyzed as a separate sample. The particular statistics employed, discriminant analysis, principal component analysis, and Mahalanobis distances were selected to avoid "pigeonholing" data into defined categories. The data sets were tested for normality and mathematical transformations were applied to data that did not exhibit a normal distribution.

Two chemically distinct groups of phyllite temper were observed on the basis of concentrations of Mg and Ca; two others were identified on the basis of variation in K and silicon (Si). Discriminant analysis showed that clay compositions of the 112 sherds corresponded well to the four raw material reference sets. Principal component analysis and Mahalanobis distances were used to assess the provenance of sherds on the basis of clay composition. Abbott et al. (2008) identified significant exchange between three pottery production locales along a 24 km stretch of Cave Creek and another at Grapevine Wash as well as the transport of manufactured vessels from them to other regions.

Non-destructive Compositional Analysis

Information about the composition of ceramic items can be obtained using a variety of non-destructive techniques. As outlined by Tite (1992), the gray level of the scanning electron microscope (SEM) (Chapter 37) image generated by backscattered electrons is related to the atomic number of the element encountered. Black areas indicate elements with low atomic numbers and white areas indicate the presence of elements with high atomic numbers. Analysis of secondary X-rays generated by the electron beam using SEM-EDS provides more quantitative information about composition. In a study focusing on the physical characteristics, rather than the chemical composition of additives, Owenby et al. (2004) used a SEM to argue intentionally crushed schist was used as a tempering material in Hohokam pottery.

Hand-held X-ray fluorescence spectrometers (Chapter 37) are available for the non-destructive analysis of surface compositions. Emerson et al. (2003) used a device called a portable infrared mineral analyzer (PIMA) to determine the mineral composition of flint clay figurines from IR energy reflected from the surface of these artifacts. This approach is basically attenuated total reflectance (ATR)-IR spectroscopy (Chapter 35) with the capacity to measure the loss of energy from sample absorption in reflected spectra for wavelengths from 1300 to 2500 nm.

Rye and Duerden (1982) recognized the potential effects of paste inhomogeneity when elemental analysis techniques that only analyze the sherd surface are employed. Using PIXE with a 3 mm diameter beam, the researchers compared the composition of sherd surfaces to subsamples that were ground to a powder and pressed into 1 cm diameter pellets. Values obtained from pressed powder pellets had much lower standard deviations than those obtained from sherd surfaces. Standard deviations of measurements on surfaces were about 5% for most elements, but approached 10% for others. Higher standard deviations for surface analyses were attributed to pottery inclusions.

Several techniques currently employed to determine the paste composition of ceramics non-destructively are described below. The main difference between older techniques used to assess surface composition and those employed today is the diameter of the particle or energy beam. A beam with a very small diameter is now used, making it much easier to avoid temper or other inclusions. Areas of the surface selected for analysis should be reasonably flat, smooth, and free of temper.

Non-destructive Compositional Analysis with PIXE and XRF

Non-destructive compositional analysis is particularly important for the examination of rare or sacred materials. Kuhn and Sempowski (2001) proposed such a method to study the movement of ceramic pipes from the Mohawk region (at the eastern extent of the territory) to well-dated Seneca sites (at the western extent) and estimate the timing of formation of the classic Five Nations League of the Iroquois. They performed elemental analysis of ceramic pottery and pipes with X-ray fluorescence (XRF) and particle-induced X-ray emission (PIXE) spectrometers equipped with energy-dispersive Si (Li) detectors (see Chapter 37).

A total of 115 potsherds and 96 ceramic pipes, or fragments thereof, from 8 Seneca sites and 121 potsherds from 2 Mohawk sites dating from ca. A.D. 1550 to 1625 were analyzed. The data from potsherds were obtained in order to establish the composition of local clay from both Seneca and Mohawk regions. Multiple measurements were taken on ten samples to demonstrate data were reasonably representative of matrix composition. Thick samples were not subjected to PIXE analysis. Concentrations of 15 major, minor, and trace elements were measured. Data were transformed into logarithms to rectify problems attributed to unequal weighting of variables by the statistical analysis software.

Principal component and cluster analysis of potsherd compositions demonstrated intersite similarity among Seneca sites and the ability to clearly separate clays from the Seneca and Mohawk regions on the basis of elemental composition. Discriminate function analysis was used to produce functions that discriminated potsherd matrices correctly 92.83% of the time. When applied to ceramic pipe data, 90 were identified as similar in composition to Seneca pottery, 5 were identified as similar in composition to Mohawk pottery, and 1 fell outside the composition range of both groups. On the basis of these results, the classic Five Nations League of the Iroquois formed between A.D. 1590 and 1605. The authors suggested that this approach could be used to establish the dates other groups joined the League or other questions about Aboriginal interaction and exchange.

This non-destructive approach enabled the analysis of irreplaceable pipes, many of which were recovered from burial contexts. Kuhn and Sempowski (2001) argued that acceptable results were possible if the surfaces analyzed were reasonably flat and smooth. In this study, areas possibly contaminated by leaching were avoided and efforts were made to avoid mineral temper. Primary tempering material was crushed rock that included plagioclase feldspar and the authors (2001:307) suggested, "even if some temper were being included in the data collection, it is doubtful that it would confound the results."

Non-destructive Compositional Analysis with LA-ICP-MS

Beck and Neff (2007) used results from a combination of colorimetric, petrographic, and compositional analyses applied to 150 sherds from three sites in the Gila Bend area to explore the interaction between Hohokam and Patayan potters. The sample consisted of large rimsherds that could be assigned to a specific ceramic type or ware. The Gila Bend area is believed to represent the western frontier of the Hohokam and the eastern frontier of the Patayan; ceramics from both traditions occur at sites in the region.

Laser ablation-inductively coupled-mass spectrometry (LA-ICP-MS) (see Chapters 30, 31, and 34) was used to sample and determine the concentration of ten elements. Areas of the surface selected for analysis exhibited the finest size fractions of the matrix and were also free of temper or other coarse-grained inclusions. Clusters or protogroups were identified using a variety of statistical methods and then refined using Mahalanobis distance calculations. Ten sherds with unusually heterogeneous compositions were excluded from the analysis. The iron oxide content of the clays was assessed on the basis of color using oxidation analysis, which involves refiring sherds at 950°C in an oxidizing environment for 2.5 h.

Oxidation analysis provided some information about the depositional context of the clay in that only leached alluvial clay oxidized to lighter colors; however, elemental analysis revealed that groupings based on oxidation colors did not correspond to groups made on the basis of chemical composition. Five compositional groups were defined; three of which (clusters A, B, and C) were based on aluminum concentration, one (cluster D) on a high tin concentration, and one (cluster E) on high antimony concentration. Of these, the clays of clusters A and B were alluvial and likely local to the Gila Bend area; the clays forming the other groups may be either alluvial or residual.

Beck and Neff (2007) found that Hohokam and Patayan potters both used local clays to make Hohokam Buff Ware and Lower Colorado Buff Ware, respectively. This implies the Gila Bend area was a shared environment and there were frequent opportunities for interaction and exchange. In addition, Patayan individuals or families may have moved into Hohokam communities to the east where they continued to manufacture Lower Colorado Buff Ware pottery.

Manufacturing

X-Radiography of Ceramics

X-radiography (Chapter 37) can be used to obtain information about the construction of ceramics that may be difficult to discern by other means. Braun (1982) demonstrated that X-radiography combined with point counting techniques provided a means of assessing the temper particle characteristics of Woodland pottery, in particular, minimum temper size, shape, and density. Pierret et al. (1996) used X-radiography to examine the porosity of ceramics. Rye (1977, 1981) focused on the preferred orientation of inclusions (temper) and voids detected by X-radiography to identify vessels formed by coiling, paddle and anvil, and wheel throwing. Berg (2008) expanded on this work with the X-radiography of 101 hand-built and wheel thrown vessels manufactured for the study by two specialists using different primary and secondary forming techniques. Most vessels were cut in half when the paste was leather hard and shrinkage had stopped, then a secondary forming technique or surface treatment was applied to one of the two halves.

Through her study of primary forming techniques, Berg (2008) found that the angle of inclusions cannot be used to indicate the speed with which a wheel-thrown vessel was raised in height. This contradicted previous hypotheses that steeper inclusion angles indicated that the walls of a vessel were raised rapidly. Comparisons were made between vessels completely formed on pottery wheels (wheel-made) and hand-made coiled vessels finished on a wheel (wheel-shaped). Particles in fully wheel-made pottery had a diagonal orientation while the horizontal particle alignment, produced through the rolling motion and associated with coiled pottery, was maintained in coiled vessels subsequently shaped on a wheel.

Rye (1977, 1981) reported that the only secondary vessel forming technique detectable with X-radiography was paddle and anvil (or beating) as inclusions become preferentially oriented parallel to the vessel walls and star-shaped cracks appear around large mineral particles. Other secondary forming techniques did not alter the internal structure of the vessel. In support of this finding, Berg (2008) found the effects of moderately vigorous scraping, performed to even the thickness of vessel walls, or turning, performed with a tool held stationary against the vessel wall to remove clay from a rotating vessel, cannot be detected with X-radiography. Knife trimming was detectable if the process resulted in differential thickness, dislodging of temper or the creation of air spaces. Surface treatments, such as burnishing and slipping, were not detectable.

Berg (2008) confirmed that marble, granite, and shell temper as well as voids produced by the loss of organic temper during firing are most visible in an X-radiograph while quartz, sand, and grog temper are difficult to detect. Regardless of temper type, inclusions are obscured when thick sherds are examined and large temper is more easily detected.

Carr (1993) described a method of using X-radiography to aid in the sorting of sherds of household-made ceramics into individual vessels for more accurate assessments of vessel frequencies, types, styles, sizes, and functions. Carr

(1993) suggested X-radiography can reveal paste inhomogeneities, such as temper clustering, which may if unrecognized, lead to inaccurate interpretations of manufacturing practices. In particular, the technique could be used to document internal structures relating to the amounts, distribution, size, and type of temper and the frequency of void spaces. Types of manufacturing techniques employed could be identified through similarities in vessel fracturing. Carr (1993) suggested X-radiography is particularly useful for sorting ambiguous vessels, such as those that are plain or minimally decorated. Using the technique to examine material from sites in Ohio dating from 500 B.C. to A.D. 1400, Carr (1993) was able to sort 3500 sherds, each measuring at least 9 cm^2 in area, into 1000 vessels. Vessels from this period are characterized by great internal homogeneity but inter-vessel variability in tempering.

Carr (1993) recommended that categorizations be based on at least four, semi-independent dimensions, including primary production, stylistic elaboration, use and post-depositional alteration. Attributes with the most power to discriminate (i.e., most variable) should be weighted more heavily than those that are least variable within vessels. Construction of a hierarchical, sequential decision tree could also facilitate the sorting process. Carr (1993) found the best, or first-order, discriminators were visually detectable and primarily stylistic or functional. The next best (or second-order) discriminators were detected with X-radiography and could subdivide previously established groups. The third-order discriminators were traits that could vary within a single vessel, such as the presence of charred residues or paste lamination. Statistical analysis should be employed for large collections of sherds (Carr 1993).

In order to obtain best results, industrial or mammography films should be used together with "softer" radiation, because lower peak accelerating kilovoltages (kVp) maximize contrast. X-ray tubes with molybdenum targets were recommended for thinner and more lightly tempered pottery. Additional measures that could improve the quality of the image include reducing scattering radiation, imaging groups of sherds with approximately the same thickness together, and minimizing the distance between the film and the sherds (Carr 1993).

Determination of Firing Conditions

Analysis with Mössbauer Spectroscopy

Archaeological applications of Mössbauer spectroscopy (see Chapter 39) typically involve analysis of the oxidation state of iron to obtain information about the conditions under which ceramics were fired. Mössbauer spectra are produced when gamma ray emissions from iron-57 (^{57}Fe), which is the decay product of cobalt-57 (^{57}Co), on a moving source were absorbed by ^{57}Fe in the stationary sample. In general, pottery fired in an oxidizing (oxygen-rich) atmosphere contains mainly ferric (Fe^{3+}) and little ferrous (Fe^{2+}) iron, whereas pottery fired in a reducing (oxygen-poor) atmosphere contains mostly ferrous iron; magnetic species of iron oxide can be produced at high firing temperatures (Feathers et al. 1998; Guangyong et al. 1989; Hess and Perlman 1974).

Mössbauer spectroscopy is usually applied as part of a suite of analytical techniques that provide information about the chemical composition of the pottery. Where possible, samples of the clay likely used to manufacture the ceramics are obtained. Reference clays are fired and changes in the chemical states of iron oxide that occur with temperature are monitored with Mössbauer spectroscopy.

In the studies described below, Hess and Perlman (1974) and Guangyong et al. (1989) both examined the conditions that cause chemically identical raw clays to fire to markedly different colors. Feathers et al. (1998) investigated possible changes in firing conditions associated with the shift from mineral to shell temper.

Hess and Perlman (1974) examined pottery from Tel Ashdod, on the southern coast of Israel. Locally available brownish red clay did not appear to contain hematite (αFe_2O_3) but it was detected when the clay was fired in an oxidizing environment and a red ceramic was produced. When fired in an argon atmosphere (i.e., under completely oxygen-free, reducing conditions), the ceramic was black and contained magnetite. Hess and Perlman (1974) determined that hematite was present in raw clay as extremely small crystals that could not be detected with X-ray diffraction; firing caused the crystals to grow in size. In order to achieve desired colors, such as gray-green, the atmosphere of the kiln must have been carefully controlled, as a "delicate balance in the oxygen supply" was required (Hess and Perlman 1974:145).

Guangyong et al. (1989) compared gray- and red-colored samples from the terracotta warriors and horses from Xi'an, Shanxi Province, China, to local clay from a nearby hill. Results from XRD, XRF (Chapter 37), and INAA (Chapter 32) showed no significant differences in the major or trace element compositions of the archaeological samples and the local clay. The Mössbauer spectra of archaeological samples showed the gray-colored samples were dominated by ferrous species, while ferric species dominated the red sherds. The red color was achieved by firing the ceramic in an oxidizing environment at a temperature of 980±50°C. The gray color was probably achieved by first firing the ceramic at high temperature (minimally 830 ± 50°C in one case and 950–1000°C in another) in an oxidizing atmosphere then switching to a reducing atmosphere at a lower temperature. This firing regime would produce the observed ceramics, which have dark gray surfaces, dominated by ferrous species, and a reddish brown core, with a lower concentration of ferrous iron.

Feathers et al. (1998) applied Mössbauer spectroscopy to archaeological sherds and reference clays from southeastern Missouri to gain information about firing conditions. The archaeological samples dated from A.D. 500 to 1400, which encompasses the shift in tempering materials from coarse quartz and feldspar sand to crushed freshwater mussel shells. A sample of 20 sherds was selected to represent four groups: early and late sand tempered and early and late shell tempered. One type of reference clay, Langdon clay, was tempered with 30% sand by weight and fired at different temperatures in reducing and oxiding environments. Prior to analysis, surface layers were scraped off to ensure the core of the sherds was examined and possible contaminants and surface treatments were removed.

Data from X-ray diffraction (XRD), differential thermal analysis, and Fourier transform infrared spectroscopy (FTIR) were used to provide a general estimate of the degree of firing as low (less than 700°C), medium (700°C), or high (800°C). The results for the reference clays were similar to those reported for other clays heated under oxidizing conditions; however, the reduction of ferric species (to ferrous) under reducing conditions was quite slow.

Both ferric (Fe^{3+}) and ferrous (Fe^{2+}) species were detected in most archaeological samples but only ferric ions were present in low- and medium-fired sand-tempered sherds and a few of the low- and high-fired shell-tempered sherds. None of the sherds appeared to have been fired far in excess of 800°C. Black cores of varying thickness were observed in both sand- and shell-tempered sherds; since ferrous ions only appear in high-fired sand-tempered sherds, Feathers et al. (1998) suggested they were likely due to the presence of remnant carbon in low- and medium-fired sand-tempered sherds. Ferrous ions were present in low-, medium-, and high-fired shell-tempered sherds and firing cores were present in all samples. The ferric to ferrous ratio (Fe^{3+}/Fe^{2+}) tended to decrease with increased firing temperature in shell-tempered sherds. Maghemite, a magnetic iron oxide, was observed in one high-fired, early sand-tempered sherd. The mineral did not form in high-fired, shell-tempered pottery, which Feathers et al. (1998) found rather puzzling. It is possible that the paste was oxidized at a temperature below 800°C. Most shell-tempered sherds were likely fired in a reducing environment to slow the decomposition of calcium carbonate.

Analysis with Infrared Spectroscopy

Infrared (IR) spectroscopy (Chapter 35) provides information about firing temper because minerals exist between finite ranges of temperatures (Eiland and Williams 2000). The IR absorption spectra from the powdered ceramic material can be obtained by incorporating a very small amount into a pressed potassium bromide (KBr) matrix. Alternatively, IR reflectance spectra can be obtained from flat, polished, and inclusion-free areas of a sherd. Analysis targets oxides present in the ceramic material as carbonates, chain and framework silicates, iron oxides, aluminum oxides, and silica (SiO_2). If firing temperatures exceed 880°C, calcite ($CaCO_3$) decomposes to lime (CaO) and carbon dioxide (CO_2); however, decomposition occurs at lower temperatures in the presence of silica. The presence of vitreous material indicates high-temperature firing; in addition, clay minerals progress through a series of transformations at different temperatures. For example, as firing temperatures increase, kaolinite dehydrates to metakaolin, then to a SiO_2 and spinel phase, and finally to mullite at about 1075°C.

Eiland and Williams (2000) used FT-IR to examine ceramics dating between 5000 B.C. to the second millennium B.C. from Tell Brak, Syria. Prior to analysis, the uppermost 100–500 μm of the surface was scraped from an area measuring less than 1 mm^2 to ensure dirt or surface alteration products were removed prior to analysis. Data obtained from IR spectroscopy were compared to results from powder X-ray diffraction (Chapter 37), scanning electron microscopy (Chapter 37), and inductively coupled plasma (likely atomic emission spectroscopy) (Chapter 34).

Ceramics from the earliest periods, Halaf and Ubaid, were fairly uniform in composition and characterized by a high degree of vitrification. Higher levels of carbonates were present in pottery from the Akkadian, period; in addition, feldspar levels were high in basalt-tempered ceramics from this middle period. Firing temperatures used throughout the Akkadian period were likely lower than the earlier periods but variability in the results suggested the presence of imported vessels. Results from pottery dating to the second millennium B.C. were more uniform. Vessels were commonly tempered with calcite and the ceramics were fired at a low temperature to avoid its decomposition.

Using a slightly different approach, Barone et al. (2004) combined data obtained by FT-IR with mineralogical identification from XRD to estimate the maximum firing temperature of transport amphorae from a site in Italy.

Paints, Slips, and Glazes

Elemental Analysis of Glaze Paints and Slip Pigments

Recent investigations of paints and slips often employ techniques able to sample and analyze materials applied to the surface of ceramics in a non-destructive or minimally destructive manner. Tang et al. (2001) showed that X-ray diffraction using synchrotron radiation could provide compositional information about the slip on Attic black gloss sherds; however, this X-ray source is not readily accessible. For the routine analysis of materials applied to the surface of pottery and other archaeological materials, laser ablation is gaining popularity (see Speakman and Neff 2005). Laser ablation provides a means of removing miniscule amounts of paints, slips, and glazes applied to the surface for analysis; however, the actual removal process must be understood, as there are implications for compositional analyses. As outlined below, Neff (2003) found that laser ablation-inductively coupled-mass spectrometry (LA-ICP-MS) (Chapters 30, 31, and 34) can assess compositions as major oxides of elements but time-of-flight laser ablation-inductively coupled-mass spectrometry (TOF-LA-ICP-MS) (Chapters 30, 31, and 34) can provide good data for elements because variations in both the stream of ablated sample and the production of ions in the plasma torch are eliminated (Duwe and Neff 2007:406).

Neff (2003) used LA-ICP-MS to examine the composition of the slipped surface of 139 Plumbate pottery sherds produced in Mesoamerica. The gray or olive-green vitrified surface of Plumbate pottery was produced when the high alumina- and high iron content clay slip was fired in a partially reducing atmosphere. Previous work using instrumental neutron activation analysis (INAA) indicated the paste of Early Postclassic, Tohil Plumbate was distinct from the pastes of San Juan Plumbate, a simpler type produced in the Late and Terminal Classic; however, both types were manufactured on the Pacific Coast within a few kilometers of the Guatemala and Mexico border.

In order to counter the effects of variable yield during laser ablation and elemental fractionation, the same area of the sherd was sampled using four or five passes of the laser at low power (after four pre-measurement ablation passes at high power to remove surface contamination). Data, as counts for each oxide of the element, were determined by subtraction of the counts from a sample blank and corrected for isotopic abundance. Amounts were standardized with respect to aluminum (oxide). Results for reference clays were internally consistent and in good agreement with data obtained by INAA but LA-ICP-MS measurements were inflated by 16%, presumably because water and a few other components are not measured (Neff 2003:25–26). Neff (2003) was able to demonstrate the existence of two Plumbate slip compositional groups and match them to raw clay sources within the region. Two groups of potters shared the Plumbate pottery making technology but the Tohil variety was more extensively traded.

Duwe and Heff (2007) used TOF-LA-ICP-MS to examine the elemental compositions of glaze paints and slips on contemporaneous pottery from Bailey Ruin in east-central Arizona. The aim of the study was to discern common recipes that may indicate vessels were produced by potting communities. The pottery examined, White Mountain Red Ware (WMRW), is characterized by a thick red slip made from yellow clay with a high content of limonite, an iron oxide which becomes a red-orange color when fired in an oxidizing atmosphere. The Pinedale types of WMRW examined included Pinedale Black-on-red and Pinedale Polychrome. The glaze paints applied to the vessels consisted of silica to which galena, which is lead sulfide (PbS), was added as a flux to lower the melting point and produce a dark black color.

TOF-LA-ICP-MS was selected for this analysis so that it would be possible to spot-sample the glaze paint and slip without moving into the vessel paste. Laser ablation only penetrates to a depth of 30 μm and the glaze paint and slip thickness ranged between 40 and 200 μm. In order to reduce analysis time, small pieces, measuring approximately 5 $\times$ 5 mm, were removed from 161 vessels and mounted to glass slides and several samples were placed inside the laser chamber at the same time. Any post-depositional surface contamination was removed (pre-ablated) with the laser; ablation was performed with short laser pulses that moved across (<1 mm) the paint and slip. Ablated particles were swept into the plasma torch and ions entered the TOF mass analyzer. As outlined in Chapter 31, ions enter TOF mass analyzers with the same kinetic energy and separate on the basis of mass as they pass through a field-free zone called a drift tube.

Data for 40 elements were obtained, although Na and Cr were discarded because they occurred in only a few samples. In order to control for archaeological time, comparisons were only made between sherds recovered from floor proveniences. Samples were carefully selected on the basis of paste characteristics and thickness to ensure all 161 samples were from unique vessels. In order to test for homogeneity across the surface of a vessel, five samples of glaze paint were analyzed from one large sherd.

Initial groups established by principal component analyses of the base-10 logarithms were refined using jackknifed Mahalanobis distances. Results indicated that, except in one case where it was very thinly applied, the composition of the black pigment was likely not affected by over-ablation, i.e., the inadvertent sampling of the underlying slip. Three major compositional groupings were identified on the basis of differing levels of iron, lead, copper, and antimony in the glaze pigments. Group 1 consisted of only 13 members and was characterized by low levels of lead and copper; the remaining samples fell into the other two, slightly overlapping, recipe groups, both of which contained higher levels of lead and copper. The analysis of slips led to the recognition of two compositional groups, which differed with respect to the amount of silica, aluminum, and rare-earth elements. Duwe and Neff (2007) suggested that the three glaze-pigment groups and the two slip groups identified on the basis of composition were evidence that specific recipes were followed in their manufacture.

Stable Lead Isotope Analysis of Glaze Paints

The use of stable lead isotopes for provenance studies is described in Chapter 13. Expanding on the findings of Habicht-Mauche et al. (2002), Huntley et al. (2007) compared the lead isotopic and elemental composition of glaze paints on Rio Grande Glaze Ware to that employed in the Galisteo Basin. The glaze paints were composed of silica to which galena (PbS) was added as a flux to lower the vitrification point. Previous studies using ICP-MS indicated the Galisteo Basin potters obtained galena from deposits in the Cerrillos Hills (Habicht-Mauche et al. 2002). Huntley et al. (2007) wished to determine the source of lead used for Rio Grande Glaze Ware manufactured at Salinas pueblos.

A total of 83 bowl rim sherds from two Salinas pueblos occupied between the 1300s to the late 1600s were selected for analysis. The reference collection consisted of a total of 97 ore samples from nine mining districts in north-central New Mexico. Habicht-Mauche et al. (2002) had previously compared sample introduction by acid dissolution to laser ablation and found ICP-MS provided superior results when glaze paints and ores samples were dissolved in acid. Samples weighing about 1 mg were removed with a surgical steel blade and dissolved in 1% trace metal grade nitric acid for 1 week; undissolved material was removed prior to analysis. Total lead concentrations were obtained using inductively coupled plasma-optical emission spectroscopy (ICP-OES) (Chapter 34) to ensure appropriate lead standards were selected. Samples were diluted to concentrations below 1 part per million then isotope concentrations were determined using high-resolution ICP-mass spectrometry (ICP-MS) (Chapters 31 and 34).

Bivariate plots of isotope ratios indicated lead ores from the different mining districts were fairly distinguishable; samples from Cerrillos North and South formed a single group but there was some overlap with Magdalena and Hansonberg ores from the southern part of the study area. Huntley et al. (2007) suggest the Salinas potters obtained most of the lead ores for their glaze paints from nearby southern mines and did not have to rely on material from the Cerrillos Hills, which was used by, and possibly under the control of, Galisteo Basin potters. In some cases, ores from the Cerrillos Hills may have been mixed with ores from other sources.

The elemental compositions of glazes used in the later part of the period (glazes E and F) on 67 sherds were determined using an electron microprobe with wavelength-dispersive detectors (Chapter 37). Cross sections of sherds were mounted on glass slides, highly polished and coated with a thin layer of carbon prior to analysis. Data were obtained for nine elements and converted to oxide weight percentages; average percentages by weight were calculated from five readings and normalized to 100%. Four compositional groups, or glaze recipes, were identified through k-means non-hierarchical cluster analysis and principal component analysis on the standardized molecular proportions of each oxide to eliminate disproportional influences from heavy elements (Huntley et al. 2007:1143). The glaze recipes appear to differ with respect to the amounts of clay, flux, and colorants added.

Trapped Charge Dating

Trapped charge dating (Chapter 10) has not been extensively applied in North America, but Feathers (2000) suggested more extensive application of thermolu-minescence (TL) dating to pottery would be beneficial, especially for materials less than 300 years old. Preliminary results reported by Lipo et al. (2005) associated with the luminescence dating of Late Precontact ceramics from the central Mississippi River Valley show promise. Their goal is to resolve the ambiguity associated with radiocarbon age estimates by determining which sites are contemporaneous as well as the durations of site occupations.

TL dating is usually applied to pottery recovered from buried contexts. Dunnell and Feathers (1994) argued that reliable dates could be obtained from surface mate-rial as long as disequilibrium effects caused by sherd porosity and leaching were recognized and corrected. When a radioactive decay series is at equilibrium, the activities or decay rates of all members are identical (Chapter 3). The loss of a member, such as radon (Rn) gas through diffusion, creates a disequilbrium. In order to recalculate the TL date, the activity of the chain was compared to that of the par-ent and the amount of disequilibrium in each sample was determined (Dunnell and Feathers 1994).

Zacharias et al. (2005) found that loss of potassium, attributed to leaching during burial, could seriously affect TL dates. Potassium is a major contributor to internal radiation (Chapter 10), so dose rate calculations are greatly impacted by potas-sium leaching. TL dates would not be affected if the leaching occurred shortly after burial then ceased, i.e., according to the early leaching model. However, if the loss of potassium occurred long after burial, i.e., according to the late leaching model, errors in dose rate estimations could be as high as 50%. Zacharias et al. (2005) cau-tioned that errors in dose rate estimations of about 20–25% could be expected under exponential and linear leaching models.

Dykeman et al. (2002) compared the accuracy and precision of TL dates on ceramics to tree ring, or dendrochronological, dates on wood from sites in Dinétah, the traditional homeland of the Navajo in northwestern New Mexico. Precise estimates of the ages of these sites could not be obtained by radiocarbon, ceramic cross-dating, or dendrochronology. Radiocarbon dates on materials from these sites in the region overestimated their ages by 150–200 years. Ceramic cross-dating could not provide fine temporal resolution because the most common type, Dinétah Gray, remained largely unchanged between A.D. 1500 and 1800. Gobernador Polychrome pottery was only produced from A.D. 1640 to 1800 but its distribution was more restricted. Navajo pueblito structures represented less than 2% of the sites in the project area, which reduced the availability of well-preserved wood samples for tree ring dating.

Comparisons of tree ring and TL dates were made with the understanding that the resulting ages represented different events. Tree ring dates provided information about the death of a tree, whereas TL dates were related to the exposure of a vessel to intense heat, likely when it was fired. Unfortunately, many of the wood samples from the study area were weathered and exterior tree rings were damaged; as a result, dates from these materials predate the actual death of the tree. Attempts to estimate the number of rings lost through analysis of the interior, dark-colored heartwood, the light-colored sapwood, and the boundary between them generated erratic results. Despite these challenges, tree ring dates from the sites were believed to be accurate within a decade. Navajo tree harvesting in the area extended from A.D. 1629 to 1750. Sherds from surface and subsurface contexts were selected for analysis. Dose rates calculated on the basis of dosimeters buried for 1 year did not significantly differ from estimates based on the composition of sediments. Only 5 of the 19 TL dates on pottery corresponded well to tree ring dates.

Dykeman et al. (2002) noted that deviations from expected dates could occur if the onset of trapped charge accumulation did not correspond to the targeted event. A TL date on an old ceremonial or other type of non-utilitarian vessel pot that was curated by site inhabitants would pre-date the targeted event; this is called the old pot effect. Exposure of pottery to high temperatures long after site abandonment would result in a date younger than the targeted event. Some of the noncorrespondent dates in this study occurred on imported vessels and likely related to the old pot effect. The long duration of site occupation was also cited as a potential source of discord as the TL date from a pot fired at the beginning of site occupation could be substantially older than a tree ring date on wood from the last structure built at the site. Correspondent dates were more often obtained from Gobernador Polychrome, which was manufactured for a relatively short period of time. Whereas the average departure from tree ring dates was 44 years on this type of vessel, the average departure for Dinétah Gray vessels was 71 years. Some of the tree ring dates related to historic wood harvesting activities that occurred after A.D. 1870.

Chapter 18
Flaked and Ground Stone Tools

Analyses of stone tools are most often conducted to obtain provenance information or determine the time of site occupation. As outlined in the following section, there is a long history of provenance studies of obsidian. Recent studies in South America show geochronological techniques can aid in the discrimination of obsidian sources with similar composition. Although the composition of chert is more variable, a recent study has demonstrated that valuable information can be garnered by applying a range of techniques. Compositional studies of andesite, dacite, basalt, and rhyolite have also been conducted. While trace element compositions are frequently used to characterize lithic materials, provenience studies of steatite and turquoise demonstrate other strategies may be more suitable for materials with highly variable compositions.

The introduction of secondary ion mass spectrometry (SIMS) to measure the thickness of the hydration rind objectively and precisely is transforming obsidian hydration dating. Morgenstein et al. (2003) have proposed a technique for dating fine-grained volcanic rocks which combines uranium series dating and weathering rind formation and subjected these materials to infrared-stimulated luminescence dating. Trapped charge dating of burnt chert (or flint) using thermoluminescence is also possible.

Examples involving analyses of lithic raw material sources and dating techniques are presented in this chapter. Analyses of organic residues related to function are described in Chapter 22. The analyses of paints and pigments applied as decoration to lithic materials are presented in Chapter 25.

Obsidian Provenance Studies

Obsidian is volcanic glass that forms when silica-rich lava cools too rapidly for crystallization to occur. Its amorphous structure makes it ideally suited for flintknapping and ancient people in many parts of the world utilized obsidian for flaked stone tools. Not all obsidian deposits provide raw material suitable for stone tool manufacturing, however. Over time, obsidian undergoes spontaneous recrystallization, or devitrification, so only relatively young sources can be exploited. Most high-quality obsidian is found in deposits that formed within the last 15 million years in North

M.E. Malainey, *A Consumer's Guide to Archaeological Science*, Manuals in
Archaeological Method, Theory and Technique, DOI 10.1007/978-1-4419-5704-7_18,
© Springer Science+Business Media, LLC 2011

America and the last 10 million years in the eastern Mediterranean (Jones et al. 2003; Pollard and Heron 1996; Williams-Thorpe 1995). Obsidian sources in South America tend to be young; all those examined by Bellot-Gurlet et al. (1999) were less than 5 million years old.

Considerable effort has been invested in provenance studies of obsidian; Shackley (2008: 199) estimates that "thousands-if not tens of thousands-of pieces" of archaeological obsidian are analyzed each year. Williams-Thorpe (1995) and Pollard and Heron (1996) agree the results stemming from decades of work to characterize obsidian in the Mediterranean and Near East amounts to an archaeological science success story. Provenance studies have been conducted on obsidian from different parts of the New World (Glascock 1994), including Mesoamerica (Glascock et al. 1998), the American Southwest (Shackley 2005), Peru (Craig et al. 2007), Argentina (Yacobaccio et al. 2004) as well as Columbia and Ecuador (Bellot-Gurlet et al. 1999, 2008). A synthesis of studies of obsidian in the Great Basin of North America and ongoing work in South America are described below.

Shackley's (1998) edited volume should be consulted for various methodological and theoretical aspects of obsidian research. Almost every technique developed for the analysis of inorganic materials has been applied to the study of obsidian (i.e., virtually all those described in Part IV, with the exception of Chapters 33 and 35). At present, the analytical techniques most frequently used to establish composition are X-ray fluorescence spectrometry (XRF), instrumental neutron activation analysis (INAA), inductively coupled plasma-mass spectrometry (ICP-MS), sometimes with laser ablation (LA-ICP-MS), and particle/proton-induced emission of X-rays (PIXE) or gamma rays (PIGE). Shackley (2008) suggests that all of these techniques provide valid and comparable results.

Other techniques that have been applied include optical emission spectrometry (OES), electron microprobe, inductively coupled plasma emission techniques (referred to as ICP, ICP-OES, or ICP-AES), strontium isotope analysis, Mössbauer spectroscopy, and atomic absorption spectroscopy. Tykot (1997) preferred electron microprobe analysis equipped with a wavelength-dispersive X-ray spectrometer for his study of obsidian from western Mediterranean because the results were more precise than those obtained by LA-ICP-MS and per-sample analysis costs were significantly less than either XRF or INAA.

Although it has been studied for decades, recent work described below indicates some assumptions about the homogeneity of obsidian sources are at least partially false. At least four subgroups occur within the Mule Creek regional obsidian source in the American Southwest (Shackley 2008). Jones et al. (2003) report that several Great Basin obsidian sources have complex compositions and three different geochemical types were found to occur within one obsidian source. The high degree of compositional variability noted in a source in Ecuador appears to be due to complete mixing of two magmas (Asaro et al. 1994; Bellot-Gurlet et al. 1999, 2008).

Great Basin Studies Using XRF

Jones et al. (2003) used compositional data on obsidian and fine-grained volcanic (FGV) rock, primarily dacite and andesite, obtained over a period of more than

25 years to trace the movement of obsidian in the Great Basin during the Terminal Pleistocene and Early Holocene. Jones et al. (2003) assumed the Paleoarchaic inhabitants of the region obtained the majority of their raw lithic materials directly and exchange played only a minor role. If so, the movement of obsidian and FGV rock was connected to the movement of people and provenance data could be used to estimate territorial ranges.

Interpretations were based on the elemental compositions of lithics from 16 sites that were occupied between about 11,000 and 8,000 years ago. X-ray fluorescence spectrometry (XRF) (Chapter 37) data from 840 obsidian specimens from eastern Nevada and 76 from the Sunshine Well locality were used to define 40 chemical types. Despite the high number of potential sources, almost 75% of the obsidian was obtained from only four deposits ranging from 150 to 250 km away from the eastern Nevada project area: Butte Mountain, Brown's Bench, Panaca Summit, and source B. Only 7 of the 40 types occurred with a frequency greater than 2%. In addition, obsidian hydration dating techniques (Chapter 11) were used to measure the rind thickness in micrometers. Instead of calculating dates, a relative chronology was established by sorting assemblages according to the mean hydration values of seven obsidian types.

A total of 177 dacite and andesite flakes associated with projectile points and production bifaces were analyzed using wavelength-dispersive XRF. Although over 20 sources are known, only 7 were considered important. Dacite, which is 62–65% silica, was utilized more frequently because it was more easily flaked (Jones et al. 2003).

Jones et al. (2003) identified discrete obsidian conveyance zones, each extending over 450 km from north to south and 150 km from west to east. During the Terminal Pleistocene and Early Holocene, little movement of source materials occurred between zones. Artifacts fashioned from obsidian sources located in western or northern parts of the Great Basin were almost completely absent from sites in eastern Nevada. Jones et al. (2003) argued the obsidian conveyance zones corresponded to foraging territories. Three territories were identified across the middle of the Great Basin; one existed in the northern Great Basin and another was identified in the Mojave Desert.

Jones et al. (2003) noted that the lack of contact between the groups was due to the low population density and the movement patterns corresponded to the distribution of significant wetlands in the area. The patterns changed when the distribution and quality of resource zones were affected by the shift to drier climatic conditions.

Use of INAA and K–Ar Dating

Yacobaccio et al. (2004) used compositional and geochronological data to examine changes in the patterns of procurement and distribution of obsidian between 2200 and 400 B.P. in northwestern Argentina. Instrumental neutron activation analysis (INAA) (Chapter 32) was employed for compositional analysis, and potassium–argon dating (K–Ar) (Chapter 9) was used to compare the ages of obsidian deposits.

The geochemistry of major compounds was also determined for many of the obsidian sources.

Multiple samples were collected from each obsidian outcrop. Co-authors who visited obsidian sources typically collected more than ten samples; between two and four samples were available from other sources (Yacobaccio et al. 2004). A total of 177 artifacts obtained from 37 different sites were also analyzed. The material was subdivided into four chronological groups: (1) 2200–1800 B.P., the time when people lived in small villages and practiced agriculture and pastorialism (14% of sites); (2) 1800–1100 B.P., a period marked by changes in social organization at specific sites and the appearance of institutional inequality (21% of sites); (3) 1100–550 B.P., during which time there was a sudden development of urban spaces with concentrated populations, social inequality, and intensification of agriculture (51% of sites); and (4) 550–400 B.P., when the Inka conquest began (14% of sites).

Compositional data for 27 elements were obtained using INAA. The sources could be discriminated with nearly 100% success using three bivariate plots: (1) thorium (Th) vs. cerium (Ce), (2) manganese (Mn) vs. lead (Pb), and (3) lanthanum (La) vs. hafnium (Hf). The elemental composition of each obsidian source analyzed was plotted as an ellipse at the 95% confidence level; artifact compositions were plotted as individual points. The chemical compositions of nearly 90% of the artifacts corresponded to eight of the ten obsidian sources analyzed. The remaining 11.3% were fashioned from material obtained from ten as-of-yet unidentified obsidian sources.

Yacobaccio et al. (2004) reported that more than 80% of the obsidian was procured from three sources: Zapaleri, Ona-Las Cuevas, and Cueros de Purulla. The distribution sphere for obsidian from Zapaleri was 350 km wide, extending over the northern part of the study area. The distribution sphere for obsidian from Ona-Las Cuevas was 340 km wide and extended over the southern part of the study area. Obsidian from Cueros de Purulla and another source was found within the Ona-Las Cuevas distribution sphere. Material from two other sources appeared in both of the two main distribution spheres.

Changes in obsidian source use through time were assessed for Zapaleri, Ona-Las Cuevas and Cueros de Purulla. More than one-half of the artifacts made of Zapaleri obsidian date to 1100–550 B.P., the third chronological period. The source was also well utilized in the first (2200–1800 B.P.) and fourth (550–400 B.P.) periods. Ona-Las Cuevas obsidian was fairly consistently exploited over the first three chronological periods. The majority (89%) of artifacts made of Cueros de Purulla obsidian date to the first and third chronological periods.

This study demonstrates the potential for elemental analysis to shed light on resource exploitation and distribution patterns over time when (1) a limited number of raw material sources exist, (2) there is clear compositional variability between sources, (3) artifacts from many sites over a broad region are analyzed, (4) there is chronological control of artifacts, and (5) artifact manufacture does not alter its chemical composition.

Although the concentrations of major oxides in all sources were very similar, the obsidian was successfully discriminated on the basis of trace elements.

Compositional homogeneity was demonstrated through the analysis of multiple samples from different outcrops of each source. Bivariate plots clearly showed the extent of overlap between sources and the degree of correspondence of artifact composition to each source.

Yacobaccio et al. (2004) were able to define two mutually exclusive distribution spheres within the study area on the basis of the occurrence of Zapaleri obsidian in the north and Ona-Las Cuevas obsidian in the south. Relative changes in the intensity of use over time for the three major obsidian sources, Zapaleri, Ona-Las Cuevas, and Cueros de Purulla, were depicted graphically but the total numbers of artifacts traced to each source for each chronological period were not presented and could not be calculated by the reader.

Use of PIXE, Fission Track, and ICP Techniques

Bellot-Gurlet et al. (1999) matched artifacts from Columbia and Ecuador to specific obsidian sources on the basis of chemical composition and age of formation.

In order to do this, destructive compositional analysis of debitage or waste and non-destructive compositional analysis of tools was performed using proton-induced X-ray emission (PIXE) (Chapter 37). These data were combined with fission track age determinations (Chapter 9) of source material and artifacts.

The collection of materials analyzed included 142 obsidian artifacts, mainly debitage, from 45 prehispanic archaeological sites and 22 obsidian samples from seven obsidian sources in Columbia and Ecuador. The largest dimension of samples typically did not exceed a few centimeters and their thickness did not exceed 1 cm. Where permitted, polished thin sections were prepared from the sample; otherwise, natural surfaces of tools were analyzed. Data from 13 elements were included in the analysis: sodium (Na), aluminum (Al), silicon (Si), potassium (K), calcium (Ca), titanium (Ti), manganese (Mn), iron (Fe), zinc (Zn), gallium (Ga), rubidium (Rb), strontium (Sr), and zircon (Zr).

Bivariate plots were used to illustrate sample compositional characteristics. On the basis of composition alone, most artifacts matched four obsidian sources, three of which were examined in this study and one unknown source. One obsidian source, Mullumica, was heterogeneous, possibly due to incomplete mixing of two magmas (Bellot-Gurlet et al. 1999).

Fission track dating revealed that obsidian sources with roughly comparable compositions could have two significantly different ages, 0.18–0.20 million years (Ma) and 0.25–0.30 Ma. Bellot-Gurlet et al. (1999) suggested another, as yet unknown and older, source with a composition similar to Mullumica must exist. Bellot-Gurlet et al. (1999) concluded that problems might occur when the source of obsidian artifacts was identified on the basis of composition alone. Previous studies of obsidian employing INAA and X-ray fluorescence would not have discriminated different aged sources with similar compositions.

Analysis with PIXE provided data for a relatively limited number of elements. The approach adopted by Bellot-Gurlet et al. (2008) more than doubled the number of elements in the database of the region from 13 to 36. In addition to analysis by PIXE, obsidian compositions were determined with either inductively coupled plasma-atomic emission spectroscopy (ICP-AES) (Chapter 34) or ICP-mass spectrometry (ICP-MS) (Chapters 31 and 34). The material analyzed included additional samples from the Chacana caldera in Ecuador, which contains the Mullumica and Callejones flows and other deposits, and the Palentará caldera in Columbia.

Samples analyzed with ICP-AES and ICP-MS were dissolved in a mixture of hydrofluoric and perchloric acids (HF-HClO$_4$), dried then redissolved in hydrochloric acid (HCl), and diluted. Trace element analysis with ICP-MS also involved thulium (Tm) spiking and rare earth element chromatographic separations. PIXE was performed on polished cross sections.

The additional analyses confirmed the complex evolutionary history of both calderas, including support for the hypothesis that the Mullumica/Callejones type represents incomplete mixing of two magmas prior to eruption. Samples from both calderas exhibit significant variations in composition due to their complex formations. Although determinations using PIXE were less precise than ICP techniques (<5–10% vs. <3–5%) and involved fewer elements, Bellot-Gurlet et al. (2008) found that the same compositional groupings were identified and the data obtained using PIXE were comparable to data obtained by others using INAA and XRF.

Sourcing Other Lithic Materials

Other materials used to make flaked and ground stone tools are not as homogeneous as obsidian, which makes sourcing studies more challenging. The examples below involve the analysis of materials from the United States, including some of which are very recent, very novel, or both.

Sourcing Chert Using INAA and UV Emissions

Lyons et al. (2003) tested the possibility that exotic grayish white chert recovered from the Lost Dune site in southeastern Oregon was obtained from the Tosawihi quarry complex, located about 300 km to the southeast in Nevada. Their multi-pronged chert analysis approach included: 1) descriptions of artifacts and hand specimens; 2) the observation, photography, and measurement of UV fluorescence emissions; and 3) determination of elemental composition using INAA. The collection of material analyzed included 121 tools from the Lost Dune site, 5 tools from the Birch Creek site, located approximately 200 km to the east, and samples of 4 regional chert sources.

The artifacts and source samples were described on the basis of color, texture and luster. Specimens were examined under short- and long-wavelength ultraviolet light in a view box; visible–UV emissions were photographed and the color identified through comparisons with pigmented chips. Ultraviolet fluorescence spectra from intact samples were measured with a recording fluorescence spectrophotometer, enabling a more precise assessment of the emissions. This instrument is similar to those described in Chapter 35 except the emission of UV light is monitored rather than the absorption.

Instrumental neutron activation analysis was performed on 32 Lost Dune artifacts, 5 Birch Creek artifacts, and samples from 4 regional chert sources; previously obtained INAA data from Tosawihi source material were included in the analysis.

Artifacts were sorted according to groupings established on the basis of color, texture, and luster. The visible emissions from grayish white Lost Dune artifacts were green under short-wave UV light, which matched those from Tosawihi source materials. Under long-wave UV light, emissions from Tosawihi source material were purple but those from Lost Dune artifacts where purple, orange, or not detectable. The visible emissions of chert from other sources were different. The short-wave UV fluorescence spectra for material from the Tosawihi Quarry contained three strong peaks in the green range of visible light, which was a unique feature of this type of chert.

Concentrations of 36 elements were determined by INAA. Principal component analysis and Mahalanobis distance calculations were performed on the log-transformed concentrations of 17 elements. The first four principal components explain more than 95% of the variance observed. Mahalanobis distance calculations, based on principal components, were used to determine the probability that a particular artifact was made from a particular type of chert.

The chert groupings made on the basis of descriptive traits largely conformed to groups generated by statistical analysis of elemental concentrations. Lyons et al. (2003) suggested that the unique UV fluorescence of Tosawihi chert might be due to elevated levels of uranium. Compositional analysis with INAA revealed that material which appeared different from recognized sources on the basis of color was chemically similar. In the case of the local Harney Lake chert, the unusual coloring (high chroma) could indicate the material was from a hot spring facies rather than the typical lacustrine deposits (Lyons et al. 2003). No artifacts were found to match the chemistry of Roma beds chert.

Statistical analyses of the elemental compositions of a few chert artifacts placed them in two different groups, Owyhee River and Red Butte. Similarities in elemental composition could indicate portions of both these sources were altered by similar hot springs systems (Lyons et al. 2003). Variation explained by the first two principal components related to the major diagenetic chert types: replacement (Tosawihi), lacustrine (Harney Lake and Rome beds), and silicified sediment (Owyhee River and Red Butte).

The identification of Tosawihi chert at Lost Dune site was further supported by the presence of Prehistoric Intermountain Tradition pottery and bisymmetrical hafted "Shoshone knives" from the Great Basin. Much of the material from

Lost Dune was actually procured from local sources. Most of the obsidian at Lost Dune came from a source about 90 km to the northeast. The sandstone and pottery originated from an area about 150 km to the east.

Elemental concentrations determined by INAA were not presented or well summarized. The discussion focused on results of principal component analysis and the Mahalanobis distance calculations made on the data. This was particularly puzzling given two statements by Lyons et al. (2003:1150): (1) "Most of the complex multi-element differences among the sources summarized by principal component analysis are also apparent in bivariate plots showing standardized element concentrations of tantalum versus sodium (not shown) and of tantalum versus iron (not shown)" and (2) "General agreement of the artifacts with the source groups can also be illustrated with bivariate element plots." Given the simplicity and visual clarity of bivariate element plots, it is unclear why Lyons et al. (2003) abandoned their use in favor of complex statistical analysis and probability of membership calculations.

Clear differences in the visible–UV fluorescence observed between the Tosawihi chert and other source materials demonstrated the effectiveness of ultraviolet fluorescence spectrophotometry; INAA and statistical analyses further support these results. Given these findings, it appears the chert from this region can be sorted with confidence on the basis of UV fluorescence results alone.

Sourcing Fine-Grained Volcanic Rock with XRF and ICP-MS

Lebo and Johnson (2007) analyzed a small sample of basalt rocks and artifacts from the Hawaiian Islands of Nihoa and Necker. A total of 18 elements were determined with XRF; concentrations of 26 elements were assessed using ICP-MS. Although a larger sample of both geological specimens and artifacts were required, their results supported the hypothesis that rock was not transported between islands.

Sourcing Rhyolite with XRD

Pollock et al. (2008) used powder X-ray diffraction (XRD) (Chapter 37) to characterize flow-banded spherulitic rhyolite from two sources in New Hampshire: Mount Jasper and Mount Jefferson. The micro- to cryptocrystalline material was light olive gray with flow bands of various color and spherules ranging in diameter from 1 to 10 mm. Flow-banded spherulitic rhyolite from various sources in New England occurred in Paleoindian, Archaic, and Woodland sites in New England and southeastern Quebec. Although there were eight rhyolitic dike outcrops at Mount Jasper, only one appeared to have been exploited as a source of raw material for tools. The Jefferson source consisted of isolated blocks of spherulitic rhyolite in till and other glacial deposits; the largest observed measured about 50 × 30 cm.

XRD was performed on samples of source materials and artifacts that had been finely ground; the mass of powdered sample analyzed was not provided. Quartz and albite were the most common minerals in the 43 Mount Jasper dike materials analyzed. Pollock et al. (2008:695) indicated it was "for the most part mineralogically homogeneous" but varied in the minor occurrence of sanidine, orthoclase, or anorthoclase. The material from the Jefferson source differed from the dike in that the flow bands consisted of thin layers of mostly closely packed spherules. The spherules in Jefferson rhyolite were variations of gray with lighter colored rims and dark colored interiors; the spherules in the Mount Jasper dike material ranged in color from red to reddish brown.

The X-ray diffraction patterns from Mount Jasper source materials were uniform and consistent. Two different diffraction patterns were observed in the material from the Jefferson blocks, one of which was identical to the Mount Jasper material. Pollock et al. (2008) concluded it was not possible to unequivocally determine the source of an artifact using X-ray diffraction.

The diffraction patterns of 10 of the 12 Late Paleoindian and Archaic artifacts were similar to weathered and unweathered material from Mount Jasper. The diffraction patterns of the remaining two artifacts resembled material from the Jefferson source but were visually different with respect to the arrangement and weathering characteristics of the spherules. In this case, X-ray diffraction appeared to provide less information than visual separation. It is possible that other analytical methods, such as INAA of the powdered samples or LA-ICP-MS of intact spherules, could provide compositional data to assist this rhyolite sourcing study.

Sourcing Steatite with INAA

Truncer et al. (1998) proposed a method of sourcing steatite, also known as soapstone, on the basis of transition metal composition. In contrast to an earlier INAA study, a large number of samples from a small number of quarries located in Pennsylvania, Maryland, and Virginia were selected for analysis. The chemical compositions of between 22 and 31 samples from each of eight steatite quarries were determined by INAA. Where possible, archaeological debris, i.e., incompletely manufactured vessels, were collected from the quarries; otherwise loose material was selected.

The data obtained on samples from steatite quarries were compared to the compositions of 133 finished steatite vessels from sites in New York, Delaware, Maryland, New Jersey, Pennsylvania, and Virginia. Samples weighing between 1 and 3 g were taken from beneath the weathered and potentially contaminated surface. Aliquots weighing 150 mg were subjected to one short irradiation followed by one count and a long irradiation followed by two separate counts. Base-10 log conversions of data from 17 of the 33 elements measured were used in the statistical analysis. Source separations were not achieved through the use of bivariate plots of element concentration or principal component analysis; however, sources were differentiated at a regional level through the use of canonical discriminant functions. The elements contributing most to the separation of sources were transition metals [Fe, Zn, scandium (Sc), vanadium (V), cobalt (Co)] rather than rare earth elements. Truncer et al. (1998) reported that differences in transition metal concentrations reflected parent material concentrations and were easier to detect than rare earth elements. Only 8 of the 42 known quarries in the region were examined; the discriminating ability might have been better had larger sample sizes been used.

The discriminant functions were used to assign steatite vessels to sources on the basis of composition. Two-thirds of the samples were judged to be different from seven of the eight quarries at the 90% confidence interval and the eighth source exhibited a high degree of variability. One unexpected result was that most vessels from New York did not appear to be formed from New England steatite, the closest known source. Truncer et al. (1998) noted that most sites had steatite vessels from more than one steatite source. Similar studies of many more steatite quarries in the Middle Atlantic region would be required in order to assess the degree of inter- and intrasource variability, but regional-level source assignments were possible.

Sourcing Turquoise with Stable Isotopes

Hull et al. (2008) conducted a study to determine if the ratios of hydrogen and copper isotopes (Chapter 13) could be used to discriminate turquoise deposits and source artifacts. Turquoise is a blue-green hydrous copper aluminum phosphate with a variable composition. Previous attempts to source turquoise on the basis of trace element and rare earth element concentrations and stable lead isotopes were not completely successful due to the wide variation in composition of this mineral.

Hull et al. (2008) reasoned that the composition of a specific turquoise deposit should be related to the water from which it precipitated. The isotopic fractionation of hydrogen isotopes in water derived from precipitation (meteoric water) at a particular location is distinct and related to the specific environmental conditions that existed at the time of its formation. Turquoise precipitates from and retains rainwater so its stable hydrogen isotopic composition should be useful for provenance studies. Copper is a transition metal and variations in its stable isotope compositions are due to abiotic and biotic processes. Abiotic processes reflect differences in reaction rates of chemical processes; biotic processes reflect the preference of microbes for lighter isotopes.

Minimally destructive solid-sample microanalysis of turquoise was performed with a secondary ion mass spectrometer (SIMS) (Chapter 31) and electron microprobe (Chapter 37). The material examined included deposit samples from 12 mines located in Arizona, Colorado, New Mexico, and Nevada. A total of 17 turquoise artifacts from sites in New Mexico were analyzed including 11 from several sites in Chaco Canyon and 6 from site ENM 848, Guadalupe Community.

All samples were polished, washed, and rinsed prior to analysis. Reflected-light photomaps were prepared to identify sampling regions of turquoise devoid of mineral impurities, such as inclusions of apatite, iron oxide, and quartz, and areas altered by weathering. Prior to conducting stable isotope analysis with SIMS, deposit samples and all artifacts were mounted in epoxy resin and sputter-coated with a thin layer (ca. 200 Å) of gold to ensure surface conductivity. A primary electron beam of oxygen ions (O^{1-}) with a diameter of about 20 μm was used for SIMS analysis. Multiple samples were made on each spot and the internal precisions of measurements were given. Elemental analysis of thin sections prepared from selected mine samples was performed with an electron microprobe equipped with a wavelength-dispersive X-ray analyzer. Oxygen and water contents of the turquoise were also determined.

Values for δD (or $\delta^2 H$) were calculated as the per mil ratio of deuterium to hydrogen in the sample and compared to Vienna Standard Mean Ocean Water (V-SMOW). Values for $\delta^{65}Cu$ were calculated as the per mil ratio of ^{65}Cu to ^{63}Cu in the sample compared to the NIST976 standard. Correction factors for instrumental mass fractionation were calculated on the basis of accepted isotopic composition of standards. A linear correlation existed between mass bias and iron content, which enabled the ratio of ^{56}Fe to H to be used as a correction factor.

Analysis using SIMS was found to be rapid and virtually non-destructive compared to traditional methods of stable isotope analysis that require 5–30 mg of powdered sample and time-consuming wet chemical processing. The stable hydrogen and copper isotopes from mine samples were relatively homogeneous within the Sleeping Beauty mine; values on samples from different mines within the Cerrillos Hill Mining District were also similar. A bivariate plot of $\delta^{65}Cu$ vs. δD showed the stable isotope values separated mines from different regions and clustered mines from the same region together. The stable isotope compositions of 13 of the 17 artifacts examined corresponded to those of turquoise sources examined in this study. Hull et al. (2008) concluded that Chaco Canyon artifacts were mined from a variety of sources while those from Guadalupe Community site ENM 848 were mainly from two sources.

Dating Lithic Material

Obsidian Hydration Dating with SIMS

While Friedman and Smith (1960) originally promoted obsidian hydration dating (OHD) as an inexpensive method of estimating the age of flaked obsidian using optical microscopy, recent studies show that more accurate and precise measurement of rind thickness is possible using SIMS (Anovitz et al. 1999; Stevenson et al. 2001) (Chapter 11). Stevenson et al. (2004) conducted an extensive comparison of the methods using material from Mound 11 and Mound 25 at the Hopewell site and Mound 13 at the Mound City site in Ohio. Of the 22 samples from Mound 11, OHD ages for 19 were previously determined using optical microscopy. OHD ages on 1 of the 7 samples from Mound 25 and 6 of the 31 samples from Mound 13 were reported previously.

Variables required to calculate the rate of hydration of a specific type of obsidian at a specific location were experimentally determined. The diffusion coefficient and activation energy range for the obsidian were measured using the hydrate and quench technique. Flakes of obsidian were suspended over water in a pressure vessel and heated to 140–180°C; after rapid cooling, the newly formed hydration layer was measured. The initial water content of the obsidian was measured by infrared spectroscopy (Chapter 35). The effective hydration temperature and relative humidity were determined using salt-based monitors, or cell pairs, buried at Hopewell site at depths of 20, 40, and 80 cm for 1 year. In addition, Stevenson et al. (2004) used induced hydration experiments from 6 to 266 days duration to confirm the thickness of the hydration rind was related to the square-root of time, as predicted by Friedman and Smith (1960).

The possibility that mound samples were altered by heat at the time of site occupation was investigated by comparing the optically measured thickness of each side of the artifact. The difference in unaltered specimens should be less than 0.25 μm, which is the error limit imposed by optical resolution (Stevenson et al. 2004). Thickness divergence in excess of 0.25 μm was observed in five samples from Mound 11 and two samples from Mound 25 at the Hopewell site; no hydration rim was visible on either side of two other Mound 25 samples. The majority of samples from Mound 13 at Mound City exhibited differences that greatly exceeded 0.25 μm; on one, the difference between the two sides was 5.38 μm. Experiments confirmed that exposure to heat caused an existing hydration rim to expand and then disappear; exposure to high temperatures (more than 750°C) prior to hydration led to the formation of thicker rims (Stevenson et al. 2004). Samples from Mound 25 at the Hopewell site and Mound 13 at Mound City were deemed unsuitable for OHD and excluded from the study.

The concentrations of hydrogen atoms at various depths, or hydration profiles, were determined on nine samples from Mound 11 at the Hopewell site, using SIMS. The point of inward water diffusion was measured at the full-width-half-maximum of the hydrogen profile with an estimated error of ±0.05 μm. One-to-one comparisons indicated that the SIMS thickness values tended to be less than those measured optically and the difference between the values exceeded 0.25 μm in more than 50% of the samples. A paired two sample t-test indicated the measurements made using SIMS were statistically different from measurements made on the same artifacts using optical microscopy.

OHD age estimates on nine samples from Mound 11 ranged between 258±119 B.C. and A.D. 607 ± 94. The OHD ages on five samples overlapped radiocarbon ages on carbon adhering to Mound 11 materials. The older OHD ages on two samples were attributed to artifact curation. Lower than expected near-surface hydrogen concentrations observed on two artifacts indicated some type of interference in the hydration history caused younger than expected OHD ages (Stevenson et al. 2004). The accepted OHD ages show the obsidian at Mound 11 had accumulated over a period of approximately 600 years. Stevenson et al. (2004) suggest this was evidence of the "ritual collecting" of items prior to their deposition in a sacred mortuary context.

Dating with U Series and Trapped Charge Dating

While trapped charge and uranium series dating of lithics is not necessarily commonplace, it is more often applied in the Old World than in North America. Truncer (2004) recently reported that initial attempts to obtain thermoluminescence (TL) dates on steatite vessels produced encouraging results. Analytical details were not provided, nor were the age estimates on steatite compared to TL dates on ceramics or to radiocarbon dates on associated organic material.

While burned flint has been successfully dated at many sites, variations in site environment over time can produce anomalous results (Richter 2007). Rink et al. (2003) encountered problems with obtaining reliable TL ages on burned flint from the Early Levantine Mousterian from the Rosh Ein Mor site, Central Negev, Israel. The site was situated on a river back terrace about 70 m above a spring. The site was excavated from 1969–1972 and all materials were interpreted as Early Levantine Mousterian, i.e., more than 200,000 years old. Three radiocarbon ages previously obtained on ostrich eggshell from the site were more than 37,000 B.P., more than 44,000 B.P., and more than 50,000 B.P.

In an attempt to improve temporal resolution, Rink et al. (2003) performed TL dating on five pieces of burned flint from the 1969–1972 excavations and uranium series dating on 8 g of ostrich eggshell recovered from the site in 1996. The outer 2 mm of the burned flint was removed. The elemental composition of the sample was determined by INAA. Carbonates were removed by acid etching prior to dating. The sample was divided into two equal portions. The TL signal was measured on one portion; the equivalent dose with correction for initial supralinear growth was determined from the second portion using the additive method. The external contribution of the annual dose was determined using two TL dosimeters that had been buried at the site. Difference between measurements from the cells was attributed to soil moisture and an average value was used.

To ensure a sufficient sample size, fragments of ostrich eggshell recovered from depths ranging from 8 to 29 cm below surface were combined (Rink et al. 2003). The shell was dissolved in 6 N nitric acid and anion exchange columns were used to separate uranium from thorium. Element concentrations were determined with thermal ionization mass spectrometry (TIMS) (Chapter 31). The sample produced the $^{230}Th/^{234}U$ ages of 200.9 +9.5/−8.7 thousand years (ka), using an early model of uranium uptake.

The TL ages on the burned flint ranged between 14 and 48 ka; exclusion of an obvious outlier produced an average age of 35 ± 8 ka. Internal radioactivity contributed 70–90% of the annual dose of four samples and 51% of the fifth. Rink et al. (2003) suggested the flint was either heated during a later, ephemeral occupation of the terrace or by natural fires that occurred when an oak-dominated forest grew in the area. The TL ages of the five flint samples were rejected. The U-series age of the ostrich eggshell was interpreted as the minimum age of the Early Levantine Mousterian materials (Rink et al. 2003).

Morgenstein et al. (2003) proposed a method of dating fine-grained volcanic artifacts, such as andesite, dacite, and basaltic trachyandesites, and applied it to materials from site 45K1464, located in the Cascade Mountains of Washington. Over time, feldspar in the igneous rocks decomposes to clay and clay-rich weathering rinds form on artifacts fashioned from these fine-grained volcanic rocks. The composition of the weathering rind of one lithic artifact, an andesite core, was examined on the basis of activities of radioactive minerals at different depths using α- and γ-spectrometry. Instead of being equal to ^{230}Th, the concentration of radium (^{226}Ra) in the rind exceeded expectations. Morgenstein et al. (2003) found that the level of excess ^{226}Ra (where excess ^{226}Ra $= {}^{226}$Ra $- {}^{230}$Th) in the weathering rind decreased with depth and then stabilized to produce a steady-state plateau.

Morgenstein et al. (2003) proposed a two-layer model to explain the distribution of ^{226}Ra. According to the model, the outer layer was diffusion-controlled and levels of ^{226}Ra were related to levels in the soil water. The inner layer was weathering controlled and ^{226}Ra levels were related to the rate of weathering and rind formation. They (2008) used calculus to calculate the rate of rind growth (w) at the boundary between the outer and inner layers as:

$$F_d = wA_d$$

where F_d is the diffusion flux of ^{226}Ra and A_d is the measured excess activity. The age of lithic artifacts and Pleistocene pebbles was then estimated using the calculated accumulation rate of the clay-rich weathering rind of $(6.6 \pm 1.6) \times 10^{-5}$ mm year^{-1} and the maximum rind thickness measured optically on a polished cross section. The values, which represent minimum ages due to a possible delay in the onset of rind development, corresponded very well with ^{14}C dates on associated sediments.

The time since last exposure to sunlight was estimated by obtaining infrared-stimulated luminescence (IRSL) ages on flakes collected at night. Poor results were obtained from large (90–250 μm) grains scraped from the surface so luminescence signals from smaller (1–8 μm) grains were used. Despite inadvertent mixing of samples and younger than expected age estimates, Morgenstein et al. (2003) reported that the technique showed promise.

Morgenstein et al. (2003) suggested weathering rind accumulation rates were particularly useful in situations where radiocarbon or alternative dating techniques were not feasible. In the case of mid-Holocene site 45K1464, there was a 5000-year long sedimentation hiatus, which ended about 7000 cal B.P., and radiocarbon dates on material that accumulated after sedimentation resumed resulted in a very broad window of cultural activities. They suggested that the rind formation rate calculated from expensive uranium series dates obtained on one artifact could be used to determine the age of other artifacts at the site. Morgenstein et al. (2003) admitted further research is required to more fully investigate the utility of these techniques.

While his research represented the results of a pilot study, the broad similarities of this technique to cation-ratio dating are somewhat concerning. The rate of weathering rind growth was established on the basis of uranium series dates on one andesite core and then applied to all other volcanic lithics at the site. While

Morgenstein et al. (2003) acknowledged that the technique is both site- and material-specific, many more uranium series dates would be needed to clearly demonstrate the validity of the technique within a single site. Uranium series dating on multiple artifacts fashioned from andesite, dacite, and basaltic trachyandesite would be required to demonstrate that the weathering rinds on different materials grow at identical rates. The researchers should also demonstrate that intra-site environmental variability, burial rate, and precipitation variability over time do not affect the growth rate of the clay-rich rind.

Chapter 19
Bones and Teeth

The analysis of bone for radiocarbon dating is central to archaeology; however, there are other potential uses. Faunal material, tooth enamel in particular, can be reliably dated by electron spin resonance (ESR) and uranium series techniques. Analysis of amino acid content using high-pressure liquid chromatography (HPLC) provides information about collagen preservation; amino acid racemization (AAR) ages can be obtained by comparing the ratio of amino acids in original form to those that have converted to their mirror images. Ancient DNA can be extracted from bone fragments and amplified using the polymerase chain reaction, enabling determination of species. Stable carbon and nitrogen analysis of bone and teeth provides information about the diet of the animal. Analysis of strontium, oxygen and hydrogen isotopes in these tissues gives information about seasonal migrations or other movements of an animal.

The success of these applications rests heavily on the preservation of the organic components, primarily collagen, and the inorganic component, apatite. This chapter begins with the methods used to detect and assess bone degradation.

Bone Diagenesis

Assessing Collagen Degradation

Assessment of collagen preservation is particularly important because this type of protein is the preferred sample material for radiocarbon dating, amino acid racemization dating, and stable carbon and nitrogen isotope analysis. Studies show that microbial activity plays a major role in collagen degradation.

Child et al. (1993) reported that over 200 organisms found in soil and feces could "subsist" on collagen alone; of these, 13 bacteria produce collegenase, the enzyme that breaks down collagen. Bacteria found in soil could preferentially attack either R- (or D-) aspartic acid (44 species) or S- (or L-) aspartic acid (38 species). Although the effect was not proven, these microorganisms could alter the ratio of S to R the inverse of D/L of amino acids. This finding supported the hypothesis that discrepancies between obtained and expected ages based on amino acid racemization of collagen might be due to microbial action.

M.E. Malainey, *A Consumer's Guide to Archaeological Science*, Manuals in
Archaeological Method, Theory and Technique, DOI 10.1007/978-1-4419-5704-7_19,
© Springer Science+Business Media, LLC 2011

Grupe et al. (2000) examined the effects of soil bacteria on bone collagen to trace the impact of microbial degradation on stable carbon and nitrogen isotope values and to assess collagen preservation in archaeological specimens. Marten femora were inoculated with three types of bacteria, then the effects on (1) non-mineral bound, non-collagenous proteins, (2) mineral-bound, non-collagenous proteins, and (3) collagens were examined after 6–9 months of storage. Amino acid profiles were determined using an HPLC-supported amino acid analyzer (Chapter 33). The $\delta^{13}C$, $\delta^{15}N$, and C:N values were determined using a mass spectrometer (Chapter 31) equipped with an elemental analyzer (Chapter 38). The $\delta^{13}C$ values of collagen exposed to microbial degradation were up to 2.4‰ more negative (depleted) and the $\delta^{15}N$ values were 3.6‰ more positive (enriched) than the control specimen.

In addition, Grupe et al. (2000) examined the collagen amino acid profiles of archaeological human bone and found that soil microorganisms preferentially consumed amino acids with the greatest number of carbon atoms. Consequently, the isotopic signals on similarly affected collagen would reflect the diagenetic history of the bone rather than the subsistence behavior of the individual. However, amino acid analysis and gel electrophoresis (Chapter 33) could be used to identify and exclude samples likely to give erroneous values (Grupe et al. 2000). Otherwise, changes in amino acid concentrations could be used to estimate the original $\delta^{13}C$ value in an unaltered collagen sample.

A study by Pfeiffer and Varney (2000) investigated the relationship between the preservation of cortical bone tissue (histological preservation) and collagen integrity using samples of human femoral bone from a military cemetery in Fort Erie, Ontario, dating to A.D. 1814. Collagen was extracted from different layers of cortical bone with preservation ranging from good to extremely poor. Midshaft cross sections of 23 femora were selected; the ratio of calcium to phosphorus in a few samples was determined using energy-dispersive X-ray microanalysis (see SEM-EDS; Chapter 37). Bone apatite was examined using X-ray diffraction (Chapter 37) and bone density was assessed on the basis of gray-level comparisons of microradiographic images (Chapter 37). The organic component from the outer (adjacent to the periosteal surface), inner (adjacent to the medullary canal), and middle layers was extracted separately. The C:N ratios of the extracted collagen were determined using a CHN elemental analyzer (Chapter 38). The amino acid composition was determined using HPLC (Chapter 33).

The collagen yield from all samples ranged between 7.9 and 21.4%, which is within the recommended range for stable isotope analysis of diet (5–25%). The C:N values of 17 of 22 samples were between the recommended range of 2.9 and 3.6, 5 poorly preserved samples had higher than acceptable C:N values. With respect to C and N concentrations, three samples exhibited values lower than the recommended minimum of 3% C and 1% N. The amino acid yield of the archaeological samples ranged from 3 to 17% of the standard by weight, well below the recommended level of 95% of the standard. There was no relationship between the C and N composition and collagen yield.

Analysis showed that little of the material extracted from the bone was actually collagen; in some cases, less than 10% of the extract actually consisted of amino acids. The amino acid profiles of the archaeological samples were generally consistent with fresh collagen; however, HPLC analysis showed the variation between the aliquots was very high, indicating the samples were not homogeneous. Removal of lipids prior to extraction improved both protein yield and elemental characteristics.

While histological preservation was not an indicator of the quantity or quality of the material extracted, the amino acid compositions of samples from the outermost layer deviated the most from expected values and exhibited the most variation. Pfeiffer and Varney (2000) recommended that external surfaces, i.e., both outer and inner, should be removed prior to analysis.

Assessing Apatite Diagenesis

Recent studies document changes in the structural and chemical composition of buried and unburied bone on the ground surface and within the burial environment. Trueman et al. (2004) examined the processes that affect the composition of unburied bone in the tropical savannah grasslands of Kenya. They found that bone exposed on the surface absorbs barium and lanthanum from the soil water; the size of bone crystallites increases and bone protein degrades. If microbial decomposition was present, the damaged bone mineral was prone to dissolution and reprecipitation. Calcite, barite, and, to a lesser extent, crandallite infilled vascular spaces of many bones; this process, called premineralization, can reduce the porosity of bones on the surface by 95% within 2 years of exposure.

Berna et al. (2004) compared the solubility of carbonated hydroxyl apatite extracted from archaeological bone to synthetic hydroxyl apatite. They found that the mineral fraction of bone is considerably more soluble than synthetic hydroxyl apatite in deionized water and neutral pH solutions. Under slightly basic conditions, pH 7.6–8.1, minerals in bone dissolved and then reprecipitated as less soluble minerals. Berna et al. (2004) noted that these conditions were common in nature and most bones will undergo similar recrystallization processes. Adderley et al. (2004) reported that dissolved bone minerals can recrystallize in the surrounding soil matrix. They suggested that amorphous material consisting of calcium, iron, and phosphate found in soil pores at an early fishing community represented the recrystallized remains of cod fish bone.

As outlined in Chapter 37, X-ray diffraction (XRD) uses the diffraction of X-rays by a mineral to establish the properties of a crystal lattice. Bartsiokas and Middleton (1992) suggested that X-ray diffraction (XRD) could indicate the degree of fossilization and provide a means of relatively dating bone. Diffraction patterns from modern, archaeological, and fossil bone were observed with a diffractometer and on film using a Debye–Scherrer camera. Cortical bone samples weighing less than 1 mg were drilled from mature animals.

The peaks produced by carbonate hydroxyapatite differed in a manner that corresponded to the age of the sample. Older fossil and archaeological specimens produced sharper peaks than modern and younger archaeological specimens. The intensity of a particular peak, the 300 reflection, was used as the basis of the proposed crystallinity index (I). This value was determined using the formula:

$$I = 10(a/b)$$

where a and b represent estimates of the local and general background, respectively, in the region of the 300 reflection. The value for a is the distance between the top of the peak and the average between valleys immediately adjacent to the 300 reflection. The value for b is the distance between the top of the peak and the X-axis.

Bartsiokas and Middleton (1992) suggested that fossilization leads to a progressive increase in crystallinity that may provide a means of relatively dating bones if the hydroxapatite growth was time dependent and spontaneous. They found that a strong relationship existed between the base-10 logarithm of the age of the bone and the crystallinity index:

$$\log_{10}(\text{age}) = 1.13 + 0.74I$$

The influence of burial environment, species, cooking, or burning and sample location on bone must be considered as well (Bartsiokas and Middleton 1992).

Lee (1995) examined mineral changes in fossil bone related to the diagenesis of carbonates using both XRD and solid-state nuclear magnetic resonance spectroscopy (NMR) (Chapter 36). The sample consisted of 10 mg of bone from a 15,000-year-old bovine knuckle without apparent microbiological degradation recovered from a cave in Spain. XRD was performed using a powder diffractometer; the diffraction pattern best matched that of a lightly carbonated hydroxyapatite.

Lee (1995) noted that, unlike XRD, NMR can be used to examine the chemical environment of targeted species whether or not that sample is crystalline (in a lattice) and the intensity is related to the concentration of absorbing nuclei. The ^{13}C cross polarization (CP) magic angle spinning (MAS) NMR signal from pure calcite was eight times stronger than that of the bone, indicating the bone was 11% calcite, by mass. A number of peaks from 0 to 50 ppm were assigned to organic material, which had not been detected by XRD. This determination was made by suppressing the signal from carbon atoms that are adjacent to hydrogen atoms. This caused carbon peaks from organic substances to disappear so remaining peaks arose from inorganic carbonates (Lee 1995). After the water detected in the proton (^{1}H) MAS NMR spectra was eliminated, a peak at 0.8 ppm remained. This peak was interpreted as the signal from a hydrogen atom in a hydroxyl group, similar to those found in hydroxyapatites. Lee (1995) concluded the techniques had different strengths. XRD indicated the presence of three inorganic components but did not detect the presence of organic material. Proton and ^{13}C NMR were unable to detect the calcium phosphate hydrate.

Blau et al. (2002) used X-ray diffraction with synchrotron radiation to test the hypothesis that wrinkling and pitting of enamel of archaeological human teeth were due to a condition known as fluorosis. Fluorosis is caused by too much fluoride and the uptake of excess fluorine could alter the structure of enamel. X-ray diffraction with high-energy, short-wavelength synchrotron radiation was selected so that the analysis would be performed in a relatively rapid manner on a small sample of enamel.

Samples weighing about 10 mg were taken from each of 42 teeth and finely ground. The material was placed in a 0.5 mm capillary tube and submitted for analysis. Suitable diffraction patterns were obtained from 30 of these and then compared on the basis of two parameters that describe, in part, the hexagonal structure of apatite. A plot of these lattice parameters resulted in two distinct scatters of data points. Most teeth containing defects occurred in one scatter, but some also appeared in the other scatter.

While the correlation was not perfect, teeth in one scatter tended to come from sites in the southern part of the United Arab Emirates and exhibited higher occurrences of enamel defects and dental caries. The teeth in the other scatter tended to come from sites in the north where the water contains lower levels of fluorine.

Wright and Schwarcz (1996) described a method that uses Fourier transform-infrared spectroscopy (FT-IR) (Chapter 35) to detect diagenetic material in bone after the application of acid treatments, such as those described by Garvie-Lok et al. (2004) (see below). Samples of bone were ground into a powder then sieved, 2 mg of the fine powder was mixed with 200 mg of potassium bromide (KBr), pressed into a pellet, and analyzed. Following Shemesh (1990), the degree of crystallization was calculated by a crystallinity index (CI):

$$CI = \{A_{565} + A_{605}\}/A_{595}$$

where A_x is the absorbance at wavenumbers 565, 605, and 595, respectively. The carbonate content was given by the ratio of absorbances of carbonate (CO_3) and phosphate (PO_4) or C/P, measured at wavenumbers 1415 and 1035, respectively. The CI is negatively correlated with both the C/P and the weight % of CO_2 produced for stable carbon isotope analysis. A bone with a high CI value and low carbonate content would contain fluoridated apatite. Highly recrystallized bone, with a CI value greater than 4.25, would also be depleted in ^{18}O. A stable isotope value obtained on highly recrystallized bone would not be connected to the dietary practices of the individual, so samples with high CI values should be excluded from these types of studies. In the karst (eroded limestone) study area at Dos Pilas, Guatemala, environmental CO_3 contamination likely occurred from dissolved bicarbonate in the soil water (Wright and Schwarcz 1996).

Hiller et al. (2006) outlined the use of small-angle X-ray scattering (SAXS) with a two-dimensional detector to examine the preservation of bone apatite. The method is similar to X-ray diffraction, but X-rays strike the sample at very small angles of incidence. Low-angle X-ray scattering detects differences in the density of electrons between a substance and its surrounding medium and provides information about the morphology (shapes, size, orientation, packing) of objects. The technique can be used to examine both bone powder and thin sections.

Data obtained using SAXS were compared to infrared splitting factors (SF) and carbonate:phosphate ratios (C:P) determined by FT-IR. SAXS was better able to detect diagenetic thickening of crystals and a direct relationship was found between crystal thickness and organic preservation indicated by nitrogen content (%N). All of the 12 archaeological bones from which amplifiable DNA (Chapter 16) was extracted were found to have thin crystals. Crystal thickening and shape changes related to thermal alteration were detected by SAXS after the bone was exposed to temperatures of 500°C for 15 min, making it a more effective screening tool than XRD which can detect changes in the mineral phase that occur at temperatures above 1000°C. Microfocus SAXS (μSAXS) using high-energy X-rays produced by a synchrotron could also be used to examine the nanotexture of bone thin sections. Hiller et al. (2006) suggested that ancient DNA is more likely to preserve in homogeneous samples with needle-like crystals with little or no diagenetic thickening.

Garvie-Lok et al. (2004) compared methods of preparing bone apatite for stable carbon isotope analysis. A commonly used method involves soaking samples in 1.0 M acetic acid for 24–36 h to remove potentially contaminated labile (adsorbed) carbonate. The potential for recrystallization and excessive sample loss tends to increase with acid concentration but a number of factors are involved; stable isotope values may also be affected.

Treatment effects of using 1.0 and 0.1 M acetic acid were compared with archaeological mammal bone less than 2000 years old and modern cow bone. Bone crystallinity was measured with FT-IR. The peak heights of wavenumbers 565, 595, 605, 1035, and 1415 were used to calculate indices of crystallinity and carbonate content. The organic content of each bone was also determined. After processing, stable carbon and oxygen isotope values were determined with a stable isotope ratio analysis mass spectrometer.

Results indicated that apatite loss was most pronounced (almost double) in samples treated with 1.0 M acetic acid and the loss occurred after only 4 h (Garvie-Lok et al. 2004). Lowering of stable carbon isotope values continued for the duration of the treatment with 1.0 M acetic acid but no sample dissolution was observed, which suggested recrystallization had occurred. The effect on oxygen isotope values was more variable but the majority of changes also occurred within 4 h. Garvie-Lok et al. (2004) recommended that 0.1 M acid treatments not exceeding 4 h duration be used for both fresh and archaeological bone. In addition, care must be taken when comparing the $\delta^{13}C$ and $\delta^{18}O$ values of bone prepared by the different methods.

Preservation of Lipids

Liden et al. (1995) investigated the effects of lipids and the procedures used to remove them on stable carbon isotope values. Lipids are a known source of error in studies of bone protein because lipids are as much as 7‰ more negative than collagen (see Chapter 13). The effects of lipid removal procedures using either organic solvents or NaOH were compared. The sample included 14 archaeological and modern unburned, demineralized bones from seven different geographical locations in Canada, Iceland, Finland, Greece, and Tonga that ranged in age from 10,000 years old to modern.

Lipids were extracted with a 2:1 mixture of choloform:methanol. The effect of lipid extraction at different stages was examined using modern horse bone from Tonga. In order to assess the effectiveness of sodium hydroxide (NaOH) extraction, the lipid content of demineralized bone was compared to the lipid content of untreated whole and demineralized bone; refluxed and filtrated bone was also examined. $\delta^{13}C$ values were determined for lipid and non-lipid extracted collagen, as well as the lipid extracts.

Lipid-extracted collagen samples were mixed with potassium bromide (KBr) and analyzed with FT-IR spectroscopy to detect the presence of lipids and/or residual extracting solvent. Effects of NaOH and ultrafiltration on the stable carbon isotope and amino acid composition were examined on collagen extracted from bison bone dating to 10,000 B.P. Amino acid composition was analyzed using HPLC equipped with an absorbance detector.

Liden et al. (1995) found that modern bones and two of the archaeological bones contained measurable amounts of lipid material, 1.0–3.5 and 0.2% respectively. NaOH treatment did not remove lipids. No lipids or solvents remained in extracted collagen but the $\delta^{13}C$ value of lipid-extracted collagen was less negative than non-lipid extracted collagen. The most obvious effect of NaOH was decreased collagen yield; however, both NaOH and ultrafiltration affected the amino acid concentrations. Regardless, the $\delta^{13}C$ values on samples processed different ways were not altered.

Liden et al. (1995) reported that lipids can remain in bone and interfere with carbon isotope analysis of bone collagen but the degree of interference depends on the amount present in the sample. For example, if lipids are not removed from fresh bone (with lipid levels of 1–2%), the collagen extraction process can concentrate them to levels as high as 18%. This would cause stable carbon isotope values to be 1.8‰ more negative than the same lipid-extracted sample. NaOH treatment removed humic acid and reduced collagen yield by several percent but did not necessarily affect the $\delta^{13}C$ value. The different collagen extraction protocols resulted in different amino acid patterns so amino acid analysis alone was not a good indicator of diagenesis. Liden et al. (1995) suggested that treatment with NaOH was unnecessary; instead, lipid extraction with chloroform–methanol should follow the demineralization step.

Evershed et al. (1995) used gas chromatography and gas chromatography with mass spectrometry to examine the preservation of lipids in both modern and archaeological bone. The composition of certain lipids, in particular phospholipids, extracted from modern bone was found to depend upon the presence or absence of marrow; however, the concentration of cholesterol did not change and appeared to be associated with the bone matrix. The lipid composition of archaeological samples was quite similar to that of modern bone lacking marrow.

In waterlogged environments, microbial reduction of cholesterol under anaerobic conditions results in the formation of degradation products, 5α- and 5β-cholestanol and 5α- and 5β-cholestan-3-one. Both cholesterol and a 7-keto compound were extracted from a sample of 70,000-year-old whalebone. Concentrations of cholesterol, as high as 46.5 μg/g, were preserved in archaeological bone from well-drained burial contexts and degradation products did not form in this environment. Evershed et al. (1995) suggested that the stable carbon isotope value of the preserved cholesterol could potentially be used as a paleodietary marker.

Stott et al. (1999) used online isotope ratio monitoring-gas chromatography/ mass spectroscopy (irm-GC/MS) (Chapters 31 and 33) to evaluate Evershed et al.'s (1995) hypothesis. Cholesterol is derived primarily from dietary carbohydrates and lipids and controlled feeding experiments with rats showed that, due to its high turnover rate, the isotopic value of cholesterol was more sensitive to dietary change than bone collagen and carbonate.

The $\delta^{13}C$ values of cholesterol were compared to $\delta^{13}C$ and $\delta^{15}N$ values of bone collagen from the skeletal remains of a Medieval coastal population ($n = 50$) and a Medieval inland population ($n = 27$) from the United Kingdom. Mean cholesterol $\delta^{13}C$ values of the populations differed by 1.7‰, which was statistically significant. Stott et al. (1999) suggested the difference is due to dissimilar proportions of marine and terrestrial foods in their diets. No significant difference was observed in the mean collagen $\delta^{13}C$ values obtained on the same sample. Enriched cholesterol $\delta^{13}C$ values were also detected in the remains of a member of an Archaic seal hunting population and two individuals from a population of Mesolithic coastal dwellers, which further supported the value of cholesterol as a dietary indicator (Stott et al. 1999).

Preservation of Blood Proteins

As noted in Chapter 15, Lowenstein (1985) announced the detection of protein preserved in 1.9 million-year-old fossil hominids specimens using radioimmunoassay. Later Downs and Lowenstein (1995) raised concerns that partially degraded proteins may react in unpredictable ways. Recently, Torres et al. (2002) used dot-blotting and quantitative dot-blotting techniques to examine material extracted from fossil bone from sites in Spain ranging in age from 0.12 to 1.6 million years old. They reported that extracts from 0.12 and 1.6 million-year-old fossil equid bone reacted more strongly to antibodies against horse immunoglobulin G (IgG) than to those against human IgG; likewise, extracts from 1.6 million-year-old fossil hominid bone reacted more strongly to antibodies against human IgG than to those against horse IgG. Material extracted from 1.4 million-year-old fossil hominid bones from another site did not react to antibodies against immunoglobulin.

Analysis of Ancient DNA from Bone

Examples of DNA analysis (Chapter 16) to determine the species of archaeological bone are becoming increasingly common. While the technique can be used to simply assess the number or proportion of animals utilized, information about procurement patterns can also be obtained. Examples of these applications are provided.

Newman et al. (2002) analyzed ancient DNA to successfully make species-specific identifications of highly fragmented cortical bone from a bison jump in Alberta, Canada, and distinguish 5 of the 14 sheep and goat bones examined from sites in northern Cameroon and Nigeria. In the former, species-specific primers targeted the cytochrome *b* gene of humans, cattle, goat, sheep, and bison and resulted in amplification products of different lengths. In the latter, universal primers were used to amplify the cytochrome *b* gene of a variety of species; after amplification, restriction endonucleases were used to sever PCR products into fragments the lengths of which were species specific. For example, the *Dde*I restriction endonuclease cut the cytochrome *b* gene of goat into 129 and 50 base pair fragments; the *Rsa*I restriction endonuclease cut the cytochrome *b* of goat into 117 and 59 base pair fragments. The lengths of the severed PCR products were determined by simple agarose gel electrophoresis.

Arndt et al. (2003) used ancient DNA to source the remains of an exotic catfish, genus *Clarias*, recovered from a site in Anatolia (modern Turkey). Differences discovered in the mitochondrial DNA (mtDNA) of modern catfish from Turkey, Syria, Israel, and Egypt indicated that the archaeological specimens, *Clarias gariepinus*, were harvested in the lower Nile.

Salmon Utilization in the Pacific Northwest

Speller et al. (2005) used ancient deoxyribonucleic acid (aDNA) analysis with the polymerase chain reaction (PCR) to identify salmon vertebra from the Keatley Creek site in British Columbia. A previous study of these remains using X-radiography (Chapter 37) concluded that the low-oil content pink salmon were associated with small, low-status house pits, and high-oil content chinook and sockeye were associated with large house pits.

The sample consisted of the remains of 60 salmon from one small- and one medium-sized residential structure and two specialized or ritual structures dating between 1500 and 1100 B.P., representing the Late Plateau and Early Kamloops Horizon. Specimens were X-rayed to determine their spawning age, which is the age of death. Ancient DNA was extracted from either one-half of a large vertebra, whole small vertebra or a fragment of head, rib, or fin elements using a method previously described by Yang et al. (2004). Samples were decontaminated with commercial bleach, 1 N hydrochloric acid, 1 N sodium hydroxide, and ultraviolet irradiation. Three sets of primers were used, starting with two primers together: Smc7/Smc8, a long primer for the D-loop control region, and CytB5/CytB6 for the cytochrome *b* gene. If the amplification failed, a shorter primer for the D-loop control region, Smc3/Smc4 was used. Previously extracted and DNA-identified archaeological salmon were used as controls; modern controls were not used to reduce the chance of contamination.

Ancient DNA sequences were BLAST compared through GenBank. Five samples had no detectable PCR amplification using three primer sets. More than 90% of samples had strong amplification of at least one of the co-amplified D-loop or cytochrome *b* fragments. Three species were identified, sockeye ($n = 55$; 81.8%), chinook ($n = 8$; 14.5%), and coho ($n = 2$; 3.6%). Re-examination by X-radiography revealed an inherent ambiguity in the method. Species identifications had been made on the basis of apparent spawning age and a second radio-opaque band at the center of the vertebra had not been recognized as a true annuli. Fish identified in the previous study as 2-year-old pink salmon were actually 3-year-old sockeye salmon. Sockeye salmon were the staple for everyone. Smaller, younger salmon more often were found at poor households while larger and fatter sockeye were associated with wealthier households. Chinook salmon were less common and only found in the larger residence and specialized structure. Speller et al. (2005) concluded that well-preserved aDNA can provide unambiguous identifications if the proper quality control procedures are in place. Subjective methods, such as radiography, are inherently less reliable.

Cannon and Yang (2006) used the techniques developed by Yang et al. (2004) to argue that a salmon storage economy existed at the site of Namu, British Columbia, as early as 5000 cal B.C. The sample consisted of 122 vertebrae with a range of sizes from spatially separated units within the site representing different temporal periods. Amplifiable DNA was successfully extracted from 116 vertebrae. Most (87.1%) positive identifications were made on the basis of two mitochondrial fragments from both the cytochrome *b* gene and the D-loop region and the remainder of the identifications were based on the presence of one of these fragments. Although five species of salmon were identified, 42.2% of the bones were from pink salmon. Ancient DNA showed that vertebrae previously identified as from 3-year-old chum on the basis of X-radiography were actually 2-year-old pink salmon with a supplementary growth check.

Cannon and Yang (2006) argued pink salmon were preferentially utilized because of their natural abundance in the region and their low fat content made them ideal for storage. Species with higher fat contents, such as sockeye and coho, were likely used fresh. The availability of high numbers of pink salmon enabled the early development of a permanent settlement with a salmon storage economy at Namu. The hiatus in occupation from 1380 cal B.C. to cal A.D. 830 was attributed to the collapse of the pink salmon fishery (Cannon and Yang 2006).

Stable Isotope Analysis of Archaeological Bone

Indicators of Human Diet

Stable isotope analyses of archaeological animal bone are frequently performed as an adjunct to studies about human diet. As described in Chapter 13, stable isotope values from different animal species are used to estimate their contribution to the human diet. Archaeological fauna are the actual remains of the food animals so the isotopic values obtained on collagen from these bones provide a more accurate reflection of ancient consumption patterns than modern fauna.

Several recent studies use this approach to examine ancient human dietary patterns. Richards et al. (2003) used data obtained from stable isotope analysis on human and animal remains from Catalhöyük (modern Turkey) to challenge the hypothesis that cattle was the main source of dietary protein. Privat et al. (2002) used stable carbon and nitrogen isotope values on human and animal remains to detect status-based differences in individuals recovered from an Early Anglo-Saxon cemetery in Berinsfield, UK. The average diets of individuals classified as "wealthy" and "intermediately wealthy" were found to be distinct from those classified as "poor"; the diet of males over the age of 30 years differed from that of younger males. Stable carbon and nitrogen analysis on human and animal bone from a monastic community on the Belgian coast indicated the proportion of marine and terrestrial foods and possible differences related to social status (Polet and Katzenberg 2003). Müldner and Richards (2005) found that the stable carbon and nitrogen values from three sites in Northern England were significantly different from those reported in other parts of Britain. Müldner and Richards (2005) attributed the combination of low-stable carbon and high-stable nitrogen ratio to a mixed diet consisting of terrestrial, freshwater, and marine foods and that the emphasis on fish reflected the adherence to Catholic fasting regulations.

While most paleodietary studies focus on stable carbon and nitrogen, other isotopes provide valuable information. Reynard and Hedges (2008) showed significant differences in stable hydrogen isotope values correspond to trophic level effects. Incremental increases, on the order of 10–30‰, were found between herbivores, omnivores, and carnivores. Geographical variation in δD was demonstrated by comparing archaeological material from the United Kingdom, Hungary, and Peru. Furthermore, it might be possible to detect manuring by comparing stable nitrogen and hydrogen values in archaeological human and cattle bone if only humans ate the nitrogen-enriched crops but grazing herbivores and humans drank the same water (Reynard and Hedges 2008).

Indicators of Environmental Change

The goal of a study by Emery et al. (2000) was to determine if stable carbon isotope values on animal bone could be used to detect changes in their environment. White-tailed deer bone collagen from six ancient communities in the Petexbatún region, Petén, Guatemala, occupied between the Preclassic and Terminal Classic periods were examined. In particular, isotopic changes consistent with increased consumption of maize or a decrease in the tropical forest canopy (i.e., become less negative) overtime were sought.

Collagen was extracted from 62 well-preserved specimens of zooarchaeological white-tailed deer bone; the effects of diagenetic processes on the collagen were assessed by C:N ratios (Chapter 38). $\delta^{13}C$ analysis of collagen was performed on bone collagen with a stable isotope ratio analysis (SIRA) mass spectrometer. Collagen extracts from 53 bones from riverine ($n = 24$) and inland ($n = 29$) environments were suitable for stable isotope analysis. These samples had C:N ratios between 2.8 and 3.8 and yields greater than 1% of dry weight. The resulting $\delta^{13}C$ values ranged from −17.45 to −22.09‰, with a mean of −20.55‰.

Emery et al. (2000) used a linear mixing model to estimate that the contribution of carbon atoms from maize was between −7 and 22%, with an average of 3%.

The negative value indicated that animals were consuming plants more depleted in ^{13}C than the C_3 plants in the reference collection. Presumably these plants were located in heavily forested habitats and the depleted values were due to the canopy effect (Chapter 13). No significant differences existed between riverine or inland environments or between different chronological periods.

The mean $\delta^{13}C$ values previously reported for wild deer in corn-growing areas of North America ranged between −16 and −21‰ vs. −21 and −23.3‰ in other areas. Based on the North American data, Emery et al. (2000) suggested that opportunistic deer browsing should also increase with corn availability in the Petexbatún region. The apparent stability in deer feeding patterns countered expectations related to agricultural field expansion and intensification of corn production. The data also countered suggestions of deer domestication in the Late Classic Maya period (Emery et al. 2000).

Emery et al. (2000) assumed that Maya period wild white-tailed deer in the Petexbatún region had the same degree of accessibility to corn crops as modern wild deer in North America. In addition, Emery et al.'s (2000) analysis of a potentially human-dependent deer with a healed midshaft fracture showed that domesticated or semi-domesticated deer might have been fed C_3 plants rather than maize.

Indicators of Seasonal Mobility

Stable isotope analysis on animal bone collagen can provide information about seasonal mobility. As described in Chapter 13, differences in the proportion of C_3 and C_4 plants in the diet are detected through the analysis of stable carbon isotopes. Proportions of terrestrial and aquatic (freshwater and marine) foods are obtained from the analysis of stable nitrogen isotopes. Information from strontium and oxygen isotope ratios are particularly valuable indicators of seasonal mobility as they reflect the local geology and drinking water.

Schulting and Richards (2002) used stable carbon and nitrogen values obtained from archaeological animal bones to evaluate differing interpretations of seasonal movement by the ancient inhabitants of two Mesolithic sites in the United Kingdom, Star Carr, and Seamer Carr. In particular, they tested the hypothesis that limestone enrichment can cause the stable carbon isotope ratios of plants and birds associated with a nearby freshwater lake to resemble that of a marine system. Stable isotope analysis of the remains of an aquatic bird from Star Carr showed no evidence of carbon-13 enrichment. Differences in the stable carbon and nitrogen values in the remains of a dog from Seamer Carr and a dog from Star Carr were consistent with the interpretation that the ancient inhabitants of Seamer Carr traveled to the coast on a seasonal basis and those of Star Carr did not.

Fenner (2008) reported that it might be possible to distinguish discrete communal hunting episodes using stable carbon, oxygen, and strontium isotope ratios on samples from the prey animal bonebeds. Since the isotopic variation for a single population should be low, higher than expected variation would suggest multiple kill events. This premise was tested using tooth enamel of modern pronghorn (*Antilocapra americana*) from several different wildlife management "hunt areas" in Wyoming as well as animals collected from a single mass death that occurred in November 1991. Statistical analysis of computer-generated virtual sites and the 1991 natural die-off showed that scaled distance measures on the stable isotope values for carbon, oxygen, and strontium were more reliable indicators of population origins than those based on two elements; use of a single isotope ratio to identify population could produce misleading results.

Sealy et al. 1991 described how the ratio of ^{87}Sr to ^{86}Sr (^{87}Sr/^{86}Sr) could be used to indicate both diet and mobility over the lifetime of an individual. As outlined in Chapter 13, the ^{87}Sr/^{86}Sr is directly related to the levels in the environment. Plants that grow in a region reflect the ^{87}Sr/^{86}Sr values of the soil and underlying bedrock, which is then passed on to consumers at different trophic levels without fractionation or enrichment. All tissues from an individual who resides within an area lacking variation in bedrock will have identical ^{87}Sr/^{86}Sr values. If an area includes more than one rock type with differing ^{87}Sr/^{86}Sr or if the individual resides in a coastal region with terrestrial ^{87}Sr/^{86}Sr different from that of the sea, then bone strontium isotope ratio measurements provide a measure of the relative importance of foods from each isotopic zone. If tissues that form at different stages of life are analyzed, the movement of the individual between the isotopic zones can be traced.

This approach takes advantage of differential timing of tissue formation by the body. The enamel of permanent teeth develops when an individual is very young but bone is constantly remodeled throughout life. The rate at which new bone is deposited and old bone resorbed, or turnover rate, varies between different bones and types of bone. The turnover rate of dense cortical bone is much slower than trabecular bone. The shaft of a long bone may require decades to completely remodel while a rib may turnover in a few years. By comparing strontium and oxygen levels in different tissues information can be gained about an individual's residence patterns. Benson et al. (2008) found that strontium isotope values on modern deer mice were suitable for assessing the value of local food animals exploited in the past, such as rabbit, turkey, and deer, because these animals obtain strontium from near-surface sources.

Late Stone Age Mobility of Pastoralists in South Africa

Balasse et al. (2002, 2003) used isotopic analysis of archaeological livestock teeth to examine the seasonal mobility of pastoralists who occupied the site of Kasteelberg, situated on the Vredenburg Peninsula in the southwestern Cape of South Africa. Balasse et al. (2002) tested the hypothesis that Late Stone Age pastoralists, who primarily kept sheep, followed the same pattern of seasonal mobility as seventeenth century cattle herders. Others indicated that the mortality profiles of sheep and fur seal remains at the site appeared to support this interpretation. It was anticipated that differences in vegetation and bedrock geology between the coast and the interior hinterland would be detected through the isotopic analysis of carbon, oxygen, and strontium.

Samples were taken from the molars of eight sheep, a cow, an eland, and two steenbok from three Late Stone Age open-air middens that overlapped chronologically. The enamel of each tooth was sequentially sampled in discrete horizontal bands measuring less than 1 mm wide. This provided up to 29 samples from a second molar with a crown height of 28 mm and 40 samples from a third molar with a crown height of 39 mm. Organic matter was removed by treating the enamel powder with sodium hypochlorite (NaOCl); diagenetic carbonates were removed using acetic acid. Prior to stable carbon and oxygen isotopic analysis with an isotope ratio mass spectrometer, enamel samples were treated with 100% phosphoric acid and subjected to cryogenic distillation. Samples selected for strontium analysis were dissolved in HNO$_3$ then purified using a cation exchange column containing Sr exchange resin and dried. The residue was redissolved in HCl, applied to single tungsten filaments, and analyzed using thermal ionization mass spectrometry (TIMS) (Chapter 31).

Balasse et al. (2002) found the variation in stable carbon isotope composition to be only about 3‰ and suggested this variation could be due to differences in the relative proportion of the local C$_3$ and C$_4$ plants the animals consumed at different seasons, or simply isotopic variation due to heat or water stress. Strontium isotope analysis revealed some of the animals lived on the coast throughout their lives but others (at least one sheep and the cow) lived in the hinterland at the time of tooth formation. Due to the absence of cyclic variation in the ^{87}Sr/^{86}Sr ratios, Balasse et al. (2002) concluded the animals were born in the hinterland, then traded to pastoralists on the coast but did not travel between the regions annually.

δ^{18}O values in tooth enamel are highest in summer and lowest in winter; while normal cyclical patterns were observed, the peak was observed in different places in the tooth enamel (Balasse et al. 2002). The peak occurred in enamel taken about 15 mm from neck of the third molar of some animals and about 8 mm from the neck in others. On the basis of this bimodal distribution in oxygen isotope ratios, Balasse et al. (2003) concluded that some sheep were born in autumn and that others were born in spring. Two birth seasons would have increased the availability of milk but limited the mobility of the Late Stone Age pastoralists. Balasse et al. (2002) further argued that just because the availability of fur seals was seasonally restricted, their presence at the site did not preclude occupation at other times of the year. The researchers suggest that the pattern of seasonal mobility described in the seventeenth century was not required for sheep herders and was only adopted once the pastoralists had significant numbers of cattle.

The Presence of Absorbed Metals in Bone

The concentration and isotopic signature of lead and trace metals are of interest to physical archaeologists as the ingestion of high levels of lead could have serious health implications. Lead traced by isotope analysis to the solder of metal cans is considered to have contributed to the loss of the Franklin Expedition (Keenleyside et al. 1996; Kowal et al. 1991).

Budd et al. (1998) examined both archaeological and modern teeth and determined that surface enamel is highly enriched in lead. Concentrations and the distribution of lead in the teeth were measured using laser ablation-inductively coupled plasma-atomic emission spectroscopy (Chapters 30 and 34) and isotope dilution-TIMS. The enrichment was found to occur in the outermost 30 μm of the enamel. Because it was observed in both modern and archaeological teeth, it is unrelated to diagenesis.

Detection of Heat Exposure

Methods able to distinguish cooked from uncooked bone are valuable as sample screening devices prior to further analysis involving bone collagen or ancient DNA. Koon et al. (2003) examined the ability of transmission electron microscopy (TEM) to detect morphological changes in mammal bone related to cooking. TEM is different from SEM (Chapter 37) in that instead of being deflected by the surface, the electron beam passes through the sample. As with fish bone, morphological changes in the collagen fibrils were observed in sheep bone after very mild heating events. The drawback was that similar changes occur in uncooked bone during burial in acidic soil (pH 3.5–4.5) after several years. The researchers suggested the technique would be of value for distinguishing between cooked and uncooked bone from the same feature or assemblage.

Hiller et al. (2006) reported that crystal thickening and shape changes related to thermal alteration were detected by small-angle X-ray scattering (SAXS) after the mammal bone was exposed to temperatures of 500°C for 15 min. This feature made it a more effective screening tool than XRD which can only detect changes in the mineral phase that occur at temperatures above 1000°C (Hiller et al. 2006).

Incremental Growth Structures

Carter (1998) proposed a method of assessing tooth development with X-radiography in sub-adult red deer so that the growth increments could be used as an indicator of season of death. The sample included 20 archaeological red deer mandibles from 17 different animals from Star Carr and 82 modern red deer samples ranging from 1 to 40 months in age.

X-radiographs of the mandibles were obtained at 50 kV and 6 milliamps (mA) for 40 s with a distance of 762 mm between the source and the bone. Three of the archaeological mandibles were found to be from adults and excluded. Six of the rami formed matched pairs, which indicated that actually 14 different animals were represented in the collection.

Results from the two youngest animals indicated they died during cold months, November–December and December–May, respectively. Data from these animals did not support the proposal that occupations at Star Carr were restricted to late spring and summer. Age estimations of older animals were not precise. Carter (1998) concluded that four animals died between August and either February or

April; another died between October and April, one died between January and April and three died between January and September.

Due to material contrasts between bone, tooth enamel, and the root, Carter (1998) showed that tooth development stages were easily assessed from the radiographs. However, large gaps existed in the reference collection and animals that died between late winter and mid summer were poorly represented. For this reason, even complete mandibles from older subadult red deer could not be aged with confidence (Carter 1998). With the exception of the youngest animal, which died between November and December, and two older animals that died between fall/early winter and mid-spring (between October/December and April), it should be noted that the data could equally support a late spring and summer occupation.

Dating Bone

Radiocarbon Dating

The processes involved in obtaining a radiocarbon age estimate on faunal remains are familiar to most archaeologists and issues related to radiocarbon dating of faunal remains were described in Chapter 8. For this reason, only one example of a particularly challenging sample is presented. If ranges overlap, age estimates obtained using radiocarbon are often compared to those made using techniques developed more recently (Rink et al. 1996, 2002).

George et al. (2005) re-examined fragments of mastodon femur from Monte Verde in south-central Chile from which age estimates of 6550 ± 160 B.P. and 11,990 ± 200 B.P. were obtained. The two pieces were originally submitted for ^{14}C dating separately; later it was determined that the long bone fragments fit together. George et al. (2005) isolated and dated the total amino acid and ultrafiltered gelatin fractions of each bone. Levels of intact collagen were found to be 20 and 31% and all fractions produced statistically identical ages, averaging 12,460 ± 30 B.P. George et al. (2005) could not ascertain the reason for the original young date of 6550 ± 160 B.P. but it would seem that the portion originally dated retained less than 5% of its original collagen and/or was massively contaminated. The sample dating 5000 years younger had eroded onto the surface in a modern creek bed at the site; consequently it is possible that partial exposure led to differential degradation of the bone fragment.

Uranium Series Dating

Uranium series dating provides an estimate of the age of a sample based on the radioactive decay of uranium and/or its daughter isotopes (Chapter 9). Tooth enamel, bone, and shell can be dated, but these materials are initially free of uranium and absorb it from the environment. Because of the need to model uranium uptake, uranium series dating of biological materials is more complicated than dating a mineral, such as calcite.

Esposito et al. (2002) used uranium series dating to reconcile the age of faunal material and calcite from Snake Cave, Cambodia. A previously determined U-series age of 130,000 years appeared to be too young. Esposito et al. (2002) determined that spleleothem calcite samples were contaminated with detritus. Ages corrected for contamination using the isochron technique indicated the main fossil layer was actually more than 160,000 years old. Esposito et al. (2002) were unable to obtain reliable uranium series ages from fossil teeth.

Combined Radiocarbon and Uranium Series Dating

Benzi et al. (2007) obtained age estimates for a nearly complete skeleton of an extinct deer, *Praemegaceros cazioti*, recovered from a cave on the Italian island of Sardinia using both liquid scintillation radiocarbon dating (Chapter 8) and two types of uranium series dating, $^{230}Th/^{234}U$ and $^{231}Pa/^{235}U$ (Chapter 9). Bone collagen was extracted for radiocarbon dating using a triple soft acid–alkali–acid treatment; about 15% of the original collagen was recovered from the bone. The CO_2 produced by combustion in chromatographic oxygen was converted to benzene for liquid scintillation counting. Although no sample sizes were given, in order to improve the accuracy a second, larger sample was analyzed and counting time was extended to about 9000 min.

Uranium series dates were obtained on whole bone; tooth enamel was separated from dentine and each component was dated. Samples underwent combustion at 800°C for 12 h, then were dissolved in nitric acid (HNO_3), and then Al nitrate carrier and $^{232}U-^{228}Th$ tracer solution were added. Uranium was separated from thorium using an ion exchange column then the elements were plated on aluminum and counted using alpha spectrometry. The ^{231}Pa content in the bone sample was not measured directly; its activity was assumed to be in secular equilibrium with, hence identical to, ^{227}Th.

Radiocarbon dates were obtained on both samples. Both the bone and the tooth dentine were found to be free of ^{232}Th contamination and acceptable uranium series dates were obtained. The tooth enamel had a very low uranium content and was contaminated with detrital ^{230}Th; an erroneous date of 52,000±13,000 years was produced.

The first attempt at radiocarbon dating produced a date of 6460 ± 130 B.P. and the second gave a date of 6790 ± 80 B.P. Calibration of the latter date with CALIB3 resulted in a range of 7650–7530 cal B.P. The uranium–thorium dates on the bone and dentine were 8500 ± 700 years and 7100 ± 500 thousand years, respectively; the $^{231}Pa/^{235}U$ date on bone was 9300 ± 800 thousand years. The researchers assert the three dating methods lead to concordant ages of about 7500 B.P., which indicated the skeleton was maintained as a closed system after burial. Benzi et al. (2007) reported the ages obtained made this *P. cazioti* skeleton the youngest example of the species in Sardinia.

While it is possible to obtain a fairly precise radiocarbon age estimate using liquid scintillation counting, Benzi et al. (2007) showed it could be a trial and error process. In order to reduce the error associated with the date, the researchers needed to obtain a second, larger bone sample and extend the counting time. The uranium–thorium age on dentine overlapped with the reported 1σ range of the calibrated radiocarbon age; the age estimate of the date on bone fell just outside this range. The calibrated radiocarbon date and the $^{231}Pa/^{235}U$ date on bone appear to overlap at the 2σ range, which for the latter is 7700–10,900 thousand years. Due to the high degree of precision offered by the technique, perhaps it would have been prudent to obtain an AMS age on bone collagen as well. Benzi et al. (2007) concluded that the skeleton remained a closed system after burial, implying that uranium absorption matched the early uptake model. There were considerable differences in the uranium content of the samples but Benzi et al. (2007) did not discuss their significance. The ^{238}U contents of the samples, given in dpm/g, were 6.8 ± 3 for bone, 11.7 ± 0.4 for dentine, and 0.060 ± 0.005 for enamel. Their finding that the tooth enamel contained thorium from the environment, but bone and dentine did not, is surprising because enamel is favored for uranium series dating since it is less likely to be contaminated.

Electron Spin Resonance Dating

Mellars et al. (1997) raised concerns about possible errors in the calculation of ESR ages made on previously excavated material. Problems may arise if fine sediments still adhering to a sample, such as a tooth, are used to estimate the dose rate. Sediments from five sites were collected, separated on the basis of particle size, and weighed. Each fraction was then finely ground; elemental compositions were determined using XRF and instrumental neutron activation analysis (INAA) (Chapter 32). Uranium, thorium, and potassium concentrations were highest in the finest fraction. Mellars et al. (1997) cautioned that if the annual beta dose rate were determined only from fine-grained sediments instead of the whole sediment, overestimations in the dose rate would cause underestimations in the age.

Rink et al. (2004) demonstrated the importance of assessing the annual dose and collecting *in situ* samples in a study of material from Tabun Cave in Israel. Prior to this, ESR dates on museum specimens were determined using attached sediments for calculation of the annual dose. Discrepancies between these and thermoluminescence (TL) dates prompted a re-evaluation of the material. The application of ESR, U-series, and combined ESR/U-series on teeth recovered *in situ* provided ages that agreed well with those obtained by TL and were much older than ESR dates on the museum pieces.

Assessing the external contribution can prove to be a challenge even when material is collected in situ. Rink et al. (2004) found that teeth from the lowest section of the central area of Hayonim Cave, Israel, yielded younger ages than material above it. These samples were located close to an erosional conformity and potassium-rich siliceous aggregate mineral assemblages that had crystallized over time. Consequently the external contribution to the annual dose had varied with time. Infrared spectroscopy (Chapter 35) aided in the identification of these mineral assemblages so that unaffected teeth could be collected for ESR dating.

Combined Uranium Series and ESR Dating

As noted in Chapters 9 and 10, the use of uranium uptake models is required to calculate uranium series and ESR dates on biological samples. Processes by which uranium was absorbed by material initially uranium-free affect the internal radiation dose, which is required for ESR age determinations, and the formation of daughter products, which affects uranium series dating.

It is possible to avoid assumptions about uranium uptake by obtaining uranium series and electron spin resonance ages on the same sample. The ages obtained on biological materials can also be compared to geological materials, such as calcite deposits in caves, which contained uranium at the time of formation. Uranium uptake is a factor in both methods; if early uptake is the correct model, the uranium series age on the sample will match the ESR age. When the concentration of uranium is very low, the early uptake and linear uptake models will also generate very similar ESR ages. At higher concentrations, linear or supralinear models of uranium uptake produces significantly different ESR and U-series ages; however, age determinations can be reconciled mathematically. The advantage of combining the two techniques is that it provides a means of crosschecking.

Because of the need to model uranium uptake, dating biogenetic material (tooth enamel, dentine and bone) initially free of uranium is inherently more complicated than dating material such as calcite. This concern applies equally to U-series and ESR dating techniques. As noted above, the external contribution to the dose must also be accurately assessed when ages are determined using ESR.

Chapter 20
Cultural Rock

This chapter examines the study of cultural rocks, including those used for ornamentation, building material, as well as thermally altered rocks associated with archaeological features or used for hot rock cooking. Techniques for determining the composition of rocks for provenance studies are considered. Methods of examining the integrity of archaeological features and human activity through the study of remanent magnetization with paleomagnetic or archaeomagnetic techniques are reviewed. In addition, dating fire-cracked or burned rock using thermoluminescence is presented.

Amber

The first application of nuclear magnetic resonance spectroscopy (NMR) (Chapter 36) to study amber, which is fossil tree resin, was by Beck et al. (1974). Beck et al. (1974) used proton (^{1}H) NMR to characterize the oil produced from the destructive distillation of Baltic amber. The source trees of Baltic amber were believed to be an extinct pine, *Pinus succinifera*, so Beck et al. (1974) compared the composition of amber to commercial resin. The soluble pyrolysis products from the 160 to 180°C fraction contained the targeted compounds, *p*-cymene and *m*-cymene. The composition of the fossil resins in Baltic amber was found to be different from those of living trees.

Lambert and colleagues employed ^{13}C nuclear magnetic resonance spectroscopy with cross polarization and magic angle spinning (^{13}C CP/MAS NMR) (Chapter 36) in provenance studies of solid amber from various parts of the world. In their 1988 study, Lambert and others compared the amber from several different parts of Europe. It had been previously reported that Baltic amber contains succinic acid; however, this characteristic alone was not sufficient for determining provenance.

Organic solvents were used to extract small molecules and oxidation products from the amber; the remaining solid polymeric matrix was analyzed. Lambert et al. (1988) found that the Baltic amber was distinct from amber from more southern parts of Europe. The ^{13}C NMR spectra of all samples of amber from locations near the Baltic Sea and North Sea, including succinites, beckerite, and gedanite, were

M.E. Malainey, *A Consumer's Guide to Archaeological Science*, Manuals in Archaeological Method, Theory and Technique, DOI 10.1007/978-1-4419-5704-7_20, © Springer Science+Business Media, LLC 2011

similar in that exomethylene peaks were present. This peak did not occur in the amber from locations in central and eastern Europe, including rumanites, simetites, schraufite, and walchowite.

Other studies focused on amber from the Dominican Republic (Lambert et al. 1985), Mexico (Lambert et al. 1989), and North America (Lambert et al. 1990). The chemical composition of ambers was found to be strongly dependent on the tree of origin. North American ambers over a wide geographic zone generated very similar spectra; however, ambers originating from trees of different genera could be distinguished. New World ambers were clearly different from Old World ambers.

JET

Jet is an organic gemstone that has been used for ornamentation for thousands of years. Like other carbonaceous fossil materials, jet was formed from organic materials in sedimentary deposits. It is similar to lignite and sub-bituminous coal with respect to levels of hydrogen and aromatic carbons (Lambert et al. 1992).

Lambert et al. (1992) utilized ^{13}C CP/MAS NMR spectroscopy to compare jet from six sources in England and two sources of Spanish jet. Portions weighing 0.2–0.3 g were powdered to facilitate packing in the sample container (rotor), although this was not absolutely necessary.

^{13}C CP/MAS NMR was able to distinguish saturated or aliphatic carbons from unsaturated carbons in double bonds or aromatic rings in the jet and the size of the peaks indicated the number of carbons at a molecular site (Lambert et al. 1992). Lambert et al. (1992) reported that the English samples tended to have larger aliphatic peaks than Spanish jet; the overall ratio of aromatic peaks to aliphatic peaks varied from 1.2 to 1.5. All English samples also had a prominent shoulder on the low-frequency side of the aliphatic resonance; this feature did not occur in the spectra of Spanish samples. These differences were confirmed when interrupted decoupling was used to remove resonances arising from carbons bonded to only one or two hydrogen atoms (CH and CH$_2$). The English jet displayed a prominent resonance from rapidly moving methyl groups (CH$_3$) but the Spanish samples did not. Lambert et al. (1992) suggested the criteria are useful for differentiating between English and Spanish jet in provenance studies.

Watts et al. (1999) explored the utility of pyrolysis gas chromatography/mass spectroscopy (py-GCMS) (Chapter 33) to distinguish jet, which is rare and precious, from more widely available materials, such as cannel coal, oil shale, and lignite. Although larger pieces could be discriminated through microscopic examination, this approach was not suitable for small artifacts, such as beads, and identifications using non-destructive X-radiography and X-ray fluorescence (XRF) were not always reliable (Watts et al. 1999) (see Chapter 37).

Watts et al. (1999) used py-GCMS to assess the chemical structure of kerogen, an insoluble macromolecule, in the jet and similar materials. Three types of kerogens were worked for making ornaments: type I, included torbanites derived from the lipids of algal remains; type II, included oil shales derived from the lipid precursors of marine plankton and plants; type III, included humic coals derived from terrestrial plant materials. The goal was to determine py-GCMS fingerprints for the different materials that related to their general depositional environment and, possibly, specific sources.

The pilot study included geological samples of lignite, cannel coal, oil shale, and two types of jet from sources in the UK. After grinding and homogenization, 2–3 mg of the powder was compressed onto the wires of a pyrolysis unit using a hydraulic press. The material was pyrolyzed at 610°C for 2 s, then transferred directly onto the GC column; separated pyrolysis products were identified with reference to published mass spectra and on the basis of relative retention times (Watts et al. 1999).

Watts et al. (1999) reported pronounced differences between the pyrolysis chromatograms of the materials with respect to the distribution and relative abundance of aliphatic hydrocarbons, aromatic compounds, and organo-sulfur. These differences related to the material from which they formed and their depositional environment. Lignite and jet, both derived from wood, could be distinguished on the basis of the distribution of normal hydrocarbons produced on pyrolysis. More significantly, the relative proportion of organic sulfur compounds, thiophenes, and alkylthiophenes provided a means of discriminating Whitby jet from jet obtained from the Kimmeridge source. Since the analysis only involved single samples of the materials, further research was necessary to assess the compositional variability of individual deposits (Watts et al. 1999).

Limestone and Marble

Provenance Studies of Limestone

De Vito et al. (2004) used information derived from the analysis of elemental and isotopic composition to determine the provenance of limestone from two sites near L'Aquila, Italy. Archaeological materials included 18 samples taken from the Ercole Curino Sanctuary, which was constructed over a major active fault between the fourth and third century B.C. Seven samples were taken from the foundation walls of a large house once occupied by a Roman family, or *domus*, dating from the first century B.C. to the second century A.D. In addition, 75 geological samples collected from limestone outcrops that were known or suspected to be quarries were analyzed.

Atomic absorption spectroscopy (AA or AAS; Chapter 34) was used to determine the elemental composition of limestone samples dissolved in a combination of perchloric and hydrofluoric acids. Only the concentrations of specific elements known to provide information about the sedimentation and diagenesis of carbonates and the paleoenvironment were determined (De Vito et al. 2004). The ratio between strontium (Sr) and magnesium (Mg) indicated the environment in which the limestone was deposited. If the Sr/Mg ratio was positive then the limestone was deposited in shallow water. If Sr and Mg were inversely correlated, the limestone was deposited in a deep, open water (pelagic) environment. The ratio between nickel (Ni) and cobalt (Co) provided data about the paleo-oxygenation conditions under which the limestone formed.

Stable carbon and oxygen isotopes were determined on 12 limestone samples from monuments and 26 from geological formations. The $\delta^{13}C$ values on samples from the Ercole Curino Sanctuary completely overlapped but the $\delta^{18}O$ values were separate. This indicated the limestone was obtained from two different sources. There was no correspondence between the stable isotope values obtained on limestone from the *domus* and any of the sampled formations.

De Vito et al. (2004) concluded that most of the limestone used to construct the Ercole Curino Sanctuary was from a local source but artistically valuable stone was obtained from the Colle Mitra outcrop about 5 km away. The source of limestone used to construct the *domus* was similar with respect to most features to a source located 3 km away but the stable carbon and oxygen values did not correspond to it.

Marble Provenance Studies Using ESR Spectroscopy

On the basis of 15 years of research, Polikreti and Maniatis (2002) were able to identify 10 parameters that are useful for establishing the provenance of white marble. These parameters, which were determined by electron spin (or paramagnetic) resonance spectroscopy (ESR or EPR) (Chapter 36), in combination with measurements of maximum grain size, enabled marble from different quarries in the eastern Mediterranean to be discriminated.

Previous studies showed that the source of white marble could not be determined using data from a single analytical technique, but partial success had been achieved through the use of nuclear magnetic resonance spectroscopy (NMR) (Chapter 36) and stable carbon and oxygen isotope analysis (Chapter 13) (Polikreti and Maniatis 2002). Building on earlier ESR studies that identified the potential utility of data obtained on manganese (Mn), Polikreti and Maniatis (2002) found that data from ESR and grain size measurements could discriminate calcitic marble sources in the Aegean Sea.

A total of 422 geological samples of marble from Greek quarries at Penteli, Hymettus, Proconnessus, and two sources on each of the islands of Paros and Naxos were studied. Measurements of maximum grain size (MGS) were made on a freshly broken surface using a stereoscopic microscope. In preparation for ESR analysis, small samples were etched in 0.5 N hydrochloric acid for 10 min, washed and dried then gently ground to avoid alteration of the ESR spectra (Polikreti and Maniatis 2002). About 200 mg of the powder was placed in a quartz sample tube and analyzed at a microwave frequency of 9.5 GHz; the magnetic field was scanned from 0 to 6000 Gauss (G). Three spectra were obtained and the total analysis time was about 10 min. Because commercially available material was not sufficiently pure, an internal laboratory standard was used to tune the EPR instrument. In order to normalize measurements to the KH1 (Kalkstein) standard from Germany, the cleanest calcite or marble standard available, a correction factor of 0.032 needed to be applied to the measurements of Mn^{2+} (Polikreti and Maniatis 2002).

The first spectrum targeted the region containing the Mn^{2+} sextet (i.e., six peaks produced by the splitting of a single large peak). The second spectrum focused on the central region around $g = 2.0000$, while the third spectrum focused on the low magnetic field region. The intensity of each peak was measured as the height of the first derivative at an amplification of 1×10^6 (Chapter 36). The parameters required for provenance studies related to the intensities of specific peaks, degree of peak overlap and width, measured as the half width at half maximum (HWHM), of a specific low field Mn^{2+} peak. The data provided information about the concentrations and physical and chemical environment of Mn^{2+}, dolomite, different CO^{2-} radicals, Fe^{3+}, and a paramagnetic center in the mineral muscovite. The EPR parameters and MGS data were found to exhibit either lognormal, normal, or other distributions. Some of the parameters were correlated, but this was expected on theoretical grounds.

Polikreti and Maniatis (2002) found that although one parameter could not discriminate all quarries, two specific quarries usually could be distinguished on the basis of a single parameter. Boxplots, showing means and ranges, were used to evaluate the effectiveness of each parameter to discriminate the six sources and select the specific parameter best able to differentiate two quarries. Polikreti and Maniatis (2002) suggested the following protocol for discriminating marble quarries. The first step involved construction of a bivariate plot between the logarithms of MGS and Mn^{2+} peak intensity. Next, marble spectrum parameters were evaluated to reduce or eliminate overlap between quarries with similar compositions.

Polikreti and Maniatis (2002) reported that a bivariate plot of the logarithms of MSG vs. the peak at $g = 14.25$ largely discriminated the two marble sources on the island of Naxos. The plot of "the ratio of MGS to width" vs. "Mn^{2+} intensity $-$ $100 \times$ the peak height at $g = 1.9998$" could almost completely distinguish a source on Paros from those on Proconnessus. If the marbles could not be differentiated on the basis of transparency and texture, sources on Paros could be discriminated from Hymettus using a bivariate plot of "Mn^{2+} intensity $-10 \times$ the peak height at $g = 2.0044$" vs. "width $-5 \times$ log of the peak height of $g = 1.9998$". Polikreti and Maniatis (2002) achieved a high rate of success when this methodology was applied to determine the provenance of approximately 120 archaeological samples from the eastern Mediterranean.

Thermally Altered Rock

Archaeomagnetic Studies of Rock Features

Gose (2000) provided examples of how the magnetization retained in burned or fire-cracked rocks could be used to examine feature integrity, determine the maximum temperature to which the rocks were exposed, and test hypotheses about ancient cooking practices.

As outlined in Chapter 11, when heated to their Curie point, minerals lose their previously acquired magnetization and acquire a new magnetization as the material cools down. If the position of the sample did not change, the heat-induced magnetization, or thermoremanent magnetization (TRM), of the minerals would correspond to the direction and intensity of the geomagnetic field at the time of cooling. If the rocks were heated to temperatures below the Curie point, a portion of the original magnetization would be retained and a partial thermoremanent magnetization (pTRM) would be acquired. Gose (2000) obtained information about feature integrity and temperature by studying the temperature-related components that contributed to the remanent magnetization of the sample. After the natural remanent magnetization was measured, different components of the magnetization were separated through progressive demagnetization studies.

Progressive demagnetization is the incremental removal of magnetization, starting with the low-temperature component. After each of the ten or more steps in the demagnetization process, the sample is placed in a zero magnetic field and then the direction and intensity of the remaining magnetization are measured. Either thermal or alternating field demagnetization can be used to obtain the data.

Gose (2000) showed that the declination and inclination measured at each demagnetization step was identical if a rock that was heated to the Curie points of its constituent minerals at the time of site occupation and allowed to cool in place. The same applied to all rocks from an intact feature and the TRMs corresponded to the direction of the geomagnetic field at the time of cooling.

If a rock were exposed to a temperature below the Curie point, magnetizations would not be completely reset. The magnetization of the rock would consist of two distinct components: a lower temperature cultural magnetization, which is the pTRM, and a higher temperature natural magnetization. The declination and inclination of the pTRM would correspond to the geomagnetic field at the time of site occupation; the high-temperature component would reflect the original geological magnetization retained in the rock. Consequently, if the rocks from an intact feature were heated to 500°C (which is below the Curie points of magnetite and hematite) they would share common pTRMs but the components above the maximum heating temperature of 500°C would be different for each rock. The point at which the magnetizations of feature rocks change from common alignments to random orientations corresponds to the maximum temperature of heating.

If a sample was heated to a very high temperature at one orientation but moved to another position relative to the geomagnetic field before it had completely cooled, the cultural magnetization would have more than one component. The higher temperature component of the TRM would reflect the magnetization acquired prior to being moved and a lower temperature component would be due to magnetization acquired after the move.

Gose (2000) used principal component analysis of the demagnetization data to separate the different components of magnetization. Magnetization is a three-dimensional vector with both direction (declination and inclination) and magnitude (intensity), so the data obtained at each step of the demagnetization process was plotted on a vector component (or As-Zijderveld) diagram. One data point was used to represent the projection of the vector onto the north–south–east–west plane and a second point showed where it lay on the up–down–horizontal plane; the intensity of the magnetization was depicted as the distance from the origin (i.e., the point where $x = 0$, $y = 0$, $z = 0$). Principal component analysis was then used to calculate the best fitting mean vector for the data.

Vector component diagrams and principal component analysis are quite difficult to explain. Think of an arrow emerging from the ground at a 45° angle and pointing in the northeast direction. The position from where the butt of the arrow intersects the ground corresponds to the point $x = 0$, $y = 0$, $z = 0$. If you shine a light directly down on the arrow, a shadow will be cast on the plane in space represented by the ground, you can mark the position of the tip of the shadow on the ground. If you shine a light horizontally at the arrow, a shadow will be cast on the plane in space represented by a vertical wall, you can mark the position of the tip of the shadow on the wall. If you do the same with arrows pointing in the same direction but one-half and another one-quarter of the size of the first arrow, the points plotted on the ground and on the wall will each be aligned and be progressively closer to the point $x = 0$, $y = 0$, $z = 0$. The length of the arrow corresponds to the magnitude of the vector. Using the positions of the marks on the wall and on the ground, principal component analysis can be used to determine that all arrows were at an angle of 45° to the ground and pointed northeast.

For a rock heated and cooled *in situ*, the data points plotted on each plane would form straight lines and the mean vector would correspond to the ambient magnetic field at the time of occupation. If the rock had two components of magnetization because it was moved while it was still hot, the points of the high-temperature component would align and the lower temperature component points would align at a 90° angle to the high-temperature component. The first and second principal components would correspond to the two different ambient magnetic fields experienced by the rock as it cooled.

Gose (2000) noted that rocks used for stone boiling could acquire multiple, possibly very complex, magnetizations. If a hot rock was moved from a hearth and placed into water, it would have two temperature components. The first, high-temperature component would correspond to its cooling during transport from the fire to the cooking receptacle and the second was acquired while it was immersed in water. The ambient magnetic field experienced by the rock would have changed if stirring moved it as it cooled.

Trapped Charge Dating of Burned Rock

Dykeman et al. (2002) compared the accuracy and precision of TL dates (Chapter 10) on burned rocks to tree-ring, or dendrochronological, dates on wood from sites in Dinétah, the traditional homeland of the Navajo in northwestern New Mexico. Three sandstone rocks were obtained from surface and buried contexts from two different sites. The age on one subsurface rock was $17,271 \pm 4361$ B.C., which predated any human occupation of the area; this was explained by insufficient heating at the time of occupation. The ages of sufficiently heated sandstone were comparable to those obtained from locally produced ceramics. The TL age on one sandstone rock corresponded to the target tree ring date at the 2σ range; only 6 of the 15 TL ages exhibited any overlap with the tree-ring date. Results from Dykeman et al.'s (2002) comparison between the accuracy and precision of TL ages on ceramics to tree-ring dates on wood from these sites is described in Chapter 17.

Chapter 21
Food-Related Pottery Residues

Techniques used to identify archaeological residues related to food and drink target lipids, proteins, DNA, or other characteristic compounds. Absorbed lipid residues extracted from the walls of pottery vessels are often analyzed using gas chromatography (GC) (Chapter 33) and gas chromatography with mass spectrometry (GC/MS) (Chapters 31 and 33). A powerful new technique that combines gas chromatography with combustion isotope ratio mass spectrometry (GC-C-IRMS) (Chapters 31 and 33) can be used to determine the stable carbon isotope values of specific fatty acids or biomarkers. In addition, analyses involving high-pressure liquid chromatography (Chapter 33), elemental analysis, optical (Chapter 35) and resonance (Chapter 36) spectroscopy, and stable isotope analysis (Chapter 13) are described.

Approaches that provide general characterizations of the residues and those that give specific identifications are presented. A newly developed process of obtaining AMS radiocarbon dates on specific compounds in archaeological lipid residues is also described.

General Characterizations of Food Residues

Phosphorus Content of Potsherds

Citing evidence that clay minerals in soil surrounding a burial become enriched with phosphates, Duma (1972) suggested similar enrichment of vessel walls might indicate a pot once contained organic substances. The phosphate content of pots, and the soil they contained, increased from top to bottom and Duma (1972:128) attributed these phosphorus enrichments to organic substances the pots once held. Furthermore, Duma (1972:129) demonstrated that clay products fired between 600 and 800°C were most efficient at absorbing phosphorus from an aqueous solution under experimental conditions. This is the expected temperature range for pottery fired outside of a kiln, such as low-fired "primitive" potteries (Duma 1972; Rice 1987).

Bush and Zubrow (1986) used energy-dispersive X-ray fluorescence (ED-XRF) (Chapter 37) to quantify the phosphorus content of sherds non-destructively and

M.E. Malainey, *A Consumer's Guide to Archaeological Science*, Manuals in Archaeological Method, Theory and Technique, DOI 10.1007/978-1-4419-5704-7_21, © Springer Science+Business Media, LLC 2011

show that vessel encrustations represented organic remains. Cackette et al. (1987) used the same technique to analyze ethnoarchaeological cooking jars from Mexico. The phosphorus level in samples taken from the exteriors of an unused control vessel, a pot used to boil water, and one used for boiling maize and/or beans were similar; however, only samples taken from the interior of the last pot showed elevated levels of phosphorus. Cackette et al. (1987) concluded the higher phosphorus levels of the cooking vessel were due to phosphate enrichment by the organic-rich liquid. Highest phosphorus levels were recorded at the interior neck area where organic substances, in the form of scum or froth, would accumulate if the vessel was used for boiling (Cackette et al. 1987:126).

Dunnell and Hunt (1990) argued that phosphorus levels were not necessarily an indicator of previous organic contents. Using ED-XRF, Dunnell and Hunt (1990) demonstrated that phosphorus content in archaeological pottery was related to (a) to composition of materials from which it was made, (b) enrichment and/or leaching of phosphorus during use, and (c) interaction with the depositional environment and its changing composition. For these reasons, reliable vessel functional inferences based on phosphorus content alone were not possible (Dunnell and Hunt 1990).

Compositional Analysis of Solid Residues

Oudemans and Boon (1991) used Curie point pyrolysis (Chapter 33) to introduce solid residue samples onto the column for analysis with GC/MS (CuPy-GC/MS). Charred vessel residue samples were taken from sites in the Netherlands, dating from the Late Iron Age to the Early Roman period. Evidence of severely denatured proteins, polysaccharides, free fatty acids, and free fatty amides were detected in two archaeological samples from the insides of pots. Denatured proteins and carbonates, but no polysaccharides, were detected in another archaeological sample.

Later, Oudemans et al. (2007) used CHN elemental analysis (Chapter 38), FT-IR (Chapter 35), and solid-state ^{13}C cross polarization/magic angle spinning nuclear magnetic resonance spectroscopy (Chapter 36) to analyze solid archaeological residues on Roman Iron Age vessels from a site in the Netherlands. One residue that was not charred was identified as calcium carbonate with a small amount of unheated proteinaceous material. The composition of soot residues was clearly different and consisted of polyaromatic hydrocarbons (i.e., cyclic carbon and hydrogen structures; see Chapter 5).

The charred residues were divided into two groups based on the ratio of aromatic carbons (i.e., those in cyclic structures) and aliphatic carbons as determined by the NMR signal strength. Residues with less than one-half of the NMR signal arising from aromatic carbons were termed "mildly condensed" and contained lipids (fats) and peptides (proteins). Residues with higher levels of carbon aromaticity were termed "highly condensed"; some of these exhibited characteristics of carbohydrates but there was little evidence of lipids or proteins.

Their findings suggested that the highly condensed residues may have been exposed to more intense heat than the mildly condensed residues. Alternatively, the residues may be of starchy materials that condense at relatively lower temperatures than lipid or proteins. Oudemans et al. (2007) suggested that this combination of techniques provided a rapid means of characterizing and classifying solid organic residues.

Stable Isotope Analysis of Carbonized Residues

Hastorf and DeNiro (1985) were the first to perform stable isotope analysis of carbonized residues (or charred encrustations) from archaeological vessels in order to identify former contents. They proposed that combined stable carbon and stable nitrogen values could be used to separate (1) legumes, (2) non-leguminous C_3 plants, and (3) C_4 and crassulacean acid metabolism (CAM) plants. Their results indicated that Early and Late Period Peruvian pots contained only boiled non-leguminous C_3 plants. In the Middle Period, nitrogen-fixing C_4 plants, such as maize, were mixed with these other plants. As there were no CAM plants in the study area, this identification could be made with confidence. No evidence that legumes were cooked in pots was found.

In an expansion of work by Morton (1989), Morton and Schwarcz (2004) compared the isotopic ratios of carbonized pot residues from sites in Northern and Southern Ontario to published stable isotope studies of human bone (Chapter 13). Stable carbon isotope values were obtained on 137 carbonized pottery residues from 50 occupations at 45 sites dating between 680 B.C. and A.D. 1725. Stable nitrogen isotope values were determined on 55 of these residues and 9 others from a total of 32 sites dating from 0 to A.D. 1725.

Residues were removed from potsherds by scraping or flaking; 5 and 15–25 mg was required for stable carbon and nitrogen analysis, respectively. Ten residues were screened and found to be negative for the presence of humic acids, so they were analyzed without pre-treatment. Organic matter in the solid residues was converted to carbon dioxide (CO_2), nitrogen (N_2), and water (H_2O) by heating the sample in an evacuated tube with cupric oxide at 900°C for 2 h. The CO_2 and N_2 gas released was analyzed with a mass spectrometer (Chapter 31); $\delta^{13}C$ and $\delta^{15}N$ values were calculated with respect to Pee Dee belemnite and atmospheric N_2 standards, respectively.

Thick mixtures of mashed fresh corn, dry kidney beans, and rainbow trout were boiled individually in cooking experiments to recreate carbonized residues found on precontact Ontario pottery. A combination of corn and beans and another of corn contaminated with charred wood were also prepared. Only the residues produced by cooking mashed beans and fish visually resembled archaeological vessel residues. The $\delta^{13}C$ values for the modern residues produced by boiling fish and beans bore the strongest resemblance to values obtained from archaeological residues. The $\delta^{15}N$ values for the modern food samples, except for cooked beans, were within the range obtained from the archaeological residue samples. In terms of both carbon and nitrogen isotopic levels, the vast majority of prehistoric residues were most similar to those produced by a thick fish soup (Morton and Schwarcz 2004).

Although maize agriculture was introduced around A.D. 600, the average $\delta^{13}C$ value of archaeological residues was 24.5‰, which was most similar to native C_3 plants and the flesh of herbivores and freshwater fish. The percentage of C_4 plants (PC4) in each residue was calculated using the equation

$$PC4 = [(\delta s - \delta 3)/(\delta 4 - \delta 3)] \times 100$$

where δs is the $\delta^{13}C$ of the sample and $\delta 3$ and $\delta 4$ are averages for the C_3 and C_4 plants, respectively (Morton and Schwarcz 2004). While an increase in PC4 values was observed in pottery residues from Southern Ontario during the agricultural period, no such increase occurred in residues from Northern Ontario. The gradual increase in the maximum $\delta^{13}C$ values of carbonized vessel residues was in sharp contrast to the dramatic increase in the $\delta^{13}C$ values of human bone that paralleled the introduction of maize to the diet, however.

The $\delta^{15}N$ values of the archaeological residues indicated the presence of significant amounts of mammal or fish protein throughout the period from 680 B.C. and A.D. 1725. Consistent with values obtained on human bone, no shift relating to the introduction of beans was detected, which showed beans did not form a significant portion of the diet. Historic accounts confirm that fish was regularly consumed, but meat was only eaten occasionally.

Morton and Schwarcz (2004) suggested the difference between the stable carbon values of the carbonized vessel residues and those on human bone might relate to seasonal variations in food consumption or vessel function. Alternatively, it could indicate that dietary maize was consumed in different forms or that the residues were significantly contaminated with C_3-based carbon at the time of use or after burial.

Hart et al. (2007) used a variety of cooking experiments to show that attempts to model the proportion of C_4 plants in residues on the basis of PC4 values might not be valid for carbonized residues. Hart et al. (2007) argued that the PC4 values underestimated the percentage of maize in the majority of residues. Different foods contained different amounts of carbon and/or released different amounts of carbon during cooking. In addition, the amount of carbon released by corn depended on whether it was prepared green or dried. The presence of fat would also lower the $\delta^{13}C$ values of carbonized residues from the vessel necks and rims. The apparent low frequency of maize in the vessels might be due to a nonlinear relationship between the proportion of maize cooked in the pot and the resultant $\delta^{13}C$ value of the residue (Hart et al. 2007).

Fatty Acid Ratios

Malainey et al. (2001) used gas chromatography (Chapter 33) to determine the fatty acid composition of residues extracted from Late Precontact period Aboriginal pottery from sites in Western Canada. Residue identifications were combined with descriptions of faunal and tool assemblages to test hypothesized settlement and subsistence strategies for the Late Precontact period. Archaeologists generally believed that both bison and humans would converge on the parklands in the late fall to avoid the inhospitable conditions of the treeless plains. This model was not consistent with 18th and early 19th century accounts from European traders and travelers indicating that the majority of winter camps and bison herds were located in the grasslands.

Sherds were selected from seven sites located on the open grasslands, six from the transition zone between the plains and parkland, two sites in the parkland proper and three sites in the southern boreal forest. Identification criteria developed from the decomposition patterns of experimental residues were applied to archaeological vessel residues (Malainey et al. 1999a, c). Relative percentage compositions were calculated on the basis of ten fatty acids that regularly appeared in ancient residues (Malainey et al. 1999c; Chapter 14).

Residues from over 200 sherds were extracted and analyzed as fatty acid methyl esters. Large herbivore products alone or combined with plants were identified in almost 85% of residues from grassland sites; about 12% of the vessels were only used for the preparation of plants. There was little evidence of medium fat content foods (such as fish or corn) in the residues, faunal recoveries of fish bone, or fishing tools at grasslands sites. The dominance of large herbivore products decreased marginally in the residues from transition zone sites and recoveries of fish bones and tools associated with a fishing economy increased. Large herbivore products were less frequent in residues from parkland sites and were not detected in the 31 pottery residues from southern boreal forest sites. At some parkland and southern boreal forest sites, the majority of faunal remains were fish.

Evidence from the vessel residues, faunal remains, and tool assemblages supported the hypothesis that mobile hunter-gatherers of the North Plains remained on the grasslands throughout the winter, close to large stable herds of bison. Communal hunting strategies continued into later winter and spring, which enabled the exploitation of fetal and newborn bison, their favorite food, when the adult females became fat-depleted. Peoples from the parkland and forest moved to the northern edge of the grassland to take advantage of wintering bison. In the spring when bison were no longer available, these people switched to spring-spawning fish and other resources.

The broad-spectrum resource use of the parkland and forest-adapted foragers enabled them to exploit the fat-rich fish that became available in the spring. By exploiting fetal bison and newborns, the plains-adapted people were able to avoid suddenly switching from their prolonged diet of lean red meat to fish. The accounts of several fur traders and travelers indicated people accustomed to a diet of lean red meat experience deleterious physical effects when they consumed fish. The symptoms experienced, including diarrhea and loss of energy, were consistent with those of lipid malabsorption. The illness suffered by those attempting to make a switch from lean red meat to fish could be at the root of the fish avoidance, commonly practiced by the indigenous people adapted to nomadic life on the plains (Malainey et al. 2001).

These types of general characterizations based on fatty acid ratios could be further refined through the detection of sterols, triacylglycerols or other lipid biomarkers using GC/MS (Malainey 2007). Compound-specific stable isotope analysis could aid in the discrimination of C_3 browsers from C_4 browsers and identification of meat from non-ruminant animals. As outlined below, the presence of maize in the vessel residues could also be demonstrated.

Specific Characterizations

Plant Waxes

Waxes are long-chain alkyl compounds formed by plants and animals as protective coatings on skin, fur, feathers, leaves, and fruit. These compounds are resistant to decay so the relative proportions of individual components can be used as "fingerprints" to identify the specific origin of the wax (Evershed 1993:85). Evershed et al. (1991) found that 70% of the volatile constituents of the organic extracts of a sherd found in a Late Saxon/Medieval settlement were epicuticular leaf waxes of plant from the genus *Brassica*. The presence of nonacosane, nonacosan-15-one, and nonacosan-15-ol, indicated that leafy vegetables, probably turnip or cabbage, were cooked in the pot. Cooking experiments later confirmed that the occurrence and distribution of these three compounds matched the leaf wax of cabbage (*Brassica olearacea*) (Charters et al. 1997; Evershed 2000).

Animal Adipose Fats

Evershed et al. (1997) and Mottram et al. (1999) proposed molecular characteristics for differentiating animal fats based on (1) distribution of fatty acids including branched and odd carbon number components; (2) composition of monusaturated fatty acids; (3) stable carbon isotope values of C16:0 and C18:0; and (4) triacylglycerol distributions. As outlined in Chapter 14, compound-specific stable isotope analysis using GC-C-IRMS (Chapters 31 and 33) provided a means of discriminating ruminant and non-ruminant fat. Fatty acid characteristics also provided valuable information.

Evershed et al. (1997) reported that medieval lamps were fuelled with the fat of ruminant animals. Residues contained more C18:0 than C16:0 and significant amounts of branched chain and odd-numbered straight chain components (C15:0, C17:0, and C19:0). Complex mixtures of positional isomers of octadecanoic acid observed in the residues were due to the biohydrogenation of unsaturated dietary fats in the rumen of animals such as sheep and cattle (Evershed et al. 1997). By contrast, the residues from dripping dishes had higher levels of C16:0 than C18:0 and only one C18:1 positional isomer, Z-9-octadecanoic acid. The presence of a single isomer was characteristic of the fats of monogastric animals, such as pigs (Evershed et al. 1997).

Marine Animal Products

As noted in Chapter 14, Hansel et al. (2004) suggested that ω-(o-alkylphenyl) alkanoic acids with 16–20 carbons together with two isoprenoid fatty acids, 4,8,12-trimethyltridecanoic acid (4,8,12-TMTD) and phytanic acid (3,7,11, 15-tetramethylhexadecanoic acid), provide evidence for the exploitation of marine fish. Evershed (2003) observed isoprenoid acids are able to serve as biomarkers because their concentrations are high in marine animals and they are either absent or occur in low concentration in terrestrial animals.

Maize Using Compound-Specific Stable Isotope Analysis

As outlined in Chapter 14, the near unequivocal identification of maize in lipid residues is achieved through compound-specific stable isotope analysis of the long-chain alcohol, n-dotricontanol, and the most common fatty acids, C16:0 and C18:0, to show they were derived from C_4 plants (Reber and Evershed 2004a, 2000b; Reber et al. 2004).

Reber et al. (2004) applied this technique to 134 sherds from 17 sites along the Mississippi Valley in order to obtain information about the preparation and consumption of maize. Stable isotope analysis of human skeletons from Emergent Mississippian sites in the American Bottom indicated that lower status individuals consumed proportionately more maize than high-status individuals. During the Mississippian period, maize formed a more consistent part of the diet, but people living along the Mississippi River were less dependent upon it than groups living in other regions.

Reber and Evershed (2004a) analyzed absorbed lipid residues from pottery manufactured during the Middle Woodland, Late Woodland, Emergent Mississippian, and Mississippian periods in order to detect changes in the abundance of maize. Absorbed lipid residues were extracted with a 2:1 by volume mixture of chloroform and methanol; trimethylsilyl (TMS) derivatives were prepared from the total lipid extract (TLE). The neutral fraction of the TLE, containing long-chain alcohols, alkanes, and sterols, was isolated by saponification (Chapters 5 and 14) and derivatized as above. Fatty acid methyl esters were prepared by adding hydrochloric acid and boron trifluoride/methanol to the TLE. All 134 samples were analyzed using GC and GC/MS; 81 residues were analyzed with GC-C-IRMS.

Maize was confirmed in slightly less than 10%, eight of the 81, vessel residues submitted for compound-specific analysis on the basis of n-dotricontanol and fatty acids from C_4 sources. Seven of the sherds were recovered from two sites in the American Bottom, Mees-Nochta and Halliday; one sherd was from the Osceola site in the Lower Mississippi Valley. Although the consumption of maize was variable, its absence from pottery residues was still surprising; Reber and Evershed (2004a) suggested that either vessels used to process maize were not sampled or that it was prepared without the use of pottery. The low levels of maize agreed with results from archaeobotanical studies and bone stable isotope results from Mississippi Valley sites, which indicated maize served as a supplement to the traditional diet of C_3 plant and fish resources.

Dairy Products

Craig and Collins (2000) proposed an ELISA immunological method (Chapter 15) for detecting the presence of dairy proteins on both mineral and ceramic surfaces. Proteins can become tightly bound to the minerals, such as clay, which makes them resistant to chemical and microbial degradation; however, it also makes them difficult to extract. Craig and Collins (2000) developed a method, called Digestion and Capture Immunoassay (DACIA), which involved digesting the mineral matrix with hydrofluoric (HF) acid in an immuno-tube for 15 h. The released protein became bound to the tube walls so that at the end of the incubation period, the HF was discarded or neutralized and the captured protein could be assayed directly.

After HF was removed, the immunoassay for dairy products involved blocking the tube walls overnight with a gelatin solution. A monoclonal antibody raised against α-casein, called anti-bovine casein Immunoglobulin M, was added to detect the presence of bovine milk proteins. The tubes were washed and a second antibody was added to detect positive reactions to the first antibody; horseradish peroxidase was used to detect positive reactions to the second antibody. Craig and Collins (2000) and Craig et al. (2005) reported that the α_{si}-casein protein served as a biomarker for milk from bovines, such as cattle, and that no cross-reactions occurred with the milk of humans, horse, goat, or sheep.

Copley et al. (2003) reported that the archaeological residues associated with ruminant dairy products could be discriminated from adipose fat using GC-C-IRMS. The $\delta^{13}C$ of C18:0 ($\delta^{13}C_{18:0}$) of dairy fats was systematically lower than the $\delta^{13}C$ values of C16:0 ($\delta^{13}C_{16:0}$) because the mammary gland cannot biosynthesize C18:0; the animal obtains C18:0 from the unsaturated fatty acid component of dietary plants.

Copley et al. (2003) found the average difference between stable carbon isotope values of carbohydrates and fatty acids in 166 plants eaten by grazing and browsing animals in Southern Britain was 8.1‰. In dairy fats, the $\delta^{13}C$ values of the plant lipid-derived C18:0 were at least 3.3‰ lower than the $\delta^{13}C$ values of carbohydrate-derived C16:0. In other words, the $\Delta^{13}C$ values, where $\Delta^{13}C = \delta^{13}C$ of C18:0 $- \delta^{13}C$ of C16:0, of ruminant dairy fats were less (more negative) than $-3.3‰$. Copley et al. (2003) produced bivariate plots of $\delta^{13}C_{16:0}$ vs. $\delta^{13}C_{18:0}$ showing these variables clearly separated reference ruminant dairy fats, ruminant adipose fats, and porcine adipose fats.

In their examination of archaeological residues from three early Neolithic sites, Copley et al. (2003) found that 54–57% of the sherds from two sites contained predominantly dairy fats. These two sites had causewayed enclosures and, based on age and sex structure of the herd, dairying activities were believed to be important. The third was a domestic site and 78% of the residues contained predominantly dairy fats. In bivariate plots, residues containing mixtures of dairy and/ or animal fats fell between the reference fat groupings.

Ancient DNA in Amphora Residues

Hansson and Foley (2008) demonstrated that the analysis of ancient deoxyribonucleic acid (ancient DNA or aDNA) (Chapter 16) could determine the former contents of ancient amphora extensively used as transport vessels in the Mediterranean region. The 2400-year-old amphorae were from a shipwreck in the Aegean Sea located off the island of Chios in 70 m of water. One amphora was of the type manufactured in Chios during the fourth century B.C., presumably for the transport of wine; the origin of the other was not known.

Ceramic scrapings (about 1 cm³) were collected from the walls of both amphorae; a sample of resin from the second vessel was also examined. Ancient DNA was extracted with proteinase K and precipitated with sodium actinium (NaAc) and ethanol then redissoved in TE buffer. Prior to amplification, the extracts were analyzed by electrophoresis (Chapter 33) to confirm the presence of genetic material. Primers were specifically designed to target short sequences unique to chloroplast plants (*Embryophyta*), rather than algae, and included regions that enabled the identification of a specific plant genus or species. After amplification, the polymerase chain reaction (PCR) products were cloned, sequenced, and then compared to reference material in the GenBank database. Since ancient DNA typically contain only short fragments, additional polymerase chain reactions were performed targeting longer segments (500–1000 bp) likely to be present if the sample was contaminated with modern plant DNA.

During the first round of PCRs, a sequence found in olive and another found in thyme, oregano, and sage were identified in the first amphora. During the second round of PCRs, the presence of olive and oregano was confirmed. The extract from the wall of the second amphora contained DNA from the *Pistacia* genus, which includes mastic and pistachio nut; the resin did not contain amplifiable DNA. Neither vessel contained evidence of grape DNA. The findings of Hansson and Foley (2008) demonstrated that fourth century B.C. amphorae from the island of Chios were not exclusively used for the transport of wine. The results from the second amphora were less conclusive as only reference sequences for mastic and pistachio were available, but not for a third *Pistacia* species, terebinth. A resin used to coat wine transport amphora was produced from terebinth; mastic was added to wine produced on the island of Chios as a flavoring agent and preservative during the Roman period (Hansson and Foley 2008).

Wine

Badler et al. (1990) used diffuse-reflectance Fourier transform infrared spectroscopy (FT-IR) to study the contents of jars from the Iranian site, Godin Tepe, dated to the mid-fourth millennium B.C. Diffuse-reflectance FT-IR is similar to ATR-FTIR

(Chapter 35) in that infrared light reflected by the sample is detected; the difference is that the beam of infrared light actually strikes the powdered sample. Since the surfaces of tiny particles are randomly oriented, the IR radiation is reflected in all directions; concave and flat mirrors are used to direct the IR radiation reflected by the powdered sample to the detector (Skoog et al. 1998).

Badler et al. (1990) extracted residues from sherds with boiling acetone, evaporated the extract to dryness, then mixed the resinous solid with potassium bromide (KBr) powder for analysis. By making comparisons with Nubian archaeological wine residues and standards, tartaric acid and its salts were identified in the residue, indicating these jars formerly contained wine.

McGovern and Michel (1996) used diffuse-reflectance FT-IR to examine the solid scrapings of red deposits from the interior and dry solvent-extracted residues from Godin Tepe jars. McGovern and Michel (1996) argued that similarities existed between the diffuse-reflectance FT-IR spectra of the powdered residue and solid extracts and the transmission FT-IR of the L-(+) form of tartaric acid stereoisomers (Chapter 6). The surface scrapings from the Godin Tepe jars were also subjected to the Fiegl test, a colorimetric test for tartaric acid, which involved treatment with β,β'-dinaphthol and concentrated sulfuric acid. The treated residue exhibited green fluorescence under ultraviolet light, which showed that tartaric acid was present in the sample. This evidence indicated that the amphora contained a grape product.

A colorimetric technique was also employed by Petit-Domínguez et al. (2003) to test for the presence of tannins in three large (40 L) and one small (0.5 L) amphora dating to, or before, the fifth century B.C. from sites in Spain. The Folin–Denis method involved the oxidation of tannins (which are phenolic compounds) by the Folin–Denis reagent, a solution consisting of sodium tungstate, phosphomolybdic acid, and phosphoric acid. The amphorae samples were washed in 5 mL of distilled water; 2.5 mL of the aqueous solution was treated with 250 μL of the Folin–Denis reagent. After 3 min, 0.5 mL of a sodium carbonate solution was added and the mixture diluted to 5.0 mL. The appearance of a blue-black color indicated the presence of soluble tannins.

The extract from one of the large amphora gave the color change demonstrating the presence of soluble tannins. To show that the color change was due to tannins rather than an interfering substance, 1 mL aliquots of the sample solution were treated with a mixture of gelatine, acid sodium chloride (made by adding concentrated sulfuric acid to a saturated sodium chloride solution), and kaolin powder. This caused tannins to precipitate out, producing a "tannin-free solution," which was retested with the Folin–Denis reagent. None of the "tannin-free" solutions gave a positive test, which confirmed the presence of tannins in the amphora. The negative reactions from the other three amphorae showed that tannins were either never present or had decomposed (Petit-Domínguez et al. 2003).

Garnier et al. (2003) proposed the use of polyphenols, i.e., linked tannin units, as an indicator of archaeological wine. In order to analyze degraded and polymerized archaeological samples with GC/MS, they employed a process called thermally assisted hydrolysis–methylation (THM), or thermochemolysis, which uses heat and a chemical reagent to release and methylate sample components.

Modern samples of grape seeds, grape skins, and wines were used to develop the method, which was then applied to archaeological grape seeds from different environments and preserved Roman wines recovered from shipwrecks. Each sample was treated with tetramethylammonium hydroxide (TMAH) in methanol; after it dried, the sample was pyrolyzed at 450°C for 5 s and the resulting products analyzed by GC-MS.

Garnier et al. (2003) found that more than 90% of the pyrolytic extract were phenols and demonstrated that the polyphenol decomposition products were very stable. Although the wine residues were treated with chloroform/methanol to remove lipids, diterpenic compounds from the resinous pitch used to coat the interior of the amphora were still present. Levels of diterpenic acids exceeded those of phenolic pyrolytic markers in one type of wine. The presence of lignin was attributed to the cork used to stopper the amphora. Important markers that were preserved in the ancient wine residues included di- and trimethoxylated benzoid compounds, which formed from condensed tannins (also called proanthocyanidins) in the fruit.

On the basis of their examination of experimental and archaeological residues, Stern et al. (2008) challenged identifications of archaeological wine residues. Analyses using the Fiegl spot test, FT-IR spectroscopy, and gas chromatography-mass spectrometry resulted in ambiguous results, which could lead to false identifications; the most reliable results were obtained using high-pressure liquid chromatography-tandem mass spectrometry (Stern et al. 2008).

Starchy Foods

Hill and Evans (1987, 1988, 1989), who were working on material from the South Pacific, spent more than 2 years developing infrared and ultraviolet standards for the heptane extracts of six cultigens: banana, sweet potato, rice, sago, taro, and yam. Hill and Evans (1987) believed these starch-producing plants with no oil composition would be difficult to detect using other methods. For each cultigen, they identified specific wavelengths that could be used to characterize the plant regardless of variety, geographic location, or method of preparation (Hill and Evans 1988). When the technique was applied to archaeological samples these cultigens were detected in three of the six residues tested: one sweet potato; one taro; and one possibly sweet potato, taro, and yam combination (Hill and Evans 1988). Two sherds with residues of banana, one of rice and one of both taro and sweet potato were identified from another site (Hill and Evans 1989).

Chocolate

Hurst et al. (1989) and Hall et al. (1990) both reported the use of high-pressure liquid chromatography (HPLC; Chapter 33) and liquid chromatography-mass spectrometry (Chapters 31 and 33) to analyze dry residues from the contents of a Mayan vessel. The presence of theobromine and caffeine in the residue of one vessel was confirmed through each analysis. As these compounds tended to be unique to cocoa, their detection strongly supported the iconographic analysis of the Mayan glyphs decorating this vessel. The application of HPLC in this instance was especially appropriate as the sample contained no fatty acids or other lipids (Hurst et al. 1989).

Compound-Specific GC Analysis and AMS Dating

Stott et al. (2003) described a method of obtaining radiocarbon dates on specific compounds in absorbed pottery residues. The approach provided a means of directly dating pottery, rather than associated materials, such as charcoal. This was considered advantageous because the composition of the absorbed lipids was directly related to the former contents of a vessel and the function of the pot. The structure of the lipids could be confirmed through GC/MS and GC-C-IR-MS (Chapters 31 and 33) and potential contaminants detected. Furthermore, the age of the lipids closely corresponded to the time of their introduction into the vessel; because they are not soluble in water, lipids are immobile and generally not affected by leaching (Stott et al. 2003).

Two fatty acids, C16:0 and C18:0, were targeted for analysis because the concentrations of the fatty acids were comparatively high, 10^{-4} to 10^{-3} g per gram of potsherd. Lipids were extracted from powdered archaeological potsherds with a Soxhlet apparatus using a 2:1 by volume mixture of dichloromethane and methanol (Chapter 14). TMS derivatives were prepared from an aliquot of the total lipid extract (TLE) and the sample was analyzed with high temperature (HT)-GC and HT-GC/MS. The remainder of the TLE was converted to fatty acid methyl esters using hydrochloric acid and methanolic sodium hydroxide then extracted with hexane. Samples included two reference compounds, n-octadecane and n-heneicosanoic acids.

Preparative capillary gas chromatography (PCGC) was conducted with a widebore column with an internal diameter of 0.53 mm. Using this approach, only a small portion of the targeted component was directed to the flame ionization detector; the majority of the isolated component was collected in a trap. Stott et al. (2003) reported that 98% of desired targeted components, C16:0 and C18:0, were directed to a preparative fraction collector, which contained six sample traps and one waste trap. Collected samples were combusted in a continuous flow CHN elemental analyzer (Chapter 38); carbon dioxide was separated from the other gases using GC and collected.

A Megabore fused silica capillary column coated with a DB-1 stationary phase was used to minimize the amounts of column degradation products, i.e., column bleed, in the sample traps (Stott et al. 2003). Tests with 120 column blanks demonstrated that only a minimal amount of column-derived carbon would appear in any compound-specific AMS analysis. The automated instrument exhibited excellent reproducibility, which enabled the trap opening times to be precisely set. The isolated fatty acids were transferred from the traps into tin capsules containing tin powder. The trapping efficiency of C16:0 and C18:0 was determined to be greater than 90%. More than 95% of the trapped fatty acids were recovered from the tin during combustion.

The organic solvents used in sample processing were found to contain less than 0.9% modern ^{14}C and were considered ^{14}C "dead" (Stott et al. 2003:5040). Analyses of the reference compounds indicated the methodology introduced a small amount of modern carbon, so a correction of 2.9 ± 0.3 μg of modern carbon was applied to date calculations. The degree of isotopic fractionation that occurred over the course of 80–100 PCGC runs was between 1.1 and 1.5%. Additional corrections were made to account for the addition of a methyl group to each of the targeted fatty acids during derivatization and for isotopic fractionation.

Archaeological sherds from Early Neolithic to Medieval sites in the United Kingdom were selected for compound-specific ^{14}C dating. Most residues were derived from animal fat so C16:0 and C18:0 were the dominant fatty acids. Four samples from medieval vessels containing between 200 and 4840 μg of preserved lipids per gram of sherd were analyzed. Three fatty acids were collected, C16:0, C18:0, and C18:1; about 100 preparative GC runs were needed to accumulate sufficient amounts of each fatty acid. The dates obtained on all three fatty acids were in good agreement with radiocarbon dates on a building timber and ages based on pottery typology (Stott et al. 2003).

Two fatty acids, C16:0 and C18:0, were collected from the residue of a Roman mortarium. The calibrated age from C18:0, A.D. 210–440, correlated well with expectations but the age from C16:0, A.D. 430–650, was too young. Age estimates on three fatty acids, C16:0, C18:0, and C18:1, collected from Mid-Late Neolithic Fengate vessel residues correlated with expectations. Ages on fatty acids from Early Neolithic vessel residues also agreed with expected values, in particular the age obtained from C18:0. Ages obtained on fatty acids from Early Neolithic vessel residues from a different site correlated well with ages from associated antler, bone, and charcoal. The age obtained on C18:0 from an Early Neolithic bowl from another site compared particularly well to the dendrochronological age of an elevated wooden walkway; the age estimate from C16:0 was too young.

Stott et al. (2003) concluded that PCGC was capable of extracting sufficient amounts of targeted fatty acids for AMS dating if the samples contained more than 200 μg of preserved lipids per gram of sherd. Almost 100 GC runs were required to accumulate sufficient amounts of fatty acids for AMS dating. While ages obtained on all fatty acids were in good agreement with expectations, ages obtained on C18:0 best correlated with ages on associated materials; ages on C16:0 ages were often slightly younger. For this reason, Stott et al. (2003) recommended only the C18:0 component of lipid residues should be selected for compound-specific dating studies.

Chapter 22
Other Food-Related Residues

Food-related residues appear in a variety of contexts, including flaked and ground stone tools, thermally altered rock, and even in the teeth of the deceased. Examples of different approaches that can be taken for the analysis of these residues are presented, including fatty acid analysis, identification of possible biomarkers and techniques that can identify both the inorganic and organic components in the residue. The analysis of lipid residues is described in Chapter 14.

Techniques used and issues surrounding the analysis of proteins are provided in Chapter 15. Examples presented in this chapter include the analysis of projectile point and ground stone tool residues using an older method of protein analysis and the blind test results of an experiment designed to test the reliability of protein identifications. Newly proposed and very rigorous protocols for protein analysis and an example of their application for the identification from stone tool residues are also presented.

An example of ancient DNA analysis of stone tool residues using the polymerase chain reaction is also described. While an attempt was made to translate the methodology into lay language, it might be beneficial to first review the descriptions of the analysis of ancient DNA and the polymerase chain reaction (PCR) in Chapter 16.

Lipid Analysis

Burned Rock Residues

Buonasera (2005) examined the fire-affected andesitic rocks recovered from the Late Prehistoric Fort Mountain rock shelter in central California to determine if they contained analyzable quantities of lipids and if these lipids were introduced by cooking activities. These findings were then used to critique Quigg et al.'s (2001) analysis of burned rock residues from the Lino site, TX. Ethnographic accounts from central California indicated acorn and pine-nut soups and mush were prepared in baskets using stone boiling techniques; heated rocks were used to bake acorn bread in earth ovens. Ceramic vessels were not used prehistorically.

M.E. Malainey, *A Consumer's Guide to Archaeological Science*, Manuals in Archaeological Method, Theory and Technique, DOI 10.1007/978-1-4419-5704-7_22, © Springer Science+Business Media, LLC 2011

Rocks were selected from the debris apron and the shelf midden; off-site rocks were collected from a forested location near Strawberry Valley, CA. The outermost 1–2 mm of exterior surfaces were removed and samples were crushed. Lipids were extracted using the Bligh and Dyer method by rocking the powdered rock in a 1:2:0.8 by volume mixture of chloroform, methanol, and citrate buffer. Lipids were extracted from intact rocks by sonication using the same solvents. Fatty acid methyl esters were prepared with 4% sulfuric acid in methanol and analyzed by gas chromatography (GC) with an HP-5 column.

Residues recovered from the archaeological and natural rocks were characterized on the basis of the relative percentages of fatty acids using criteria developed by Marchbanks (1989) and Malainey et al. (1999a–c). Using the criteria developed by Marchbanks (1989), residues were identified as plant ($n = 4$), fish ($n = 4$), and land animal ($n = 1$). According to the criteria developed by Malainey et al. (1999a–c) residues were identified as beaver ($n = 7$), fish, or corn ($n = 1$); two residues did not conform to any category. Based on the degradation patterns of experimental residues from Texas, Quigg et al. (2001) reported that the decomposed residues of moderate and high fat content seeds, and hence nuts, had high levels of C18:1 isomers and would be identified as beaver. The category was later subdivided based on levels of C18:1 isomers in the sample (Malainey 2007).

Buonasera (2005) found that three of the archaeological samples had lipid concentrations two to six times greater than off-site samples from a forest. The off-site rocks contained substantially higher concentrations of the polyunsaturated C18:2 than archaeological rocks. Buonasera (2005) noted that firing pottery at high temperatures gave the vessels a clean slate and suggested this also could apply to heated rock. Evidence of this exists in the data presented, but was not discussed by Buonasera (2005); the concentration of fatty acids in residues extracted from archaeological burned rocks FMRS 6 and FMRS 24 contained about 58 and 82% of those reported for natural rocks.

It must be noted that the criteria developed by Malainey et al. (1999a–c) were for the identification of decomposed archaeological residues. The purpose of analyzing lipids from non-cultural rocks should be to assess the natural lipids that occur in the environment and could potentially contaminate a sample. The presence of high levels of polyunsaturated fatty acids in an archaeological residues is an indicator of contamination (Quigg et al. 2001). In addition, only data from cultural rocks with significant quantities of lipids should be incorporated in to archaeological interpretations. Quigg et al. (2001) excluded residues extracted from 1 ground stone and 11 burned rock samples from the Lino site on the basis that insufficient lipids were present to permit a reliable identification. Measurable amounts of fatty acids were actually detected in 9 of the 11 burned rock residue samples excluded. Residues extracted from two Lino site burned rocks were devoid of fatty acids, which clearly indicated heating completely eliminated any natural lipids formerly present in the rocks. The context of recovery is another consideration as the Lino site rocks were only selected from archaeological features, primarily burned rock dumps.

Betel Nut Residues in Dentitions

Oxenham et al. (2002) used GC/MS to investigate the origin of dark reddish brown stains observed on the dentitions of 23 of 31 individuals from the Bronze Age site of Nui Nap in northern Vietnam. In particular, Oxenham et al. (2002) wished to test the hypothesis that the stains were from betel nut, which is frequently chewed by the inhabitants of Southeast Asia and the Pacific today.

Betel nut is known to contain amino acids, tannins, gallic acid, and tetrahydropyridine alkaloids, such as arecoline, iarecaidine, guvacine, and guvacoline. After an epoxy cast was made of the tooth, the stained surface was ground off with a dental drill. Lipids were extracted from the tooth powder and a fresh betel nut, serving as a modern reference, by sonication in a mixture of chloroform, methanol, and water. Following the preparation of trimethylsilyl (TMS) derivatives, the extracts were analyzed by GC/MS.

While tetrahydropyridine alkaloids were detected in the fresh betel nut extract, these compounds did not appear in the residue extracted from the tooth. A compound interpreted as a tannin derivative appeared in both the archaeological residue and the betel nut, however. Oxenham et al. (2002) considered this evidence that betel nut juice was intentionally applied to the teeth, possibly for aesthetic purposes. Examination of the cast with scanning electron microscopy (SEM) indicated the tooth may have been etched in preparation for the betel nut staining.

Analysis of Surface Residues from a Milling Stone

Babot and Apella (2003) used several analytical techniques to examine the solid residue on a milling stone from a rockshelter in northwestern Argentina. The whitish-gray residue was recovered from the pecked working surface of a magmatite milling stone that had been placed upside down over a bedrock mortar.

The powdered residue was determined to be a mixture of burned bone and maize starch. Semi-opaque bone particles, starch grains, silica phytoliths, and microcharcoals were observed using a petrographic microscope. Maize starch granules were cracked and broken from the milling process. The presence of hydroxyapatite was detected through the analysis by powder X-ray diffraction (XRD) (Chapter 37) using a diffractometer. The presence of hydroxyapatite was also identified using Fourier transform-infrared spectroscopy (FT-IR) (Chapter 35), performed with the residue powder mixed with potassium bromide (KBr) and pressed into a disc. No absorption bands consistent with the presence of carbohydrates or fatty acids were detected. The chloroform-soluble components of the residue were examined with GC-MS. Babot and Apella (2003) reported the distribution of fatty acids in the residue resembled that observed in animal fat; cholesterol was also detected. Babot and Apella (2003) suggested that the milling stone was last used to grind burned bone and the presence of small amounts of maize were remnants of earlier uses. Maize bran was rarely detected, which suggested the kernels were peeled and winnowed prior to grinding.

Protein Analysis

Analysis of Milling Implements

Yohe et al. (1991) used immunological methods (Chapter 15) to identify residues extracted from a selection of ground stone tools from two sites in southern California. In particular, evidence was sought to confirm ethnographic accounts that small mammals were processed using stone implements. The artifacts analyzed included six cobble manos, three metate fragments, a hammerstone and five soil samples from one site, and a basket hopper mortar and pestle from another site.

Residues were extracted from the ground working surfaces with 5% sodium hydroxide (NaOH) with occasional agitation. The solutions were concentrated by lyophilization (i.e., freeze-drying), then reconstituted in a 5% ammonia solution. Testing of the residues against pre-immune serum yielded three positive results. Pre-immune serum is the blood of the animal in which the antisera were raised; a positive test shows the residues would produce a false-positive result to immuno-logical tests (Chapter 15). These residues were re-tested after treatment with the non-ionic detergent, Tween; one residue produced a second positive result and was eliminated from the study. Crossover immuno-electrophoresis (CIEP) was used to test sample residues against antisera for deer, rabbit, rat, guinea pig, and mouse.

Three positive results were obtained on cobble manos from one site, two residues tested positive to rodents (mouse and rat) and one residue tested positive to non-specific proteins. None of the soil samples from this site yielded positive results. The mortar and pestle from the other site both produced positive reactions when tested against rodent antisera (mouse and weakly rat). Yohe et al. (1991) concluded that the findings support ethnographic accounts of processing small animals with milling implements. Since the antisera were prepared from Old World rodents, it is not possible to clarify the identifications further. The wood rat would be an example of a New World member of the rat family; the white-footed and the pocket mouse would be examples of New World mice.

Analysis of Experimental Residues

Leach and Mauldin (1995) and Leach (1998) conducted a series of blind tests of CIEP using simulated stone tool residues. In the first test, Leach and Mauldin (1995) prepared 54 experimental stone tool residues by processing flesh from ten animal species. Freshly knapped artifacts were used to cut or scrape one type of animal flesh for several minutes. The residues were allowed to dry, then visible hair was removed by rubbing the tool with a piece of cloth or sand, but blood was still visible. Some samples were also exposed to heat. The tools were placed in sealed plastic bags and stored in a refrigerator for between 1 and 2 months prior to testing. "Blank" tools without residues and fire-cracked rock from a hearth used to cook rabbit were also submitted for analysis.

CIEP was conducted using polyclonal antibodies for a variety of mammals, two birds, and six plants. Only 20 residue identifications were deemed to be correct or partially correct; another 9 were considered partially wrong or wrong; no blood was detected on 25 tools that were used to process animal flesh. Some misidentifications could be attributed to cross-reactions between closely related animals, but others were well beyond the level of family. The success rate of CIEP was judged to be 37%, but approached 50% if samples exposed to heat were excluded.

In the second test, Leach (1998) devised a similar blind test to verify the ability of CIEP to identify plant residues. Freshly knapped tools were used to scrape, grind, or pulverize living plants. Afterward, the tools were dried in sunlight for several hours; visible plant remains were then removed by rubbing with a cloth or sand. The 19 tools were stored for 1–2 months in sealed plastic bags in a refrigerator prior to testing. Solid residues from vegetables boiled dry in a ceramic pot were also submitted for analysis.

One plant residue was correctly identified by CIEP and another identification was judged as a partial success; other samples were identified as negative or incorrectly identified, mainly as human. Leach (1998) indicated that archaeological residues previously submitted to this lab had included plant identifications. The approximately 200 archaeological tools previously submitted had been extensively handled and only one had been identified as human, so contamination should not be a problem in this case. Leach and Mauldin (1995) and Leach (1998) both treated negative results (i.e., no protein detected) on tools that were used to process animal flesh or plants as identification errors. This is contrary to standard practice as these results relate to the sensitivity of the technique, not its reliability.

New Protocols for Protein Residue Analysis

Shanks et al. (1999) proposed new CIEP protocols for protein analysis in order to improve confidence in residue identifications (Chapter 15). Recommendations included the use of only forensic quality antisera obtained from professional labs. The concentrations at which a serum could accurately identify a known amount of protein, or its sensitivity, should be established by serial dilution. Cross-reactivity analysis should be conducted to identify inconsistencies in antisera specificity so that cross-reactions could be distinguished from true positives. The antisera should be tested with whole blood from animals in the same family taxon in order to establish the range of concentrations at which cross-reactions occurred.

The effectiveness of the new protocols was tested using 15 experimentally generated artifacts coated with the whole blood of different animals. Some of the artifacts were exposed to direct sunlight for 14 days while others were stored uncovered on a laboratory bench. Residues were extracted and identified with 100% accuracy.

Shanks et al. (1999) implemented the new CIEP protocols for the analysis of residues on stone tools from the Bugas-Holding site in northwest Wyoming. The site is located on the alluvial floodplain of Sunlight Creek, near the eastern border of Yellowstone National Park. A total of 46 artifacts were selected for residue analysis at the time of excavation and handled with gloved hands; 18 soil samples were also collected and analyzed.

Biological residues were extracted with 5% ammonium hydroxide using ultra-sonication followed by gentle agitation on a wave table. The samples were vacuum dried, then stored in phosphate buffer saline (PBS) at $-20°C$. CIEP was used to test each tool extract against a series of known antisera and pre-immune controls. Each gel plate also included one positive and one negative control. If any immunological activity was detected the entire test was repeated.

Shanks et al. (1999) reported a success identification rate of 15.2% without any inconsistencies. Positive reactions identified fauna indigenous to the Rocky Mountains including bovine ($n = 4$), bear ($n = 3$), sheep ($n = 2$), and human ($n = 1$). The residues were extracted from seven tools including three bifaces, two retouched flakes, one endscraper, and one utilized flake. Five tools had only one type of residue, one had two, and one had three. No positive reactions were generated by the soil extracts; therefore, the residues extracted from the tools related to artifact uses (Shanks et al. 1999).

DNA Analysis

DNA Analysis of Flaked Stone Tool Residues

Shanks et al. (2005) tested 24 flaked stone tools from the Bugas-Holding site in northwest Wyoming for the presence of ancient deoxyribonucleic acid (ancient DNA or aDNA). Comparisons were also made between DNA preservation in bone and stone tools from the same stratigraphic context. The analyses were performed according to the protocols recommended for the study of ancient DNA (Chapter 16). Prior to the initiation of the study, 20 no-template PCR reactions were performed to test the reagents for contamination. A mock DNA extraction with buffer alone was performed alongside every extraction; only one genuine and one mock extraction were performed each day. After each extraction, work areas were cleaned with bleach and ultraviolet light; equipment was cleansed with hydrogen peroxide and ethylene oxide gas. Between 4 and 10 no-template reactions containing purified water were performed for every positive reaction.

The protocol for DNA extraction was identical to that described above for protein residues by Shanks et al. (1999). After the vacuum drying step, residues undergoing DNA analysis were digested in a protein extraction buffer consisting of 5 mM ethylenediaminetetraacetic acid (EDTA) with 0.5% sodium dodecyl sulfate (SDS) and incubated with proteinase K at 56°C for 6–12 h. The solution was extracted with water-saturated phenol, followed by a 24:1 part mixture of chloroform and isoamyl alcohol. The aqueous phase, which would contain the DNA, was concentrated and then purified using a silica extraction method.

The PCR amplification was performed using a protocol similar to the one described in Chapter 16 using two sequential 40-cycle PCR reactions. The primers used targeted the cytochrome *b* region. Extracts that contained amplifiable DNA or visible surface residues were amplified with additional primers targeting (1) a 191 base pair (bp) region of the mitochondrial 16S ribosomal ribonucleic acid (RNA) gene, which is present in members of the cat (Felidae), deer (Cervidae), and sheep (Ovidae) families; (2) a 137 bp region of the mitochondrial 12S ribosomal RNA gene found in birds in the Phasianinae subfamily; and (3) a 292 bp region of the mitochondrial D-loop sequence that can distinguish wolf from dog (Shanks et al. 2005). All amplifications were repeated.

PCR products were cloned using a plasmid vector and introduced (transformed) into *Escherichia coli* bacteria (see section "Special Topic: DNA Cloning" in Chapter 16). The resulting DNAs were sequenced by the dye-terminator [i.e., Sanger (dideoxy) method with color-labeled nucleotides] using an automated sequencer (see section "Special Topic: DNA Sequencing" in Chapter 16). The nucleotide sequences obtained from the stone tool residues were compared to those in computerized databases.

PCR sensitivity was monitored using modified templates with 20–40 bp insertions or deletions, which could be distinguished from products amplified from wild-type DNAs. Additional processes were used to detect rare amplicons. PCR products were inserted into a plasmid vector, then transformed into *E. coli* bacteria and cloned. The number of cloned products was increased by PCR amplification or colony hybridization.

Of the 24 tools examined, 14 were collected without being touched by human hands. Amplifiable DNA was recovered from nine stone tools, three of which were not handled by humans, including five bifaces, two side scrapers, one end scraper, and one utilized flake. Canid DNA was detected on all tools; one extract also contained domestic cat DNA, and another extract also contained both domestic cat DNA and mule deer DNA. No DNA was detected on three flakes lacking use-wear, which were included as negative controls. No animal DNA was detected in sediment samples. The degree of protein preservation in six bison bones from the site was generally good and four contained amplifiable bison DNA.

The test for rare amplicons yielded positive results for pig, human, and cow. Tests for contamination in the PCR reagent (see Chapter 16) showed 94% of reactions were negative; positive reactions indicated the presence of cow, guinea pig, human, and chicken DNA.

Based on the abundance of big horn sheep and bison bone in the faunal assemblage, the absence of DNA from these animals in the extracts was somewhat unexpected. Shanks et al. (2005) noted that the site was occupied from October until March or April. It is possible that big horn sheep residues from animals butchered in the late fall or early winter were removed when the tools were resharpened. Debitage produced by tool resharpening could possibly retain big horn sheep DNA. The canid residues either indicated dogs were butchered in the spring at a time of resource stress or that dogs simply introduced their own DNA by licking the tools (Shanks et al. 2005).

Tool Residue Analysis Using Multiple Techniques

Loy and Dixon (1998) applied a wide variety of techniques in their analysis of residues from fluted projectile points from nine sites in northern Alaska and northwest Yukon. During the last glaciation, this region would have represented the eastern side of the Bering Land Bridge or eastern Beringia.

The collection of artifacts examined included 34 fluted projectile points from museum collections and two morphologically related projectile points from the same region. Protein analysis of the projective point residues involved the application of seven independent techniques, which are described in Chapters 15, 16, and 33. Tools were first screened for the presence of residues using a low-power stereobinocular microscope. This process led to the elimination of 15 specimens that lacked visible residues and six specimens that lacked sufficient residue for analysis.

Red blood cells (RBCs) were detected in eight residues in three size ranges between 4.5 and 11 μm. Variations in size were attributed to distortions of red cell membranes and most RBCs did not appear to carry hemoglobin. Residue RBCs were compared to RBCs from mammoth tissue and the dried blood smears of musk ox, sheep, caribou, bear, and Indian elephant. Loy and Dixon (1998) reported that the sizes of RBCs in three residues were large and matched the size range of RBCs in mammoth tissue and modern Indian elephant blood. RBCs in the medium size range were observed in two residues, which matched the size of RBCs in bear blood. RBCs in the small size range were observed in five residues, including two residues that also had medium-sized RBCs. RBCs of similar size occurred in the blood of musk ox, sheep, and caribou.

In order to screen for the presence of heme (Chapter 15), 10–20 μL of the residues washed from the tools with ultrapure water was applied to a Hemastix colorimetric dip-stick. After 1 min, the reactions were read; all 15 residues were judged to exhibit positive reactions. All residues were positive for the presence of mammalian immunoglobulin G when tested with the broad-spectrum antibody, staphylococcal protein A (Loy and Dixon 1998).

As described in Chapter 33, the behavior of a protein in an electric field is related to the properties of its constituent amino acids. At a particular pH, called its isoelectric point (pI), internal neutralization occurs and the protein becomes electrically neutral. Isoelectric focusing is the separation of proteins with different isoelectric points using electrophoresis. A pH gradient is established within the gel slab so that one end is basic (pH greater than 7) and the other end is acidic (pH is less than 7). When the electric field is applied, the protein will carry an electric charge and move through the pH gradient. At its isoelectric point, the protein becomes electrically neutral and stops moving.

Loy and Dixon (1998) used isoelectric focusing (IEF) to compare the distribution of isoelectric points of tool residue components to those in the blood of control species. The stone tool residues were placed in individual gel lanes and allowed to separate under a constant power of 8 W for 45 min or until the visible pI standards reached a stable configuration. The positions of separated components in the residues and control references were determined using a densitometer to record the differential absorbance of light along each gel lane. Based on the position of components in the standard, the absorbance peak locations of the separated components were calibrated to pI values. The coefficient of match between the pI values of tool residue components and those of control species blood was then calculated.

Four residues examined were selected for IEF analysis. The components of two tool residues, including the one exhibiting the highest similarity coefficient score of 0.58, most closely match the components in the mammoth tissue extract. Two tool residues appeared to contain blood from more than one species (Loy and Dixon 1998).

Following Loy (1983)'s procedure, which is outlined in Chapter 15, hemoglobin crystals were grown from the tool residue extracts. All but one residue produced at least one type of crystal; these were identified as the hemoglobin crystals of six different animal species. Loy and Dixon (1998) reported the presence of caribou crystals in eight residues and mammoth crystals in five residues. Bison and sheep crystals each appeared in two residues; bear and musk ox crystals each appeared in two residues.

DNA analysis using the polymerase chain reaction was performed on one tool residue and compared to material extracted from mammoth tissue and bone but the results were not yet available (Loy and Dixon 1998).

As outlined in Chapter 15, the reliability of some of the techniques employed by Loy and Dixon (1998) has been questioned. Given these concerns, perhaps Loy and Dixon (1998) could have begun by proving the tool residues were actually protein. For example, elemental analysis could have shown that the carbon–nitrogen ratios matched those of protein (see Nelson 1993). Loy and Dixon (1998) admitted that the applicability of IEF to prehistoric residues has not been evaluated. In addition, there has been much criticism of hemoglobin crystallization on theoretical grounds, although no one appears to have rigorously tested the methods Loy and Dixon (1998) employed. Finally, very strict protocols are recommended for the analysis of ancient DNA and it is unclear whether or not they were followed.

Chapter 23
Other Organic Residues

Approaches used to analyze the wide variety of archaeological organic residues that do not directly relate to foods are presented in this chapter. The presence and/or distribution of organic compounds that occur in perfumes and cosmetics, beeswax, resins, tar, and pitches and other materials can be used to identify them. These identifications can reveal the former contents of a vessel, the function of an artifact, and how natural materials were processed. Lipids and proteins preserved in soil can also provide information about past human activities and human health. Further information about the approaches used to analyze these residues appears in Chapters 14 and 15.

Perfumes and Cosmetics

While absorbed residues are often extracted from powdered sherds, Gerhardt et al. (1990) found it was possible to non-destructively extract residues from intact Corinthian figure vases by simply washing the interior with non-polar organic solvents. Gerhardt et al. (1990) were able to identify over 40 compounds in the extracted residues using gas chromatography (GC; Chapter 33) without producing methyl or trimethylsilyl (TMS) derivatives. One type of vase contained a series of straight-chain hydrocarbons in the C_{14-27} range, characteristic of flowers and other waxy plant parts. As well, an oleoresin of pine, cypress, or juniper was found, indicating the presence of perfumed oil in these vases. The major components of another vessel were cedrol and cedrene, found in cedarwood oil.

Pérez-Arantegui et al. (1996) used X-ray diffraction (Chapter 37), Fourier transform-infrared spectroscopy (FT-IR) (Chapter 35), and proton-induced X-ray emission (PIXE) (Chapter 37) to characterize the inorganic component of solid residues from two Roman Age glass bottles known as *unguentaria*. The residues from both vessels largely consisted of gypsum, calcite, and clay minerals. Sulfur (as SO_3) was a major component of the pink residue from an *unguentarium* with a globular body. Pérez-Arantegui et al. (1996) determined that hematite was used to give the residue from a dove-shaped receptacle its red color. An organic component was present in the latter residue; further analysis with proton nuclear magnetic resonance spectroscopy (NMR) (Chapter 26), thin layer chromatography (Chapter 33),

M.E. Malainey, *A Consumer's Guide to Archaeological Science*, Manuals in Archaeological Method, Theory and Technique, DOI 10.1007/978-1-4419-5704-7_23, © Springer Science+Business Media, LLC 2011

and IR spectroscopy suggested the presence of a cholesterol by-product, an aliphatic alcohol, and, possibly, amino groups. Pérez-Arantegui et al. (1996) suggested both residues represent cosmetic powders consisting of oils and resins added to the solid base consisting of gypsum, calcite, and clay. It is unclear why gas chromatography was not employed to analyze the organic components of these residues.

Beeswax

Much research has been devoted to the study of ancient beeswax, which was used for a variety of purposes. Beeswax is resistant to water and has a high degree of plasticity. It was believed to have healing powers and was employed in the manufacture and protection of everyday objects and in sacred rituals (Regert et al. 2001).

Evershed et al. (2001) and Regert et al. (2001) indicated beeswax residues could be characterized using gas chromatography on the basis of the distribution of long-chain organic compounds. Identifications were based on the distribution of odd-numbered n-alkanes containing from 23 to 33 carbons and even-numbered long-chain palmitic acid wax esters containing from 40 to 52 or 54 carbons. Long-chain alcohols containing between 24 and 34 carbons were produced when wax esters degraded through diagenetic hydrolysis (Evershed 2000). It was more difficult to identify highly degraded beeswax used in lamps because shorter chain components were missing.

Regert et al. (2001) traced the effect of temperature on the degradation of modern beeswax through heating experiments at 60 and 100°C over a period of 7 months. Samples were analyzed using FT-IR, GC, and gas chromatography-mass spectrometry (GC/MS; Chapter 33). FT-IR was used to follow the evolution of functional groups throughout the experimental degradation process; the molecular constituents of beeswax were identified using GC. The composition of raw beeswax from different regions was similar and consistent with previously published data. Regert et al. (2001) reported that the beeswax lost volatile degradation products after 3 days of heating at 100°C; this occurred after 6 months when beeswax was heated at 60°C. The composition of the white deposit, which condensed on Petri dishes covering the heated beeswax, was similar to beeswax; but an orange deposit consisted of aromatic compounds, specifically phenolic derivatives. Only a few milligrams were lost through volatilization, so beeswax composition was largely unchanged; however, alcohols formed through the hydrolysis of wax esters. Examination of archaeological beeswax using both GC and GC/MS showed that the profile of wax esters was very stable over time but the long-chain hydrocarbon profile was altered by the loss of the lowest molecular weight compounds (Regert et al. 2001).

Evershed et al. (2003) suggested analysis of n-alkanes and n-alkanols in suspected beeswax residues with compound-specific GC-C-IRMS (Chapter 31 and 33) could confirm that the wax was not derived from plants. The δ^{13}C of these components in ancient beeswax residues from Greek combed ware pottery used as beehives ranged between −24.5 and −26.9‰ for n-alkanes and between −24.5 and −25.7‰ for n-alkanols.

Beewax and Tallow Candles

Frith et al. (2004) examined white, flaky organic deposits taken from iron and lead candleholders recovered from Fountains Abbey, Britain. Small samples, weighing 5–10 mg, were sonicated with a 2:1 mixture of dichloromethane and methanol, filtered through glass wool and dried. After TMS derivatives were prepared, the samples were analyzed using GC and GC/MS.

Analysis demonstrated the residues on three of the candleholders consisted of tallow, which was animal fat collected during rendering. Triacylglycerols were detected and the presence of C15:0 and C17:0 suggested a ruminant animal source. Dicarboxylic acids, or diacids, were detected, which is also consistent with a ruminant animal source. Based on the historical records, the likely source was sheep (Frith et al. 2004).

Residues from two other candleholders were a mixture of beeswax and tallow. Frith et al. (2004) indicated that pure beeswax candles, which produced brighter light and emitted a pleasant honey fragrance, were normally only used in the church. This could be the evidence that used church candles were recycled in domestic contexts. Another possibility was that tallow candles were coated in beeswax to give them additional strength and improve the smell. Ketones were only detected in the thermally altered animal fat residues of tallow and beeswax mixtures. The higher temperature of these mixed composition candles may have contributed to the formation of products associated with the thermal degradation of unsaturated fatty acids (Frith et al. 2004).

Resinous Compounds

Pine Resin

Shackley (1982) used gas chromatography to examine resinous crusts, believed to represent the former vessel contents, from a sixth century storage jar from a site in Israel. GC analyses indicated the major constituents of the residue were dehydroabietic and 7-oxo-dehydroabietic acid, the known oxidative decomposition products of tree resins. While the presence of a *Pinus*-type resin was confirmed, firm conclusions about the nature of the deposit were not possible. Historically, similar resins were used (1) in the production of unguents, (2) for caulking and waterproofing, and (3) in resinated wine (Shackley 1982).

Heron and Pollard (1988) used GC/MS to analyze visible and absorbed residues taken from imported trade amphorae in the Museum of London reference collection. The major component in each of the eight samples was the resin degradation product, dehydroabietic acid. The amphorae originated from a variety of places including France, Rhodes, Spain, and Italy, indicating the widespread use of conifer wood extracts to seal the interiors of amphorae (Heron and Pollard 1988).

Copal

Stacey et al. (2006) analyzed modern and ancient Mesoamerican resins referred to as copal. While copal was burned as incense during ceremonies, the samples analyzed were from mosaics, where copal was used to glue turquoise, shell, lignite, pearls, and malachite onto wood. In most cases, copal appeared as a dark brown adhesive but the resin was also colored red with hematite and used as inlay material.

Small samples (2 mg) taken from damaged areas of the mosaics were extracted with a 2:1 by volume mixture of chloroform and methanol. Trimethylsilyl derivatives of the lipids were analyzed using GC/MS. The compositions of the ancient resins were compared to reference materials from botany collections and commercial Mexican resins purchased at a market.

On the basis of diterpenoid composition and natural distributions of the trees in the region, Stacey et al. (2006) identified the mosaic adhesive as pine (*Pinus*) resins, but the precise species could not be determined. Resins from two Burseraceae family trees, *Protium* and *Bursera* species, also appeared on the mosaics, sometimes mixed with pine resin. The *Protium* resin was only detected on one mosaic; this type of resin was more frequently used as incense (Stacey et al. 2006). The commercial samples purchased for this study consisted mainly of pine and *Bursera* species resins. One of the copal samples included *Boswellia* resin, also known as frankincense, from trees that are not native to Mesoamerica (Stacey et al. 2006).

Frankincense

Frankincense is a pentacyclic (i.e., the molecules each contain a five-carbon ring structure) triterpenoid compound derived from trees of the genus *Boswellia*. Evershed et al. (1997) identified frankincense on materials from Qasr Ibrîm in Egyptian Nubia, from which pine resin was also recovered. The presence of these resins suggests incense-burning ceremonies took place at the site. The resins must have been imported because neither tree type grows in the region (Evershed et al. 1997).

Pistacia Resin

The resin collected from trees of the genus *Pistacia* was used for a variety of purposes in ancient times, including a flavoring agent, water-proofing, varnish, incense, and during the embalming process (Hansson and Foley 2008; Stern et al. 2003). These studies and others show that characteristic triterpenoids detected with GC/MS or fragments of DNA preserved in residues can establish the presence of the resin. Stern et al. (2003) found that the most stable triterpenoid biomarkers were moronic acid and 28-norolean-17-en-3-one. Oleanonic acid appeared frequently in archaeological residues but isomasticadienonic acid and masticadienonic acid readily degraded. Unfortunately, the compounds shown to be the most reliable indicator of burning or charring could not be identified because they emerged from the GC column (co-eluted) with other components (Stern et al. 2003). Two unidentified components appeared in seven of ten locally produced bowls used as incense burners from a Late Bronze Age site in Egypt but were absent from the residues of Canaanite amphorae used to transport resin. The presence of these unknown compounds, with a mass spectral base peak at *m/z* 453, demonstrated the resin was used as incense (Stern et al. 2003).

Tars and Pitches

Spectroscopic techniques were first used to provide a general characterization of archaeological tars and pitches more than 40 years ago. Hadži and Orel (1978) used nuclear magnetic resonance (NMR) spectroscopy (Chapter 36) to show that three archaeological pitches were derived from birch, supporting the results obtained using infrared (IR) spectroscopy (Chapter 35). Sauter et al. (1987) used both ^{1}H and ^{13}C NMR to identify betulin in archaeological tar from an Austrian site, demonstrating the Iron Age tar was produced from birch.

Robinson et al. (1987) used ^{1}H NMR spectroscopy to verify a common *Pinus* origin for tars and pitches obtained from a Tudor shipwreck, called the Mary Rose, and an Etruscan shipwreck. Robinson et al. (1987) demonstrated general similarities between pitch and tar samples from the Mary Rose, the Etruscan wreck, and Stockholm tar with IR spectroscopy. However, GC and GC/MS provided the most informative data and showed that the major components were diterpenoid compounds (Robinson et al. 1987). Since then, the approaches have become the analytical tools of choice for archaeological tars and pitches.

Embalming Products

By combining their knowledge of geochemistry and extensive sample processing with GC and GC/MS analysis, Maurer et al. (2002) were able to determine the chemical composition of embalming material from four Egyptian mummies dating to the fourth century A.D. Three of the samples were lumps of dark material from the body cavities and the last was a resin-soaked cloth.

Lipid extraction with dichloromethane/methanol (99:1 v/v) was performed by ultrasonication. In order to simplify the analysis, Maurer et al. (2002) added an excess of *n*-hexanes, which caused asphaltenes (aromatic components of crude oil), to precipitate out of the lipid extract. Lipids soluble in *n*-hexane were further separated by medium-pressure liquid chromatography into three groups prior to analysis with GC/MS: non-aromatic hydrocarbons, heterocompounds, and aromatic hydrocarbons.

Maurer et al. (2002) found that the major components of the embalming material were associated with fossil hydrocarbons. These included steroid hydrocarbons, including steranes, sterenes, and aromatic hydrocarbons, which formed when steroids decompose, and polycyclic terpanes. Fatty acids from contemporary plant material and tricyclic diterpenoids from coniferous resins (i.e., abiatic acid and related compounds) were also present. The *n*-alkanes in the residues were from both fossil hydrocarbons and higher land plants.

The overall composition of two of the lumps was similar in that the insoluble portion contained carbonized plant material and the remains of beetles. The solvent extractable portions were dominated by asphaltenes and identical with respect to lipid biomarkers. The fourth lump and the residue from the cloth resembled Dead Sea asphalt with respect to sterane and terpane biomarker distributions; the cloth residue could also contain beeswax (Maurer et al. 2002).

Smoke Condensates

Oudemans and Boon (1991) used Curie point pyrolysis to analyze solid residues from pottery from a site in the Netherlands, dating from the Late Iron Age to the Early Roman period. Curie point pyrolysis with mass spectroscopy (CuPy-MS) (Chapters 31 and 33) and discriminant analysis of the results provided a broad initial assessment of 33 samples. Four of these were then selected for analysis with Curie point pyrolysis-gas chromatography with mass spectrometry (CuPy-GC/MS), to characterize the sample at a molecular level. One sample from the outside of the pot consisted mainly of polycyclic aromatic hydrocarbons, which are compounds found in smoke condensates of wood fires. Their presence indicated the pot had been suspended over an open fire.

Nicotine

Smoking pipes predate archaeobotanical evidence for tobacco in eastern North America by more than a thousand years, so Rafferty (2006) used GC/MS in an attempt to identify organic molecules that could provide evidence for early tobacco use. Demonstrating tobacco use was complicated because nicotine is water soluble, which makes it vulnerable to loss through groundwater leaching and post-excavation washing. In addition, nicotine is present in plants native to eastern North America, such as milkweed, swallow-wort, and false daisy (Rafferty 2002). One could argue that tobacco was the most likely source of nicotine or its degradation products detected in archaeological pipe residues but the assertion could not be proven.

The validity of the methodology was demonstrated by analyzing a nineteenth century A.D. ceramic pipe from Ghana. The presence of nicotine was indicated by the retention time of a chromatographic peak, but only the identity of decay products was confirmed by GC/MS (Rafferty 2002). Similar results were obtained for residue from an Adena pipe from the Cresap Mound site in West Virginia dating from 730 to 375 B.C. Rafferty (2006) examined smoking pipes from the Boucher site, an Early Woodland cemetery in Vermont, with calibrated radiocarbon dates ranging from 1036 B.C. to 49 B.C.; only one pipe retained sufficient solid residue for analysis. Alkaloids were extracted with methylene chloride in a Soxhlet apparatus, concentrated down to 1 μL under nitrogen, then injected into the GC/MS. Based on previous research with modern reference material, the presence of nicotine was indicated by a small chromatographic peak with retention time of 12.9 min (Rafferty 2002). Unfortunately, the abundance of this component was too low to permit confirmation by GC/MS. A chromatographic peak appearing at 7.6 min could indicate the presence of anabasine, which is a decay product of nicotine (Rafferty 2002).

As a word of caution with respect to sample handling, Wischmann et al. (2002) found that porous material initially free of nicotine could be contaminated by modern tobacco smoke. The researchers (2002) placed a human bone from a Bronze Age individual in an office where tobacco smoking was permitted for a period of 6 weeks. One-half of the bone was washed prior to analysis and the other half was not; the outer 1.5 mm was discarded and the inner bone was analyzed for the presence of nicotine. Wischmann et al. (2002) reported that the alkaloid was detected in both halves of the bone but the concentration of nicotine was three times higher in the portion washed with water compared to the unwashed portion. Washing had carried nicotine deposited on the surface into the inner regions of the bone.

Archaeologically Significant Soil Lipids

It has long been recognized that human activity can introduce lipid residues into soils. Morgan et al. (1983) examined the fatty acid composition of waxy material found in the soil from a Thule site on Herschel Island in the Western Arctic. Four samples from a midden showed the material consisted of animal fats that had partially converted to adipocere. Polyunsaturated fatty acids had decomposed, but Morgan et al. (1983) suggested that the fatty acid compositions of the residues most closely resembled harbor seal. Based on the bone recovered from the site, seal and whale were likely sources.

Sterols in archaeological soil can provide evidence about human activities. Davies and Pollard (1988) analyzed lipid extracts from soil from an Anglo-Saxon grave using various techniques. Spectra obtained by ^{1}H NMR analysis provided little information due to the small amount of residue present and the complexity of the mixture; IR spectroscopy (Chapter 35) only indicated the presence of organic matter (Davies and Pollard 1988). Using GC/MS, Davies and Pollard (1988) showed that the concentration of cholesterol in soil surrounding the grave was three times higher than in control soil samples. Davies and Pollard (1988) suggested that decomposing human bodies deposit cholesterol residues in the surrounding soils and these residues could potentially be used to detect burials in environments where bone has completely disintegrated.

Evershed et al. (1997, 2001, 2008) found that specific lipid biomarkers enabled the identification of ancient manuring practices. The presence of human and animal fecal matter could be detected by the presence of 5β-stanols and bile acids. These compounds are formed by the microbial reduction of cholesterol, found in animals, and campesterol, stigmasterol, and sitosterols, which are the phytosterols found in plants (Evershed et al. 1997, 2001, 2008). Examination of the proportions of the different 5β-stanols enabled manure from humans or pigs (porcine) to be discriminated from ruminant manure. The 5β-stanols from ruminants are only derived from phytosterols. The distribution of four different bile acids provided further confirmation of sources. Hyodeoxycholic acid is produced only in the porcine gut. Deoxycholic acid and lithocholic acid are produced in the gut of bovines; these two bile acids and cholic acid are produced in the human gut.

Archaeologically Significant Soil Proteins

Mitchell et al. (2008) used enzyme-linked immunosorbent assay (ELISA) (Chapter 15) with monoclonal antibodies to detect evidence of microorganisms (protozoa cysts) known to cause dysentery. Historical accounts indicate dysentery was a common problem among the crusade armies, but the enteric parasites responsible were difficult to detect microscopically. Following the successful application by others, Mitchell et al. (2008) tested medieval latrine soil with modern clinical stool sample kits capable of detecting whole cysts, cyst wall fragments, and proteins excreted by the microorganisms.

The latrines were located in the city of Acre, in modern Israel, which was within the crusader kingdom of Jerusalem during the twelfth and thirteenth centuries A.D. The first set of samples was collected from a 35 stone seat latrine block associated with the Hospital of St. John. Members of the religious order and the ailing crusaders and pilgrims they treated would have used the latrine. Control soils were collected from a thirteenth century storage hall and from a soil layer dating to the eighteenth century located above the latrine. The second set of samples was collected from a cesspool that was likely used by the city inhabitants; control soils were also collected. Radiocarbon dates on charcoal in the latrine and coins from the cesspool confirmed the facilities were used during the crusader period.

The soil samples were disaggregated in a 0.5% aqueous solution of trisodium phosphate for 72 h. Pre-made commercial kits were used to test the disaggregated soil for the presence of microorganisms that cause dysentery. The kits specifically tested for *Entamoeba histolytica*, *Giardia duodenalis*, and *Cryptosporidium parvum*. Multiple cysts were required to give a positive result, from 10–15 for *Giardia* to 78–156 for *Cryptosporidium*. A positive test was indicated by the change of the well color to a deep yellow quantified with a visible light spectrophotometer at a wavelength of 450 nm (Chapters 2 and 35) and through comparisons with positive and negative controls. Mitchell et al. (2008:1850) suggested, "If the tests avoid giving false positive results with commensal gut microorganisms they are also likely to avoid false positives with commensal soil organisms." The kits were more likely to miss a positive test than to produce a false negative (Mitchell et al. 2008).

Mitchell et al. (2008) reported six of the eight soil samples from the hospital latrine tested positive for *E. histolytica*. One of the eight eighteenth century control soils also tested positive but all eight of the thirteenth century control soils tested negative. One latrine sample tested positive for *G. duodenalis* and all 16 control soils tested negative. All latrine soils and controls tested negative for *C. parvum*. All samples and controls collected from the cesspool tested negative for the three microorganisms.

Mitchell et al. (2008) suggested that the high rate of negative results (seven of eight hospital latrine samples) for *G. duodenalis* could relate to the threshold requirement of 10–15 cysts for a positive result. The negative result for the cesspool could indicate that individuals who primarily used it were free of dysentery or that the protozoa were not preserved. The results confirmed the presence of two microorganisms known to cause dysentery in the crusader kingdom of Jerusalem (Mitchell et al. 2008). Microscopic analysis revealed the presence of a variety of parasites including roundworm, tapeworm, and whipworm in the solution used to disaggregate the soil.

Protein Pigment Binders

Scott et al. (1996) used a variety of techniques to analyze the composition of a black pigment cake made by the Chumash Indians who lived on the south-central coast of California. Ethnographic accounts indicated pigment cakes were made by mixing charcoal or soot with deer bone marrow until it had the consistency of dough. The pigment cake was then pressed into a container for storage.

The solid pigment was analyzed with FT-IR by pressing approximately 2 μg with barium fluoride (BaF_2) into a disc. Proteins were recognized in the IR spectrum and it compared well to the reference spectrum for dried blood. The pigment was treated with 6 N hydrochloric acid vapor in order to hydrolyze the protein and produce free amino acids. After derivatization with ethyl chloroformate, 14 volatile amino acids were detected through GC analysis. The pigment was found to have a protein content of 1% by weight and the amino acid composition was in good agreement with the reference sample of blood (Scott et al. 1996).

Immunological techniques were then used to determine the source of the protein in the pigment cake. The portion of the pigment that was soluble in ethanol was subjected to cross-over immunoelectrophoresis (Chapter 15). The extract did not react to pre-immune serum, i.e., the blood of the animal in which the antisera was raised, but positive reactions were obtained in tests against the antisera of deer, pronghorn, and human. Scott et al. (1996) concluded that pronghorn was the source because of the species-specific reaction to pronghorn antisera. Since pronghorn lived on the open plains, not near Santa Barbara where the pigment cake was found, the pigment cake may have been traded into the region (Scott et al. 1996).

Chapter 24
Organic Artifacts

Techniques used to determine the composition, structure, biological source, and age of organic artifacts are described in this chapter. As these types of artifacts are rarely preserved, minimally destructive techniques, such as plasma oxidation, are of particular interest.

Compositional Analysis

Stable Isotope Analysis for Textile Sourcing

Benson et al. (2006) investigated the possibility of using stable isotope analysis (Chapter 13) and trace metal concentrations to determine the source of plant materials, such as willow and tule, used in the construction of baskets and matting. The eastern Sierran drainage system, which includes the Carson, Truckee, and Walker rivers, Pyramid Lake, Winnemucca Lake, and the Carson sink, has lower $^{87}Sr/^{86}Sr$ ratios than the Humboldt River, which flows east from the Ruby Mountain metamorphic complex to the Humboldt Sink. The $\delta^{18}O$ values of rivers and streams are similar to that of precipitation and snowmelt (meteoric water), but values of water in wetlands, basins, and sinks are higher due to evaporation of the lighter isotope. Willows typically grow along rivers because they are not salt tolerant and cannot survive permanent flooding. Tule is a wetland emergent plant requiring a saturated, oxygenated root zone. Consequently, the $\delta^{18}O$ values of plants that grew along rivers would be lower than plants that grew in lakes (Benson et al. 2006).

Western Great Basin sites in the Lake Lahontan Basin and Dixie Valley of Nevada contain Lovelick Wickerware manufactured from willow (*Salix* sp.) using a semi-flexible, paired weft and rigid warp rod. Twined weaving was used to manufacture baskets and matting from tule (*Schoenoplectus* cf. *acutus*) where a pair of weft elements were twined or twisted around a stationary, semi-flexible warp.

Water samples were collected from rivers; synthetic water was produced by leaching sediments taken from dry sinks and basins with acid. Samples of modern willow and tule were also collected. Prior to isotopic analysis, potential contaminants were removed from archaeological textiles using two different methods. The stable isotope values of samples that underwent "standard cleaning" were compared to those that underwent "deep cleaning." Modern and "deep cleaned" samples were wet-ashed by an acid digestion process; the other archaeological samples were dry-ashed. Trace element and stable isotope analyses were performed using thermal ionization mass spectrometry (TIMS), inductively coupled plasma-mass spectrometry (ICP-MS), ICP-atomic emission spectroscopy (ICP-AES), and an elemental analyzer interfaced to a stable isotope ratio mass spectrometer; these techniques are described in Chapters 31, 34, and 38.

Based on a mean $\delta^{18}O$ value of $-1‰$ on water from Pyramid Lake, which is hydrologically closed, oxygen isotope values of sinks and shallow water lakes in the Great Basin were estimated to range from -1 to $-3‰$ (Benson et al. 2008). Assuming the fractionation between water and plant was approximately 34‰, plants that grew in these environments would have $\delta^{18}O$ values of 31–33‰. The average $\delta^{18}O$ values from tule and willow harvested from the Carson, Humboldt, and Truckee rivers were found to be 22.8 and 22.7‰; values on archaeological textiles ranged from 17.4 to 25.4‰. Benson et al. (2006) reported that $\delta^{18}O$ values of modern tule and willow samples were 3.3–4.8 ‰ higher than precontact values due to recent irrigation and water diversion.

Although strontium isotope values on synthetic water samples overlapped, there was a clear separation between the river water from the Sierra and the Humboldt drainages. The $^{87}Sr/^{86}Sr$ values obtained on archaeological textiles processed by "standard cleaning" fell into one or the other category; however, deep cleaning was found to dramatically alter the samples. In two of the three cases, the $^{87}Sr/^{86}Sr$ value on the portion subjected to standard cleaning was associated with the Sierra drainage but deep cleaning shifted the value to the range for the Humboldt River drainage. Contamination was especially problematic for tule samples because the plant has very thin walls, which make it difficult to clean. Benson et al. (2006) recommended that whole (un-split) sections of tule and willow be selected for analysis. Trace metal concentrations could not distinguish the different rivers flowing from the Sierra Nevada or the plants that grew along them.

Analysis of Paper Using ESR Spectroscopy

While archaeological applications of electron spin resonance (ESR) spectroscopy (Chapter 36) more often relate to dating, Attanasio et al. (1995) used the technique to examine the composition of paper dating from the fifteenth to the eighteenth centuries. Paper is a combination of cellulose, water with small amounts of organic and inorganic impurities, and additives. Previous studies with proton nuclear magnetic resonance spectroscopy (NMR) (Chapter 36) indicated paper degradation was due to structural changes of cellulose, including loss of bound water and the transformation of fibrous, highly crystalline molecules into amorphous material. The study by Attanasio et al. (1995) investigated the possible role of transition metal ion impurities as reaction catalysts. These impurities are paramagnetic (Chapter 1) species that can be detected by ESR spectroscopy.

Eighteen samples of unprinted paper, weighing about 10 mg or measuring about 1.5 cm^2, were taken from books produced in Italy, Germany, France, and Sardinia. Some samples were taken from different parts of the same book that exhibited differential degradation. Spectra were obtained at low temperature, 110 K, which is $-163°C$.

The ESR spectra of the paper samples were similar and contained peaks arising from rhombic and octahedral iron (Fe), manganese (Mn), and occasionally copper (Cu). The signal from Cu indicated that some of the atoms were loosely bound (adsorbed) and others appeared to be tightly bonded to cellulose molecules. Attanasio et al. (1995) found that the preservation of paper was strongly correlated to the presence of Cu and rhombic iron. While not all samples of poorly preserved paper contained Cu, all samples that contained even small amounts of Cu were badly degraded. Attanasio et al. (1995) concluded that copper was the most effective catalyst of the degradation process.

Structural Analysis of Organic Artifacts

Structural Analysis of Textiles Using X-Radiography

Yoder (2008) described the use of "soft" X-rays to examine the internal structure of 35 Anasazi woven yucca fiber sandals. The sandals were made from about 500 B.C. to A.D. 1350 using construction techniques that included twining, braiding, and plain weaving. Rather than higher energy X-rays required to examine bones, Yoder (2008) employed low-energy X-rays typically used to image soft body tissues; the analysis was performed at the mammography unit of a hospital. The voltage used to produce soft X-rays was about 30 kV and the energies of the resulting X-rays ranged between 15 and 30 keV. Analysis revealed internal construction elements otherwise undetectable without damaging the sandal. In particular, characteristics of the warp, including arrangement, ply, and twist, and remains of the tie system were discernable.

Analysis of Plant Fibers with SEM and SEM-EDS

Plant fibers employed for the construction of archaeological textiles contain siliceous phytoliths, calcium oxalate crystals, and calcium carbonate cystoliths that may aid in their identification. Jakes and Mitchell (1996) described a method of preparing these inorganic inclusions for analysis using cold plasma ashing. Commonly employed processing techniques involve high temperatures and/or hazardous chemicals. High-temperature procedures could distort the shape and chemical composition of the inclusions and strong chemicals could pose safety hazards. Jakes and Mitchell (1996:151) described their procedure as "nondestructive and non altering of phytolith morphology." It is important to note that this statement referred only to the inclusions; the surrounding organic constituents were destroyed.

Samples of paw paw and basswood, which are fibrous plants used in textiles, cordage, or baskets, were mounted onto a carbon planchette and placed in a sample chamber which was evacuated and filled with an oxygen atmosphere. Oxygen plasma, consisting of ions, electrons, and neutral particles, was created by exciting the molecules with radio frequency energy at 13.56 MHz. After an ashing period of 30 min, the materials were examined with a scanning electron microscope with and without the use of energy-dispersive analysis of X-rays (SEM and SEM-EDS) (Chapter 37).

Jakes and Mitchell (1996) reported this method enable the examination of both silicon- and calcium-based structures with minimal alteration of the inclusions. Previously undocumented crystal structures were observed and these could aid in the identification of specific plant genera. The composition of the materials was determined *in situ* through the use of X-ray microanalysis and this information could contribute to the understanding of phytolith formation.

Analysis of Ancient DNA

The two studies described in this section used the analysis of ancient deoxyribonucleic acid (DNA) to identify the specific source of processed hides from domestic animals. The high success rates of these studies indicates it was likely not a factor, but it is important to remember that the contamination from modern cow, pig, chicken, or goat DNA occurred in about 5% of PCR reagents tested by Leonard et al. (2007) (Chapter 16).

Analysis of Ancient DNA Preserved in Parchment

Poulakakis et al. (2007) analyzed ancient DNA recovered from six Greek manuscript parchments dating between the thirteenth and sixteenth century AD in order to determine the species of origin. Parchment is typically prepared by removing hair and soft tissue from animal skin; after the hide has dried it is scraped. Small samples, each measuring approximately 0.5 cm^2, were taken from each manuscript. The extraction process was performed five times on each sample using primarily phenol–chloroform methods, but silica-based techniques were also employed.

Poulakakis et al. (2007) selected two pairs of primers, 150 base pair (bp) and 194 bp in length, to produce an overlapping 323 bp sequence of the cytochrome *b* gene. After 60 cycles of PCR amplification, products were isolated and reamplified an additional 30 cycles. These products were inserted into plasmids and cloned with *Escherichia coli*; sequencing was performed after the sizes of the inserted (recombinant) DNA in positive colonies were confirmed.

Amplifiable DNA was recovered from three of the six manuscripts. Statistical analyses were performed to compare the parchment sequences to nine species of goat (*Capra* spp.), two species of sheep (*Ovis* spp.), cow (*Bos taurus*), pig (*Sus scrofa*), and human (*Homo sapiens*) DNA. The ancient DNA sequences from each of the three parchments corresponded to positions 64–386 of the *Capra hircus* sequence.

DNA Preservation in Leather

Vuissoz et al. (2007) examined the preservation of nuclear DNA (nuDNA) and mitochondrial DNA (mtDNA) in ancient and modern leather. Modern samples included rawhide as well as vegetable-tanned and mineral-tanned leather. Archaeological samples from medieval leatherworking sites in Denmark and Norway were also examined. Vuissoz et al. (2007) believed that the ancient leather was produced using a vegetable tanning process. The preservation of the leather at both sites was due to wet burial environments.

In order to detect ancient mtDNA, a 177 bp sequence in the mitochondrial control region of cattle was targeted. To detect ancient nuDNA, a 79 bp sequence of the single copy Cyp19 gene that appears in all members of the tribe Bovini was targeted. All amplifications failed due to the presence of PCR inhibitors until the DNA extracts were filtered and diluted up to 100-fold. Vuissoz et al. (2007) successfully amplified mtDNA from all samples of modern and medieval leather; amplifiable nuDNA was not recovered.

The transformation of animal hide to leather generally involves several processes including (1) removal of water, (2) isolation of the dermis (middle skin layer) through the removal of hair, the epidermis (outer skin layer), and subcutaneous tissue (inner skin layer), (3) pickling of the skin in acid, and (4) tanning (Vuissoz et al. 2007). In order to determine when damage to nuDNA occurred, experimental vegetable tanning processes were developed and the hide was sampled at each of the 19–22 steps. Vuissoz et al. (2007) found that nuDNA degradation occurs when lime was added to remove hair and during the first tanning bath if caustic soda was not added.

Dating Techniques

AMS Radiocarbon Dating Using Plasma Oxidation

Steelman et al. (2004) used plasma oxidation to remove microscopic amounts of carbon from organic artifacts for accelerator mass spectrometer (AMS) radiocarbon dating (Chapter 8). The organic artifacts were associated with an infant bundle burial recovered from Hind's Cave, which is a dry cave on the Pecos River in southwest Texas. Organic artifacts selected for dating included grass, a woven mat, a wooden stick from a desert ash tree, a stalk of sotol and sotol twine; all but the sotol stalk were also submitted for routine combustion AMS dating.

Acid–base–acid pretreatments were applied to all samples prior to combustion AMS dating. The plasma oxidation AMS ages on the sotol and one sample of the woven mat were obtained without any pretreatment. The remaining samples were treated with water, 0.1 or 1 M sodium hydroxide (NaOH), or the standard acid–base–acid prior to plasma oxidation. The radio frequency-generated oxygen plasma converted organic carbon into carbon dioxide at a temperature of 50°C. The carbon dioxide (CO_2) was collected, dried in liquid nitrogen, and then used to make a graphite target for AMS dating.

Age estimates obtained by plasma oxidation AMS were very similar to those obtained by combustion AMS. Both techniques found that the desert ash stick was about 1200 years younger than items from inside the bundle burial. The weighted average of 10 dates obtained by plasma oxidation AMS, including one made on bone and skin from the mummified infant, was 2137 ± 13 B.P. The weighted average of the three combustion AMS dates was 2128 ± 20 B.P.

Steelman et al. (2004) suggested that plasma oxidation is non-destructive in that no observable physical or chemical changes result. Since plasma oxidation only removed organic carbon, the first acid step in the chemical pretreatment process was unnecessary. Comparisons of dates obtained on treated and untreated specimens showed that the items in the burial were virtually free of contamination. Until a non-destructive method of removing humic and fulvic acids is developed, base treatment to remove these organic soil contaminants and the final acid wash to remove adsorbed carbon dioxide from the base solution would be required for most samples (Steelman et al. 2004).

Van der Plicht et al. (2004) showed that AMS radiocarbon ages could be obtained from organic artifacts associated with bog burials believed to date from the Late Iron Age/Roman Period, from the second century B.C. to the fourth century A.D. In a comprehensive study, van der Plicht et al. (2004) obtained more than 100 dates on 40 bog bodies and associated materials from 35 findspots.

Van der Plicht et al. (2004) reported the effect of humified water in peat bogs on biological remains was analogous to a tanning process. Physical pretreatment to remove visible contaminants and the acid–alkali–acid (AAA) chemical pretreatment specifically developed for bog bodies were required to obtain reliable dates. Similar to the process described above, the first acid removes carbonates, resins, and fulvic acids from the sample; alkali removes tannic acids and lignin; and the final acid treatment removes carbon dioxide absorbed during the alkali treatment.

The organic artifacts analyzed included fur capes, wool, textiles, leather, and wood. The full treatment was recommended because partial treatment or no treatment could produce ages that are too young or too old. For especially delicate items, van der Plicht et al. (2004) suggested that pretreatment be conducted at room temperature with dilute (less than 1%) hydrochloric acid for a shorter duration.

Dating Shell Artifacts

Rick et al. (2002) reported the results of AMS radiocarbon dating on eight single piece shell fishhooks from the Channel Islands and mainland coast of Santa Barbara,

CA. The hooks were circular or j-shaped and usually made of red abalone or Norris top shell. They ranged in height from 2 to 8 cm and occasionally were barbed.

The eight shell fishhooks were obtained from a site on the mainland coast ($n = 2$), a site on the Southern Channel Island of San Nicholas ($n = 1$), and four sites on two Northern Channel Islands, Santa Rosa ($n = 1$) and San Miguel ($n = 4$). Samples weighing less than 1 g were removed from areas that had broken previously. All were etched with dilute hydrochloric acid and rinsed with distilled water. When dry, samples were converted to CO_2 through a reaction with 85% phosphoric acid (H_3PO_4); then the CO_2 was converted to graphite for dating. A regional correction (ΔR) of 225 ± 35 was used to compensate for local upwellings but Rick et al. (2002) acknowledged that upwelling intensity at the time of shell formation could affect the date (see Chapter 8).

The timing of the introduction of the single piece shell fishhook seemed to correlate with a regional intensification of marine fishing and population growth. Most of the previously reported dates for these fishhooks were made on associated material, rather than on the artifacts themselves. While some suggested they first appeared about 5000 years ago, one previously reported AMS age was much younger. Rick et al. (2002) reported the oldest age, 2450 cal B.P., was obtained on a j-shaped red abalone fishhook, which was consistent with the previously reported direct date on shell. These results indicated that single piece shell fishhooks appeared around 3000 years ago and were in wide use by 2500 cal B.P.

Rick et al. (2005) reported anomalous dates on purple olive and red abalone shell artifacts from sites on San Miguel Island. Two *Olivella* spire-lopped shell beads, from a collection of 25 strung on woven sea grass, were submitted for AMS dating. The age of 7610–7480 cal B.P. on one bead was consistent with expectations, while the other produced an age of $30,900 \pm 100$ B.P. A similar bead from a nearby site was nearly 1000 years older than other material from the cave. Age estimates of red abalone from a site occupied during the protohistoric and historic periods were between 100 and 500 years older than associated marine shells representing food remains.

Rick et al. (2005) suggested the beads were constructed from "old shell" obtained from fossil deposits, previously abandoned sites or the beach. Use of this material might have been restricted to the manufacture of shell beads. No anomalous dates were obtained on fishhooks or other tools, which suggested more durable, unweathered new shell was more suitable for utilitarian artifacts (Rick et al. 2005).

Use of Strontium Isotope Dates to Source Shell

Vanhaeren et al. (2004) used strontium isotope dating as a means of sourcing *Dentalium* shell beads from La Madeleine rockshelter, located east of Bordeaux, France. Four of the beads were among the 1314 associated with the burial of a child dated to $10,190 \pm 100$ B.P. and the other was among the 39 recovered from the occupation layer. The primary purpose of the study was to determine whether the shells were collected from a nearby Miocene outcrop or from beaches on the coast.

Differences in the strontium isotope values of ocean water would clearly distinguish shell that formed during the Miocene epoch from shell formed 10,000 years ago.

A red pigment, determined to be magnesium and iron oxides by energy-dispersive (XRF) spectroscopy (Chapter 37), was removed from the shells prior to analysis. Shells were placed in an ultrasonic bath with 0.6 N hydrochloric acid, which resulted in the removal of about 50% of the exterior coating. Strontium isotopes were chemically separated with a cationic chromatography column and then analyzed with thermal ionization mass spectrometry (TIMS; Chapter 31).

The $^{87}Sr/^{86}Sr$ ratios of two of the shell beads, both with smooth exteriors, from the burial and the bead from the occupation layer were close to the present-day value for seawater, which is 0.7092. The strontium isotope date on the third bead from the burial ranged between present and 1 million years ago (Ma); the exterior surface of this bead was also smooth. The strontium date on the final bead from the burial ranged between 0.7 and 1.9 Ma; the exterior of this bead was longitudinally striated.

Vanhaeren et al. (2004) concluded that three of the beads with close to present-day strontium values were collected from beach sources about the time of site occupation. It was possible that the early date on one bead with a smooth surface was due to post-depositional diagenesis, which would have altered the original aragonite to calcite. Insufficient material remained after acid etching to submit beads for both strontium isotope analysis and X-ray diffraction (XRD) (Chapter 37), which could have detected the presence of diagenetic calcite (Vanhaeren et al. 2004). The age of the bead with longitudinal striations suggests it was obtained from an early Quaternary outcrop; the nearest known potential source was located 200 km away. None of the beads were collected from the nearby Miocene outcrop.

Vanhaeren et al. (2004) reported that the beach was located 35 km farther from the site at the time of occupation due to lower sea levels, so the Miocene outcrop would have been about 50 km closer. Examination of the Miocene *Dentalium* shells revealed that the maximum size of the opening was less than 1.9 mm, whereas the thickness of 20 sewing needles recovered from La Madeline rockshelter ranged from 1.9 and 4.4 mm, with an average of 2.7 mm. The opening of the beads associated with the child burial ranged between 1.6 and 2.9 mm. Shells from the Miocene outcrop were too narrow to pass the needles through, so it was necessary to collect *Dentalium* with wider openings from the coast (Vanhaeren et al. 2004).

Chapter 25
Paints, Pigments, and Inks

The compositions of paints, pigments, and inks have been analyzed using a variety of techniques. While more destructive techniques can be applied for the compositional analysis of raw materials, completely non-destructive techniques are typically required to examine the ink on ancient and sacred manuscripts. Examples of various approaches taken to address these problems and issues related to dating rock art using charcoal-based pigments are presented in the chapter.

Elemental Analysis of Ochre with Instrumental Neutron Activation Analysis

As the starting point for provenance studies, Popelka-Filcoff et al. (2008) examined the elemental composition of iron-rich red pigments collected from three geological ochre sources near Tucson, Arizona, using instrumental neutron activation analysis (INAA). The iron oxide formations in the soil of Beehive Peak formed just above the bedrock. The iron oxides at Rattlesnake Pass were deposited near the surface by hydrothermal fluids in a region known for copper ore deposits. Materials from Ragged Top occurred at the interface of iron-rich sedimentary rock formations and iron-rich veins produced by volcanic activity. The iron oxides from Ragged Top were darker in color than material from Beehive Peak and Rattlesnake Pass, which were both described as light orange-red.

Intrasource and intersource variability was examined by collecting five samples from each of 22 locations; 10 samples found to be material other than ochre were excluded. The ochre was dried at 100°C overnight to remove moisture prior to and after grinding and crushing. Samples weighing about 60 mg and references were subjected to two irradiations, one lasting only 5 s and another lasting 24 h. Following the short irradiation, gamma emissions were counted for 720 s after a 25 min decay period. Following a 24 h irradiation and 7-day decay period, samples were counted for 2000 s, then again for 10,000 s after an additional decay period of 3 weeks.

Elements reliably measured in the majority of samples were included in the statistical analysis. A Pearson's correlation was used to discriminate elements that were correlated with iron (Fe) from those associated with surrounding material. The 17

M.E. Malainey, *A Consumer's Guide to Archaeological Science*, Manuals in
Archaeological Method, Theory and Technique, DOI 10.1007/978-1-4419-5704-7_25,
© Springer Science+Business Media, LLC 2011

elements associated with iron oxide were considered to represent the "Fe-oxide" signature of the material. The amount of each element was computed as a ratio to the amount of Fe in the sample; the base-10 logarithmic ($\log_{10}$) transformations of these values were subjected to principal component and canonical discriminant analyses.

Much of the observed variance was due to concentrations of transition metals and rare earth elements in the ochres. Bivariate plots showed the three source areas could easily be distinguished by plotting $\log_{10}$ ratios of antimony and iron ($\log_{10}[\text{Sb/Fe}]$) against arsenic and iron ($\log_{10}[\text{As/Fe}]$). Intersite variability was revealed in plots of $\log_{10}$ ratios of arsenic and iron ($\log_{10}[\text{As/Fe}]$) against manganese and iron ($\log_{10}[\text{Mn/Fe}]$). Statistical outliers exhibited high levels of calcium (due to the presence of gypsum), As, or Mn. Canonical discriminant analysis verified the three source areas formed statistically different groups. These results indicated that future studies involving elemental analyses using INAA could provide a means of assigning ochre, ochre-stained, or ochre-painted site materials to sources.

Compositional Analysis of Paint Using SEM-EDS, CHN, and FT-IR Spectroscopy

Fralick et al. (2000) examined the composition of red, green, black, and gray paints on gypsum plaster masks and an effigy head from San Lázaro Pueblo, New Mexico, using a variety of techniques. The occupation of the pueblo began in the twelfth century A.D. and extended up to about A.D. 1680, but the pueblo was abandoned for much of the sixteenth century. The masks, representing bear or badger, and the head effigy, likely used in religious performances, were recovered from a back room. Samples taken from previously broken pieces of the objects were analyzed with the permission of Native American representatives. The composition of the paint was compared to potential sources including local clays, copper-mineralized sandstone, and carbonaceous shales recovered at the site and from nearby coal deposits.

The composition of paints was determined using scanning electron microscope-energy dispersive X-ray spectrometry (SEM-EDS) (Chapter 37), carbon–hydrogen–nitrogen (CHN) elemental analysis (Chapter 38), and Fourier transform-infrared (FT-IR) spectroscopy (Chapter 35). Polished cross sections prepared from the samples were examined with SEM-EDS. An elemental analyzer was used to determine CHN content of the combustion products generated when samples were heated to 1000°C. Transmission FT-IR was conducted by incorporating 100 mg of the finely ground sample into 2 g of potassium bromide (KBr) and pressing the material into a disc. Spectra could not be obtained using attenuated total reflectance (ATR-) FT-IR because good contact could not be maintained between the ATR element and the sample. The ritual objects were made of gypsum ($CaSO_4$) plaster, so any carbon detected by CHN and/or FT-IR analyses would have arose from organic sources.

Fralick et al. (2000) reported that the green paint on the sample contained copper-bearing minerals, probably malachite, which is copper carbonate. The sample of copper mineralized sandstone contained both malachite and azurite. Traces of hydrocarbons were detected using FT-IR but their source was not clear. Variations detected in hematite concentration in the red paint may have been intentionally produced by the artist in order to alter the shade of red. Gypsum was mixed with hematite to create a lighter shade of red on one mask. Absorptions in the FT-IR spectra indicated the presence of an organic binder in the red paint, possibly derived from plant materials. High levels of carbon were detected in the black paint using CHN and FT-IR; similarities between the paint and the carbonaceous shale recovered from the site indicate the shale or similar material were possible sources. This interpretation was not clearly supported by SEM-EDS data, indicating another carbon source, possibly charcoal, coal, or carbonaceous shale from a different source, was employed. The gray-colored areas were created by combining amber tinted and white gypsum with carbonaceous material (Fralick et al. 2000). Results from microstructural analyses related to the construction and decoration of these plaster objects are described in Chapter 29.

Vermillion et al. (2003) used a similar, but different, approach to identify pigments on Moorehead phase Ramey knives from a site near Cahokia. Ramey knives are typically large, finely crafted tools fashioned from Mill Creek chert. Red and green pigments appeared on both knives under investigation, but in opposite places. Transmission electron microscopy with energy dispersive spectroscopy (TEM-EDS) was used to determine the composition of the pigments. Unlike SEM (Chapter 37) where the beam is deflected off the sample surface, the TEM electron beam passes through a thin film of sample.

In order to prepare the film, small samples of pigment were removed, crushed in a mortar and pestle, and then immersed in water; drops of the particle suspension were placed on TEM grids for analysis. Samples were analyzed after drying overnight on a glass slide.

The red pigment was found to contain iron oxides or iron hydroxides, likely from ochre; copper was detected in the green pigment. Green copper minerals such as malachite would not bind to the mineral surface as well as oxides, which could account for the relatively low amount of green pigment on the knives (Vermillion et al. 2003).

Non-destructive Analysis of Pigments with Raman Spectroscopy

David et al. (2001) used both Fourier Transform-Raman (FT-Raman) spectroscopy and visible Raman spectroscopy (Chapter 35) to examine pigments encrusted on the tools of ancient artists that were recovered from Tell el Amarna in Middle Egypt. The royal city was constructed, occupied, and then destroyed between about 1350 and 1334 B.C., during the reign of Amenhotep IV (Akhenaten). Raman spectroscopy was selected because both inorganic minerals and organic dyes and binders could be identified with the wavenumber range employed, 50–3500 cm^{-1}.

Pigments of various colors still adhering to artists' palettes and stone mortars were analyzed without processing. FT-Raman spectroscopy was conducted with a Nd/YAG laser at a power of 100–200 mW and sampling diameters of either 100 or 8 μm. Visible Raman spectroscopy was performed with a helium–neon (He–Ne) laser at a power of 0.7 mW focused on spots measuring 1–2 μm in diameter. Samples were identified on the basis of comparisons to reference materials consisting of pure minerals.

Samples colored red, red-brown, or yellow all contained α-hematite in combination with other materials, including sand, resin used as an organic binder, and goethite. Two different forms of carbon were also detected in the latter two pigments. The variety of goethite in the yellow-brown pigment was different from the one in the yellow pigment. A combination of two arsenic (III) sulfides, pararealgar and realgar, was detected in the orange-yellow pigment. One of the two blue samples was a synthetic pigment called Egyptian blue; the other consisted of the mineral azurite. The identity of the green pigment was confirmed as malachite (David et al. 2001).

Clark and Gibbs (1998) used portable and remote laser Raman microscopy to non-destructively analyze pigments on Qazwīnī manuscripts dated to the sixteenth century A.D. With the remote laser device, the laser light is directed to the sample through an optical fiber to a probe; the Raman scattered light returns to the spectrometer through a larger optical fiber. Although the quality of the spectra obtained was typically poorer, the same six pigments were detected by both techniques: vermilion, red lead, lead white, lapis lazuli, carbon black, and Indian yellow.

Non-destructive Analysis of Ink Using X-Ray Fluorescence and X-Ray Diffraction

Nir-el and Broshi (1996) used energy-dispersive X-ray fluorescence (XRF) to analyze red ink on four fragments of the Dead Sea Scrolls; one of these was also subjected to X-ray diffraction (XRD). The analysis was conducted without any preliminary sample treatment. The ink was very well preserved in three samples; parchment preservation on two fragments was good but in one case it had transformed into gelatin and another was affected by the spread of fungus. An area with red ink and another without red ink were examined on each fragment.

Analysis with XRF used an incident X-ray beam angle of 7° and spectral measurements were taken for a period of 200 s. The one sample subjected to XRD was both flat and small enough to be accommodated in the sample chamber; it was simply attached to a microscope slide for analysis. A wavelength (λ) of 1.5405 Å [1 angstrom (Å) $= 10^{-10}$ m $= 0.1$ nm] was used. Results from XRF indicated that mercury was the main component of the red ink; minor amounts of chlorine (Cl), potassium (K), calcium (Ca), titanium (Ti), manganese (Mn), iron (Fe), and strontium (Sr) were also present. Main peaks detected in the areas without red ink were from Ca, Fe, bromine (Br), and lead (Pb), with minor amounts of Cl, K, Ti, Mn, and Sr; a trace amount of mercury (Hg) was also present.

Results from XRD showed that the red ink was mercury sulfide (HgS), which is also called cinnabar. The cinnabar was very pure; only a very small amount of lead contamination was detected, 0.2%. Cinnabar was used in antiquity as a pigment but was very expensive and rarely recovered in archaeological materials. The high purity of the Dead Sea scrolls red ink showed the cinnabar was not adulterated with hematite, a practice known to have occurred. Possible sources of the cinnabar were Spain, from where Romans acquired the mineral, or Ephesus in Western Anatolia. An unassigned peak in the area without ink may be due to something in the parchment, the black ink, the paper on which the fragment was mounted or the adhesive used for that purpose (Nir-el and Broshi 1996).

Radiocarbon Dating of Charcoal-Based Pigments

In order to secure the chronology of known rock art sites in Missouri, Diaz-Granádos et al. (2001) collected pigment samples from pictographs from Picture Cave for accelerator mass spectrometer (AMS) dating (Chapter 8). The site was selected because of the abundance of black pigment on the rock art and the presence of diagnostic motifs, including depictions of supernatural beings.

Small samples of charcoal pigments were taken from five pictographs with precautions to avoid contamination, including wearing latex gloves and using new scalpel blades and clean aluminum foil. Treatment with hydrochloric acid (HCl) was avoided because calcium oxalate could be present in the sample (Diaz-Granádos et al. 2001). Humic acid removal involved treating samples with sodium hydroxide (NaOH) and ultrasonication for approximately 1 h at about 50°C. The sample was recovered by filtration using binder-free borosilicate glass filters that had been baked at about 600°C to remove organic contamination and dried.

A plasma extraction system was used to take carbon from the charcoal pigment while leaving the substrate rock and carbonate/oxalate accretions intact. The sample was loaded into the plasma chamber, air was removed (evacuated), and then a high-purity argon atmosphere was introduced. Adsorbed carbon dioxide (150°C) was eliminated using low-power argon plasmas then the chamber was evacuated once more. After it was verified that no significant leaks existed through which atmospheric CO_2 could enter, the chamber was filled with ultra-high purity (also called ultra-pure) oxygen. Radio frequency-generated low temperature ($\sim$150°C), low pressure ($\sim$1 torr) oxygen plasmas converted (oxidized) carbon in the sample into gaseous CO_2. Water was removed by freeze-drying the gas in liquid nitrogen at a temperature of -194°C; then the dry CO_2 was sealed in a borosilicate glass tube and sent for AMS dating. The CO_2 extraction technique was verified using ^{14}C-free samples and previously dated materials.

Dates obtained on the pigment were earlier than expected, ranging from 950 $\pm$ 100 B.P. to 1090 $\pm$ 90 B.P.; the weighted average of four dates on three pictographs was 994 $\pm$ 42 B.P. Fragments of St. Clair Polished pottery found on the cave floor were associated with the Moorehead phase, which dated from A.D.

1150–1250 in Illinois. The authors (2001) note that radiocarbon dating of charcoal pigment provided an age estimate for the charcoal, not the pictograph. As such, the dates represented the maximum age of the drawings.

Using basically identical processes, Armitage et al. (2001) obtained AMS dates on hieroglyphic texts from Naj Tunich cave, Guatemala. Taking care to avoid contamination, tiny samples of black pigment and the underlying limestone were removed from three drawings containing Maya calendar dates. The four radiocarbon dates from the three drawings were found to be statistically indistinguishable; the average of 1443 ± 47 B.P. produced a calibrated 2σ range of A.D. 540(630)670 (Armitage et al. 2001).

The Maya calendar dates ranged between A.D. 738 and 771; when compared to the mean of A.D. 630, the discrepancy ranged between about 110 and 140 years. The difference could indicate that the charcoal was produced from old wood or that the drawings were rendered with old charcoal. Alternatively, oxalates or carbonates from the underlying rock and accretion material may have contaminated the sample. Isotopic alteration of the wood due to the recycling of carbon dioxide in the tropical forest environment, i.e., the canopy effect (Chapter 13), would only account for about 12–25 years of the discrepancy (Armitage et al. 2001).

Chapter 26
Metal and Glass

Metal

Compositional analysis of metals and their alloys is used to gain information about provenance, manufacturing processes, and artifact distributions at sites. The examples presented in this section show that elemental compositions of artifacts fashioned from unrefined native metal can be used to relate them to ore sources. Isotope ratios are more commonly used to assess the composition of refined metals, although concerns about changes related to manufacturing processes have been raised. Furthermore, it may not be possible to adequately demonstrate the homogeneity and normality of ore sources if too few samples are analyzed. Finally, a method for dating iron artifacts on the basis of carbon preserved in rust is presented.

Compositional Analysis of Metal

Compositional Analysis of Bronze with GD-OES

Ingo et al. (1997) demonstrated the compositional analysis of copper–tin bronze coins using the minimally destructive technique, glow discharge-optical emission spectrometry (GD-OES) (Chapters 30 and 34). Glow discharge devices are used to simultaneously remove and atomize, or sputter off, a small portion of a sample. The resulting excited atoms are suitable for analysis by a number of techniques; in this case, OES was selected. Bulk composition can be determined by analyzing the inner portion of the sample. Depth profiling involves taking samples from the same spot repeatedly, which removes ions from progressively deeper parts of the sample for analysis. Provided the material is homogeneous, quantitative results can be attained in a straightforward manner.

Ingo et al. (1997) compared data obtained by GD-OES with results from ICP-AAS, which involves the analysis of samples atomized with an inductive coupled plasma torch (ICP) by atomic absorption spectroscopy (AA or AAS) (Chapter 34). With atomic absorption spectroscopy, the concentration of the targeted element is determined by the amount of characteristic light absorbed by the excited atoms.

M.E. Malainey, *A Consumer's Guide to Archaeological Science*, Manuals in Archaeological Method, Theory and Technique, DOI 10.1007/978-1-4419-5704-7_26, © Springer Science+Business Media, LLC 2011

More commonly, the emissions from the atomized sample are analyzed, as in ICP-AES and ICP-OES. The results were also compared to those obtained by energy-dispersive X-ray fluorescence (XRF), which is described in Chapter 37.

Five Punic bronze coins dating to 350 B.C. were selected for analysis. The glow discharge device removed material from areas of the sample measuring 0.4 cm^2. The sputtering was performed in an argon atmosphere and the depth of the resulting crater was measured using a surface roughness instrument. Analyses by GD-OES and XRF were minimally destructive. ICP-AAS involved cutting the coin in half to obtain a 0.8 mm thick slice of bright metal from near the center, which was dissolved in acid for the analysis. Calibration curves for GD-OES and ICP-AAS were developed from pure metal standards. Ingo et al. (1997) found GD-OES provided good results for copper, tin, iron, zinc, and lead. Depth profiling enabled comparisons between the compositions of oxide layers and the bulk of the sample.

Analysis of Metal Using XRF and XRD

Dungworth (1997) analyzed copper alloys in 1163 samples dating to the Roman period from a wide variety of sites in Northern Britain with energy-dispersive X-ray fluorescence (ED-XRF). Elemental concentrations in the samples were determined by comparing count rates for each element detected to 21 standards of known composition. In some cases, the composition was determined on metal after corrosion was removed. In other cases, a drill sample, weighing up to about 5 mg, was taken. Data were obtained for nine elements but only values for seven were reported: zinc (Zn), lead (Pb), tin (Sn), iron (Fe), nickel (Ni), manganese (Mn), and arsenic (As).

The results enabled a general categorization of the alloys and detection of changes in the proportions at sites dating between the first and fourth centuries A.D. Brass was initially very common, forming up to 37% of first century alloys, but levels dropped to only 4% by the fourth century A.D. At this time, artifacts fashioned from leaded bronze and leaded gunmetal accounted for 64% of the alloys. About 20% of the alloys at sites influenced by Romans were brass; however, the relative proportion of brass at small, isolated rural farmsteads was about 40%. Significantly lower levels of brass were observed at caves and hillforts. Dungworth (1997) suggested that coppersmiths re-using scrap metal took care to control the mixing of metals to achieve the desired composition.

Lutz and Pernicka (1996) reported that energy-dispersive X-ray fluorescence (ED-XRF) (Chapter 37) performed with portable instruments could provide non-destructive multi-element analysis of archaeological metals. Results were compared to values from reference alloys obtained using other techniques. For most elements relevant to the study of ancient copper alloys, Lutz and Pernicka (1996:316) found that XRF data were accurate "in the order of 5% or better at concentrations above 1%, and in the order of 10–15% at concentrations between 0.1 and 1%."

The information obtained by non-destructive analysis of archaeological metals using portable XRF devices depended upon the degree of patination. Corroded areas could be depleted or enriched in different elements compared to unaltered portions of the sample. If an area of original metal was not available for analysis, it was usually possible for the major component of the alloy to be determined by XRF. Lutz and Pernicka (1996) analyzed more than 1000 drill samples from archaeological metal from Mesopotamia using XRF. About 100 of these were subjected to instrumental neutron activation analysis (INAA) (Chapter 32) and about 20 were analyzed with AA. Some of the variation in results between XRF and INAA may have been due to the presence of partially corroded material in the XRF sample. Lutz and Pernicka (1996) found that, in general, analysis of drill samples provided more accurate data than surface analysis. For certain elements, results were comparable to AA.

Friedman et al. (2008) used high-energy synchrotron X-rays to perform non-destructive XRF and XRD analysis of seven bronze bangles from Tell en-Nasbeth in northern Judah. The energy of the synchrotron X-rays enabled them to penetrate much deeper into the metal than is possible with conventional X-rays. For this reason, data on both the surface and the bulk composition of metal objects were obtained non-destructively.

Depending on the thickness of the item, 6–12 steps were required to analyze the compositions of the bangles from the surface to the center. Corrosion was restricted to the outer 1 mm; three phases were recognized in the bulk region of five of the seven bangles. The lead (Pb) phase was distributed homogeneously throughout the bulk of the bangle, but copper–tin separated into intermetallic and solid solution phases. The intermetallic phase was usually restricted to the outer 3–4 mm of the bangle, which suggested it was the product of rapid cooling. Two bangles contained only two phases, copper and lead, due to low levels of tin.

Compositional Analysis of Copper with INAA

Mauk and Hancock (1998) used INAA to assess the variability in composition of native copper from the White Pine mine in northern Michigan. A total of 85 samples were analyzed including pure vein samples ($n = 15$), sheet samples, which contained some impurities ($n = 50$), and disseminated samples ($n = 27$), which had copper grains evenly dispersed in the rock. Seven of the samples analyzed in duplicate showed good reproducibility. Only about 12 of the 22 elements targeted were detected in most samples. Of these, significant variability was measured in concentrations of silver (Ag), sodium (Na), antimony (Sb), scandium (Sc), and arsenic (As).

The mineral deposits at the White Pine mine formed as a result of two discrete hydrothermal events, so the trace element geochemistry of the ore was highly variable. Regardless, the data supported previous studies showing that nickel (Ni), indium (In), cadmium (Cd), and cobalt (Co) usually did not appear in native copper and that levels of gold (Au) and Sb were consistently low. Mauk and Hancock (1998) contend the latter four elements (Cd, Co, Sb, and Au) along with As, which occurred in particularly high concentrations in disseminated samples, provided a means of discriminating between native copper and European trade copper.

Levine (2007) used INAA to examine the provenance of copper artifacts in northeastern North America to test the assumption that all raw metal for these items was obtained from deposits around Lake Superior. Although native copper deposits occurred in easily procurable contexts throughout New England, the Middle Atlantic, Appalachia, and eastern Canada, they were not incorporated into archaeological models. In addition, ethnohistoric accounts contained direct observations of indigenous populations in these regions procuring and working native copper (Levine 2007).

Reference materials consisted of 103 geological samples from northeastern North America amassed by the author and a database containing approximately 1400 geological samples of native copper housed at the Michigan Technological University.

Ore samples and artifacts were irradiated for 120 min and counted for 1 h after a cooling period of 7 days and again after 16 days. Forty-two elements were initially targeted but only ten were included in the statistical analysis: silver (Ag), chromium (Cr), iron (Fe), mercury (Hg), antimony (Sb), zinc (Zn), arsenic (As), gold (Au), lanthanum (La), and tungsten (W).

Results from 270 native copper samples were used to define trace element fingerprints of each deposit. Thirteen discrete sources were recognized including seven from Lake Superior and six from regions east of Lake Superior in New Jersey, Nova Scotia, and Pennsylvania. Insufficient material was collected from sources in Connecticut to define other trace element fingerprints.

After mathematical transformation of the raw data to improve its approximation of a normal distribution, trace element compositions of the 270 reference samples were subjected to linear discriminant analysis. The resulting trace element signatures were described as excellent, good, useful, or poor depending on their reliability. When geological samples could be traced back to their known source more than 75% of the time, deposits were deemed to have excellent trace element signatures. Eleven of the signatures were considered excellent or good. The remaining two signatures were ranked as useful; they provided correct group assignments about 4.5 times (35%) and 5.7 times (44%) more often than expected by random chance (7.69%).

The trace element compositions of 54 artifacts from 18 different archaeological sites in eastern Canada and New England were compared to the 13 sources. A total of 23 metal artifacts were obtained from nine Late Archaic sites; of these, 18 were traced to deposits in the Lake Superior region. By contrast, 20 of the 31 artifacts from nine Early Woodland sites were probably fashioned from copper procured from deposits in Nova Scotia. The analysis showed that although copper from the Lake Superior region was still employed in the Early Woodland, metal ore from closer sources was favored (Levine 2007).

Compositional Analysis with ICP-MS

Young et al. (1997) and Young and Pollard (2000) recommended the use of inductively coupled plasma-mass spectrometry (ICP-MS) as a means of determining both elemental concentrations and isotopic abundance ratios of metal artifacts. Young et al. (1997:381–882) suggested that ICP-MS has a distinct advantage over INAA, ICP emission (ICP-AES and ICP-OES), and other techniques in that a very large number of elements can be analyzed simultaneously.

ICP-MS is performed on sample solutions so the metal must first be dissolved. Iron and copper-based metals were usually dissolved in warm concentrated nitric acid (HNO_3). A combination of HNO_3 and hydrochloric acid (HCl) was used for artifacts that contained gold, silver, tin, and lead; if tin was present, it might be necessary to add a small amount of hydrofluoric acid (HF). The trace elements present in a sample solution were determined, then the sample was diluted so that the concentrations of major and minor elements could be measured. When solid sample material was introduced by laser ablation (LA) (Chapter 30), Young et al. (1997) reported that ICP-MS did not experience the same spectral overlap and matrix interferences that occurred when XRF and PIXE were employed for surface analyses. Single or multi-element standards must be analyzed together with the unknowns in order to obtain quantitative concentration data (Young and Pollard 2000; Young et al. 1997).

In addition, ICP-MS could provide information about lead isotope ratios in metals. While thermal ionization mass spectrometry (TIMS) offers a higher degree of precision than ICP-MS with a quadrupole mass analyzer, data from ICP-MS with a magnetic sector mass analyzer matched or outperformed TIMS. Another advantage was that isotopic fractionation did not occur, although there might be preferential discrimination of heavy isotopes (Young et al. 1997).

Analysis of Native Copper Using INAA, ICP-MS, and LA-MC-ICP-MS

Cooper et al. (2008) used INAA and ICP-MS techniques to analyze copper artifacts and source material from south-central Alaska and southwestern Yukon Territory to determine if sources could be discriminated on the basis of trace element composition or lead isotope ratios. Primary deposits of copper occurred in basaltic and sedimentary rocks where it precipitated out of copper-bearing hydrothermal solutions. Due to weathering and erosional processes, secondary deposits occurred downslope and downstream of the primary sources. At least 46 discrete sources were known to exist in the region, including a 120 km long contact belt that was extensively mined in the early twentieth century.

Source materials were collected from three stream drainages exploited by indigenous peoples at the time of EuroAmerican contact (Cooper et al. 2008). These native copper sources, Kletsan Creek, Dan Creek, and Chititu Creeks, were all located about 200 km southeast of the site. Twenty-four of the 170 copper artifacts selected for analysis were from the Late Precontact Gulkana site, situated on a tributary of the Copper River.

Forty-one samples from source locations and 24 artifacts, weighing from 80 to 160 mg, were analyzed with INAA. After cleaning, samples were irradiated once for either 4 or 6 h; gamma ray emissions were counted for between 15 and 30 min after a decay period of about 1 week and again for 6–24 h after a decay period of about

4 weeks. Prior to analysis with a quadrupole ICP-MS, 20 source materials and 10 artifact samples weighing between 50 and 100 mg were dissolved in concentrated HNO_3, which was then diluted. Source materials, standards, and fragments of copper artifacts were subjected to lead isotope analysis using laser ablation-multiple collector (MC)-ICP-MS. Portions of the sample removed by laser ablation were directed to the ICP torch; ratios of the four lead isotopes in the sample and two thallium (Tl) isotopes in the reference solution were mass analyzed simultaneously using multiple collectors, specifically six Faraday cups (Cooper et al. 2008).

Although ICP-MS provided data for more elements, Cooper et al. (2008) found that elemental compositions obtained using INAA were most useful for matching artifacts to copper sources because data for mercury (Hg) could be obtained. As the copper was not heated to its melting point during manufacture, the artifacts retained Hg. Mercury cannot be measured by ICP-MS because it occurs in the argon (Ar) used as both the carrier gas and the fuel for the plasma torch.

Cooper et al. (2008) found that all but one of the artifacts were probably fashioned from copper from a single source, the composition of which most closely resembled the deposits at Dan and Chititu Creeks. Native copper in the region contains very low levels of Pb, less than 0.1 parts per million (≤ 0.1 ppm), which reduced the utility of lead isotope analysis. The results obtained by Cooper et al. (2008) supported of the findings of others and showed that copper from sources in the Arctic and Subarctic had fairly uniform trace element compositions. Cooper et al. (2008) noted that additional samples from sources within the region were needed to better assess the intraregion variability, but the remoteness of certain copper sources represented a significant challenge.

Lead Isotope Analysis of Metals

Lead isotope analysis is used for provenance studies because it is not possible to match the refined metal in an artifact to its source on the basis of trace elements (Gale and Stos-Gale 1992; Pollard and Heron 1996; Rehren and Pernicka 2008). The processes used to refine the metal modify the distribution of trace elements, which might not have been homogeneous within the original ore deposit. As described by Henderson (2000), ancient metallurgy involved a number of steps that separated the metal from other matrix materials and increased the level of purity. The extracted ore underwent beneficiation processing, during which it was crushed, washed, and sorted. The ore then was roasted to convert metal sulfides into sulfur dioxide and carbonates into carbon dioxide. Pollard and Heron (1996) noted that volatile elements, such as arsenic, might be lost at this time.

During the smelting process, the chemical composition was further altered by the separation of the metal from the parent rock or vein material (gangue). The ore was heated to a high temperature in a reducing (oxygen free) environment to ensure the pure metal was recovered rather than a metal oxide (Henderson 2000). Flux agents might have been added to lower the melting point of the impurities and ensure their removal as slag, a silica-rich liquid. Further refining would alter the trace elements to a greater degree. The trace element composition of an artifact formed from the resulting metal were not homogeneous and might also be affected by corrosion (Henderson 2000).

As described in Chapter 13, the isotopic composition of an ore deposit is related to its geological history. Ore bodies that formed at about the same time by similar processes in similar geological environments should have similar isotope ratios. Although Barnes et al. (1978) found that the lead isotopic ratio of metal is unaltered by the metal-refining techniques or any subsequent corrosion, Budd et al. (1996) suggested that the limitations of trace element analysis of metals apply equally to lead isotope studies.

As described in Chapter 12, Baxter et al. (2000) raised concerns about the distribution of isotopes in an ore body. Researchers who conduct provenance studies tend to assume rather than demonstrate that the lead isotope field for a deposit is normally distributed. Baxter et al. (2000) noted that more than twice the recommended number of 20 ore samples was actually needed to test an ore deposit for normality. Between 60 and 100 ore samples would be required to characterize a deposit with a non-normal lead isotope field.

Radiocarbon Dating Iron Artifacts

Cook et al. (2003) revised the process developed by van der Merwe (1969) and obtained accelerator mass spectrometer (AMS) radiocarbon ages on iron artifacts from carbon preserved in rust as cementite (Fe_3C). Carbon is added to iron as a necessary part of the smelting process. Typically steel contains about 2% carbon and cast iron contains somewhat more. Prior to the advent of AMS dating, about 50 g of these metals were required to obtain a radiocarbon age; now less than 300 mg of steel or cast iron is needed.

Cook et al. (2003) extracted CO_2 for AMS dating from rusty iron using a process called sealed tube combustion. Iron samples were sealed inside a prebaked 6 mm quartz tube with copper oxide (CuO), without pretreatment with 10% nitric acid. The 6 mm tube was then sealed inside a prebaked 9 mm quartz tube in case of breakage. The sample was baked at 1000°C for a minimum of 10 h to release the carbon.

The process was tested on ten cast and bloomery iron samples, weighing from 45 to 932 mg, from the United States, Scotland, and China that were dated by van der Merwe (1969). Six European and five American artifacts considered too small to be dated previously were also analyzed. In addition, an iron artifact that had rusted under the ocean and another that had rusted while buried in soil were examined. Sufficient carbon for AMS dating was extracted from all but one of the previously dated samples. The ages of eight of the nine previously dated samples were within one standard deviation of those obtained by van der Merwe (1969). The lack of agreement between the ages obtained on the final sample was attributed to an inhomogenous distribution of charcoal and coal in the metal.

Earlier than expected dates on four of the six European samples indicated ^{14}C-depleted coal was used in the steel-making process, instead of charcoal; the other ages obtained were plausible. The ages of the samples that rusted in the ocean and in soil overlapped with ages on associated material within 2σ.

Carbon yields from the sealed tube combustion process often exceeded those previously reported by van der Merwe (1969). Cook et al. (2003) suggested that either the new extraction process was more efficient or that rusting had increased the relative concentration of carbon in the samples. Correspondence with previously dated material indicated that rusting did not introduce new carbon into the sample. Cook et al. (2003) emphasized that age estimates obtained from metal were based on the amount of ^{14}C remaining. If coal, or a mixture of coal and charcoal, was used, the age of the artifact would be too old. Cook et al. (2003) noted this factor, and the possible effects of metal mixing or recycling, may account for the older than expected age estimates on the European artifacts.

Glass

Compositional analysis of glass is often conducted as part of provenance studies. Examples of techniques that could be employed for this purpose are presented in this chapter. Recently, Baxter et al. (2006) described some of the problems that may be encountered when attempting to combine glass compositional data sets. While the statistical analysis of separate data sets could reveal similar patterns, problems could arise if the data were merged into a single large data set.

Baxter et al. (2006) attempted to merge a data set acquired in the 1990s with one obtained more recently using the same technique within the same facility. They found some samples excluded as outliers in the earlier analysis of 15 glass cups prevented the recognition of a group that was more obvious when 63 vessels were analyzed. Good agreement between old and new measurements was observed for only 2 of the 11 metal oxides, MgO and CaO. For this reason, statistical analysis of the combined data set would only highlight methodological differences rather than compositional differences.

Systematic differences between the data sets were reduced by applying data adjustment factors; individual factors were determined by the ratio of the old and new measurements for each metal oxide. Measurements on two variables, P_2O_5 and PbO, were eliminated from both data sets. Re-analysis of the merged data set revealed potential problems because it was impossible to conclusively determine if a grouping consisting mainly of cups analyzed in the 1990s was a true cluster or if it was a product of inadequate data set reconciliation (Baxter et al. 2006).

Compositional Analysis of Glass

Non-destructive Analysis of Trade Beads Using INAA

Sempowski et al. (2001) used INAA to determine the chemical composition of seventeenth century opaque red glass trade beads, known as redwood beads, recovered from archaeological contexts, including a glass bead factory in Amsterdam and 17 Aboriginal sites in New York, Ontario, and Quebec. Usually samples submitted for INAA are powdered prior to analysis, but in this case, each of the 221 trade beads was analyzed whole.

If necessary, the beads were cleaned in an ultrasonic bath prior to analysis. Because the beads were so small, weighing between 5 and 10 mg, they were simply placed in individual polyethylene sample vials. The first count, lasting 5 min, was made after a 5 min irradiation and cooling period lasting between 5 and 7 min. A second count lasting between 5 and 33 min was made the next day. Gamma ray emissions were counted twice using a hyper-pure germanium detector based spectrometer. The concentrations of 12 elements were obtained as percentage amounts or parts per million (ppm). The first count provided data for cobalt (Co), tin (Sn), copper (Cu), sodium (Na), aluminum (Al), manganese (Mn), chlorine (Cl), and calcium (Ca). Concentrations of longer lived radioisotopes of sodium (Na), arsenic (As), antimony (Sb), and potassium (K) were measured in the second count.

Sempowski et al. (2001) reported that all redwood beads were soda-rich mixed alkali–lime–silica glasses that varied with respect to the amounts of Cu, Sn, and Sb present. Four mutually exclusive groups were identified on the basis of chemical composition. Two groupings were recognized among beads manufactured prior to 1655 and two groupings among beads manufactured after 1655. The amount of Cu found in certain uncored beads manufactured prior to 1655 was higher than other beads made at that time. High levels of Sb and slightly elevated levels of Sn were detected in cored beads manufactured after 1655, but this was not observed in either round or circular uncored redwood beads manufactured at the same time (Sempowski et al. 2001).

A significant aspect of this study is that Sempowski et al. (2001) were able to characterize the opaque red trade beads using a non-destructive approach. Another important feature was that one short irradiation followed by two gamma ray counts provided sufficient data to establish differences in elemental composition. A second, long duration irradiation and associated gamma ray counts after days or weeks of decay were unnecessary.

Compositional Analysis Using LA-ICP-MS

Shortland et al. (2007) used LA-ICP-MS to compare the trace element composition of glass working debris and finished objects from sites in the Middle and Near East. Colorless and blue glass from four roughly contemporaneous Late Bronze Age sites in Egypt and Mesopotamia (modern Syria and Iraq) were examined. Samples were mounted in resin blocks and laser ablation was performed on a spot measuring 80 μm in diameter. The concentrations of potassium and phosphorus were difficult to determine using this technique but more than 30 other elements were detected below the parts per million level with a precision of better than 10%.

By comparing blue and colorless glass at each site, Shortland et al. (2007) were able to distinguish elements associated with the use of cobalt as a colorant from those colored with copper. The remaining elements could be detected in either type.

Shortland et al. (2007) divided the elements into three groups: those derived from colorants, those appearing in all types of glasses, and elements whose appearances are variable. In Egypt, blue glass was produced through the addition of either cobalt or copper; in Mesopotamia, only copper was used as a colorant.

The set of elements common in all glasses from Egypt differed somewhat from those common to all glasses from Mesopotamia, which suggested regional glassmaking. Mesopotamian glass was more uniform in composition and "purer" in that it contained fewer trace elements (Shortland et al. 2007). The elements exhibiting the greatest interregional variation were titanium (Ti), zirconium (Zr), lanthanum (La), and chromium (Cr) and, due to geochemical factors, the differences likely relate to the original material. In particular, examination of Zr/Ti and Cr/La ratios provided a robust means of discriminating the regional sources.

These four elements were not associated with colorants; Shortland et al. (2007) proposed they were somehow related to the plant ash added as flux, the clay vessels used in glass production or clay accidentally added with the quartz pebbles. Shortland et al. (2007) considered the best explanation might be that zirconium occurred in the quartz pebbles from Egypt and the mineral chromite occurred in the ultra-mafic rocks of Mesopotamia.

Isotope Analysis Using TIMS

Degryse and Schneider (2008) used isotopic analysis with thermal ionization mass spectrometry (TIMS) to examine the provenance of ancient glass. Natron glass was the predominant type of glass in the Mediterranean and Europe from about the middle of the first millennium B.C. until the ninth century A.D. It was made in primary workshops by combining soda-rich mineral matter with quartz sand; the glass was then supplied to secondary shops where it was shaped into objects. Degryse and Schneider (2008) wished to determine whether primary glass workshops from the first to the third century A.D. were restricted to Egypt and the Levant or if natron glass was manufactured in other locations, as indicated by Pliny the Elder.

Degryse and Schneider (2008) suggested the isotopic concentration of strontium (Sr) and neodymium (Nd) in ancient glass would match that of the raw materials from which it was made because these elements do not undergo fractionation. Lime-bearing materials were the primary source of Sr; if Holocene seashells were added during glass production, the Sr value would be low and match modern seawater. The addition of limestone would result in high Sr values that closely matched the deposit. The Nd value of glass would relate to the heavy mineral content of the silica from which it was made.

Samples were dissolved in a heated mixture of concentrated hydrofluoric (HF) and nitric (HNO_3) acids, dried, and dissolved in *aqua regia*. *Aqua regia* is a three-to-one by volume mixture of concentrated hydrochloric acid (HCl) and HNO_3 (Skoog and West 1982). Aliquots spiked with ^{84}Sr and ^{150}Nd tracers were prepared along with aliquots that were not spiked in order to determine both the concentration and the isotope ratios in the samples. Chromatography was used to separate the Sr from

Nd and then isotope ratios were determined using TIMS. Values for ^{143}Nd/^{144}Nd ratios represented the ratio measured in the sample relative to the amounts in the Chondritic Uniform Reservoir (CHUR) and were given as parts per ten thousand, εNd. Calculations of strontium isotopes are described in Chapter 13.

Isotopic compositions of 27 natron glass vessels from different parts of the Roman Empire, including the Netherlands, Turkey, Slovakia, and Belgium, were compared to values obtained on sands from Italy, the Levant in modern Israel, Egypt, and Belgium. The Sr content of the ancient glass was highly variable, but all glass samples from three primary production centers in Italy, the Levant, and Egypt had ^{87}Sr/^{86}Sr values similar to present-day seawater. Very low levels of Sr were detected in the sand samples and the Sr isotopic values ranged widely. Isotopic composition and Nd content varied in both the ancient glass and the sand samples. The εNd values for sand and glass from the eastern Mediterranean were very similar and higher than -6.0; whereas values from western Mediterranean were lower than -7.0.

The Sr–Nd isotope signature of ten samples of ancient glass closely matched values from known glass production centers in the Levant and Egypt. Values for 14 samples of archaeological glass differed markedly from the eastern Mediterranean values. The primary production centers of samples with low εNd values were probably located in the western Mediterranean; samples with intermediate values could result from recycling or mixing of material from different primary sources. Degryse and Schneider (2008) concluded that isotopic analysis of Sr and Nd isotopes enabled ancient glass manufactured in eastern parts of the Mediterranean to be discriminated from glass produced elsewhere and clearly demonstrated that the account of Pliny the Elder was correct.

Chapter 27
Plant Remains

Instrumental techniques are routinely employed in the study of archaeological plants; the most common application is radiocarbon dating. In addition, a variety of techniques are used to identify, source, and monitor the domestication of plant materials. The composition of ancient plant materials may also reflect changes in their growing environment resulting from natural events or human activity. Examples of these applications are presented in this chapter.

Analysis of Microflora with SEM and SEM-EDS

Silicon- and calcium-containing inclusions present in plant fibers can often be used to identify the source plant. Siliceous phytoliths, calcium oxalate crystals, and calcium carbonate cystoliths occur in plants eaten as food. Jakes and Mitchell (1996) described a method of preparing phytoliths and inorganic inclusions using cold plasma ashing as an alternative to techniques that employ high temperatures and/or hazardous chemicals. High-temperature procedures could distort the shape and chemical composition of phytoliths and strong chemical could pose safety hazards. Jakes and Mitchell (1996:151) described the procedure as "nondestructive and non altering of phytolith morphology," but this referred only to the inclusions; other sample components were converted to ash.

Samples of dietary plants, including winter wheat, cholla, and prickly pear, were mounted onto a carbon planchette and placed in a sample chamber, which was evacuated and filled with an oxygen atmosphere. Radio frequency energy (13.56 MHz) was used to excite the oxygen molecules, creating an oxygen plasma consisting of ions, electrons, and neutral particles. After an ashing period of 2 h, materials were examined with a scanning electron microscope with and without the use of energy-dispersive analysis of X-rays (SEM and SEM-EDS) (Chapter 37). The method enabled the examination of both silicon- and calcium-based structures with minimal alteration of the phytoliths. Previously undocumented crystal structures were observed; Jakes and Mitchell (1996) reported that these could aid in the identification of specific plant genera. The composition of the materials was determined *in situ* through the use of X-ray microanalysis; this information could contribute to the understanding of phytolith formation.

M.E. Malainey, *A Consumer's Guide to Archaeological Science*, Manuals in Archaeological Method, Theory and Technique, DOI 10.1007/978-1-4419-5704-7_27, © Springer Science+Business Media, LLC 2011

Using SEM as a very high-power microscope can also enhance the analysis of ancient plant remains. Lee et al. (2004) reported that, because macrofossils are often not recovered, the use of wild rice by the precontact inhabitants of nearby archaeological sites in Ontario could not be confirmed. The researchers used SEM to compare modern wild rice to other wetland grasses and to compare modern and fossil wild rice. The pollen of modern wild rice could be differed from other wetland grasses on the basis of size. Although the size of fossil pollen was not identical to modern pollen, the sculpturing of wild rice was the same. Lee et al. (2004) concluded that fossil wild rice can be identified on the basis of its micromorphology using SEM.

DNA Analysis of Plants Remains

DNA Analysis of Ancient Wood

Deguilloux et al. (2006) wished to determine whether the genetic diversity of the trees in Europe related to the post-glacial expansion or to anthropogenic impacts on the forest. To address this question, they examined the genetic relationship between ancient wood and modern trees through the maternally inherited chloroplast genome (cpDNA) of oaks. Samples of oak wood dating from the Neolithic period to the eighteenth century were collected from various sites in Europe. Following protocols recommended for the analysis of ancient DNA using the polymerase chain reaction (Chapter 16), the researchers successfully extracted and amplified ancient DNA from 5 of 51 specimens of archaeological oak wood. The chloroplast PCR products were first cloned, then the nucleotide sequences were analyzed and compared to a molecular database.

Wood samples that contained amplifiable DNA were from sites in France and Italy dating from the Late Imperial Roman period to the eighteenth century. Four different oak cpDNA haplotypes were identified among the five samples and the interspecific variation was largely consistent with studies of modern oak. Samples containing amplifiable DNA were recovered from waterlogged environments or were preserved in clay. Failure to recover amplifiable DNA from 46 oak wood specimens could have been due to the presence of ellagitannins, which are PCR inhibitors (Deguilloux et al. 2006). Results showed that a genetic continuity existed between the ancient and the modern European oaks, despite the human impact on the forests.

DNA Analysis of Ancient Maize

An overview of research into the domestication of maize by Jaenicke-Després and Smith (2006) appears in the recent compilation, *Histories of Maize*, edited by Staller, Tykot, and Benz. The domestication of maize (*Zea mays* ssp. *mays*) from teosinte (*Zea mays* ssp. *parviglumis*) began between 6300 and 10,000 years ago. Unconscious and deliberate selection of traits resulted in observable modifications, including the reduction of branching and numerous changes to the morphology of seeds. Analysis of ancient DNA enabled the study of genes that cause physical modification as well as biochemical changes in the properties of starch and protein, which affect the nutritive value of the plant.

Domestication reduced the variability in plants, first because it was unlikely that all possible alternative forms of a gene (alleles) were present in the population and second, because intentional selection by humans increased the frequency of alleles responsible for desirable traits and decreased the frequency of undesirable traits. Characteristics related to plant structure and the biochemical properties of starch and protein are associated with three different genes: *teosinte branched-1* (*tb1*), the *prolamin box-binding factor* (*pbf*), and *sugary-1* (*su1*). Whereas teosinte is a bush-like plant with many long side branches each with a small grain spike containing 10–12 kernels, maize has one large main stalk on which two or three cobs, holding up to 500 kernels each, are located. The *tb1* gene is responsible for about 50% of this difference in plant architecture and the allele found in modern maize, Tb1-M1, also appears in the 4400-year-old cobs analyzed by Jaenicke-Després and Smith (2006). Two *pbf* alleles that affect protein storage occur with similar frequency in the 4400-year-old ancient maize DNA and the DNA of modern maize. Two *su1* alleles affecting the quantity and quality of starch that often appear in modern maize, but are rare in teosinte, were detected in ancient maize DNA dated at 1900–1800 B.P. An allele exclusively found in teosinte was also detected in ancient maize DNA, which may indicate that the selection process for starch traits was still in progress or that rare alleles had not yet been eliminated from the population.

The argument for human selection is particularly strong with respect to the frequency of the *tb1* and *su1* alleles, because they directly affect the properties relating to harvesting and the suitability of maize for tortilla manufacture, respectively. Changes in protein storage associated with the *pbf* gene provide no adaptive advantage to the plant so were unlikely to have occurred in the absence of human selection. Jaenicke-Després and Smith (2006) argued that similarities between the frequencies of *su1* alleles strongly suggest that Northern Flint maize from the northeastern United States was derived from maize from the American Southwest.

DNA Analysis of Ancient Wheat

Using wheat as an illustration, Brown (1999) outlined the potential of molecular genetics, specifically phylogenetics, to determine how many times a plant or animal was domesticated. If a species was only domesticated once, modern domesticates should display limited genetic diversity compared to wild specimens because the initial domestication process operated like a genetic bottleneck. The analysis of ancient DNA might lead to the identification of the wild progenitor of a modern species but not necessarily the site of domestication. This is because a wild plant or animal is readily available at the core of its natural range; the development of domesticates was more beneficial to humans living at the edge, or completely outside, of its natural range.

Phylogenetic comparison of genetic data obtained by aDNA analysis enabled the examination of the spread of agriculture and agricultural practices. Through this type of work, a wild population of einkorn wheat from the Karacadağ Mountains of southeast Turkey was identified as a likely progenitor of all cultivated einkorn wheat; it was also determined that einkorn wheat was domesticated just once. The research of Brown (1999) showed that high molecular weight (HWM) glutenin alleles could be used to explore the genetic relationship between wild emmer (*Triticum dicoccoides*) and the free-threshing tetraploid wheat and hexaploid wheat developed from it. Brown (1999) reported that the Glu-D1-1b allele, which is good for bread-making, was identified in a 3000-year-old Bronze age grain of wheat from Assiros, Greece.

The best preservation of DNA occurs in plant remains that were carbonized or desiccated. Charring causes the plant DNA to become extensively fragmented and chemically damaged but decay was less severe if the grain was not exposed to temperatures above 200°C. Brown (1999) estimated that DNA might be preserved in only 1 in 20 seeds. Threadgold and Brown (2003) conducted studies into the effects of charring on a type of hexaploid wheat, *Triticum aestivum*. Weight loss through the expulsion of water and more substantial organic changes were observed during heating. While the decay of DNA leveled off after 2 h in seeds exposed to temperatures up to 200°C, exposure to higher temperatures caused DNA to decay to undetectable levels more rapidly. DNA preservation was better in seeds heated in low-oxygen environments compared to those heated in aerobic environments.

Recent studies revealed some potential problems with the strategies currently employed for analysis of ancient DNA sequences. In order to simulate damage observed in ancient DNA, Banerjee and Brown (2004) exposed wheat DNA to elevated temperatures for extended periods of time. A specific portion of mito-chondrial DNA, a 181 long base pair segment of the *atpA* gene, was examined in samples heated at 95°C for periods ranging between 2 and 21 days. Instead of randomly distributed alterations, 107 of the 124 cloned sequences displayed the same change: adenine was replaced by guanine. Another type of damage was observed in 15 cloned sequences from six of the eight samples. Banerjee and Brown (2004) cautioned that if non-random nucleotide variations resulting from diagenesis are a widespread phenomenon, they could be mistaken for genuine sequence features in ancient DNA.

Compositional Analysis of Plant Remains

Analysis of Ash to Determine Material of Origin

Pierce et al. (1998) wanted to determine if fuel depletion played a role in the abandonment of Pueblo III settlements in southwest Colorado near the end of the thirteenth century A.D. As an alternative to macrobotanical analysis, they developed a technique to reconstruct ancient fuel use from the chemical composition of ash.

Compositions of modern and archaeological wood, modern and ancient ash, and sediments were determined with inductively coupled plasma-atomic emission spectroscopy (ICP-AES) (Chapter 34). Samples were oven-dried, then reduced to ash in a muffle furnace at 600°C for 12 h, and weighed. The material was dissolved in *aqua regia*, which is a 1:3 mixture of nitric acid (HNO_3) and hydrochloric acid (HCl), at 125°C; undissolved material was removed and the solution was diluted prior to analysis. Analysis of seven samples using a scanning electron microscope with energy-dispersive spectrometry (SEM-EDS) showed the undissolved material consisted largely (>80%) of silica (Si) together with aluminum (Al) and potassium (K).

Of the 30 elements targeted by ICP-AES, 14 produced reliable results and were included in the analysis. Results on modern wood indicate that it may be possible to distinguish wood from bark on the basis of elemental composition. All six wood taxa were discriminated using discriminant analysis on seven elements: potassium (K), manganese (Mn), strontium (Sr), titanium (Ti), thallium (Tl), yttrium (Y), and zinc (Zn). Weathering and leaching altered the composition of ancient ash samples to the point that they could not be identified using the discriminant functions; however, sources of ash produced from ancient charcoal could be differentiated. In some cases, regression analysis could determine each taxon present and its relative amount of mixtures of modern ash; however, results on ancient ash did not correspond well to plant macrofossil data.

Pierce et al. (1998) concluded that the taxa and tissue type of fuels could be distinguished on the basis of composition, but post-depositional processes prevented categorizations created from modern data. Large challenges were associated with the identification of mixed taxon ash samples through regression analysis. Further work was required before the method could be used with confidence.

Analysis of Tree Rings to Detect Short-Term Climate Change

Pearson et al. (2005) investigated the possibility that the chemical composition of annual tree-rings can be altered by short-term environmental changes. In particular, the researchers were interested in detecting climatic perturbations of 1–3 years duration related to volcanic eruptions. The first hypothesis tested was that a chemical fingerprint directly related to the eruption can be identified in individual tree rings. The second hypothesis was that volcanically induced changes in the acidity of the environment would alter the relative availability of certain elements to the trees.

Previous studies indicate that trees growing in poor, shallow, well-drained soils would be more responsive to changes in the environment compared to those rooted in deep, fertile ground. Cores taken from a *Pinus sylvestris* tree that grew in a marginal environment in Sarikamiş, Turkey, was used to test the hypotheses. Laser ablation-inductively coupled plasma-mass spectrometry (LA-ICP-MS) (Chapters 30, 31, and 34) was used to provide multi-element analysis of the tree rings because long sequences could be analyzed with minimal preparation and little impact on the sample. The minimum diameter of the ablation crater from the Nd:YAG laser was expected to be 30 μm. Element concentrations were given as ratios to ^{13}C in the sample.

The initial analysis involved early and late wood from A.D. 1805 to 1818, then the entire range from A.D. 1788 to 1828 was examined. Glass wafers spiked with 61 elements were used as calibration standards and to monitor changes in the instrument output. Of the 30 elements analyzed, only Al, Mn, nickel (Ni), Zn, copper (Cu), Sr, cadmium (Cd), barium (Ba), lanthanum (La), and rubidium (Rb) were above the detection limits of the instrument. Sharp increases in the levels of Al, Zn, Cu, and Ca were observed in the rings around A.D. 1815, when the Tambora eruption occurred. In contrast with the rest of the sequence between A.D. 1805 and 1818, the chemistry of the wood formed during this period was very heterogeneous and measurements of elements had wide error bars. The contrast was more pronounced when only values on wood from the beginning of the growing season (early wood) were considered. When the variation across the period from A.D. 1788 to 1828 was examined, the difference between values from A.D. 1815 and other years was not statistically significant but still appeared as a prominent change in tree ring chemistry.

In order to reduce the effects of variability within the tree, Pearson et al. (2005) recommended averaging compositions of samples taken from different heights and along different radii. Although some limitations were recognized, the technique showed great promise to provide rapid, high resolution and minimally destructive multi-element analysis of tree ring sequences (Pearson et al. 2005).

Analysis of Maize as a Possible Method of Sourcing

Benson et al. (2008) evaluated the possibility of matching the soil on which ancient maize was grown to the plant using strontium isotope values (Chapter 13) on modern deer mice and the trace element composition of modern plants and soils. Deer mice were used to estimate biologically available strontium levels in the environment; because they integrate strontium from a large area (up to 3000 m^2), the deer mice could provide a better average isotopic value for a field than discrete soil samples. Since isotopic fractionation does not occur, the ^{87}Sr/^{86}Sr ratio of the field would be reflected in the soil water, the plants, and the mice bones. Two deer mice were collected from each of 12 sampling sites; bones recovered from the animals were dissolved in nitric acid and diluted. The chemically separated strontium component was analyzed using thermal ionization mass spectrometry (TIMS) (Chapter 31); trace element analysis was performed by inductively coupled plasma-atomic emission spectroscopy (ICP-AES) and ICP-mass spectrometry (ICP-MS) (Chapters 31 and 34). Comparisons with soil collected at depths ranging from 0 to 55 cm indicated that deer mice only obtained strontium from the uppermost soil layer. The isotopic values on the animal bones were not good indicators of the ^{87}Sr/^{86}Sr values for maize, which has a root system that extends downward about 1.5 m. The trace element composition of deer mice bones could not be related to that of the maize grown in the fields where the animals foraged either.

Instead of directly comparing the trace elements in soil water to those in plants, a trace element distribution coefficient (K_D) was employed to account for the biological availability of the chemical species. Ratios of two trace metals were used to negate the effect of changes in soil water concentration on the concentrations of individual dissolved trace elements (Benson et al. 2008:913) and enabled the use of synthetic soil solutions produced by leaching a soil with a weak acid. Five of the 155 Native American maize landraces associated with the Acoma, Hopi, and Zuni were sampled for this study. A 4 cm wide portion from the center of each cob (without kernels or cupules) from the five varieties, as well as kernels from two varieties, were homogenized and freeze-dried. A small sample was dry-ashed in a muffle furnace; the maximum temperature of 450°C was held for 16 h. The ash was then dissolved in a mixture of nitric (HNO_3), hydrochloric (HCl), and hydrofluoric (HF) acids, evaporated, then redissolved, and re-evaporated twice before dilution in deionized water and analysis.

Metal ratio values were determined and then distribution coefficients were calculated. Systematic distribution coefficients were observed between soil water and cobs (soil water:cob) in four metal pairs (Ba/Mn, Ba/Sr, Ca/Sr, and K/Rb) and between soil water and kernels (soil water:kernel) in four metal pairs (Ba/Rb, Ba/Sr, Mg/P, and Mn/P). Systematic partitioning was also observed in rare earth element pairs, including Eu/Gd and Y/Yb. While these distribution coefficients could provide a means of relating maize to the ground in which it was grown, data were obtained from plants grown under ideal conditions. Benson et al. (2008) cautioned that until results are replicated with plants grown under the same conditions as prehistoric plants, the distribution coefficients should be applied with caution. Until then, strontium isotope analysis remains the best tool for sourcing maize.

Radiocarbon Dating of Food Plant Remains

Although it had been common practice to obtain age estimates on charcoal associated with maize, Little (2002) obtained accelerator mass spectrometer (AMS) and conventional radiocarbon ages (Chapter 8) on maize recovered from sites in New England. Often little or no overlap existed between the calibrated ages on maize and those on associated material at the 95% confidence interval ($\pm 2\sigma$). Instead, the radiocarbon age estimates on maize coincided with AMS ages reported on beans recovered from sites in New England, showing that the two cultigens rapidly spread throughout the region after cal A.D. 1250. Little (2002) noted that soils in New England tend to be acidic (Chapter 6), so both plant yields and preservation of botanical remains were generally poor; this reduced the archaeological visibility of maize. Little (2002) suggested that in order to improve yield, people began growing maize and beans together on calcareous floodplains or material from old shell middens was used to fertilize fields. Adoption of the practice of growing beans and maize together on alkaline soils could account for the sudden increase in archaeological visibility of the cultigens.

Chapter 28
Matrix and Environmental Deposits

Studies of sediments from archaeological sites tend to either target compositional changes associated with past human activities or focus on establishing site chronology. Analysis of soil phosphorus content is particularly relevant to the former and is discussed in some detail. Trapped charge and/or uranium series dating techniques are typically used to obtain age estimates, with optically stimulated luminescence applied to sediments previously exposed to sunlight, and electron spin resonance and uranium series dating applied to cave deposits. The use of X-radiography to examine soils by Butler (1992) is an exception. Analyses of organic residues introduced into soil by human activity are presented in Chapter 23.

X-Radiography of Soil

Butler (1992) demonstrated that X-radiography (Chapter 37) of peat deposit cores and soil sections could be used to elucidate stratigraphic information not readily detectable by other means. The coefficient of absorption of organic material is low so X-rays are transmitted; these areas appear light on X-radiographs. Mineral inclusions absorb X-rays so these areas appear dark. Fine stratigraphic and structural details, otherwise not easily detected, are clearly visible on the X-radiographs (Butler 1992). When combined with radiocarbon ages, sedimentation rates were more accurately estimated because boundaries between sediment types were precise. Mineral inclusions within an organic layer were also readily apparent. Soil sections can range between 1 and 5 cm thickness but the thickness of individual slabs must be uniform. X-rays generated at low voltages are appropriate for examination of soils.

Soil Composition Changes from Campfires

Werts and Jahren (2007) proposed a method of detecting the location and estimating the minimum temperature of archaeological campfires using stable carbon isotope analysis (Chapter 13) and measurements of organic content. Tests were conducted

M.E. Malainey, *A Consumer's Guide to Archaeological Science*, Manuals in
Archaeological Method, Theory and Technique, DOI 10.1007/978-1-4419-5704-7_28,
© Springer Science+Business Media, LLC 2011

on mature and undeveloped soils located 15 m apart. Comparisons of incinerated and unheated soil indicated heating resulted in increased $\delta^{13}C$ values in the upper O and A soil horizons by 2.3–3.5‰, but a decrease of 1‰ was observed in deeper soil from the B and C horizons. Organic carbon loss occurred at temperatures between 200 and 400°C. Changes in organic content and stable carbon values were not related to either the initial organic content or the clay content of the soils. Given that the analytical error associated with $\delta^{13}C$ values is less than 0.1‰, Werts and Jahren (2007) suggested the hypothesized locations of ancient low-temperature fires could be confirmed by comparing organic carbon and stable carbon isotope values on soil to those obtained on a same age soil from an unburned location.

Phosphorus in Soil

Holliday and Gartner (2007) reviewed methodologies associated with soil phosphorus analysis as an indicator of human activity, with the stated goal of clarifying issues of terminology and the appropriateness and meaning of individual procedures. Their terminology and perspectives are adopted in this section.

Human waste, organic refuse, burials, ash from fires, and the wastes associated with animal husbandry were common sources of anthropogenic phosphorus prior to the Industrial Revolution. Phosphorous (P), usually found in the environment as phosphate, is used as an indicator of human activity because, instead of being lost through leaching or other processes, it quickly bonds with iron (Fe), aluminum (Al), or calcium (Ca) to form stable compounds. These compounds are not overly susceptible to loss so P tends to accumulate in soil at the site of deposition.

The total amount of phosphorus (Ptot) in soil is the sum of the inorganic P (Pin), occluded P, and organic P (Porg) forms. Both Pin and Porg may be dissolved in soil water or chemically or biologically bound (sorbed) to soil particles. Occluded P is physically or chemically trapped within particles, such as clays. One type of soil P is able to transform into another. The available pool (Pav) consists of soluble and weakly adsorped P that is highly mobile and can be easily transformed. Phosphorus that is more tightly bonded to soil particles forms the active P pool (Pact). The relatively stable P pool (Psta) consists of immobile and tightly bound occluded and organic P. Solution P is already dissolved in soil water and labile (i.e., available to plants); soluble P could dissolve but is not in solution.

The form P takes in soil depends upon several factors including organic matter, soil pH, soil moisture, particle size, and mineral content. The addition of organic matter influences the relative amounts of Pin, Porg, and occluded P, as well as P mobility. However, its effect is variable and does not necessarily result in elevated levels of soil P. Soil pH affects the binding capacity of P; P binds with Fe and Al in acidic (low pH) soils but binds with Ca in neutral and alkaline (high pH) soils. The mobility of P is highest in neutral soils.

The relationship between P and soil moisture is complex because moisture affects many different variables. In general, well-drained soils tend to retain more Pin than poorly drained soils. Particle size is an important variable because finely textured (clayey) soil has a higher P sorption capacity than coarsely textured (sandy) soil. The phosphorus in organic matter added to soils rich in limestone and gypsum is quickly mineralized because it binds to Ca. The amount and types of P in soil tend to vary over time; however, P accumulation does not occur unless suitable soil P receptors are present. The retention of P tends to be quite high in cold climates.

General correlations exist between human activities and soil P levels. In particular, areas associated with food preparation and ash accumulation have elevated levels of P, whereas storage, sleeping, and workshop areas have lower levels of P. The P levels in ritual spaces tend to be low unless organic material was introduced into the soil through feasts or blood offerings.

Holliday and Gartner (2007) reported that 30 of the 50 methods developed to measure phosphorus in soil have been applied for the analysis of archaeological materials. Methods developed by agricultural scientists target available P (Pav). While this is likely not the best indicator of archaeological activity, the technique appears to work well in dry environments, such as the American Southwest. Spot tests that measure Pav may also be suitable in situations where spatial patterning of P is simply used to develop an excavation strategy.

Chemical digestion techniques that target total P (Ptot) produce more quantitative and replicable results than methods measuring Pav. Both Ptot and total Pin are measured so Porg can be determined by simple subtraction, where

$$Porg = Ptot - Pin$$

Holliday and Gartner (2007) cautioned that methods that extract Ptot should not be applied in areas where phosphorus is present in the parent material as the human contribution will be overwhelmed by naturally occurring P. Fractionation studies that target different types of P are very labor intensive and not widely used. Methods that measure Porg are less suitable for archaeological sites where Pin from animal bone is the primary source of phosphorus.

Phosphorus concentrations are usually not measured directly. Instead, reagents are added which result in the formation of molybdophosphoric compounds, which produce a blue color when reduced in an acid environment. A spectrophotometer (Chapter 35) is used to measure the capacity of the sample to transmit or absorb a specific wavelength of light. In general, light will pass through samples with low concentration of compounds and be absorbed by samples with high concentrations.

As outlined below, Entwistle and Abrahams (1997) found that P levels could not be determined with ICP-MS at the same time as trace and rare earth elements due to the large differences in the concentrations. Holliday and Gartner (2007) found that although measurements obtained by colorimetric methods were not identical to those obtained using ICP-based techniques, similar trends were observed.

Analysis of Soil Chemistry

Soil Chemistry Analysis with ICP-MS

Entwistle and Abrahams (1997) used inductively coupled plasma-mass spectrometry (ICP-MS) (Chapters 31 and 34) to determine concentrations of trace and rare earth elements in soils from historic settlements in Scotland. Samples were collected with a manual auger, then dried and sieved. Subsamples of the fine-earth fraction (<2 mm) were powdered and homogenized prior to acid dissolution using a nitric–perchloric (HNO_3–$HClO_4$) digestion. Diluted samples were analyzed semi-quantitatively using ICP-MS. Measurements of 32 trace and rare earth elements were determined simultaneously; analysis time was about 4 min per sample. Concentrations of major elements, such as potassium, calcium, and magnesium, were determined with atomic absorption and flame (i.e., optical) emission spectrometry (Chapter 34). Due to spectral interferences and differences in concentration between trace and rare earth elements, levels of phosphorus, another major element, were quantitatively determined with ICP-MS in a separate run.

Phosphate Measurement as a Survey Technique

Rypkema et al. (2007) reported a method for rapidly assessing the relative phosphorus content in samples to detect the location of an eighteenth century independent leased farmstead in a large tract of "manor land" measuring 71 ha. Instead of allowing soil samples to dry overnight, an average water content of $16 \pm 3\%$ was determined on the basis of nine samples collected within the region. Iron-bound phosphorus was targeted for extraction using the Melich III method. Instead of 1 h, samples were digested for only 3 min; Rypkema et al. (2007) reported a phosphorus yield of 75% was possible after 5 min when using this approach. Solids were removed from the solution using a syringe filtration system. The extract solution was treated with two reagents, the first created phosphomolybdate, and the second caused the molybdophosphoric compound to form a complex with malachite green, a cationic dye. Depending on the concentration of phosphorous in the sample, the color could range from light yellow-green to dark green. A portable spectrophotometer was used to measure the transmission of light through the samples; phosphorus content was assessed as high, medium, or low.

The phosphate analysis was conducted at 50 m intervals as part of a shovel testing survey. The survey crew consisted of four people, two of whom dug the pit and screened for artifacts, one person was in charge of phosphate analysis and one person provided assistance for both procedures. When abnormally high phosphorus readings were encountered, the interval between tests was reduced to 10 m. In general, more artifact-positive shovel tests were recorded in areas with abnormally high phosphorus levels.

Soil Chemistry of Ethnoarchaeological Arctic Fish Camps

Knudson et al. (2004) conducted an ethnoarchaeological study of the chemical composition of soils associated with seasonal fish camps using inductively coupled plasma-atomic emission spectroscopy (ICP-AES) (Chapter 34). The fish camps included butchering areas, drying racks, smokehouses, and living quarters. The aim of the study was to determine if camp activities could be identified through chemical signatures in the soil. Soils from abandoned and active seasonal fish camps near Chevak, Alaska, were examined. In total, 38 samples from activity areas and features and 14 from off-site locations were analyzed. Soils were dried, pulverized, and screened and then elements were extracted by soaking in hydrochloric acid (1 M HCl) for 14 days. Elemental analysis of samples, blanks, and standards was performed with ICP-AES.

High concentrations of all elements were expected because of the cold climate and poor drainage. The soil at the abandoned camp, which had been seasonally occupied for 30 years, had high levels of manganese (Mn), phosphorus (P), strontium (Sr); elevated levels of barium (Ba), calcium (Ca), potassium (K), and sodium (Na) were detected at the former site of the covered drying rack. The sodium level was attributed to the practice of dipping the fish in salt water prior to drying. Contrary to expectations, evidence of ash was not detected in the area around the smokehouse. Because tents were erected on platforms, no anthropogenic signatures were associated with living quarters.

Although the second camp had only been used as a fish camp for one season, elevated levels of Ca, K, Mg, Na, and P were detected in the fish processing area; these same elements as well as Fe and Sr were elevated in the soil of the covered drying rack. Evidence of ash was not detected in the smokehouse because fires were prepared in steel drums, but the level of Mg was elevated. Comparisons between activity areas and off-site controls confirmed that the differences in the soil chemistry represented an anthropogenic signature. There was no evidence that animal droppings, in particular contributions from migrating waterfowl, increased levels of P or Ca at either site.

Soil Chemistry of Maya Floors

Terry et al. (2004) compared the soil chemicals in the floors of an occupied guardhouse to those of Classic Maya structures that had been rapidly abandoned. The guardhouse was selected for the ethnoarchaeological study because of its proximity to the archaeological site at Aguateca, Guatemala. Within the site, soil samples were collected from two structures that were likely elite residences and two others that appeared to be either lower-status residences or workshops. Control soils were taken from areas outside of the site including an old growth forest, uncultivated land, a corn field, and from wetlands near a lagoon.

A total of 630 samples representing the upper 4 cm of soil and sections of floor were collected on a grid and dried. Phosphorus was extracted using the Mehlich

II procedure and concentrations were determined with a colorimeter. Holliday and Gartner (2007) noted this is a dilute acid procedure that extracts Pin. Heavy metals were extracted from soils with diethylene–triaminepentacetic acid (DTPA) and concentrations were determined with ICP-AES. Concentration in control samples exceeded those from the floors so the average concentration of the 158 samples (25%) with the lowest concentrations of extracted elements was used to estimate background.

Concentrations of phosphorus in the occupied guardhouse were highest in the kitchen and associated disposal area. Elevated levels of phosphorus were attributed to organic materials and the phosphorus-rich soap used to clean the dishes. High concentrations of iron were observed along the porch where the guards sharpened their machetes. The highest concentrations of copper, magnesium, and lead were observed in social areas. High levels of zinc corresponded to the battery disposal area.

At the archaeological site, high phosphorus concentrations confirmed the identification of food preparation, consumption, and storage areas in one elite residence. Low P concentrations were observed in reception, sleeping, and craftwork areas. High concentrations of Fe in the soil in the southeast corner of this structure were likely due to the production of pyrite (FeS) mirrors and ornaments. A tree root likely added P to the floor of the second elite residence and the highest concentration of trace elements were associated with a midden.

High levels of P west of a lower status house corresponded to the location of a midden. Low concentrations of trace elements were found inside but elevated levels occurred outside the doorways of both lower status residues. Material swept from the structures may have accumulated in these areas. High levels of heavy metals in one residence were due to the presence of red hematite pigment.

Compositional groupings produced by cluster analysis corresponded to (1) samples from archaeological middens, (2) archaeological mineral ore processing areas, (3) modern and archaeological food processing, consumption, and disposal areas, (4) modern and archaeological work areas, (5) two areas with low concentration of trace elements, and (6) an area that was likely used for the disposal of both mineral pigments and organic waste. Terry et al. (2004) suggested analyses of soil chemistry provided confirmation of activity area interpretations, lead to the recognition of activity areas without artifacts and prevented the misinterpretation of areas containing artifacts but had simply been used for storage.

Dating Techniques

Trapped Charge Dating

Different approaches can be used to determine the number of trapped charges in order to obtain an age estimate on archaeological matrix materials (Chapter 10). Optically stimulated luminescence (OSL) uses light, typically green, in the visible region of the electromagnetic spectrum (Chapter 2). When infrared light is employed, the technique is called infrared stimulated luminescence (IRSL). If heat is used, the process is called thermoluminescence (TL).

Smith et al. (1990) described the optical dating of archaeological sediments and reported their initial results compared well to dates obtained by thermoluminescence and other techniques. While Smith et al. (1990) suggested sediments last exposed to light 100 years ago could be dated, Rees-Jones and Tite (1997) recommended the widespread use of optically stimulated luminescence (OSL) dating for archaeological sediments ranging in age from 1000 to 300,000 years. A study of seven sites indicated that age estimates on fine-grained sediments were less scattered, but sometimes only coarse-grained material was available. Insufficient bleaching was found to be a problem when sediments were rapidly deposited, but this could be detected by comparing the OSL signals from quartz and feldspar grains.

Burbidge et al. (2001) applied OSL dating to sediments at Old Scatness, a site in the UK that had been regularly occupied since the Bronze Age. The coarse grain quartz fraction (90–155 µm) of buried soils from the agricultural infield area was examined. Samples were collected from a test pit using sharpened steel tubes that were inserted horizontally into the walls. Usually two samples were collected for each soil layer so that one could be used for luminescence and the other for dose rate determination. Direct measurements of gamma rays were taken for four samples and the water content of each sample was determined. After sieving, the soil fraction was treated with acids to removed fluorides, non-quartz components, and the outer surface of quartz grains. Prior to taking OSL measurements, sample aliquots were pre-heated to 180°C to remove electrons from unstable traps. Luminescence was stimulated by a green (420–550 nm) light from a tungsten–halogen lamp and detected with a photomultiplier. The single-aliquot regenerative protocol was used to establish the equivalent dose (D_e). After the natural intensity was measured, the aliquot was exposed to several rounds of progressively higher artificial doses of beta radiation and the resulting OSL was remeasured. The D_e was interpolated from the resulting dose–response curve.

Burbidge et al. (2001) reported that age increased with depth but high aliquot-to-aliquot variation occurred in some samples. The highest variability was associated with the deepest soil, from which ages ranging between 2700 and 4600 years ago were obtained; ages from another soil exhibited a 1000 year range. The standard deviations associated with three other soil layer ages was significantly lower, ranging from 25 to 188 years. In general, the OSL ages obtained on soil were consistent with archaeological recoveries from the site.

Zhou et al. (2000) examined luminescence retained in red "terra rossa" sediments from several open-air sites in Greece. Blocks of soil collected from the sites were wrapped in aluminum foil; after removal of the outer layer, samples were processed with acids. Measurements were made on the fine silt grain fraction (4–11 µm) using either IRSL or TL. The dose rate was determined from the sediments; both the additive and the regeneration methods were used to measure D_e. Most of the sediments appear to have been well bleached but a sample of coastal sand was not. Differences in D_e determination made by additive and regeneration methods were on the order of 15%. Most ages obtained by IRSL overlapped TL ages at the one-sigma range, but IRSL was deemed more appropriate for sediments deposited under water. In general, luminescence ages on sediments provided support for a Middle Paleolithic occupation of these sites.

Feathers et al. (2006) used OSL to estimate the ages of Paleoindian and Archaic sites on the Southern High Plains of western Texas and eastern New Mexico. The sites investigated were occupied from the Late Glacial to the Middle Holocene and situated on either dry tributaries of rivers (draws), sand dunes or small, seasonally drained lake basins (playas). Samples were collected by driving light–tight cylinders into exposed profiles, except in one case, where a soil coring rig was used. Radioactivity was determined either in the lab from sediments or in the field with dosimeters.

Measurements were made on three quartz grains ranging from 90 to 125 μm simultaneously in a sample holder designed for larger (180–212 μm) grains. Feathers et al. (2006) reported only a fraction of the grains, perhaps 25% or less, emitted a measurable signal. Feathers et al. (2006) suggested the chance that more than one of the three grains in the holder would emit luminescence that was less than 18%, so single gain resolution was attainable using this technique. Multi-aliquot IRSL ages of two samples were determined using the slide method (Chapter 10).

Equivalent doses (D_e) were determined by the single aliquot regenerative (SAR) dose method and on multi-grain aliquots. A narrow, normal distribution of D_e was expected from well-bleached, unmixed assemblages of grains. In mixed or disturbed contexts, grains with smaller D_e values were used to provide a minimum age estimate for the deposit. A variety of different models were used to calculate single grain D_e distributions including the central age model, which uses a weighted mean, and the common age model, which assumes a single true value age. A dose recovery test showed that the SAR procedures applied to these sediments were valid. Sample aliquots consisting of 400 grains were fully bleached, a 17.5 Gy dose was applied, and then SAR was used to estimate the dose. Between 22 and 88 grains from each of the five samples produced estimations of D_e within one or two standard errors of the administered dose.

Measurements were obtained on at least 900 grains from each sample; the goal of obtaining at least 100 acceptable grains was not met in 9 of the 21 samples. The amount of variation that could not be accounted for through statistics was reported as the over-dispersion (σ_b) values. While well-bleached, unmixed single grain distributions should have over-dispersion values of less than about 20%, only 4 of the 20 samples met this criterion, 8 had average σ_b values of less than 22%. Six samples with high σ_b values were subjected to linear modulated OSL and results indicated post-depositional mixing was more likely the cause of the over-dispersion than partial bleaching.

OSL ages of sites from draws and playas produced good correspondence with radiocarbon ages while those from dune sites generally did not. The distribution of D_e values at one site with a very high σ_b value was found to be bimodal, the reason for this was not known; however, Feathers et al. (2006) suggested the sample tube may have partially intersected a rodent burrow. Feathers et al. (2006)

recommend use of the single-grain technique (which actually involved three grains, in this case) as it provided a means of evaluating ages; multi-grain aliquots would have made it more difficult to detect problems due to mixing.

Uranium Series Dates on Calcite

A discussion of estimating age on the basis of sediment and faunal materials appears in Chapter 19. Shen et al. (2004) obtained uranium series dates on 12 pure, well-crystallized calcite deposits from the New Cave site, which is located about 70 m south of Zhoukoudian, China. Material was collected from the uppermost flowstone that capped all fossil and artifact-bearing deposits, a second layer situated about 1–1.5 m below, and a third layer, which was the lowest accessible. Measurements of ^{230}Th and ^{234}U were obtained by ICP-MS and compared to previous measurements using alpha spectrometry.

The range of age estimates obtained from the capping flowstone indicated it consisted of sublayers deposited over a 70,000-year period. Ages on the second layer show it was deposited between 248,000 and 269,000 years ago. The age results for samples from the lowest layer were consistently about 300,000 years. Shen et al. (2004) concluded that human activity at New Cave occurred between 300,000 and 120,000 years ago, which is broadly contemporaneous with other sites in the region.

Chapter 29
Other Materials

This chapter includes examples of the analysis of materials that do not fall under other categories. Techniques used to determine the structure of stucco from buildings and ritual masks made of plaster are presented. Dating techniques applied to coral offerings in Hawaii and baked clay features in Bulgaria and Switzerland are also described.

Analysis of Stucco Samples

Goodall et al. (2007) used a variety of analytical techniques to investigate temporal and spatial variability in the physical and chemical composition of stuccos and pigments used to decorate buildings during the Early, Middle, and Late Classic periods at Copan. Due to the practice of erecting new buildings directly on top of partially demolished structures in Copan, stucco samples were available for buildings constructed over a 400-year period beginning about A.D. 450.

Fragments exfoliating from wall surfaces were collected from the buildings. Structures examined included the Early Classic Clavel building, the Middle Classic Ani structure, and the Late Classic Structure 10L-22. Analyses were performed with Raman spectroscopy (Chapter 35) and Fourier Transform (FT)-Raman microspectroscopy (Chapter 35). Micro-attenuated total reflectance infrared spectroscopy (micro-ATR-IR) (Chapter 35) and environmental scanning electron microscopy with microenergy-dispersive X-ray fluorescence analysis (ESEM-EDX also known as environmental SEM-EDS) (Chapter 37) were performed on polished cross-sections of samples encased in resin.

Raman spectroscopy was performed directly on stucco fragments, using a microscope to select the area of analysis. Light, with wavelengths of 632.8 and 785 nm, from two lasers were optically filtered to prevent thermal degradation of iron oxide minerals. FT-Raman microspectroscopy was performed with a 1064 nm laser light beam focused down to a diameter of 100 µm. Cross-sections for micro-ATR-IR and ESEM-EDX analysis were prepared by setting stucco fragments in resin. After sectioning, cross-sections were polished with diamond paste. Micro-ATR-IR spectroscopy was performed with the ATR crystal in direct contact with the polished surface using light with wavenumbers between 4000 and 700 cm^{-1}, which is in

M.E. Malainey, *A Consumer's Guide to Archaeological Science*, Manuals in
Archaeological Method, Theory and Technique, DOI 10.1007/978-1-4419-5704-7_29,
© Springer Science+Business Media, LLC 2011

the mid-IR region. ESEM-EDX was performed without coating the sample with graphite or another conducting material.

Examination with an optical microscope under 10 and 20× magnification revealed evidence of stucco and paint layering. The mineral composition of pigments was established using spectroscopic techniques. Each stucco/paint assembly consisted of a coarse stucco layer covered by a fine stucco layer and then by pigment. The stucco consisted of calcite, quartz, and charcoal from the lime-production process. One stucco/paint assembly covered the Early Classic Clavel building; two stucco/paint layers covered the Middle Classic Ani building. The Late Classic Structure 10L-22 was covered with 15–20 stucco/paint layers.

Analysis of the Microstructure of Plaster Objects

Fralick et al. (2000) used a scanning electron microscope with electron-dispersive X-ray spectrometry (SEM-EDS) to examine the construction of a plaster mask, representing a bear or badger, recovered from the San Lázaro Pueblo, New Mexico. The occupation of the pueblo began in the twelfth century A.D. and extended up to about A.D. 1680, peaking between A.D. 1300 and 1500; the pueblo was abandoned for much of the sixteenth century.

Samples were taken from previously broken pieces with the permission of Native American representatives. Microstructural analysis was performed by backscattered electron (BSE) images on prepared polished cross-sections. The layers or zone interfaces were found to vary. The zone between the first and second layers was sharp and relatively flat. Inclusions in both these layers were preferentially oriented parallel to the interface. The next layers of plaster were applied before the second layer had completely dried. In some cases, the plaster from an upper layer penetrated into the layer below due to voids or cracks.

The mask was formed by applying multiple layers of plaster to an armature constructed from plant matter. The primary foundation of the mask was a bulrush cylinder to which a piece of carved wood was attached for the snout. The innermost plaster layer was relatively thick and included clasts and voids. In most cases, after the first layer was dry, a thick second layer of plaster, which was free of inclusions, was applied. The third layer, which was similar in texture to the second layer, may have been applied to fill small cracks. The fourth, and final, plaster layer, which was relatively thin and fine-grained, was interpreted as a smoothing layer to which paint was applied. The third and fourth plaster layers were applied before the previous layer had dried completely. Both the fourth plaster layer and the final paint layer were likely applied with a brush.

Areas of the mask painted red differed from those painted with other colors in that only two layers of plaster were applied. The basal layer was covered with a thin coat of plaster with fewer inclusions and red paint was applied while this second layer was still wet.

Results from compositional analyses of the paint applied to the plaster objects are described in Chapter 25.

Uranium Series Dating of Branched Coral

Weisler et al. (2006) used thermal ionization mass spectrometry (TIMS) (Chapter 31) to measure the ratio of uranium (^{238}U) to thorium (^{230}Th) (Chapter 9) in coral associated with structures and features on the leeward side of the island of Moloka'i, Hawaii. A period of significant cultural change began at about A.D. 1100 and lasted to the time of European contact, which included the movement of people from the wet, windward side of the islands to the dry, leeward zones. In an effort to more precisely determine the ages of structures and features than is possible using radiocarbon dating, Weisler et al. (2006) developed a ^{230}Th chronology based on coral.

Platforms, residences, shrines, enclosures, and other architectural features on the Hawaiian Islands were constructed of dry-laid stone. Offerings of branch corals (*Pocillopora* spp.) were often associated with these structures and middens and provided a means of precisely dating the features. Living corals absorb U from seawater but do not absorb Th; the ^{230}Th present in the sample was produced by decay of radioactive uranium. Age estimations were based on the ^{230}Th/^{238}U and ^{234}U/^{238}U activity ratios normalized to the measured ratios of the secular-equilibrium HU-1 standard (Weisler et al. 2006). The concentration of ^{232}Th contamination was minimal so corrections for the presence of non-radiogenic ^{230}Th typically reduced the ages by less than 2 years.

Branch corals harvested fresh from the reef by the former occupants of the sites were selected for analysis. These corals exhibited fine sculptural details, which were sharp to the touch, and were not affected by post-depositional weathering. Only the top 1–2 cm of each branch from the most pristine samples were submitted for analysis.

Twenty-one ^{230}Th ages were obtained on coral from eight structures or features and ranged from A.D. 1417 ± 3 to 1726 ± 3. Age estimates of A.D. 1604 ± 3 and 1605 ± 3 on three pieces of coral from a major temple and nearby residential complex showed they were occupied within the same year. Age estimates on two coral samples from a small shelter and nearby fishing shrine were A.D. 1725 ± 2 and 1726 ± 3, respectively. Ages obtained on eight samples of coral from a midden indicated it accumulated between A.D. 1540 ± 3 and 1620 ± 2. Age estimates on associated coral indicated a large fishing shrine was used over a period extending from A.D. 1417 ± 3 to 1709 ± 2. Some stratigraphic mixing occurred between the uppermost levels at the large fishing shrine and at the coastal midden, which contained numerous postholes and pits (Weisler et al. 2006).

Archaeomagnetic Dating of Baked Clay Features

Kovacheva et al. (2004) used archaeomagnetic dating techniques to obtain ages for four potter's kilns and the well-baked walls of two rectangular trenches from one site in Bulgaria and three in Switzerland.

Magnetic remanence was measured with three different types of magnetometers, spinner, astatic, and fluxgate. Secondary components of the natural remanent magnetization were removed through stepwise alternative field demagnetization. The unstable viscous remanent magnetization was removed by storing the samples in a field-free space for 2 weeks. The magnetic susceptibility of the sample was measured after each heating step during thermal demagnetization. The stability of the minerals during heating was also monitored.

Samples were collected from two kilns at the Early Byzantine site of Serdica, Bulgaria, to determine if they were contemporaneous and the time of their last use. Seven oriented samples were collected from one partially preserved kiln; of these, the archaeomagnetic results from two were accepted. A total of 25 oriented samples were collected from the other kiln; the results from seven samples were accepted. Some of the rejected samples were bricks from the first kiln that were not sufficiently heated to erase the thermoremanence acquired at the time of their manufacture. The temperature of this kiln did not exceed 400°C in the past. Some samples collected from the second kiln were rejected because of mineral alteration in the laboratory at high temperature. Measurements of inclination, declination, and intensity generated three possible age solutions; the most probable date on the first kiln ranged between A.D. 1513 and 1583. Two age solutions were obtained for the second kiln; the earlier range of A.D. 566–638 was accepted on the basis of coins recovered from the site.

A total of the 18 oriented baked clay samples were collected from the walls and between the stones of a medieval kiln from the Reinach site, in Basle (Basel), Switzerland. Due to missing data in the paleointensity dating curve, the archaeomagnetic date was only based on directional parameters, declination and inclination. The final heating of the kiln, determined at the 95% confidence level, occurred between A.D. 753 and 901.

The archaeomagnetic dating of a late Iron Age potter's kiln from the site of Voltastrasse in Basle, Switzerland, was complicated by the proximity of a tramline and earthmoving equipment. A sun compass could not be employed as samples were collected on a cloudy day. Many of the 18 samples were found to be unsuitable; the archaeomagnetic date, determined on the basis of five measurements on four samples, ranged between 268 B.C. and A.D. 122.

The ages of two rectangular trenches with well-baked vertical walls and few associated artifacts from the La Beuchille site, Delémont, Switzerland, were also determined. Twenty-five samples were taken from one trench and 23 were taken from the other. The mean directions from the samples indicate the two trenches were fired at the same time. The data were combined in order to determine an archaeomagnetic date and two age solutions were obtained, 415–163 B.C. and A.D. 1890. An uncalibrated radiocarbon date of 175 ± 70 B.P. was obtained, which corresponded to four separate ranges of calendar dates between A.D. 1650 and 1950. For this reason, the later archaeomagnetic date was accepted.

Part IV
Instrumentation

Chapter 30
Sample Selection and Processing

Processing procedures are specific to the analytical technique utilized but most involve a degree of sample modification. Many techniques were originally designed to examine samples in solution. If this is not possible or otherwise undesirable, a variety of other options exist. Solid sample introduction techniques covered in this chapter include: laser, spark, and arc ablation, microwave digestion, plasma oxidation, pyrolysis and thermal ionization. The precision and reproducibility of the analysis may be diminished when these alternatives are employed. Different approaches to sample selection and processing are described and issues surrounding these decisions will be presented. Issues related to sample selection and processing are discussed in general books on analytical techniques, such as Khandpur (2007), Patnaik (2004), Skoog and West (1982), Skoog et al. (1998), and Willard et al. (1988). Leute (1987), Moens et al. (2000), and Young and Pollard (2000) consider the topic with respect to different archaeological materials.

Sample Solutions

Organic materials are usually dissolved in solvents prior to analysis with most spectroscopic techniques including infrared, ultra violet, Raman, and nuclear magnetic resonance (Chapters 35 and 36). Photoacoustic techniques, such as IR-PAS spectroscopy, can be applied to solids and samples that do not completely dissolve (Chapter 35). High-pressure liquid chromatography is also used to examine substances dissolved in organic solvents (Chapter 33).

Gas chromatography (Chapter 33) involves the analysis of sample components as gases, yet many substances have high boiling points or decompose when exposed to high temperatures. In order to facilitate analysis, it is usually necessary to modify the chemical composition of the sample through a process known as derivatization. Derivatives of lipids are usually formed by replacing the hydrogen of a hydroxyl ($-OH$) group with either a methyl group ($-CH_3$) or a trimethyl silyl (TMS–) group (Chapter 14). The resulting methyl esters or TMS esters and ethers are dissolved in an organic solvent prior to injection.

M.E. Malainey, *A Consumer's Guide to Archaeological Science*, Manuals in Archaeological Method, Theory and Technique, DOI 10.1007/978-1-4419-5704-7_30, © Springer Science+Business Media, LLC 2011

Pyrolysis techniques can be used to introduce samples with low volatility onto a GC column. Pyrolysis is heat-induced decomposition and is typically used for a sample dissolved in an organic solvent or acid. The solution is applied to a heating element or Curie point needle; as the solvent evaporates, a thin film of the sample is deposited. The process is repeated several times to increase the thickness of the sample layer. The heating element or Curie point needle is then inserted into the injection port and heated rapidly. When Curie point needles are used, the sample can be heated by induction to its Curie point in 20–30 ms. The Curie point for nickel is 358°C; for iron it is 770°C and for a 40:60 iron–nickel alloy it is 590°C. The Curie point is the temperature at which a ferromagnetic metal becomes paramagnetic.

Acid Dissolution of Inorganic Material

Atomic absorption (AA or AAS; Chapter 34) and inductively coupled plasma (ICP; Chapter 34) techniques are typically used to determine the composition of crystalline solids that have been dissolved in acid. Only a small portion of the solid is required and it should be examined to ensure that it is free of contaminants prior to dissolution. The acid selected depends upon the material examined. Nitric acid (HNO_3) can be used to dissolve a variety of substances. Tin, tin alloys, lead, lead alloys, pewter, gold, and silver may require a mixture of nitric acid and hydrochloric acid (HCl). Hydrochloric acid alone efficiently dissolves some metals, bone, and paint. Hydrofluoric acid (HF) is required for other minerals, ceramic, or glass but care must be taken when using this powerful acid. Hydrofluoric acid will damage laboratory glassware and the quartz plasma torches in ICP instruments. Young and Pollard (2000) recommend drying down samples dissolved with HF then redissolving the residue in another acid prior to analysis.

During analysis, the acid solution is converted into tiny particles, or aerosol, through a process known as nebulization. The sample droplets are carried into the flame, in the case of atomic absorption, or the plasma torch, in the case of inductively coupled plasma, where atomization takes place.

Analysis of Solid Samples

It is possible to perform compositional analysis of a solid without acid dissolution. The simplest approach is to use one of the many techniques capable of analyzing solids. These include instrumental neutron activation analysis, solid-state nuclear magnetic resonance spectroscopy, X-ray diffraction, X-ray fluorescence, particle-induced X-ray emission, scanning electron microscopy with energy-dispersive X-ray fluorescence, and electron microprobe analysis (Chapters 32, 36, and 37). Samples often only need to be finely ground or powdered prior to analysis.

Other available options depend upon the analytical technique selected. In general, elements in a solid must be converted into gaseous atoms and ions, which is a process known as atomization. If the atomization process is an integral part of the analytical process, a method that simply introduces the solid into the instrument as fine particulate matter and vapors can be used. Otherwise, a technique that simultaneously vaporizes and atomizes the sample must be employed.

Sampling Without Atomization

These sampling techniques simply reduce solid samples to fine particles or vapors without altering the chemistry of the substance. The particulates and/or vapors produced are then carried to an atomizer in a stream of inert gas where they are converted into gaseous atoms and elementary ions. Spark, arc, and laser ablation techniques are commonly used for this purpose; electrothermal vaporization is another option. By using these techniques, it is possible to analyze solid samples using atomic absorption or inductively coupled plasma spectroscopy.

The term "ablation" simply refers to the removal of a substance by mechanical means. With laser ablation, a highly intense focused beam of light precisely removes a tiny portion from the surface of the solid. A laser beam is capable of vaporizing most materials. It is suitable for both conducting and non-conducting solids, so even organic materials can be sampled. Spark and arc ablation techniques use either a spark, produced by a sudden electrical discharge, or an electrical arc bridging two electrodes to remove a portion of the solid. Spark and arc ablation can only be used to sample solids that conduct electricity. Non-conducting solids can be sampled if they are powdered and mixed with a conducting material, such as graphite.

While electrothermal vaporization is typically used for sample solutions, it can be used for certain solids. The sample is placed on or within a rod made from a conducting material such as carbon/graphite or the metal tantalum. When an electric current is passed through the conducting rod, the solid is vaporized by heat. A stream of inert gas carries the vapors to the atomizer.

Ablation and electrothermal vaporization techniques are performed in sample chambers directly connected to an atomizer. In cases where the sample is too large to fit into the chamber and removal of a portion is not an option, laser ablation can be performed using an O-ring to create an airtight miniature sample chamber on a small area of the object (Young and Pollard 2002).

Sampling with Atomization

The techniques described in this section simultaneously remove and atomize a portion of a solid sample. Spark and arc atomizations are direct sample introduction techniques that convert the sample into atoms and ions suitable for analysis. Metal samples can be shaped to form one or both electrodes used to create the spark or arc. To analyze a non-conducting sample, it is mixed with graphite and placed in a cup-shaped electrode. In spark source mass spectrometry, a double focusing mass spectrometer is used to perform mass analysis of the resulting ions. Arc and spark source emission spectroscopies are used to detect energies emitted by excited atoms and ions as they return to ground state.

Glow discharge devices simultaneously introduce and atomize a sample. A glow discharge device is similar to a hollow cathode lamp (Chapter 34), except that the sample serves as the negatively charged cathode. Consequently, the sample must conduct electricity; if it does not, it can be powdered, mixed with graphite or copper, and then shaped into a pellet. The process takes place in a low-pressure argon gas atmosphere. When a dc potential between 250 and 1000 V is applied, the argon gas breaks down into positively charged ions (cations) and electrons. The argon cations accelerate toward and strike the negatively charged sample with sufficient force to dislodge (sputter off) neutral sample atoms. The sample atoms form a vapor suitable for analysis using atomic absorption and mass spectrometry. In addition, optical emission techniques can be used to analyze the energy emitted by excited species returning to ground state.

Laser microprobes can be used to simultaneously remove and atomize a tiny portion of the sample. As described previously, the laser beam vaporizes a minute amount of material from the surface. When using emission spectroscopy, the cloud of sample atoms, ions, and/or molecules is excited between a pair of electrodes. The energy emitted as the excited species return to ground state is analyzed. As described in Chapter 31, the atomic cloud can also be directed into a spectrometer for mass analysis.

Special Topic

Laser: Light Amplification by Stimulated Emission of Radiation

This discussion of how lasers work is largely based on descriptions by Halliday and Resnick (1981) and Skoog et al. (1998). The process of light amplification by stimulated emission of radiation generates an extremely intense beam of highly monochromatic light that is less than 1 μm in diameter; the term "laser" is actually an acronym of the process. The beam is coherent, which means the light is stable and in phase. There is constructive interference and the light forms a wave front that extends across its entire width. Laser light is parallel and can be tightly focused.

Lasers are classified on the basis of the substance, or lasing material, used to generate the light. A variety of lasing materials can be used, including solid crystals, semiconductors, gases, and even organic dye solutions. The common property of these materials is their metastable excited state, which means that an atom can remain in an excited state much longer than average. Most atoms remain in an excited state for about a hundred millionth of a second (10^{-8} s) before spontaneously emitting the excess energy and returning to ground state. Materials with metastable excited states remain so for about a thousandth of a second (10^{-3} s), which is 100,000 times longer. These materials are capable of accumulating and storing excess energy.

Stimulated emission occurs when an external energy source causes an excited atom to emit its excess energy before it would do so spontaneously. The energy of the photon emitted as the excited atom returns to ground state is exactly the same as the triggering (or stimulating) photon. The emission has the same wavelength, is precisely in phase with, and travels in the same direction as the electromagnetic wave that triggered its release. Amplification occurs because the amplitude of the resulting electromagnetic wave is double that of the stimulating photon.

At room temperature, the majority of atoms in a substance are at ground state. If exposed to radiation, absorption occurs and the majority of atoms move into excited states; this is termed a population inversion (Fig. 30.1). Laser light is produced by first creating a population inversion then triggering the release of excess energy in an avalanche of photons. A process called optical pumping is used to create population inversions in the lasing material. Ground state atoms absorb energy from an electrical discharge, electrical current, or an intense, continuous spectrum source and the majority move to short-lived, highly excited states. They release thermal energy and the atoms drop to the metastable excited state.

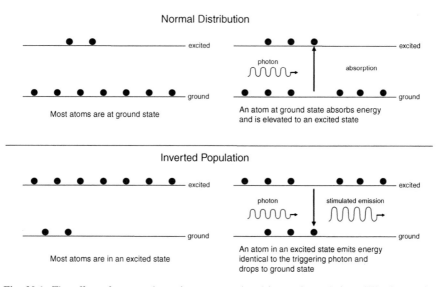

Fig. 30.1 The effect of energy absorption on normal and inverted populations ("Fundamentals of Physics", Second Edition Extended, David Halliday and Robert Resnick, copyright 1981, Reproduced with permission of John Wiley & Sons, Inc.)

Laser light can be produced using either a three- or a four-level laser scheme (Fig. 30.2). Under the three-level scheme, atoms at ground state (E_0) absorb energy from optical pumping and move to various short-lived excited states (E_n). They release thermal energy and move to the metastable excited state (E_2). This causes a population inversion with more atoms in the metastable excited state than at ground state. Collisions with photons having energy equal to the difference between the metastable and the ground states ($E_2 - E_0$) cause the excited atoms to relax. The

(a)   Three-Level Laser Scheme

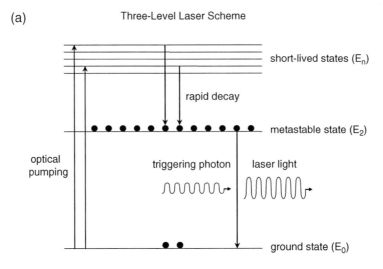

(b)   Four-Level Laser Scheme

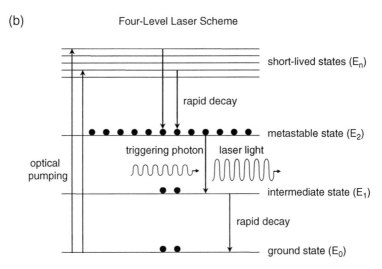

Fig. 30.2 Three- and four-level laser schemes ((**a**) "Fundamentals of Physics", Second Edition Extended, David Halliday and Robert Resnick, copyright 1981, Reproduced with permission of John Wiley & Sons, Inc. (**b**) Adapted from SKOOG. Principles of Instrumental Analysis, 5E. © 1998 Brooks/Cole, a part of Cengage Learning, Inc. Reproduced by permission. www.cengage.com/permissions)

resulting laser light is a combination of both the triggering photons and those from stimulated emissions. In the four-level laser scheme, triggering photons cause atoms to drop from metastable excited state (E_2) to a short-lived intermediate state (E_1) before reaching ground state. Laser light is produced as long as a population inversion exists between E_2 and E_1. Since the energy differential between E_2 and E_1 is smaller than that between E_2 and E_0, less energy from optical pumping is needed to maintain the population inversion.

A common type of solid-state laser consists of neodymium ion in a host crystal of yttrium aluminum garnet and is called a Nd:YAG laser. Helium/neon is a common type of neutral atom gas laser; ion, molecular, and other types of gas lasers are also available. Dye lasers use solutions of organic compounds that fluoresce light in the ultraviolet, visible, or infrared regions.

Chapter 31
Mass Spectrometry

Mass spectrometry is a versatile analytical technique used to determine fundamental properties of both organic and inorganic materials. The technique can be used to establish elemental or molecular composition, the structure of molecules, and isotopic ratios of specific elements. The information desired determines sample preparation procedures prior to analysis, the method of sample introduction into the mass spectrometer, the manner in which it is converted into ions, and how ion masses are analyzed. Detailed descriptions of mass spectrometry can be found in Barker (1999), Becker (2007), Beynon (1960), Ebsworth et al. (1987), and Watson and Sparkman (2007). More general outlines are available in books on analytical techniques, such as Hammes (2005), Khandpur (2007), Patnaik (2004), Pasto and Johnson (1979), and Skoog et al. (1998). Basic theoretical principles of mass spectrometry are described in general chemistry and physics texts, such as Halliday et al. (2005) and Mortimer (1986). General descriptions and examples of archaeological applications of mass spectrometry combined with chromatographic (Evershed 1992a, 1993b, 1994, 2000; Hites 1997) and inductively coupled plasma (Young and Pollard 2000) techniques are also available.

Mass spectrometers are widely used for the analysis of a variety of archaeological materials. The technique is employed when the abundance of radioactive isotopes in a sample must be known to obtain an age estimate. Accelerator mass spectrometer (AMS) radiocarbon dating is described in Chapter 8; potassium–argon, argon–argon, and uranium series dating are described in Chapter 9. The composition of a wide variety of archaeological materials can be determined, as element concentrations or isotope ratios, and provide information about diet, migration patterns, provenance, trade patterns, artifact function, climate, and metallurgy (see Chapters 12 and 13). Using the depth profiling capability of secondary ion mass spectrometry (SIMS), it is possible to monitor changes in composition that occur between the surface and bulk of a sample. This feature enables the precise measurement of the thickness of the hydration rind for obsidian hydration dating (Chapters 11 and 18) and minimally destructive stable isotope analysis for provenance studies (Chapters 12 and 18).

By employing solid sample introduction and ionization techniques, such as thermal ionization, secondary ion, or laser ablation, a minuscule amount is removed

M.E. Malainey, *A Consumer's Guide to Archaeological Science*, Manuals in
Archaeological Method, Theory and Technique, DOI 10.1007/978-1-4419-5704-7_31,
© Springer Science+Business Media, LLC 2011

from an artifact and sent directly into the mass spectrometer. Mass spectrometers can be combined with other instruments in ways that enhance the sensitivity and selectivity of the analyses. Mass spectrometers linked to inductively coupled plasma torches (ICP-MS) provide highly sensitive elemental analyses of samples dissolved in acids; by using laser ablation (LA-ICP-MS), acid dissolution of solid samples is avoided. When linked to either a gas chromatograph (GC-MS) or high-pressure liquid chromatograph (LC-MS), complex mixtures of organic compounds are separated into their individual components prior to their introduction into the mass spectrometer. Hybrid instruments have recently been developed that enable the ratio of stable carbon isotopes (GC-C-IRMS) and the amounts of radioactive ^{14}C in individual components separated by gas chromatography to be determined.

Analysis of a sample using mass spectrometry involves three steps: ionization, mass analysis, and ion detection. The specific process employed at each of these steps depends upon the nature of the sample and goals of the analysis.

Ionization

In order for a sample to be analyzed, it must first be converted into ions (charged particles). These ions may represent a single atom with an electrical charge (elemental ion), a single molecule with an electrical charge (molecular ion), or fragments of the molecular ion. The ions are sorted according to their mass-to-charge ratio (m/z) and the number of each is determined. The m/z of ions carrying a single positive charge is its mass. Mass is measured in atomic mass units (amu) or Daltons (Da), which is defined as precisely 1/12th the mass of a ^{12}C atom. To enhance the stability of ions during analysis, mass spectrometry must be performed in a vacuum.

Techniques for Molecular Mass Spectrometry

Molecular mass spectrometry is primarily used to study organic matter. Organic compounds largely consist of carbon and hydrogen; oxygen and nitrogen are often present and other elements may also appear. Properties of an organic substance are determined by the structure of its molecules, which is the arrangement of atoms within the compound. Mass spectrometry can be directly performed on pure substances. Chromatography is often used to separate different components in complex mixtures of substances prior to analysis (see Chapter 33).

Molecular mass spectrometry involves the study of molecules and the fragments formed when molecules break apart. Molecular ions and fragments of molecular ions are created within the ion source of the instrument using either hard or soft ionization techniques. Hard ionization causes most molecules to break apart. Soft ionization involves giving the molecule a positive or negative charge without destabilizing it; the molecular ion remains intact so that its mass can be determined.

Either hard or soft ionization techniques can be employed for the analysis of components separated using gas chromatography. The linkage of these instruments is straightforward because a gas chromatography column can be directly inserted into the mass spectrometer. Separated sample components are carried from the gas chromatograph into a mass spectrometer in a stream of helium gas. The combination of high-pressure liquid chromatography and mass spectrometry is more complex and described below. When analyzing molecular ions and fragments thereof, the relative abundance of each ion is given with respect to the most abundant ion present. Identifications are based on the relative proportions of the molecular ions and the different fragments that form.

Electron Impact

The most common ionization process is electron impact (EI), which is a type of hard ionization (Fig. 31.1). The sample is first heated to a high temperature, creating a molecular vapor. Electrons emitted from a metal (often tungsten) filament in the ion source are accelerated toward a positively charged anode to energies between 50 and 100 V; 70 V is commonly used. The molecular vapor passes through the ion source and individual sample molecules collide with high-energy electrons at right angles. The energy of each electron in the beam is sufficiently high to strip an electron from

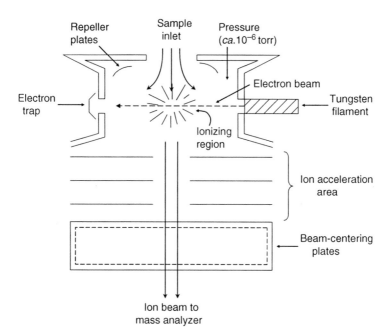

Fig. 31.1 Electron impact ionization (Reprinted from James Barker, Figure 2.1a, "Mass Spectrometry", 2nd ed. copyright 1999, with permission of Wiley-Blackwell)

the sample molecule, which results in the formation of a molecular ion. A molecular ion, which is also called a radical cation, is usually unstable and breaks into smaller fragments:

$$M + e^- \rightarrow [M]^+ + 2e^- \rightarrow A^+ + B^+ + etc.$$

The fragments carrying a single positive electrical charge and any surviving molecular ions are accelerated into the mass analyzer where they are sorted according to their m/z. Since molecules fragment in a consistent manner, it is often possible to deduce the structure of the compound through the nature of the fragments produced. Sample spectra are then compared to those in computerized databases or derived from reference materials. The statistical probability of a certain fragmentation pattern occurring forms the basis of identification. In the case of pure substances, identifications can be made with a very high degree of certainty.

Soft Ionization Techniques

The mass and information about the structure of the molecule itself can be determined using soft ionization techniques that produce stable molecular ions. Positive or negative molecular ions are formed through reactions of the sample with ionized reagent gases or electrons from the reagent gases. The production of positive molecular ions is called positive ion chemical ionization (PCI) or simply chemical ionization (CI); the production of negative molecular ions is called negative ion chemical ionization (NICI).

Positive chemical ionization typically involves reagent gas ions formed from methane, ammonia, or isobutene. The reagent gas is introduced into the ion source where it is bombarded with a beam of electrons. Occasionally, the impact removes an electron resulting in reagent gas ions carrying a single positive charge. Reagent gas ions can react with sample molecules in different ways to produce positively charged sample ions of different masses. Reactions involving the transfer of protons increase the mass of sample molecules by 1 amu $(M + 1)^+$. Removal of hydrogen atoms reduces the mass of sample molecules by 1 amu, $(M - 1)^+$. The addition of reagent gas ions to sample molecules substantially increases its mass. For example, the transfer of $C_2H_5^+$, produced from methane, to a sample molecule increases its mass by 29 amu, $(M + 29)^+$.

Negative chemical ionization involves either negatively charged reagent gas ions or slow-moving, low-energy (thermal) electrons produced from the reagent gas. Reactions with sample molecules cause the formation of negatively charged molecular ions. Ammonia is commonly used as the reagent gas.

Techniques Used for Liquid Chromatography with Mass Spectrometry (LC-MS)

Mixtures of compounds with high boiling points or damaged by temperature are not suitable for gas chromatography; but separations can be achieved through the use of high-pressure liquid chromatography (HPLC). The sample dissolved in the organic solvent mobile phase is forced through a column filled with solid particles of the stationary phase under high pressure. Recently developed atmospheric ionization techniques are the best way to ionize sample components separated by high-pressure liquid chromatography. They involve the removal of excess solvent, accommodation of the dramatic pressure change and sample ionization. Electrospray ionization, atmospheric pressure chemical ionization and atmospheric photoionization are described.

With electrospray ionization, the separated sample component and solvent mobile phase (eluent) exit the column under high pressure and enter a chamber that is at atmospheric pressure (Fig. 31.2). The pressure difference causes the sample component/solvent mixture to spray into tiny droplets (or nebulize), which are then exposed to a strong electrostatic field (4000 V) and a heated drying gas. Droplets become highly charged, creating both positively and negatively charged ions. As solvent is lost through evaporation, the size of each droplet is reduced so the distance between ions is smaller. Ions carrying the same charge are repelled, which causes sample ions, carrying one or more positive charges, to be ejected (or desorbed) from the droplet. These ions are then drawn into the mass spectrometer for mass analysis.

With atmospheric pressure chemical ionization (APCI), mobile-phase solvent is used for the chemical ionization of sample molecules. The eluent is heated to a

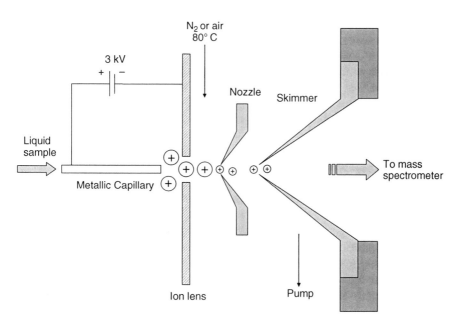

Fig. 31.2 Electrospray ionization (Reprinted from Johanna S. Becker, Figure 2.36, "Mass Spectrometry: Principles and Applications", copyright 2007, with permission of Wiley-Blackwell)

very high temperature as it enters the chamber, which causes the solvent to vaporize. Electrons, from a corona discharge needle, ionize the solvent gas, which in turn ionizes the molecules of the separated sample component. With atmospheric pressure photoionization, the eluent is also vaporized as it enters the chamber. Sample component molecules are then ionized by photons emitted from an ultraviolet lamp. The energy of the photons is specifically selected to maximize ionization of sample molecules and minimize ionization of the solvent.

Techniques for Atomic Mass Spectrometry

Atomic mass spectrometry is used to analyze elemental ions to precisely determine which elements are present and their exact proportions. The technique is primarily used to establish the composition of inorganic matter. Ratios of different isotopes of an element can also be determined using atomic mass spectrometry.

The sample is converted to atoms (atomized) and then ionized to elemental ions, usually carrying a charge of +1. The production of elemental ions often occurs outside of the mass spectrometer. This can be achieved through thermal ionization, secondary ionization, and the use of spark sources and laser microprobes. As described in Chapter 34, inductively coupled plasma torches are excellent ion sources and ICP-MS is among the most sensitive instruments ever developed.

Secondary Ion Mass Spectrometry (SIMS)

Secondary ion mass spectrometry is used to determine the atomic and molecular composition of solid samples. The advantage of this approach is that solids can be analyzed without first being powdered, which is required for instrumental neutron activation analysis, or dissolved in acid, which is required for atomic absorption and standard inductively coupled plasma techniques. SIMS can be used to examine the composition of the outermost surface of the sample or to study changes in composition with depth.

A narrow beam of high-energy ions from an ion gun is used to atomize and ionize the sample. These primary ions, typically Ar^+, Cs^+, N_2^+, and O_2^+, are formed within the ion gun by electron impact, which was described earlier. A high dc potential accelerates the 0.3–5 mm diameter beam of primary ions to energies of 5–20 keV. Most of the sample atoms removed (sputtered off) are neutral but a small fraction occurs as positively or negatively charged secondary ions. These charged particles are directed into the spectrometer for mass analysis.

The composition of the surface is determined by analyzing the uppermost layers of atoms from the sample. In order to obtain information about changes in composition with depth, the sputtering process is continued. A small crater or pit forms as subsequent layers of atoms are removed and analyzed from one spot. The record of changes in elemental composition is called a depth profile.

A more sophisticated (and hence more expensive) variation of this technique involves the use of an ion microprobe. The beam of primary ions can be more tightly focused, down to a point only 1–2 μm in diameter. The position of the beam is adjustable and can be moved across the surface about 300 μm in the *X*- and *Y*-directions. A microscope is used to precisely select the area of the sample to be analyzed.

Laser Microprobe Mass Spectrometry (LMMS) or Laser Microprobe Mass Analysis (LAMMA)

Two slightly different names are used to describe the technique that uses a laser to remove and ionize material from a solid for mass analysis. A low-power laser is used to select a point on the surface of the sample for analysis. The diameter of the pulsed high-power laser beam is less than 1 μm, so sample inhomogeneity can affect the analysis. For this reason, either a homogeneous portion of the matrix or a particular inclusion within the matrix is targeted.

The high-power energy beam, often from a Nd:YAG laser, is used to sample the solid specimen. A small amount of material undergoes ionization and volatilization, which creates a plume of atoms, ions, and molecules. The ions are drawn into the spectrometer for mass analysis. This technique is suitable for both organic and inorganic materials. Lasers are described in Chapter 30.

Thermal Ionization Mass Spectrometry (TIMS)

Thermal ionization mass spectrometry uses temperature to ionize certain inorganic samples for mass analysis. The sample is first dissolved then the solution is applied to a metal filament (or ribbon), usually made of either rhenium (Re) or tungsten (W). As the solvent evaporates, a thin layer of solid sample material is deposited on the filament. The filament is placed within the ion source and evacuated. An electric current is passed through the filament that heats the applied sample to a very high temperature (typically 1000–2500 K for Re), which causes the sample to evaporate and ionize without affecting the filament. The sample ions generated are subjected to mass analysis.

The method described above employs a single metal filament; however, double filament thermal ionization can be performed. In this arrangement, the sample is applied to one of the filaments, which is heated to a temperature just high enough to cause the sample to evaporate. Ionization occurs when the sample vapors interact with the other filament, which is heated to a higher temperature. Both the evaporation and the ionization processes are optimized when two filaments are used.

Mass Analysis

Ions are charged particles that can be separated according to mass through the use of magnetic and electric fields. In the absence of these fields, ions possessing the same kinetic energy can be separated because lighter ions move faster than heavier ones. The ability of a mass spectrometer to separate ions of different masses is called its resolution. A mass spectrometer with a resolution of 500 can separate ions with *m/z* values of 500 (or 50.0 or 5.00) from those with *m/z* values of 501 (or 50.1 or 5.01). By contrast, mass spectrometers with a resolution of 500,000 could separate ions with *m/z* values of 50.0000 from those with *m/z* values of 50.0001. Magnetic sector, quadrupole, ion trap, and time-of-flight mass analyzers are described in this section.

Magnetic Sector

The simplest means of mass separation is achieved through the use of magnetic fields alone (Fig. 31.3). If the strength of a homogeneous magnetic field is constant, all ions having the same charge, mass, and energy follow identical circular paths. The precise trajectory of the ions depends upon their *m/z* value. Because they are deflected more by the magnetic field, the path of a lighter ion is an arc with a smaller radius; the path radius is larger for heavier ions. If the strength of the magnetic field is increased, the radii of all ion paths decrease. Separation achieved on the basis of differential deflection utilizes the prism effect of magnetic fields.

The focusing effect of a magnetic field on an ion beam is similar to that of a convex lens. Ions with the same mass emerging from a narrow entrance slit will diverge and then refocus at another point; the precise position of the focusing point depends on the strength of the magnetic field. By varying the magnetic field strength, the lens effect is used to direct ion beams separated on the basis of mass through a fixed exit slit and to the detector. Mass spectrometers that rely on magnetic fields alone to separate ions are called single-focusing magnetic sector instruments. Many stable isotope ratio analysis (SIRA) mass spectrometers are elemental analyzers (Chapter 38) linked to single-focusing magnetic sector instruments.

Good mass separation is achieved when the separated ion beams are tightly focused; this is only possible if all ions enter the magnetic field with the same energy. In a double-focusing magnetic sector instrument, resolution of the mass

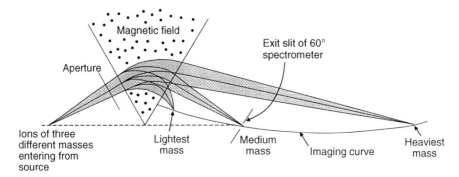

Fig. 31.3 The movement of ions in a magnetic field (Reprinted from Hermann Kienitz, Abb. B 51, "Massenspektrometrie", copyright 1968, with permission of Wiley-VCH Verlag as modified by Johanna S. Becker, Figure 3.3, "Mass Spectrometry: Principles and Applications", copyright 2007, with permission of Wiley-Blackwell)

spectrometer is increased by the incorporation of the electrostatic analyzer (or electric sector field) to ensure that ions enter the magnetic field with a narrow range of energy. The prism and lens effects of a radially symmetrical electrostatic field are similar to a homogeneous magnetic field. If a beam of ions passes through an entrance slit into an electrostatic field, they will separate on the basis of their kinetic energy. The circular path of ions with the least kinetic energy will have the smallest radius; the circular path of those with the most kinetic energy will have the largest radius. Each separated ion beam diverges as it enters the electrostatic field and is then refocused. Since the ions enter the magnetic field with a narrow range of energies, their paths are governed by mass alone.

Double-focusing magnetic sector instruments usually have either a Nier–Johnson or Mattauch–Herzog configuration. The ion beam entering a mass analyzer with the Nier–Johnson geometry is bent in the same direction by both the electrostatic analyzer and the magnetic sector before entering the detector (Fig. 31.4). The ion beam entering an instrument with the Mattauch–Herzog geometry is bent in one direction by the electrostatic analyzer and bent in the opposite direction by the magnetic sector.

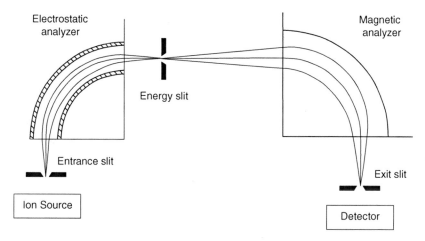

Fig. 31.4 Double-focusing mass spectrometer with Nier–Johnson geometry (Adapted from SKOOG. Principles of Instrumental Analysis, 5E. © 1998 Brooks/Cole, a part of Cengage Learning, Inc. Reproduced by permission. www.cengage.com/permissions)

Configurations of accelerator mass spectrometers are more complex and can vary. One arrangement involves two magnetic sectors, one low and the other high energy, separated by a tandem accelerator; the ion beam is passed through an electrostatic analyzer before reaching the detector. In a slightly different configuration, the beam of sample ions emerging from the ion source is passed through an electrostatic analyzer prior to entering the low-energy magnetic sector field (see Chapter 8).

Quadrupole

Quadrupole mass analyzers consist of four parallel rods, the ends of which form the shape of a diamond, spaced equal distances from a central point (Fig. 31.5). The ends of one pair of rods are oriented horizontally in the same xz plane; the ends of the other pair are oriented vertically in the same yz plane. All sample ions are accelerated through the channel between the rods, but only those ions with specified m/z values are allowed to reach the detector. Ions outside of the specified m/z range collide with the rods and are converted to neutral molecules. The upper range of ions able to pass through the quadrupole is set by the low-pass mass filter; the lower range is set using the high-pass mass filter.

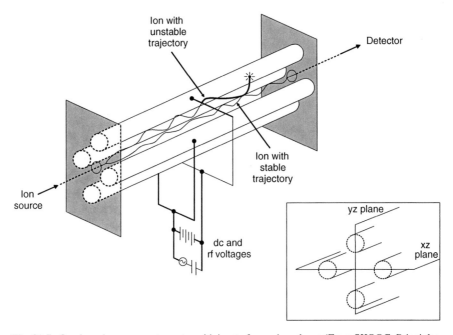

Fig. 31.5 Quadrupole mass spectrometer with inset of xz and yz planes (From SKOOG. Principles of Instrumental Analysis, 5E. © 1998 Brooks/Cole, a part of Cengage Learning, Inc. Reproduced by permission. www.cengage.com/permissions)

The four parallel rods of the quadrupole are electrically connected in two pairs. One pair is connected to the positive terminal of a variable direct current source and the other connected to the negative terminal. Radio frequency alternating currents are also applied to each pair of rods 180-degrees out of phase. The voltage of the applied direct current is constant while the alternating current oscillates. The trajectories of ions moving through the quadrupole are completely controlled by adjusting the ac and dc potentials. By varying the electric field, ions of different m/z values are allowed to pass into the detector.

Alternating currents have both positive and negative cycles. During positive cycles, positively charged ions are repulsed and move toward the center of the channel between the rods. During negative cycles, positively charged ions are attracted and deflected toward the rods. The pair of rods connected to the positive dc terminal acts as a high-pass mass filter for positively charged ions traveling in the *xz* plane. Heavier ions are more difficult to deflect and pass between the rods while lighter ions collide with the rods during the negative cycle of the alternating current. The pair of rods connected to the negative dc terminal acts as a low-pass mass filter for positive ions traveling in the *yz* plane. Light ions are more easily manipulated during the positive cycle of the alternating current and pass through the space between the rods while heavy ions tend to collide with the negative rods.

Triple quadruple mass spectrometers consist of three quadrupoles arranged linearly and are used to gain more information about the sample. Ions selected on the basis of mass by the first quadrupole collide with a gas in the second quadrupole and undergo collision-induced dissociation. The different ions created as a result of these collisions undergo mass analysis in the third quadrupole prior to entering the detector.

Ion Trap

Ion traps are compact devices that use constant and oscillating electric fields to control the movement of ions. The ion trap consists of two end cap electrodes and a doughnut-shaped ring electrode to which variable radio frequency voltage is applied (Fig. 31.6). Ions enter the trap through a hole in one-end cap and circulate within the electric field in stable three-dimensional orbits. A voltage increase is used to destabilize the orbit of ions with specific *m/z* values, which are then ejected through the other end cap to the detector.

Selected ion storage (SIS) is possible with an ion trap mass analyzer. This allows the analyst to selectively accumulate or store a limited number of ions with the trap. To maximize the number of desired ions, unwanted ions are ejected from the trap. It is possible to select ions having one or more specific masses using either single-ion or multiple-ion modes. Alternatively, ions within specific mass ranges can be selected by defining low-mass and high-mass cut-offs.

Time of Flight

Time-of-flight (TOF or ToF) mass analyzers separate ions on the basis of the difference in time required for them to travel through a 1–2 m long drift tube connecting the ion source and accelerator to the detector (Fig. 31.7). No electrical or magnetic fields are applied to the drift tube so it is referred to as the field-free region. If all the ions begin the trip through the drift tube at precisely the same time and possess the same amount of energy, the difference in time will depend only on ion mass. The lightest ions will arrive first followed by progressively heavier ions.

In order to control the timing of the start of the journey, the production of ions is not continuous. Instead, a short ionization pulse from an electron (or other particle)

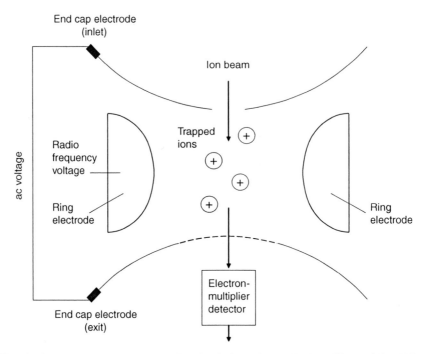

Fig. 31.6 Ion trap mass analyzer (Reprinted from James Barker, Figure 3.6a, "Mass Spectrometry", 2nd ed. copyright 1999, with permission of Wiley-Blackwell)

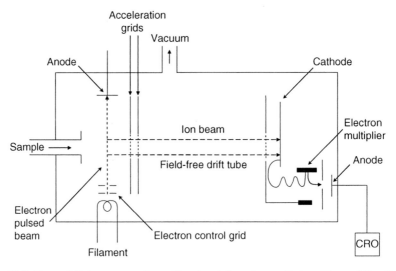

Fig. 31.7 Time-of-flight mass analyzer (Reprinted from James Barker, Figure 3.7a, "Mass Spectrometry", 2nd ed. copyright 1999, with permission of Wiley-Blackwell)

gun is used to produce a small packet of sample ions, which is then accelerated by an electric field pulse. Upon exiting the ionization and acceleration region, the kinetic energy of the ions is virtually identical. In a linear TOF instrument, lengthening the drift tube and reducing the kinetic energy ions possess as they enter the tube improves mass resolution.

Another way to improve the resolution of TOF instruments involves using an ion reflector (or reflectron) to alter the paths of sample ions with different masses. The reflectron generates an electric field that slows down sample ions in the drift tube and then deflects and accelerates them toward the detector. Heavier sample ions penetrate deeper into the reflectron than lighter ones before being deflected, which lengthens the distances heavier ions travel and increases the separation time between ions of different mass.

Ion Detection

The detection of ions emerging from the mass analyzer generally involves either counting the positive charges or converting them into negative charges and multiplying the number of electrons to amplify the signal.

Faraday Cup

A Faraday cup is a device that records impacts of positive ions emerging from the mass analyzer on a collector electrode (or collision surface). Both the collector electrode and its surrounding cage are connected to ground potential through a resistor (Fig. 31.8). Whenever a positive ion is neutralized, a voltage drop occurs across the resistor and the resulting signal is amplified and recorded.

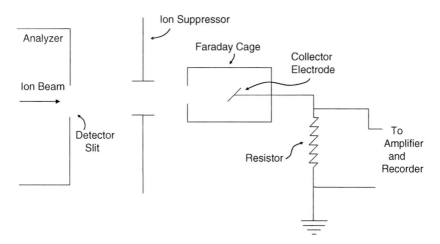

Fig. 31.8 Faraday cup detector (Reprinted from G.R. Waller, ed., Figure 2.2, "Biochemical Applications in Mass Spectrometry", copyright 1972, with permission of John Wiley & Sons, Inc.)

Positive ions strike the collision surface at an oblique angle so that reflected ions and any secondary electrons produced are directed toward the walls of the cage. An ion suppressor is positioned between the Faraday cup and the mass analyzer to prevent secondary electrons from escaping.

Electron Multiplier

Two types of electron multipliers are used, discrete dynode and continuous dynode (Fig. 31.9). In both cases, positive ions emerging from the mass analyzer are converted to electrons, the numbers of which are exponentially multiplied and then the signal is directed to a recording device.

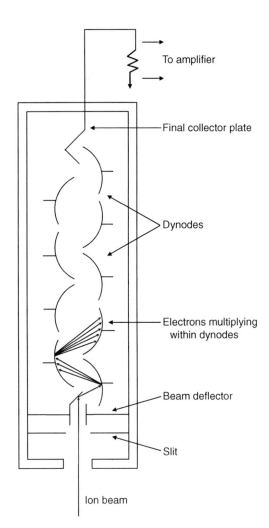

Fig. 31.9 Discrete dynode electron multiplier. (Adapted from Barker (1999) and Becker (2007), with permission. Reprinted from James Barker, Figure 4.1a, "Mass Spectrometry", 2nd ed. copyright 1999, with permission of Wiley-Blackwell. From Johanna S. Becker, Figure 4.2, "Mass Spectrometry: Principles and Applications", copyright 2007, with permission of Wiley-Blackwell.)

A discrete dynode type is similar to the photomultiplier detectors used in optical spectroscopy (see Chapter 35). This type of electron multiplier consists of ten or more dynodes, either coated with copper–beryllium or constructed of lead-doped glass, held at progressively higher voltages. The first is a "conversion dynode" that emits electrons when struck by the positively charged ions, which are accelerated toward the second dynode. Secondary electrons are generated when the electrons emitted by the conversion dynode strike the second dynode and all are accelerated to the third dynode. This "cascading effect" causes the total number of electrons to increase with each successive dynode.

Continuous dynode electron multipliers, which are also called channel electron multipliers or channeltrons, employ a single horn-shaped dynode constructed of lead-doped glass connected to a power supply (Fig. 31.10). Positive ions enter the wide end of the horn; when they strike the walls, the electrons emitted are reflected deeper into the horn. Secondary electrons are emitted each time the electrons strike the wall and all continue deeper into the detector. As the horn narrows, the frequency with which the electrons strike the walls increases, resulting in a million-fold amplification.

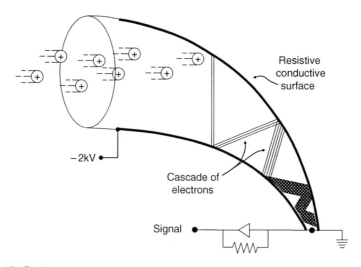

Fig. 31.10 Continuous dynode electron multiplier (Reprinted from J. T. Watson and O. D. Sparkman, Figure 2-57, "Introduction to Mass Spectrometry: Instrumentation, Applications and Strategies for Data Interpretation", 4th edition, copyright 2007, with permission of John Wiley & Sons, Inc.)

Faraday cups are considered to be precise and accurate detectors, but are less sensitive than electron multipliers, while electron multipliers are more sensitive but less precise. Some detectors combine a Faraday cup and electron multiplier. Multiple ion collection systems use several Faraday cups or electron multipliers to detect different ion beams simultaneously.

Chapter 32
Instrumental Neutron Activation Analysis (INAA or NAA)

Instrumental neutron activation analysis (INAA) or simply neutron activation analysis (NAA) is used to detect and quantify elements in inorganic substances. The technique is extensively employed for provenance studies of archaeological lithics, ceramic, and glass (Speakman and Glascock 2007). By targeting emissions from specific isotopes, simultaneous analysis of major, minor, and trace elements is possible. The detection limits for most elements are in the order of parts per million; sample sizes range from about 30 mg to a few grams. Usually samples are powdered prior to analysis but small archaeological samples have been analyzed whole.

Samples are placed in vials which then undergo neutron irradiation in the core of a nuclear reactor, where they are converted into specific radioactive isotopes or radionuclides. After a period of cooling, gamma ray emissions from the decay of these radionuclides are measured with a germanium (-lithium) detector. Specific elements are detected by either the frequency or the energy of these gamma rays. The strength of the signals is related to their concentration in the sample. Irradiations of short duration are required to study isotopes with short half-lives; isotopes with longer half-lives must be irradiated for a longer period.

Descriptions of INAA are presented in some books on analytical techniques, such as Skoog et al. (1998) and Verma (2007). Several overviews of the technique directed to archaeologists and geologists are available (Dostal and Elson 1980; Gibson and Jagan 1980; Haskin 1980; Neff 2000).

Radionuclide Formation and Emissions

Instrumental neutron activation analysis is used to determine the elemental composition of a sample by converting stable (non-radioactive) atoms into radioactive isotopes with unstable nuclear configurations. These newly formed radionuclides decay and emit gamma rays with energies or frequencies characteristic of that element. The number of emissions is related to the concentration of the element in the sample.

In order to produce radioactive isotopes, the sample is usually irradiated with neutrons. Nuclear reactors produce neutrons during the controlled fission of uranium (^{235}U). Within its core, a reactor can have a neutron flux between about 10^{11} and 10^{14} neutrons/cm^2/s. Since there is such an abundance of neutrons, radioactive isotopes are formed from a wide variety of sample elements. The detection limit of neutron activation analysis is very high because gamma ray emissions from many different elements can be monitored simultaneously. If a source of neutrons is not available, it is possible to perform this type of analysis using small charged particles, such as hydrogen, deuterium, and helium ions, from a particle accelerator (Chapter 3).

Neutrons within the reactor core are free and highly reactive with energies of about 2 MeV. These "fast" neutrons are too energetic for neutron activation. Collisions between fast neutrons and nuclei occur with such force that alpha particles or protons are ejected. The loss of protons results in the formation of new elements with lower atomic numbers. Only a low-energy collision is required for neutron activation. In order to make them suitable, the highly energetic neutrons are passed through a moderator, consisting of water, paraffin, or graphite. Fast neutrons are slowed by colliding with atoms of the moderator and become low energy, slow, or thermal neutrons with energies of less than 0.5 eV. A few neutrons with moderate energy (0.5 eV to 1 MeV), called epithermal neutrons, are also formed; these account for only a small portion of neutron activations.

Prepared samples undergo irradiation by thermal neutrons within the nuclear reactor. Since they do not carry a charge, neutrons approach sample atoms without electrostatic repulsion. The nuclei of sample atoms are able to capture the neutral particles. Most of the radioisotopes used in activation analysis are produced by neutron–gamma (n, γ) neutron capture reactions, where the capture of a neutron results in the emission of gamma rays. The newly formed isotope has the same atomic number but mass number is greater by one:

$$^{23}_{11}\text{Na} + {}^{1}_{0}\text{n} \rightarrow {}^{24}_{11}\text{Na} + \gamma\,(\textbf{prompt})$$

Certain neutron–proton (n,p) reactions, where neutron capture results in the emission of a proton, are also monitored.

The process of neutron capture adds 8 MeV of energy, so newly formed radionuclides are highly excited. Almost immediately, they expend excess energy, or relax, through the emission of prompt gamma radiation. Prompt gamma rays from the highly radioactive samples are usually not studied; instead, samples exiting the reactor core are directed into a lead container. After the prompt radiation dies down, the gamma ray emissions associated with the decay of radionuclides are counted.

Newly formed isotopes are radioactive because neutron capture reactions upset the ratio of protons to neutrons in the nucleus. Balance is restored by converting a neutron into a proton through the emission of a beta particle. In order to reach ground state, excess energy is also released through the emission of gamma rays. The gamma rays emitted have characteristic energies, which establish the presence of particular elements.

Sample Preparation and Analysis

Steps involved in instrumental neutron activation analysis include sample preparation, irradiation, cooling, and counting. Unlike techniques such as atomic absorption (AA or AAS) or inductively coupled plasma-atomic emission spectroscopy (ICP-AES) (Chapter 34), instrumental neutron activation analysis is used to study solids. While it is necessary to grind them into a fine powder, samples are not dissolved. Only a small sample is required, typically ranging from less than 100 mg to a few grams. Geological samples consisting mainly of silicon, aluminum, and oxygen are especially well suited for analysis. Neutrons easily penetrate the matrix and radionuclides freely emit gamma rays. The matrix of an artifact containing gold, silver, or a high amount of boron and cadmium, inhibits both neutron irradiation and gamma ray emissions, making it difficult to analyze with INAA.

A sample removed from an artifact is first inspected under a microscope to ensure it is free of contaminants, especially if it was taken with a metal drill bit. After grinding, the powder is precisely weighed then sealed within a small capsule made of either high-density polyethylene or high-purity quartz.

Samples and standards are carried inside the core of the nuclear reactor for irradiation by pneumatic tubes or another delivery system. The time of irradiation depends upon the species of interest but is generally three to five times the half-life of the targeted elements. It can range from a few seconds to several hours. When they emerge from the reactor, samples are highly radioactive, due to prompt gamma ray emissions. Depending upon the irradiation time, cooling periods may range from a few minutes to several days. During cooling, short-lived interferences decay away and health hazards to analysts are reduced. If desired, various sample components can be separated prior to counting.

Counting involves the use of a radiation detector to analyze gamma radiation emitted by the sample (Fig. 32.1). Each gamma ray that strikes the detector produces a pulse of energy, which is counted. A spectrometer determines the energy or frequency of the gamma ray. Germanium (or germanium–lithium) detectors have been used for this purpose for more than 30 years.

Samples with complex compositions generally require two separate irradiations and three counts. The first irradiation is used to target isotopes with short half-lives, ranging from a few minutes to several hours. Samples are irradiated for a very short time, perhaps only a few seconds, so only a brief cooling period is required. The second irradiation is much longer, usually lasting several hours or a full day. Samples are extremely radioactive and require extended cooling periods. Two counts are made to measure isotopes with longer half-lives.

For example, samples that have undergone a 5-s long irradiation can be counted for 720 s (12 min) after 25 min of cooling (Neff 2000). Neff (2000) indicates that the presence and amount of aluminum, barium, calcium, dysprosium, potassium, manganese, sodium, titanium, and vanadium can be determined using this procedure. Following a 24-h long irradiation and week-long cooling period, the first count period targets isotopes with half-lives between about 24 h and 11 days. Elements analyzed during this count, which lasts about one-half hour, include

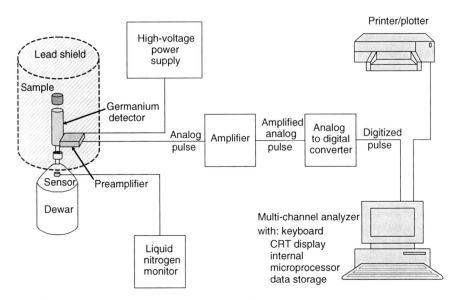

Fig. 32.1 Germanium–lithium detector (From SKOOG. Principles of Instrumental Analysis, 5E. © 1998 Brooks/Cole, a part of Cengage Learning, Inc. Reproduced by permission. www.cengage.com/permissions)

arsenic, lanthanum, lutetium, neodymium, samarium, and ytterbium. Four weeks after irradiation, a second count, with a duration of about 2.75 h, is made. Emissions from antimony, cerium, cobalt, chromium, cesium, europium, strontium, tantalum, terbium, zinc, and zirconium are detected; half-lives of these elements range from about 1 month to 30 years (Neff 2000).

Identification and Quantification of Elements

Important factors for neutron activation analysis include the size of the sample, the abundance of targeted elements within the sample, and neutron flux in the reactor. As these increase, the probability of neutron capture reactions involving targeted elements also increases. Not all elements in the periodic table can be analyzed through neutron activation analysis, however. In order for an element to be detected, neutron capture must result in the formation of a radioactive isotope that decays with a half-life suitable for monitoring and its gamma ray emission must be relatively free of interference from other sources.

Neutron capture reactions involving the most abundant isotopes of certain elements, such as potassium, produce stable isotopes. When this happens, it is necessary to monitor emissions from less abundant radioactive isotopes. Calculations of the amount of element in the sample are adjusted according to the isotopic abundance of the radionuclide used. In some cases, no radionuclide suitable for neutron activation analysis is produced by neutron capture reactions.

The appearance of peaks in a gamma ray spectrum depends upon radionuclide formation and varies with irradiation and decay times. Isotopes with half-lives less than about 2 min long are not suitable due to the extent of their decay prior to the counting period. Isotopes with half-lives longer than 30 years are not suitable for neutron activation analysis because they are less likely to decay and produce detectable gamma ray emissions during the counting period. Lengths of irradiation, cooling, and counting periods are selected to optimize the detection of targeted elements.

The decay of radionuclides may be simple or complex. The simplest decay schemes involve the formation of a stable daughter product through emission of a single beta particle and associated gamma ray with a characteristic frequency. Complex decay schemes may involve daughter products in different excited states that emit gamma rays of different frequencies. In these cases, usually one gamma ray frequency will most accurately reflect the amount of element and be used to calculate its concentration in the sample.

Germanium–lithium [Ge(–Li)] detectors are typically used to monitor gamma ray emissions. Gamma spectra consist of peaks indicating the energy of detected emissions, plotted along the X-axis, and the number of emissions counted, plotted along the Y-axis. Energies monitored usually range between 0 and 3300 keV; if 8192 channels are available, each channel is 0.403 eV wide (Fig. 32.2). Even so,

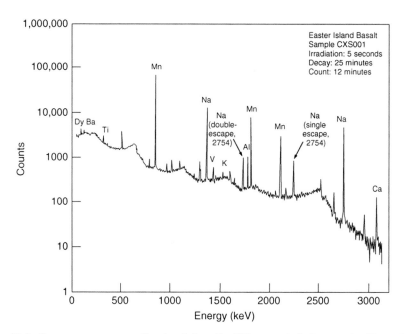

Fig. 32.2 Gamma ray spectrum (Reprinted from E. Ciliberto and G. Spoto, eds., Figure 5.1, "Modern Analytical Methods in Art and Archaeology", copyright 2000, with permission of John Wiley & Sons, Inc.)

gamma ray emissions from different isotopes may completely or partially overlap. It may be necessary to use a different gamma ray emitted by the targeted radioisotope to determine the amount present or precisely calculate and subtract the amount of interfering element present. If not recognized, signals representing the sum of two overlapping emissions could lead to inaccurate assessments of elemental composition. Calibration standards of known elemental composition are used to simplify the determination of absolute amounts of elements present. Correction factors must be applied to samples containing high levels of uranium as the irradiation can cause ^{235}U to fission and produce radioactive isotopes.

Using the two irradiation and three-count approach, more than 33 elements can usually be detected. The advantages of INAA include its high sensitivity and minimal sample processing and pre-treatment. During each count, multiple elements are determined simultaneously. While a small amount of the sample is taken and crushed, the powder itself is not consumed during the process. Neff (2000:103) has suggested that INAA is still the best technique for achieving high precision and high sensitivity with easy, relatively error-free sample processing, compared to ICP-AES and ICP-mass spectrometry (MS). Recently developed solid sample introduction and ionization techniques now enable compositional analysis with high-resolution mass spectrometry without sample dissolution in strong acids.

The main disadvantage of INAA is the requirement of a neutron generator, which is usually a nuclear reactor. Compared to other types of analysis, the distribution of facilities available for INAA is very sparse. The cost of analysis is generally more expensive than alternative techniques that may be more readily available. In addition, high concentrations of certain elements can obscure the measurement of trace elements.

Chapter 33
Separation Techniques

Chromatography and electrophoresis are used to separate the individual constituents of complex mixtures of organic compounds. The sample, dissolved in a liquid or carried by a gaseous mobile phase, is passed through or over a stationary phase, which is a solid, gel, or liquid. Stationary phases exploit differences in the physical and chemical properties of individual components of a mixture to isolate them. The stationary phase may simply act as a filter to physically separate components on the basis of their size. In other cases, the interaction between the sample components and the stationary phase is chemical. Sample components may become partially dissolved in the stationary phase. Components with lowest affinities for the stationary phases move through the system the fastest, while those with highest affinities move the slowest. On the basis of their relative affinity for the stationary phase, individual sample components migrate through the system at different rates and separate.

Separation techniques are commonly employed and descriptions are available in general books on analytical techniques, such as Hites (1997), Khandpur (2007), Patnaik (2004), Pasto and Johnson (1979), and Skoog et al. (1998). Several works address the use of gas chromatography alone (Christie 1982; Evershed 1992b) or combined with mass spectrometry (Evershed 1992a, 1993b, 1994, 2000) for the analysis of lipids.

Chromatography

The different types of chromatography are defined by the stationary or mobile phases employed. The use of thin layer and high-pressure (or high-performance) liquid chromatography techniques for examining archaeological materials has been limited. Gas chromatography alone, and in conjunction with mass spectrometry, is extensively employed. The volatility of individual components is important for gas chromatography, as those with lower boiling points enter the column as a gas before those with higher boiling points.

M.E. Malainey, *A Consumer's Guide to Archaeological Science*, Manuals in
Archaeological Method, Theory and Technique, DOI 10.1007/978-1-4419-5704-7_33,
© Springer Science+Business Media, LLC 2011

Thin Layer Chromatography (TLC)

With thin layer chromatography, a sample is applied to one end of a slab of stationary phase, which is then immersed in solvent (Fig. 33.1). Capillary action causes components dissolved in the solvent to rise vertically through the stationary phase. While identifications can be made through comparisons with standards, TLC is more often used to separate components of a complex mixture for further analysis. The separated sample components are adsorbed onto the stationary phase in discrete spots and can be easily recovered.

Fig. 33.1 Thin layer chromatography (Original Figure by M. Malainey and T. Figol)

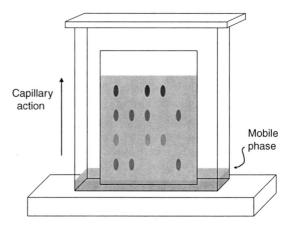

Capillary action

Mobile phase

The stationary phase is usually silica gel with a controlled pore size between 1 and 25 μm. Silica gel plates can be purchased ready made or made in the lab by uniformly spreading the gel on a plate of glass. A solution of the sample is applied to the bottom edge of the silica. When the solvent has evaporated, the plate is placed vertically in a tank with its sample marked, lower edge in the liquid mobile phase. The mobile-phase solvent rises up the silica gel by capillary action carrying dissolved sample components. The components separate according to their affinities for the silica and the polarity of the solvent. When the desired degree of separation is achieved, the plate is removed from the tank and the solvent is allowed to evaporate. The separated components may be visible as dark spots under ultraviolet light and identified through comparisons with standards. If further analysis is desired, selected components can be scraped off the plate and washed out of the silica with a suitable solvent.

High-Pressure (Performance) Liquid Chromatography (HPLC)

In its earliest incarnation, the process was simply called liquid chromatography. A liquid containing the sample was poured into the top of a vertically oriented column and separated components were collected at the bottom of the column. Gravity carried the solution through the column of particles made of, or coated with, the stationary phase. Fine component separations were not possible because the particles of the stationary phase were relatively large in size. With only gravity, separation times were also very long due to the slow flow rates. These problems were largely overcome by the introduction of pumps that force the liquid phase through the column under high pressure, up to 6000 pounds per square inch (psi) (Fig. 33.2). With high-pressure or high-performance liquid chromatography, it is possible to use very small particles of the stationary phase to achieve excellent component separation.

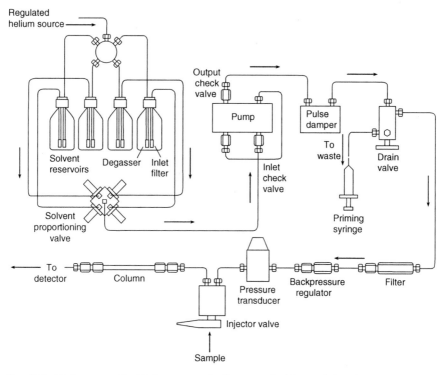

Fig. 33.2 High-pressure liquid chromatograph (Image used with permission from PerkinElmer, Inc., Waltham, MA)

The mobile phase carrying the sample is a liquid, usually an organic solvent or combination of solvents. When only one solvent of constant composition is employed, the process is referred to as isocratic elution. With gradient elution, two or more solvents are used to move sample components through the column. The ratios between the different solvents can be programmed to vary over the course of the separation.

Separation of sample components is achieved through surface interactions between sample components and the solid particles of the stationary phase. HPLC columns consist of a metal cylinder containing very small particles of stationary phase, ranging from 3 to 10 μm in diameter. The particles are porous and composed of silica, alumina, ion-exchange resin, or some other substance. Column selection depends upon the goals of the analysis, in particular the targeted component or components.

Once the separated component emerges (elutes), it can be collected in a sample vial or analyzed using any one of a variety of detectors. Ultraviolet absorbance detectors are most common, but infrared and fluorescence detectors are also used. UV absorption detectors are often highly simplified versions of those used for UV spectroscopy (see Chapter 35). After passing through the column, the separated component (eluent) enters a small detector cell where it is exposed to filtered UV light of a single wavelength, such as 254 nm from a mercury lamp. More sophisticated devices consist of scanning spectrophotometers that allow the analyst to select different wavelengths in the UV and visible range. If the eluents are sufficiently separated, a different wavelength can be used for each component. While the eluent is still only exposed to a single wavelength, the detector can be programmed to use the optimal one for each component. A high-pressure liquid chromatograph can be linked to a mass spectrometer (LC-MS); this instrument is described below.

The main advantage of HPLC is that separations are usually performed at room temperature. Samples capable of being dissolved in organic solvents can be analyzed using this technique. This is of benefit when examining samples with low levels of component volatility and thermal stability. Samples with low component volatility are not easily transformed from a liquid to a gas. Those with low thermal stability decompose when exposed to heat. In this respect, samples that are not suitable for analysis with gas chromatography can be analyzed with HPLC.

Liquid Chromatography with Mass Spectrometry (LC-MS)

The combination of high-pressure liquid chromatography and mass spectrometry enables precise identifications of components based on molecular weight and structure. During liquid chromatography, the sample is carried through the column under high pressure by a liquid mobile phase, usually an organic solvent. By contrast, mass spectrometry is performed on individual sample ions under vacuum. Using recently developed atmospheric pressure ionization techniques, removal of the solvent, pressure change accommodation, and sample ionization take place in the interface between the instruments. These LC-MS ionization techniques are described in Chapter 31.

Gas Chromatography (GC)

Gas (or gas liquid) chromatography is used to separate complex mixtures of organic compounds, or derivatives of these compounds, that have relatively low boiling points and are not decomposed by heat. The mobile phase is a carrier gas, typically helium or hydrogen; the stationary phase is liquid coated on the interior of a fused silica column. The column is situated in an oven and separation is based on component volatility and degree of interaction with the stationary phase.

Although other methods of introduction are possible, most samples are injected onto the column dissolved in an organic solvent. Initially, the oven temperature is relatively low and then gradually increased. First, the solvent evaporates and deposits the sample as a liquid residue. As the oven temperature rises, sample components with progressively higher boiling points change from liquid to gas and begin moving through the column in the stream of the carrier gas. Differing degrees of interactions with the stationary phase serve to separate components with similar boiling points. Once they emerge from the column, or elute, sample components enter a detector (Fig. 33.3).

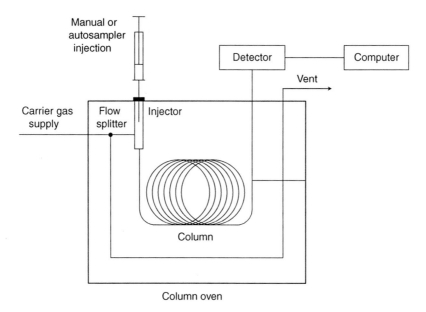

Fig. 33.3 Gas chromatograph (Adapted from SKOOG. Principles of Instrumental Analysis, 5E. © 1998 Brooks/Cole, a part of Cengage Learning, Inc. Reproduced by permission. www.cengage.com/permissions)

The term, gas chromatography, is generally applied to analyses with a flame-ionization detector. When combined with a mass spectrometer, the hyphenated form, GC-MS, is used. The term, GC-C-IRMS, refers to a system where the sample emerges from the gas chromatograph, undergoes combustion, and then is analyzed by an isotope ratio mass spectrometer.

Gas chromatography columns are generally long wall-coated open tubular (WCOT) or capillary columns made of fused silica. A column can be thought of as a very long flexible thin-walled glass straw with an extremely small diameter. The stationary phase selected should have the physical properties, specifically the degree of polarity, to achieve the best separation of sample components.

Samples are most often injected onto the column in solution. In order to compare chromatograms, the volume injected, the point of introduction, and initial temperature must be reproducible. This procedure is automated on gas chromatographs equipped with autosamplers. Prepared samples are placed in small glass vials. A syringe is used to inject a small amount of the sample solution, usually 1–5 μL, onto the column. Autosamplers are robotic devices that clean the syringe, inject the desired amount of sample onto the column, and then clean the syringe again; when one run is complete, the next sample is injected.

Gas chromatographic analysis of non-volatile solid samples is possible using pyrolysis, which is heat-induced decomposition. A portion of the sample is dissolved and the solution is applied to a special heating element or Curie point needle. When the solvent evaporates, a thin layer of the sample is deposited onto the element. This procedure is repeated several times until a film of sample coats the element. The element is then inserted into the injection port and heated rapidly to a high temperature. Temperatures up to about 800°C are reached in 20–30 ms, completely burning the sample. All pyrolysis products of the sample enter the column.

Component Separation

As mentioned above, gas chromatography uses the volatilities of individual sample components to separate them. Some separations are conducted with the oven held at one temperature, which is called an isothermal analysis. In many cases, however, separation of sample components is enhanced by gradually increasing the oven temperature over the course of the run. A typical run may begin with a short initial isothermal hold close to the boiling point of the solvent, then the temperature is ramped up a few degrees each minute; a second isothermal hold occurs when the final temperature is reached. As the temperature increases, progressively less volatile components evaporate and travel through the column in the stream of the carrier gas. Temperature programming usually results in better separations and shorter run times.

Sample constituents begin moving through the column at different times, depending on their volatilities. The rate at which they pass through the column depends on their affinities to the stationary phase. Components with a low affinity to the stationary phase move through the column very quickly, whereas components with a high affinity to the stationary phase progress through it more slowly. Differing levels of interactions between individual components and the column coating act to further separate them. Under ideal circumstances, a complex mixture can be completely separated into its individual components.

Detectors

Detectors monitor the column effluent and measure variations in its composition due to the presence of separated components. When the mobile phase alone is passing through, the detector registers a zero signal. Flame ionization detectors (GC or GC-FID) and mass spectrometers (GC-MS) are most commonly used. Using GC-C-IRMS, the isotope ratio of the separated components can be monitored. As they emerge from the column, the separated component undergoes combustion and enters an isotope ratio mass spectrometer. The ratio of stable isotopes, often ^{13}C and ^{12}C, in selected components is then determined. More information about this process can be found in Chapter 31.

Flame ionization is the most common method of detection (Fig. 33.4). It has a high sensitivity to all organic compounds and little or no response to H_2O,

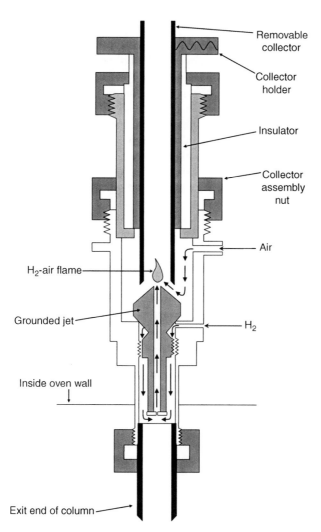

Fig. 33.4 Flame ionization detector (© Agilent Technologies, Inc. Reproduced with Permission, Courtesy of Agilent Technologies, Inc.)

CO_2, etc. When a separated component emerges from the column, the detector records the time lapsed since the sample was injected. This is called the retention time. The component passes by a flame where it undergoes combustion. The resulting thermal fragmentation and free radical reactions cause ionization to occur:

$$CH\cdot + O\cdot \rightarrow CHO^+ + e^-$$

The ions pass by a collector electrode where a current is observed; the strength of the current is related to the concentration of charged species. Identifications of sample constituents are based on retention times and comparisons made against known standards run under identical conditions.

Mass spectrometry is an established analytical technique often used to aid in the identification of components separated with high pressure or gas chromatography (see Chapter 31). Sample components enter a mass spectrometer as they emerge from the column where they are converted to ions and sorted according to their mass-to-charge ratios.

A very common method of generating ions is electron impact, which is a "hard ionization" technique. Ions are produced by bombardment with a beam of high-energy (usually 70 eV) electrons. The electron beam strips an electron off the sample molecule (M) producing a molecular ion (also called a radical cation) $[M\cdot]^+$. Molecular ions are usually highly unstable and overcome the forces holding them together and fragment:

$$M + e^- \rightarrow [M\cdot]^+ + 2e^- \rightarrow A^+ + B^+ + \text{etc.}$$

Positively charged fragment ions are sent through a magnetic or electrostatic field to the mass analyzer. The mass analyzer sorts and collects these cations on the basis of their mass/charge ratio, m/z. Since molecules fragment in a consistent manner, it is often possible to deduce the structure of the compound through the nature of the fragments produced. Component identification is made by comparing its spectrum to those in a computerized database of known compounds. A more detailed description of mass spectrometry can be found in Chapter 31.

Electrophoresis

Electrophoresis is similar to thin layer chromatography but an electric field is applied to facilitate the separation of the molecules. A typical set-up involves placing the sample solution into wells cut into one end of an agarose or polyacrylamide gel slab. The negatively charged pole (cathode) is applied to the "sample end" of the gel slab and the positively charged pole (anode) is applied to the other, producing an electric field. Negatively charged molecules migrate through the stationary medium toward the anode and separate. If necessary, stains are used to color the sample molecules so that their location can be established.

Electrophoresis of Proteins

Electrophoresis is useful for examining proteins (Chapter 5) because their behavior in an electric field depends upon the properties of their constituent amino acids. In aqueous (water-based) solutions, all amino acids form ions. The amino group forms the $-NH_3^+$ ion and carries a positive charge; the carboxyl group forms the $-COO^-$ ion and carries a negative charge. If the solution is neutral (pH 7.0) and its R group does not carry a charge, the positive charge on the amino group balances the negative charge on the carboxyl group so the amino acid is an electrically neutral "zwitterion." The amino acid would not be drawn to either the anode or the cathode, but remain stationary.

Under acidic conditions, the amino group $(-NH_2)$ is basic and reacts with the acid to form $-NH_3^+$ ions. Consequently, when the pH value is lower than 7, amino acids are no longer neutral but cations carrying a positive charge. In an electric field, cations move toward the negatively charged terminal or cathode. Under basic conditions, the carboxyl group is acidic and reacts with the base to form $-COO^-$ ions. When the pH value is higher than 7, amino acids are anions carrying a negative charge. In an electric field, anions move toward the positively charged terminal, or anode.

The R groups of some amino acids are also ionized in water so the amino acid is not electrically neutral at pH 7.0. Additional amino groups are present on basic amino acids; these carry a net positive charge. Acidic amino acids have additional carboxyl groups; these carry a net negative charge. At pH 7.0, acidic amino acids move toward the anode, and basic amino acids move toward the cathode. However, at some pH below 7.0, acidic amino acids are electrically neutral and at some pH above 7.0, basic amino acids are electrically neutral.

Isoelectric Focusing Point (pI or IEP)

Proteins are combinations of the 20 different amino acids arranged in long chains. The behavior of a protein in an electric field is related to the properties of its constituent amino acids. At a particular pH, called its isoelectric point, internal neutralization occurs and the protein becomes electrically neutral. If there are many basic amino acids, the isoelectric point of the protein will be above pH 7.0. If many acidic amino acids are present, the protein will have an isoelectric point below pH 7.0. The isoelectric point of serum albumin occurs at a pH value of 4.9, for γ-globulin it is 6.6 and it is 6.8 for hemoglobin.

Isoelectric focusing is the separation of proteins with different isoelectric points using electrophoresis. A pH gradient is established within the gel slab so that one end is basic and the other end is acidic. At any pH value below its isoelectric point, a protein will have a net positive charge and move toward the cathode. At pH values above its isoelectric point, a protein carries a net negative charge and will move toward the anode. At its isoelectric point, the protein is electrically neutral and does not move.

Separation by Molecular Weight

Electrophoresis can be used to separate molecules, such as DNA and proteins, on the basis of their molecular weights. DNA naturally carries a negative charge; when proteins are analyzed, the sample is dissolved in a sodium dodecyl sulfate (SDS) solution to give the molecules strong negative charges. When applied to the gel and placed in an electric field, the molecules all migrate toward the positively charged anode. The rate at which they move through the gel depends only upon their molecular weight. Small molecules move through the gel faster than large ones.

Chapter 34
Atomic and Emission Spectroscopy

As outlined in Chapter 1, atoms are able to absorb and emit characteristic energy. The techniques described in this chapter utilize this property to identify the elemental composition of an unknown substance. More detailed descriptions are available in books on analytical techniques, such as Khandpur (2007), Patnaik (2004), and Skoog et al. (1998). Rice (1987), Young and Pollard (2000), and Young et al. (1997) review applications of these techniques to archaeological materials.

Atomic Absorption

Atomic absorption spectrometry (AA or AAS) is used to test for the presence of certain elements; if present, the concentration of that element in the sample can be precisely determined. It is based on the principle that, under certain conditions, the atoms of an element absorb light of characteristic wavelength. Atomic absorption spectrometers consist of three components: (1) a sample atomizer, (2) a light source, and (3) a detector. Within a flame or electrothermal device, the sample is exposed to very high temperatures that break it down into atoms, a process called atomization. These atoms are exposed to light characteristic of a particular element. If that element is present, a finite amount of the light is absorbed and the detector measures the loss of light. The amount of light absorbed is directly related to the quantity of that element in the sample. If light characteristic of multiple elements is directed at the sample atoms, the presence and amounts of all those elements can be determined simultaneously.

Sample Atomization

Samples undergoing atomic absorption spectrometry are usually dissolved in strong acids. For flame atomization, a nebulizer is used to introduce the sample solution. The liquid is drawn into a stream of compressed gas, which converts it into a mist or aerosol. The aerosol is sprayed directly into the flame, which causes the acid to evaporate quickly. The solid sample is reconstituted then converted directly

M.E. Malainey, *A Consumer's Guide to Archaeological Science*, Manuals in
Archaeological Method, Theory and Technique, DOI 10.1007/978-1-4419-5704-7_34,
© Springer Science+Business Media, LLC 2011

into a gaseous state. A portion of the sample dissociates into atoms and elemental ions. In this hot, gaseous state, atoms are capable of absorbing light of specific wavelengths.

Flame atomization involves the combustion of a fuel in the presence of a source of oxygen or oxidant. The temperatures achieved depend on which fuel and oxidant combination are used. Commonly used fuels include natural gas, hydrogen, and acetylene. The hottest temperatures are attained when these fuels are burned in pure oxygen, followed by nitrous oxide and air. The combination of natural gas and air produces flame temperatures between 1700 and 1900°C, which is capable of atomizing samples that readily decompose. Much higher flame temperatures, over 3000°C, are achieved with the combination of acetylene and oxygen.

Electrothermal devices use increasing current to first remove the solvent from the sample solution, leaving the sample in solid form; second, the solid sample is turned to ash; finally it is vaporized. Electrothermal atomization takes place within a graphite furnace. A small amount of sample solution, 0.5–10 μL, is introduced into a graphite tube or cup and an electric current is applied. Solvent evaporation takes place at low temperature; the resultant solid is transformed into ash at a higher temperature. A large, abrupt increase in current produces a rapid temperature rise to as much as 3000°C, which causes the solid to atomize. This method is generally slower than flame atomization but has a higher sensitivity. The graphite tubes are sometimes coated to reduce their porosity.

One of the potential drawbacks of atomic absorption is the requirement that solid samples be dissolved prior to analysis (see Chapter 30). Samples that are difficult to dissolve can be analyzed in solid form using glow discharge atomization. This process requires that the sample either conduct electricity or be powdered, mixed with conductive material, such as graphite or copper, and formed into a pellet.

Light Sources

Special lamps (or radiation sources) are used to produce the visible or near ultraviolet light characteristic of particular elements. The most common source is the hollow cathode lamp (Fig. 34.1). The cathode is a deep, cylindrical cup constructed entirely of the element of interest or has a layer of that element applied to it. When an electric current is applied, the cathode holds the negative charge. The anode, usually made of tungsten, is positively charged. The lamp is filled with either argon or neon gas at a pressure of 1–5 torr. When about 300 V is applied, the inert gas ionizes to form either Ar^+ or Ne^+. These positively charged cations strike the negatively charged cathode with sufficient energy to dislodge atoms from the surface. This process is called sputtering and causes some of the dislodged atoms to absorb energy and become excited. Atoms in an excited state release the excess energy by emitting light characteristic for that element as electrons return to lower energy levels. The wavelengths of emitted light are in a narrow range, so appear as one color or monochromatic. Alternatively, electrodeless discharge lamps may be used.

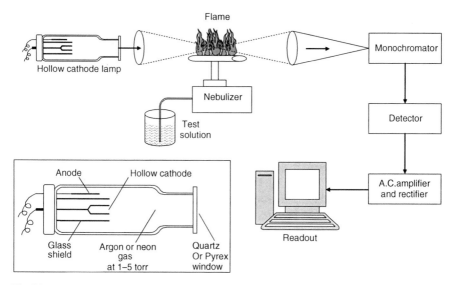

Fig. 34.1 Atomic absorption with hollow cathode lamp (From SKOOG. Principles of Instrumental Analysis, 5E. © 1998 Brooks/Cole, a part of Cengage Learning, Inc. Reproduced by permission. www.cengage.com/permissions)

Microwaves or radio frequencies are used to ionize the inert gas. The light from these lamps is more intense than that produced by hollow cathode lamps.

The monochromatic light from the radiation source is shone through the gaseous atomized sample while it is in the flame or graphite furnace. If atoms of that particular element are present in the sample, they will absorb the energy. This causes electrons at lower energy levels to move to upper levels. In general, the number of gaseous sample atoms increases with temperatures. The use of higher temperature flames will usually improve the spectra.

Detection

The amount of light lost through absorption is measured by a monochromator. According to Beer's law, the absorbance, A, is the product of three variables: the absorptivity of the element, ε, the path of the light beam, l, and the number of atoms (given as its molar concentration), c: $A = \varepsilon c l$. The general form of linear relationships is $Y = aX + b$, where Y and X are variables, a is the slope of the line, and b is the Y-intercept. Consequently, there is a linear relationship between of light absorbance and the concentration of that element in the sample. A plot of absorbance, A, vs. molar concentration, c, results in a line with a slope equal to the product of absorptivity and the path of the light beam (εl). The amount of light absorbed by the acid alone must also be determined by analyzing an acid blank. The concentrations of most metals can be measured using atomic absorption spectrometry, each with its own lamp. Each element of interest is measured individually; it is necessary to switch lamps and adjust the wavelength monitored by the monochromator. Lamps with hollow cathodes consisting of several metals are also available for simultaneous measurement of multiple elements. A photomultiplier is used to measure the amount of transmitted light, which is the light not absorbed by the sample. The difference between the light produced by the lamp and the light detected is the amount absorbed by sample atoms.

Advantages and Disadvantages

The instrument itself is quite inexpensive and widely available. Its operation is simple so minimal training is required. The technique is highly sensitive, capable of measurement to 10 ppm. The determinations also have a high degree of precision, on the order of $\pm$ 2%. Some claim 50 elements are detectable; others state satisfactory results for only 22 are possible. When using electrochemical atomization, liquid samples can be directly introduced into the graphite furnace.

Its ability to analyze only one element at a time is considered a disadvantage of this technique. To counter these problems, hollow cathode lamps consisting of a mixture of several metals have been developed. This permits the determination of more than a single element at a time. A polychromator is required to analyze light of different wavelengths simultaneously. Furthermore, analysts are likely only to test for elements that are expected to appear in the unknown; unusual elements in the unknown may be missed.

Another disadvantage is that most instruments are designed to handle samples in solution. Dissolving a solid sample may involve the use of strong acids or other reagents and also increases the potential of contamination. If the sample, or a constituent of the sample, is difficult to dissolve, elemental concentrations may become diluted. Care must be taken to ensure that the rate of sample introduction, the concentration of the sample in the acid solution, and solution viscosity do not vary throughout the analysis.

Solid sample introduction methods, such as glow discharge atomization, have been developed. The results are less reproducible, however. Problems with calibration, sample conditioning, precision, and accuracy have also been encountered.

Optical Emission Spectroscopy (OES)

This technique is based on the principle that when an element is converted into atoms and ions (atomized), light characteristic for that particular element is emitted; recognition of this phenomenon led to the early development of the technique. Optical emission spectroscopy (OES) was first applied to archaeological materials in the 1930s. In order to perform this type of analysis, it is necessary for the sample to be either a liquid or in solution; solid samples are often dissolved in a strong acid. The next step is nebulization, where the sample solution is converted into an aerosol and then the tiny droplets are sprayed into a heat source, which is often a flame. Depending upon the oxidant and fuel used, flame temperatures between 1700 and 3000°C can be achieved. At these relatively low temperatures, a portion of the sample is atomized; furnaces, sparks, and arcs can be used to produce higher temperatures than flames. Sample atoms are excited through collisions and outer electrons are temporarily lifted to higher energy levels. In order to return to ground state, energy characteristic of the particular element is released. This energy is a well-defined wavelength in the near ultraviolet or visible light range. The light is dispersed by a diffraction grating which is then detected either by a photographic plate, where individual wavelengths appear as a series of black lines, or by a slit and photomultiplier.

Only semi-quantitative analysis based on intensity of light emitted is possible using this technique. The amount of an element is determined by the blackness of the line on film or on the basis of the photomultiplier output, with an accuracy of about 10%. Only small samples are required for this type of analysis, ranging from 5 to 100 mg. If a large amount of a metallic sample is available, it could be shaped and used as the electrode for arc and spark excitation methods. The technique is also sensitive as 30–40 metallic elements can be detected at levels down to 100 ppm. The advantages of this technique are that it is widely available, rapid, and only a small amount of sample is required. The main disadvantage is its low degree of precision, in that it can only provide a semi-quantitative analysis. At lower temperatures, only a small population of sample atoms are atomized and only a relatively low level of excitation is possible. Even so, samples with complex compositions give complex spectra, which are difficult to interpret. The technique fell into disuse after the advent of X-ray fluorescence and atomic absorption spectroscopy, both of which have higher sensitivity and precision.

Inductively Coupled Plasma-Atomic Emission Spectroscopy (ICP-AES)

The basic principles of ICP-AES are identical to OES. The major differences between the two techniques rest with the thermal energy source and the manner in which emitted light is detected and quantified. Sample atoms are atomized, ionized, and excited by thermal energy; as the excited ions return to lower energy levels, light characteristic of that element is emitted, and analyzed (Fig. 34.2). The plasma torch used in ICP-AES is capable of producing temperatures significantly hotter than a flame; consequently, a larger population of the sample undergoes atomization and ionization and a higher number of excitation states are produced. The light emitted is related to the de-excitation of the ions of an element rather than its atoms. Using a multichannel analyzer, emitted light of many different energies or frequencies can be detected and quantitatively analyzed simultaneously, which facilitates the analysis of samples with complex compositions.

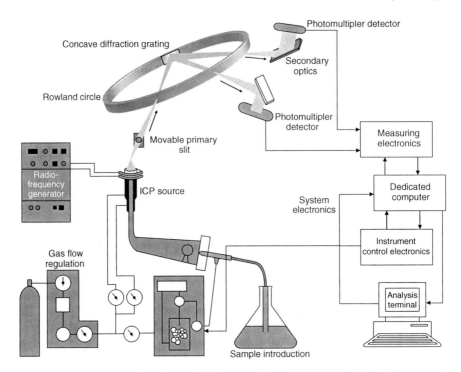

Fig. 34.2 Inductively coupled plasma polychromator (From SKOOG. Principles of Instrumental Analysis, 5E. © 1998 Brooks/Cole, a part of Cengage Learning, Inc. Reproduced by permission. www.cengage.com/permissions)

Sample introduction of ICP-AES is more flexible. Liquid samples or solutions are converted into an aerosol in a nebulizer (Fig. 34.3a). Techniques that vaporize solids, including furnaces capable of electrothermal vaporization and different ablation techniques, including laser ablation can also be employed Chapter 30. The sample, in the form of either tiny droplets or particulate matter, is carried to the plasma torch in a stream of gas, usually argon.

Plasma is a gaseous mixture containing significant concentrations of cations and electrons. While the net charge of the gas approaches zero, atoms are present in an ionized state. Plasma conducts electricity and the movement of plasma ions can be manipulated by a magnetic field. Resistance to this movement produces very high temperatures.

A plasma torch consists of three concentric quartz tubes; the largest of which is about 2.5 cm in diameter (Fig. 34.3b). Argon flowing through the tubes is ignited by a spark from a Tesla coil, leading to the formation of plasma consisting of argon cations and electrons. The torch is surrounded by a water-cooled induction coil powered by a radio frequency generator. According to Faraday's law of induction, a changing magnetic field induces an electric field and the reverse is also true. When an electric field is changed, a magnetic field is produced. The induced magnetic

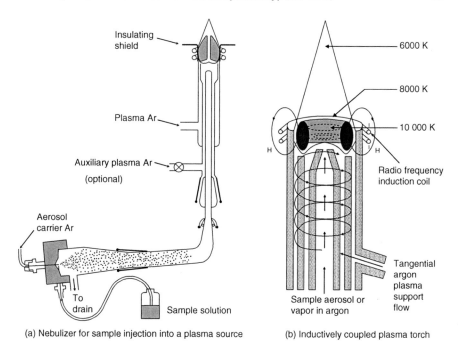

(a) Nebulizer for sample injection into a plasma source   (b) Inductively coupled plasma torch

Fig. 34.3 Inductively coupled plasma nebulizer (**a**) and torch (**b**) (From SKOOG. Principles of Instrumental Analysis, 5E. © 1998 Brooks/Cole, a part of Cengage Learning, Inc. Reproduced by permission. www.cengage.com/permissions)

field causes the ions and electrons in the plasma to move in annular paths or spirals. Resistance to this movement produces ohmic heating and the resistivity of plasma increases with temperature. Plasma ions absorb this heat and maintain a high temperature, which leads to further ionization. The plasma can be sustained indefinitely and temperatures as high as 10,000 K can be attained. While in operation, the torch consumes between 5 and 20 L of argon each minute, depending on its design.

The torch is protected from the high temperature of the plasma by flows of argon. Argon flowing in a spiral between the outer tube and the middle tube centers the plasma and serves as insulation between the plasma and the outer tube. The inner quartz tube is for sample delivery. Argon flowing through this tube carries the sample into the plasma. Since argon is an inert gas, the sample is protected from oxygen attack and the effects of oxidation en route to and within the plasma torch.

Plasma has a non-transparent core or fireball topped by a flame-like tail that extends a few millimeters above the tube. Spectral observations for atomic emission spectroscopy are made 15–20 mm above the induction coil in a region where the plasma is optically transparent. Temperatures in this area range between 4000 and 8000 K, which is about two to three times higher than the hottest flame. The high temperatures of the plasma torch lead to more complete atomization of the sample. The temperature is also more uniform than flame, arc, or spark.

Plasma emission spectrometers encompass the entire UV and visible spectrum, from 170 to 800 nm (Fig. 2.2). Light from the source is filtered out. Simultaneous multichannel instruments are designed to measure the intensities of emission lines for a large number of elements (Fig. 34.4a). Up to 50 or 60 elements can be analyzed simultaneously and analysis time is about 30 s. Sequential instruments are less

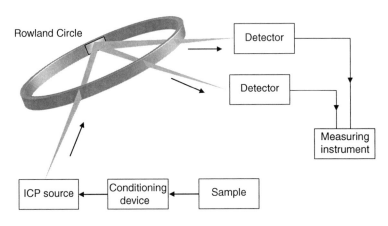

(a) Simultaneous Multichannel Detector

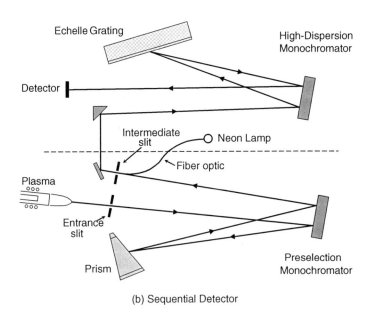

(b) Sequential Detector

Fig. 34.4 Simultaneous multichannel (**a**) and sequential detectors (**b**) ((a) Adapted from SKOOG. Principles of Instrumental Analysis, 5E. © 1998 Brooks/Cole, a part of Cengage Learning, Inc. Reproduced by permission. www.cengage.com/permissions (b) Image used with permission from PerkinElmer, Inc., Waltham, MA)

expensive but analyze only one wavelength at a time (Fig. 34.4b). This lengthens the analysis time and increases the amount of sample and argon consumed.

ICP-AES provides excellent quantitative and qualitative analysis of all metals. Several non-metals can also be analyzed in a vacuum, including boron, carbon, phosphorous, nitrogen, and sulfur. The detection limits of ICP-AES are very high; 41 elements are detectable at levels equal to or less than 10 parts per billion, compared to 16 for OES and 15 for AAS. ICP-AES is more expensive compared to the older techniques, however.

ICP-MS

The inductively coupled plasma torch is very efficient in producing positive ions (cations) of most elements. For this reason, it is an excellent ionization source for mass spectrometry and enables discrimination of various elemental isotopes by mass.

In order to transfer the ions into the mass spectrometer for analysis, significant cooling is required. The plasma torch typically operates at an air pressure of 1 atmosphere (atm), which is about 760 torr. The hot plasma gas first enters a low-pressure chamber, at 1 torr, through a sampling cone. The dramatic change in pressure causes rapid expansion of the gas and it cools. A fraction of the ions enter the mass spectrometer, which operates at 10^{-4} torr, through a skimmer cone. The positively charged ions are separated from electrons, accelerated, focused, and mass analyzed. Standards and acid blanks must be run with the sample.

Over 90% of the elements in the periodic table can be analyzed; however, those most commonly found in organic compounds (carbon, hydrogen, oxygen and nitrogen) as well as argon, neon, and fluorine cannot be examined. Interference from atmospheric gases makes it difficult to measure sulfur, chlorine and silicon as well. The detection limits are between 0.1 and 10 ppb and ICP-MS has an accuracy of 2–4%. The mass spectra generated by ICP-MS tend to be easier to interpret because there are few spectral lines.

In general, the technique works well for most metals but matrix effects can occur. The mass of most common isotopes of trace elements can be similar to major elements in the sample; in these cases, measurements must be based on less common isotopes. This approach is required to measure zinc in copper samples and antimony in samples containing tin.

If it is necessary to examine major, minor, and trace elements in the sample, they must be analyzed separately. Major and minor elements can be measured in a solution diluted to the optimal detection range of the instrument; trace elements are measured in a concentrated solution. For some applications, such as the analysis of lead isotopes, thermal ionization mass spectrometry is a superior technique.

Chapter 35
Optical Spectroscopy

Infrared (IR), ultraviolet (UV), visible (VIS), Raman spectroscopy, and related techniques subject sample molecules to radiation in the middle of the electromagnetic spectrum (Chapter 2). The absorption of energy by the sample molecules in solution is indicative of certain properties that aid in its identification. IR and UV spectroscopy use these absorptions to ascertain the presence of functional groups and types of molecular bonds. Fourier transformation is a data-processing technique used to improve the quality of the IR spectrum. Photoacoustic spectroscopy (PAS), specifically IR-PAS, is used to obtain infrared absorption spectra from solids. Attenuated total reflectance-Fourier transformation infrared spectroscopy (ATR-FTIR) can be used to examine liquid or solid samples. Raman spectroscopy uses the scattering of radiation by sample molecules to gain information about their chemical structure.

Optical spectroscopy techniques are widely used for the analysis of organic compounds. Basic descriptions are available in organic chemistry texts (Solomons 1980; Solomons and Fryhle 2004); books on analytical techniques can be consulted for more information (Harris and Kratochvil 1981; Khandpur 2007; Pasto and Johnson 1979; Patnaik 2004; Skoog et al. 1998). Coates (1998), Hammes (2005), Workman (1998a,b), Workman and Springsteen (1998) and the product literature from PerkinElmer, Inc (2005) provide more detailed information about the instruments and theoretical aspects. Examples of archaeological applications of these techniques are considered by Bacci (2000), Cariati and Bruni (2000), Rice (1987), and Smith and Clark (2004).

Ultraviolet (UV) and Visible (VIS) Spectroscopy

As described in Chapter 1, atomic orbitals represent regions in space around a nucleus where the probability of finding an electron is highest. Similarly, molecular orbitals represent regions in space where the probability of finding bonding electrons within a molecule is highest. Covalent bonds are formed when two atoms attain stable configurations by sharing valence electrons in their outermost shells. The orbitals of the two atoms overlap and the distance between the nuclei is reduced. The

M.E. Malainey, *A Consumer's Guide to Archaeological Science*, Manuals in
Archaeological Method, Theory and Technique, DOI 10.1007/978-1-4419-5704-7_35,
© Springer Science+Business Media, LLC 2011

mutual attraction for electrons exceeds the forces of repulsion between the positively charged nuclei. The lowest energy configuration occurs when the shared electrons are in the space between the two nuclei. Since electrons are in constant motion and their positions cannot be precisely determined, this condition exists when the electron probability density between the nuclei is large. This ideal electron configuration forms a low-energy bonding molecular orbital. This is the lowest electronic energy, or ground, state of the molecule.

Two overlapping atomic orbitals also produce a high-energy antibonding molecular orbital. The electron probability density between the nuclei is small since both of the shared electrons tend to occur outside this area. Electrons in antibonding orbitals destabilize a molecule because the force of repulsion between the positively charged nuclei is strong. Electrons of a covalent bond can occupy antibonding orbitals if a molecule is in an excited state, however. The bonding and antibonding molecular orbitals of the hydrogen molecule are shown in Fig. 35.1. The single covalent bond formed by the overlap of two $1s$ atomic orbitals is called a sigma (σ) bond; the sigma antibonding molecular orbital is indicated by σ^*. Double bonds occur when adjacent atoms share two pairs of electrons; these are called pi (π) bonds. The symbol π^* is used to designate pi antibonding molecular orbitals. Absorption of energy can result in the movement of an electron from a bonding to an antibonding molecular orbital. Excitation of an electron from a sigma bonding orbital to a sigma antibonding orbital is called a $\sigma \rightarrow \sigma^*$ transition. The shift of an electron from a pi bonding molecular orbital to a pi antibonding molecular orbital is a $\pi \rightarrow \pi^*$ transition. Electrons in non-bonding (n) pairs around an atom can also move into sigma and pi antibonding orbitals; these are designated as $n \rightarrow \sigma^*$ and $n \rightarrow \pi^*$ transitions, respectively.

The range of wavelengths in the ultraviolet (UV) region extends from 100 to 350 nm. The visible (VIS) region of the electromagnetic spectrum includes wavelengths from 350 to 800 nm. Light in the UV and VIS regions can move an electron from a low energy bonding molecular orbital to a high-energy antibonding orbital. Ultraviolet light with wavelengths from 100 to 200 nm produces $\sigma \rightarrow \sigma^*$ and $\pi \rightarrow \pi^*$ transitions in many organic compounds, including

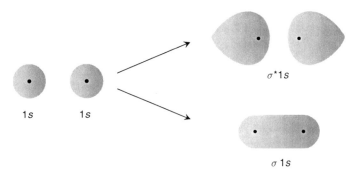

Fig. 35.1 Sigma bonding and antibonding orbitals of hydrogen (From SKOOG. Principles of Instrumental Analysis, 5E. © 1998 Brooks/Cole, a part of Cengage Learning, Inc. Reproduced by permission. www.cengage.com/permissions)

alkanes, alkenes, alkynes, and simple aldehydes and ketones. These absorptions occur in the far or vacuum UV region and are absorbed by air, which complicates their detection.

Routine UV and VIS spectroscopy uses wavelengths exceeding 200 nm and is especially useful for examining molecules containing conjugated bonds, carbonyl compounds, and aromatic groups, which absorb energy with longer wavelengths. Conjugation occurs when adjacent carbon atoms connected by a single bond are connected to other carbons through double or triple bonds (Fig. 35.2). Molecules with two double bonds separated in this manner are called conjugated dienes; those with more than two double bonds are conjugated polyenes. Carbonyl compounds contain a carbon–oxygen double bond; aromatic compounds contain ring structures with charge delocalization.

Fig. 35.2 The conjugated double bonds of 1,3-butadiene

$$CH_2 = CH - CH = CH_2$$

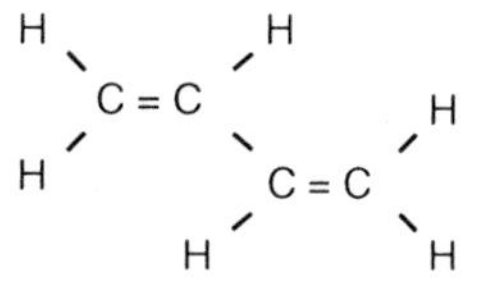

Double and single bonds alternate

Since electrons are excited from lower to higher energy levels, UV and VIS spectra are referred to as electronic spectra. Absorption bands are broad because each electronic energy level is associated with multiple vibrational and rotational states. The absorption of energy promotes electrons at any one of several possible vibrational and rotational states at the lower energy level to any one of several possible vibrational and rotational states at the higher energy level.

Ultraviolet and UV/visible spectrometers monitor the absorption of light by the sample. The amount and wavelength of light absorbed provide information about the structure of sample molecules. Typically, a beam of electromagnetic radiation in the UV and/or VIS range is split into two parts. One part passes through a cell containing the sample in solution; the other, a reference beam, passes through a cell containing the solvent alone. Specific groupings of atoms, or chromophores, in the sample absorb light at specific wavelengths; these subtractions of energy reduce the intensity of the sample beam.

The relationship between absorbance at a given wavelength (A_λ) and the intensities of the sample beam (I), and reference beams (I_O) is defined by the equation:

$$A_\lambda = -[\log(I/I_O)]$$

A graph of absorbance against wavelength is called an absorption spectrum.

The combined work of Bouguer, Lambert, and Beer established that a linear relationship exists between the amount of light absorbed and the concentration of a molecule in solution. As described by Beer's law, $A = \varepsilon c l$, the amount of energy absorbed, A, depends upon three factors. The first variable, ε, is the strength of the absorption, which is called the molar extinction coefficient or molar absorptivity, given in liters per molescentimeters. The second is the concentration, c, of the molecule in solution in moles per liter (mol/L). The third is the length, l, of the path through the cell holding the sample solution in centimeters (cm). Quantitative analysis is possible using UV−VIS spectroscopy when sample solutions analyzed follow Beer's law.

Instrumentation

Instrument components include a light source, filter or monochromator, sample cell, photodetectors and signal processors, and readout devices (Fig. 35.3). Tungsten filament lamps are commonly used to provide light with wavelengths ranging from 350 to 2500 nm. Xenon arc lamps emit light with wavelengths from 200 to 1000 nm. Helium or deuterium lamps produce light with wavelengths between 160 and 375 nm.

Filters or monochromators are used to select the light passing through the sample. Prisms or gratings further restrict the light so that only a narrow range of wavelengths can reach the sample at any given time. The sample solution is contained in cells, or cuvettes, constructed of materials that do not absorb ultraviolet or visible light in the region of interest. Quartz and fused silica sample cells are transparent

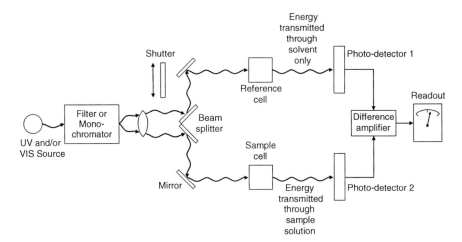

Fig. 35.3 Double beam instrument for absorption measurement (From SKOOG. Principles of Instrumental Analysis, 5E. © 1998 Brooks/Cole, a part of Cengage Learning, Inc. Reproduced by permission. www.cengage.com/permissions)

in the UV and visible range. Silicate glass is used for wavelengths between 350 and 2000 nm. Plastic sample cells can be used for light in the visible range. It is critical that the sample cells be kept free of fingerprints, grease, or other materials that can affect the transmission of light.

The amount of light reaching the detector at a particular wavelength is measured by a photodetector, usually either a photomultiplier or a photodiode. These devices are radiation transducers that convert light into an electrical signal. A photomultiplier consists of a tube containing a series of dynodes. Light passing through the sample strikes the first dynode; every photoelectron that hits it causes the emission of several electrons. These are accelerated toward the second dynode that emits several electrons for each one that strikes it. If the tube consists of nine dynodes, each photon striking the first dynode will generate a million or more electrons. A photodiode is a silicon chip that only conducts electricity when it is struck by light. The amount of current passing through is proportional to the amount of light that hits it. A multichannel detector can simultaneously monitor transmitted light consisting of a range of wavelengths so that polychromatic light sources can be used. The resulting signal is amplified and processed.

The amount of light absorbed by a sample at a given wavelength cannot be directly measured. It is determined by calculating the difference between the amount transmitted by the sample solution and the amount transmitted by the solvent alone. A single-beam instrument can only hold one cuvette at a time so the sample and reference must be analyzed separately. A double-beam instrument shown in Fig. 35.3 holds both the reference and the sample cells. The light from the source can be split to pass through both cells simultaneously or a sector mirror can be used to direct the light through one cell then the other. Once the difference in light transmission between the sample and reference cells has been calculated, an absorption spectrum is generated where absorbance is plotted against wavelength.

Infrared Spectroscopy

Organic and certain inorganic compounds absorb infrared (IR) light. This portion of the electromagnetic spectrum is divided into near, middle and far infrared regions; light in the middle IR region, with wavelengths ranging from 2.5 to 15 μm, is most often used for analyses (Skoog et al. 1998). IR radiation is less energetic than ultraviolet or visible light and causes shifts between vibrational energy levels within electronic energy levels.

Atoms connected by covalent bonds are not motionless; instead, they vibrate and rotate. Positions of atoms change relative to each other, but the molecule itself does not move. The natural vibrational frequency of a specific grouping of atoms, or chromophore, falls within a discrete energy range and different rotational and vibrational energy levels are quantized, not continuous. If exposed to radiation that exactly matches the natural vibrational frequency of a chromophore within the molecule, energy is absorbed and the vibrational energy level increases to the next higher level.

IR spectroscopy provides information about molecular structure because specific chromophores are associated with a narrow range of natural vibrational frequencies.

The difference in energy between one vibrational level and another is small compared to that between electronic energy levels. At ground state, the vibrational energy level of a molecule is zero. The energy gap between vibrational states is equally spaced so that the energy required to increase the vibrational level of a chromophore from 0 to level 1 is identical to the amount required to increase it from level 2 to 3. Whether the change is from 0 (ground state) to level 1 or from level 3 to level 4, the vibrational quantum number increases by 1.

There are two general categories of molecular vibrations: stretching and bending (Fig. 35.4). Stretching deformations change the distance between two atoms and occur along the axis of the bond; the movement is similar to that of two masses connected by a spring. The stretching may be symmetrical or asymmetrical. Bending deformations are perpendicular to the axis of the bond. They include in-plane scissoring, in-plane rocking, out-of-plane wagging, and out-of-plane twisting. Not all vibrations of covalent bonds absorb IR radiation. Molecules containing only pure covalent bonds where electrons are equally shared, such as H_2, O_2, Cl_2, and N_2,

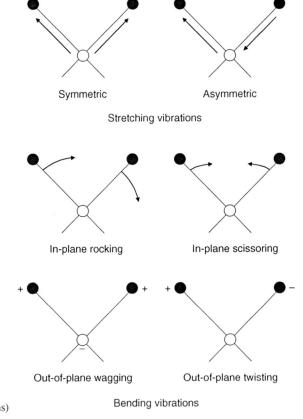

Symmetric   Asymmetric

Stretching vibrations

Fig. 35.4 Stretching and bending molecular vibrations. *Arrows* indicate movement of atoms within the plane of the page; (+) and (−) indicate movement toward and away from the reader (From SKOOG. Principles of Instrumental Analysis, 5E. © 1998 Brooks/Cole, a part of Cengage Learning, Inc. Reproduced by permission. www.cengage.com/permissions)

In-plane rocking   In-plane scissoring

Out-of-plane wagging   Out-of-plane twisting

Bending vibrations

do not produce IR absorption spectra. Molecules with polar covalent bonds absorb IR radiation when the vibration produces a net change in the dipole moment of the molecule. The dipole moment of a polar covalent bond is related to the bond distance and charge on the atoms. Symmetrical stretching of bonds of the linear molecule carbon dioxide, $O{=}C{=}O$, does not absorb IR radiation because the effects cancel; however, asymmetrical stretching and bending of these bonds generate IR absorption spectra (Fig. 35.5).

Absorption bands directly arising from stretching and bending vibrational motions, called the normal modes, usually have the strongest intensity. These are the fundamental absorptions of a molecule and reflect an increase in vibrational energy to the next higher level. Other types of weak intensity bands may also appear in IR spectra. Overtone bands are produced by energy absorptions that elevate the vibrational level by more than one. These bands can appear at whole digit multiples of the fundamental absorption frequencies. First overtones of intense bands are most likely to fall within the IR region and appear at frequencies twice that of fundamental absorptions. Second overtones appear at frequencies three times the fundamental absorptions. Combination bands can appear at frequencies representing the sum or difference of fundamental absorptions. Coupling bands are due to interactions between different absorption bands arising from the same portion of a molecule. The interaction causes a shift in the position of the bands out of the expected region for each chromophore. Fermi resonance occurs when an overtone or combination band is enhanced or split by a fundamental absorption appearing close to the same frequency.

Infrared absorption spectra are usually plotted as the intensity of the absorption (Y-axis) against the wavelength (λ) and/or wave number ($1/\lambda$) of the radiation (X-axis). The IR region with wavelengths from 2.5 to 15 μm corresponds to wave numbers ranging between 4000 and 670 cm^{-1}. The absorptions between 2000 and 670 cm^{-1} are particularly useful for identifying chromophores so the horizontal scale in this region is expanded. Bonds connecting heavy and light atoms absorb IR radiation at relatively higher frequencies and shorter wavelengths than those between atoms of similar mass.

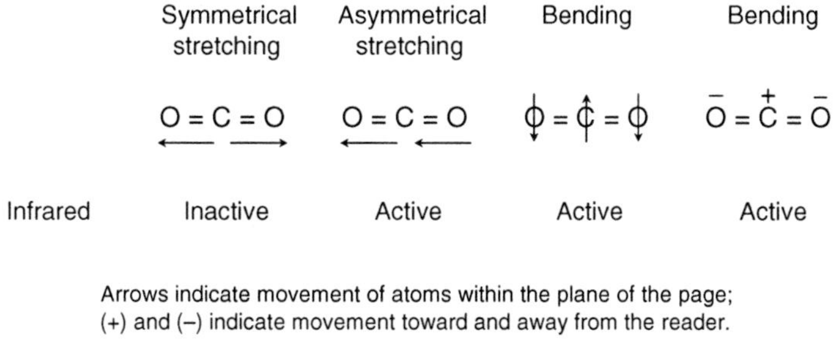

Arrows indicate movement of atoms within the plane of the page;
(+) and (−) indicate movement toward and away from the reader.

Fig. 35.5 Molecular vibrations of carbon dioxide (PASTO. DANIEL J.; JOHNSON, CARL R., LABORATORY TEXT FOR ORGANIC CHEMISTRY: A SOURCE BOOK OF CHEMICAL AND PHYSICAL TECHNIQUES, 1st edition, © 1979 pp. 130. Adapted by permission of Pearson Education, Inc., Upper Saddle River, NJ.)

Absorptions due to changes in vibrational and rotational energy levels appear as closely spaced lines in the IR spectra of freely moving gas molecules; the IR spectra of liquid and solid samples consist of broad peaks. The positions of absorption bands in certain regions provide unambiguous structural information about a molecule, such as the presence of particular chromophores. Due to the large number of vibrational interactions that occur, IR spectra of pure compounds are unique and can be used to identify an unknown substance with a high degree of certainty. While IR spectroscopy provides excellent qualitative information, the intensity of absorptions are affected by certain operational parameters of an instrument (slit width) and cannot easily be used for quantitative analyses.

Sample Preparation and Instrumentation

Many substances absorb radiation in the IR region, so only materials that transmit this energy can be used. Water is not a suitable solvent; instead, carbon tetrachloride and carbon disulfide are used. Optical glass absorbs IR radiation, so mirrors are used to focus the radiation rather than lenses. Sample cells, windows, and prisms must be constructed from materials such as sodium chloride, potassium bromide, and calcium fluoride.

Common infrared radiation sources are Nernst glowers and Globars, which are solids that can be electrically heated to temperatures between 1500 and 2200 K. A Nernst glower is a rod or hollow tube constructed of rare earth oxides, such as cerium, zirconium, thorium, or yttrium oxide; a Globar is a silicon carbide rod. Tungsten filament lamps can be used as a source of near-infrared irradiation.

Infrared irradiation is heat, so older detectors typically record increases in temperature. Thermal transducers and thermocouples are capable of detecting temperature changes of less than a thousandth of a degree. Newer Fourier transform infrared spectrometers (described below), generally use either photoelectric or photoconducting transducers. Photoelectric transducers contain dielectric crystalline materials that become polarized in an electric field and retain the polarization after the field is removed. The residual polarization is temperature dependent so infrared radiation changes the charge distribution across the crystal. Photoconducting transducers are made from materials that experience a change in conductivity when exposed to infrared energy. The electrical resistance of the semiconductor is reduced when infrared energy promotes a non-conducting valence electron to a higher energy conducting state. Semiconductor materials used in photoconducting transducers include lead sulfide, mercury/cadmium, telluride, and indium antimonide.

Fourier Transform IR Spectrometers

Most FT-IR spectrometers are based upon the Michelson interferometer (Fig. 35.6), which provides very accurate wavelength measurements. The beam of infrared radiation from the source is divided in two. Often a half-silvered mirror is used as the

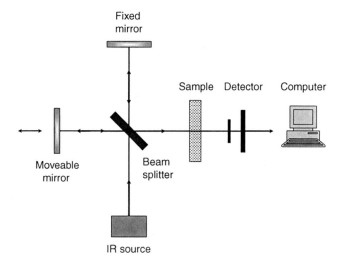

Fig. 35.6 FT-IR spectrometer with interferometer (Reprinted from "Applied Spectroscopy", Jerry Workman, Jr., page 9, Copyright 1998, with permission of Elsevier)

beam splitter; 50% of the light is reflected and 50% is able to pass through (transmitted). One-half of the beam is reflected by a fixed mirror back through the beam splitter, through the sample, and to the detector. A different mirror moving at a constant speed reflects the other half of the incident beam back through the beam splitter, sample, and to the detector. The path lengths of the two beams are not the same, which causes either constructive or deconstructive interference when they are recombined. The power of the IR fluctuates in a predicable manner depending upon the position of the movable mirror, which changes as a function of time.

When a sample is in place, absorption by chromophores alters the infrared radiation reaching the detector. The interferogram produced is a plot of light intensity (Y-axis) against the mirror position (X-axis). The sample is scanned repeatedly and many interferograms are collected. Fourier transformation of the raw data is required to produce an IR spectrum of the sample. Fourier transformation involves repeatedly scanning a specific wavelength region, then averaging the scans to improve the signal to noise ratio, and increase sensitivity; this process is described in Chapter 36.

Photoacoustic Spectroscopy (PAS)

Ultraviolet, visible, and infrared spectroscopies are typically applied to samples in solution. The amount of energy absorbed at each wavelength is determined indirectly by measuring the amount that passes through the sample. In order to measure transmitted energy, the sample must be clear. By using photoacoustic spectroscopy, certain opaque solids, semi-solid materials, and solutions containing suspended solids can be analyzed. The sample is irradiated with pulses of energy that induce audible pressure fluctuations in the gas that surrounds it; the magnitude of the pressure fluctuations is directly related to the amount of energy absorbed by the material.

Photoacoustic spectroscopy is performed with the sample inside a chamber containing a gas that does not absorb the irradiating energy. Whereas conventional UV, VIS, and IR spectroscopy involves exposing a sample to a continuous beam of radiation, PAS employs a monochromatic beam that is rhythmically interrupted or chopped. The material heats up as it absorbs the periodic energy then releases heat to the surrounding gas, causing the gas to expand. A sensitive microphone is used to detect the resulting pulses of gas pressure.

If a single-beam instrument is employed, a spectrum from the light source alone is obtained and subtracted from that of the sample. A double-beam instrument uses a beam splitter to simultaneously irradiate both the sample and the reference chambers. Finely divided carbon, called carbon black, is used as the reference material. The sample and chamber must be thoroughly purged with dry gas prior to measurement to remove all water vapor. The scan rate must be slow enough to allow sufficient time for sample heating and heat transfer to the gas to occur. The incorporation of Fourier transform techniques with infrared photoacoustic spectroscopy is usually required to obtain acceptable signal-to-noise ratios.

Attenuated Total Reflectance

Attenuated total reflectance (ATR) can be used to obtain an IR spectrum (usually FT-IR) from a liquid or solid sample. The sample is placed directly on the surface of the optically dense internal reflectance element (IRE) or ATR crystal, which is usually made of zinc selenide (ZnSe), germanium, or diamond. The infrared beam is directed at the IRE at a particular angle of incidence so that it is internally reflected; however, the radiation extends beyond the surface of the crystal as an "evanescent wave" and a few microns into the sample (Fig. 35.7). Absorption of IR energy by the sample reduces (attenuates) the intensity of the energy that reaches the detector. The absorption spectrum is produced by comparing the energy of an IR beam passing through a clean IRE (infrared background) to that of a beam reflected by the ATR crystal while in contact with the sample.

ATR-FTIR can be performed on most infrared absorbing liquids and solids as long as the refractive index of the sample is significantly lower than the IRE so that the IR beam is reflected not transmitted. No additional preparation is required for liquids; solid samples are typically powdered. In order to achieve good surface contact, powdered samples must be pressed against the crystal. Diamond crystals are well suited for the analysis of solids because they are scratch resistant.

The number of times the IR beam is reflected before it reaches the detector depends on the length and thickness of the crystal. Depending on the angle of the IR beam, it can be reflected up to ten times. Single-reflection ATR typically uses a

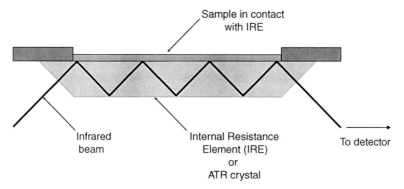

Fig. 35.7 Interaction between the sample and totally reflected infrared radiation in ATR spectrometry (Image used with permission from PerkinElmer, Inc., Waltham, MA)

very small diamond crystal, about 2 mm across. Due to the reduced surface area, a significant amount of pressure can be applied to ensure the sample is in good contact with the IRE.

Raman Spectroscopy

Raman spectroscopy involves the analysis of monochromatic light scattered by a molecule. After being exposed to radiation, a molecule usually returns to its initial state; the energy it releases is identical to the incident beam. A few molecules do not return to their initial state, but move to a different vibrational energy level. This new energy level may be higher or lower than the initial state. The energy scattered by these molecules will be at a slightly higher or lower frequency than the incident beam. This phenomenon is called Raman scattering, after the physicist who first described it, C.V. Raman.

Raman spectroscopy is similar to infrared spectroscopy in that it involves changes in vibrational energy rather than electronic transitions. A major difference is that symmetrical covalent bonds, which do not absorb energy in the infrared region, can be examined with Raman spectroscopy.

Most collisions between a molecule and a photon of light are elastic; the frequency of the incident beam matches that of the scattered radiation. This is called Rayleigh scattering. Raman scattering occurs when a photon of light has an inelastic collision with a molecule. The energy transferred in the inelastic collision alters both the vibrational energy of the molecule and the energy of the scattered photon. The shift in vibrational energy level is related to the chemical structure of the molecule. The intensity of the scattered light is a measure of how easily electrical charges can be shifted in a molecule to make it more polar, or the polarizability of the molecule. At most, the intensities of the Raman lines are 0.001% that of the source. Raman spectroscopy requires extremely intense incident light, so lasers or arc lamps are used. The scattered radiation is usually detected at right angles to the incident beam (Fig. 35.8).

A Raman spectrum is a plot of the intensity of radiation (Y-axis) against the wavenumber (cm^{-1}) difference between the source and the scattered energy (X-axis). As described earlier, the wavenumber is the reciprocal of wavelength. Three different types of radiation can appear in Raman spectra: Rayleigh scattering, Stokes lines, and anti-Stokes lines (Fig. 35.9). Rayleigh scattering due to elastic collisions gives rise to an extremely intense line with an energy difference of zero because it is identical to the incident radiation.

Stokes lines are fairly intense lines that appear at wavenumbers lower than the incident radiation. These lines occur when a molecule initially at a low-vibrational state is elevated to a higher vibrational state as a result of an inelastic collision. This can occur when a molecule initially at vibrational state $\upsilon = 0$ is raised to vibrational state $\upsilon = 1$ after the collision. The energy difference between the incident radiation and the scattered radiation corresponds to the energy difference between the two vibrational states.

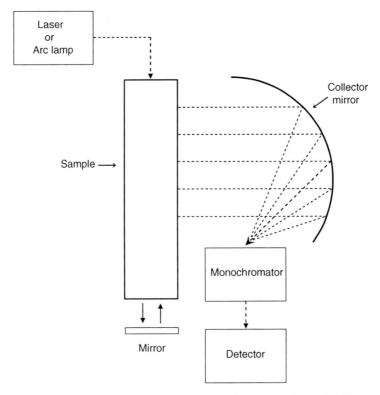

Fig. 35.8 Raman spectrometer (Reprinted from Gordon G. Hammes, Figure 5.2, "Spectroscopy for the Biological Sciences", copyright 2005, with permission of John Wiley & Sons, Inc.)

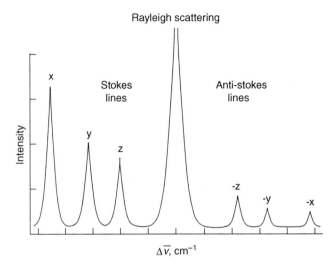

Fig. 35.9 Generalized Raman spectrum (From SKOOG. Principles of Instrumental Analysis, 5E. © 1998 Brooks/Cole, a part of Cengage Learning, Inc. Reproduced by permission. www.cengage.com/permissions)

Anti-Stokes lines appear at wavenumbers higher than the incident radiation and with significantly lower intensity than Stokes lines. These lines arise when a molecule at a higher vibrational state drops to a lower vibrational state after an inelastic collision. This can occur when a molecule initially at vibrational state $\upsilon = 1$ falls to vibrational state $\upsilon = 0$, or ground state, after the collision. Anti-Stokes lines exactly mirror the pattern of the Stokes lines. Usually only the Stokes lines are presented in a Raman spectrum.

The peaks in a Raman spectrum usually appear in an infrared spectrum, but their intensities are reversed. A chromophore that produces a weak IR band will generate an intense Raman peak in the same position. Conversely, one that produces a strong IR band results in a low intensity Raman peak. Raman spectra are usually simpler than IR spectra because overtone and combination bands do not appear. Sample analysis is also somewhat easier than IR spectroscopy because Raman spectra can be obtained from aqueous solutions, crystals, and films; sample cells can be constructed of glass and quartz.

As described earlier, energy in the visible and ultraviolet region of the electromagnetic spectrum is used to elevate electrons from low-energy to high-energy molecular orbitals. Resonance Raman spectroscopy involves the use of incident light with wavelengths near or coinciding with the absorption band for electronic transitions. Raman line intensities resulting from this type of incident radiation can be enhanced by a factor ranging from 100 to 1 million. This amplification permits the analysis of target molecules with extremely low concentrations (10^{-8} M).

Chapter 36
Resonance Spectroscopy

Resonance spectroscopy involves the manipulation of sample nuclei or electrons with electromagnetic radiation in a magnetic field. While nuclear magnetic resonance spectroscopy is typically used for compositional analyses of organic substances, electron spin (or paramagnetic) spectroscopy is more often employed as a means of obtaining an age estimate on crystalline archaeological materials. Basic descriptions of nuclear resonance spectroscopy are available in organic chemistry texts (Solomons 1980; Solomons and Fryhle 2004) and one or both techniques appear in books on analytical techniques (Harris and Kratochvil 1981; Khandpur 2007; Pasto and Johnson 1979; Patnaik 2004; Skoog et al. 1998). Descriptions of resonance techniques directed to researchers in the biological sciences by Brown and Uğurbil (1984), Hammes (2005), and Kosman (1984) are more detailed, yet comprehensible. Many books on resonance techniques, such as Derome (1987), Gordy (1980), and Sanders and Hunter (1993), are specifically targeted toward chemists and physicists.

Nuclear Magnetic Resonance Spectroscopy

Nuclear magnetic resonance (NMR) spectroscopy is used to gain information about the chemical structure of a pure substance by observing the behavior of certain nuclei. Analysis involves placing the sample in a strong magnetic field, which causes the nuclei to align themselves in its direction. The material is then subjected to pulses of energy in the radio frequency (rf) range perpendicular to the magnetic field. Depending upon the precise frequency of the pulse employed, the nuclei of certain elements absorb rf energy and flip to become aligned against the magnetic field. The frequency of energy released by nuclei as it flips back depends upon its neighboring atoms and types of bonds between them. Samples in solution or solid state can be examined.

NMR spectroscopy is possible because the nuclei of atoms have magnetic properties. Nuclei are not static; they spin on an axis. Since a nucleus carries a positive charge, the spinning motion creates a magnetic field. This magnetic field is like that of a bar magnet, with a north and south pole, aligned along the spin axis of

M.E. Malainey, *A Consumer's Guide to Archaeological Science*, Manuals in
Archaeological Method, Theory and Technique, DOI 10.1007/978-1-4419-5704-7_36,
© Springer Science+Business Media, LLC 2011

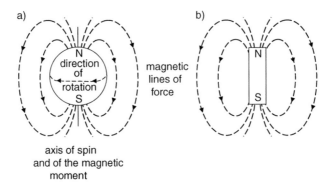

Fig. 36.1   The magnetic field associated with a spinning proton – The spinning proton (**a**) resembles a tiny bar magnet (**b**) (Reprinted from T.W. Graham Solomons, Figure 13.4, "Organic Chemistry", Second Edition, copyright 1980, with permission of John Wiley & Sons, Inc.)

the nucleus (Fig. 36.1). If placed in the core of a strong magnet, a nucleus will align itself so that it is parallel or antiparallel to the externally applied magnetic field. The parallel alignment is the low-energy state; the antiparallel alignment is the high-energy state of the nucleus.

The number of spin orientations, or spin states, a nucleus can assume depends on the configuration of subatomic particles in the atom. If both the atomic number and the atomic mass numbers are even, the nucleus has only one spin state and its spin quantum number is 0. It is not possible to examine these atoms with NMR spectroscopy. If either the atomic number or the mass number of the atom is odd, multiple spin states exist. These atoms, including ^1H, ^{13}C and ^{31}P and about 200 others, have non-zero spin quantum numbers and can be detected using NMR. Most archaeological applications have utilized either proton and/or ^{13}C NMR spectroscopy.

Nuclear magnetic resonance spectroscopy is used to obtain information about the chemical environment of nuclei by inducing transitions from one spin state to another. The amount of rf energy required to cause a transition ranges from 4 to 900 MHz. In a constant magnetic field, the frequency at which these transitions occur depends upon the magnetic moment and the spin quantum number of the nuclei. Consequently, different nuclei can be selected for analysis by simply adjusting the radio frequency.

Absorption of energy by a nucleus, a process called magnetic resonance, is quantized and only occurs at values that correspond to the difference between the low- and high-energy states. If it were possible to examine isolated nuclei stripped of electrons, the amount of energy would be identical for all nuclei of an element. Within a compound, variations in neighboring nuclei and electrons cause the magnetic resonance of different nuclei of the same element to occur at slightly different energies. For this reason, NMR provides information about the structure of molecules.

Instrumentation and Sample Analysis

Components of a nuclear magnetic resonance spectrometer include a powerful magnet and a radio frequency generator, rf detector, rf coil, and a computer for data processing (Fig. 36.2). High-resolution NMR spectrometers use magnetic fields that typically range from about 1.4 to 14 T, although instruments with strengths exceeding 20 T have been developed. A single rf coil surrounding the sample within the magnet can be used to both apply the energy pulse and acquire the signal for the detector; alternatively, separate coils are used. During analysis, the sample tube is spun on its longitudinal axis to reduce the effects of magnetic field inhomogeneities. An air-driven turbine is used to rotate the sample tube between 1200 and 3000 revolutions per minute.

Upon being placed in the magnet, sample nuclei align so that the bulk magnetization is in the direction of the magnetic field. A short pulse, usually lasting less than 10 μs, of rf energy is directed at right angles to the magnetic field. The targeted nuclei absorb energy and flip, so that their magnetization is aligned against the magnetic field. At the end of the pulse, nuclei emit energy and flip back to their original orientations. This emitted energy is detected as a small electric current in the rf coil. With Fourier-transform NMR, the free-induction decay (FID) signal is acquired in the time domain (seconds). It is initially strong then rapidly decays to zero. Another rf pulse is then applied and the resulting signal acquired by the detector. The pulse-acquire sequence is repeated many times and signal averaging is performed to improve the signal to noise ratio.

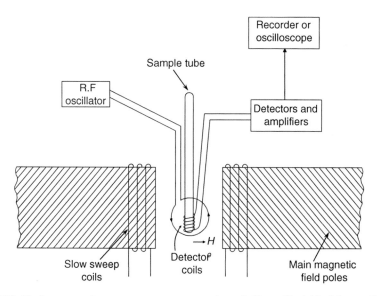

Fig. 36.2 Nuclear magnetic resonance spectrometer schematic (Pasto. Daniel J.; Johnson, Carl R., Laboratory Text for Organic Chemistry: A Source Book of Chemical and Physical Techniques, 1st edition, © 1979 pp. 188. Adapted by permission of Pearson Education, Inc., Upper Saddle River, NJ.)

Nuclear magnetic resonance spectroscopy is usually applied to samples in solution. Sample molecules move rapidly and randomly in isotropic solutions. By simply spinning the sample, magnetic field inhomogeneities are reduced and excellent spectra can be obtained from pure samples.

Proton NMR Spectra

All of the targeted nuclei in sample molecules contribute to the NMR signal. In the case of ^{1}H NMR, each hydrogen nucleus, which is a single proton, provides some information about the overall structure of the molecule. The presence of specific functional groups and their positions in the molecule can be established.

The number of peaks appearing in the spectrum depends upon the chemical environment of each hydrogen atom. In the case of methane, CH_4, the chemical environment of each hydrogen atom is identical and it gives rise to a single signal or peak. Likewise, the hydrogen atoms in both ethane, $CH_3–CH_3$, and dimethyl ether, $CH_3–O–CH_3$, are equivalent and give rise to a single signal. There are two kinds of hydrogen in propane, $CH_3–CH_2–CH_3$, which gives rise to two signals. The chemical environments of the six hydrogens on the two methyl groups are equivalent and give rise to one peak. The chemical environments of the two hydrogens on the central carbon are also equivalent and give rise to a different peak. The area under the peak arising from the hydrogen atoms on the methyl groups is three times greater than the peak arising from the other two hydrogen atoms.

Chemical Shift

The position of a peak on the X-axis, which is called its chemical shift, is a function of the chemical environment of the nuclei that generated the signal. This is affected by the species and/or functional groups to which the nuclei are bonded. The electrons that surround a nucleus create small magnetic fields. This induced magnetic field partially shields or deshields nuclei from the full effect of the external magnetic field. The frequency at which magnetic resonance occurs and the ^{1}H nuclei flip is influenced by the induced magnetic field.

Peak positions are measured relative to a standard, such as tetramethylsilane (TMS); the TMS signal is set to zero at the right side of the spectrum. The 12 equivalent hydrogen atoms on TMS are very highly shielded. Peaks from nuclei that are more highly shielded appear close to TMS. Hydrogen atoms on carboxyl groups, carbonyl groups of aldehydes, or benzene rings are highly deshielded. The electrons that surround these nuclei reinforce the effect of the external magnetic field. These peaks appear relatively far from the TMS signal.

Signal (or Spin–Spin) Splitting

In proton NMR, signal splitting provides additional information about the chemical environment. It results from the magnetic influences of hydrogen nuclei adjacent to those producing the signal. Equivalent protons do not interact with each other but affect neighboring protons in different chemical environments. A peak arising from one or more equivalent hydrogen nuclei is split into two by one hydrogen on an adjacent carbon, producing a doublet. A triplet is formed when a peak is split into three parts by two adjacent hydrogen atoms that are chemically equivalent. A quartet occurs when a signal is split into four parts by three chemically equivalent hydrogen atoms.

Carbon-13 NMR

Whereas all hydrogen atoms in a molecule contribute to the NMR signal, only the ^{13}C isotopes, about 1% of carbon atoms, undergo magnetic resonance. Procedures involved in ^{13}C NMR analyses of isotropic solutions are similar to those for proton NMR. Once the rf frequency of the applied field is adjusted according to the magnetic field strength, narrow peaks corresponding to the signal from each carbon in the molecule appear in the spectra. Spin–spin splitting does not occur in ^{13}C NMR spectra.

Solid-state ^{13}C NMR is more complex because molecules are rigidly held. Even in a finely ground powder, their fixed orientations produce individual dipolar interactions between atoms. Once they flip, it may take several minutes for nuclei to return to ground state. Instead of the sharp, narrow peaks of solution NMR, line broadening causes wide peaks to form. The chemical shifts of ^{13}C nuclei are generally larger than those for hydrogen nuclei, so even broad peaks of solids are separate. Dipolar decoupling, magic angle spinning (MAS), and cross polarization (CP) are used to persuade the molecules of a solid to behave more like those in solution.

Dipolar decoupling and magic angle spinning are applied to reduce the line broadening. Dipolar decoupling reduces interactions between carbon and hydrogen atoms in molecules. The sample is irradiated with rf frequencies that protons absorb while the ^{13}C data are acquired. Magic angle spinning involves rotating the sample very rapidly (about 2000 Hz or 2 kHz) at an angle of 54.7°.

With Fourier transformation NMR, the next pulse is applied only after the effects of the previous pulse have disappeared and all nuclei have relaxed. Relaxation time is usually very long in solids. It can take several minutes for ^{13}C nuclei to return to equilibrium ground state. Cross polarization is used to reduce the time required for ^{13}C nuclei to return to ground state. The technique causes the magnetic field of ^{13}C nuclei to interact with the magnetic field of protons attached to it. The excess energy is diffused enabling the ^{13}C nuclei to return to equilibrium ground state faster.

Fourier Transformation

The output measured by an instrument during sample analysis arises from two sources. One part is the signal produced by the species of interest, which is the actual target of the analysis. The other part of the output is due to random noise. The amount of noise present in a system determines the smallest concentration of the species of interest that can be accurately measured. Signal to noise ratio is a measure of the ability of an instrumentation system to discriminate between signals and noise. Fourier transformation is a method of mathematically enhancing the signal to improve the signal to noise ratio.

In frequency-domain spectroscopy, the output is recorded as a function of frequency. For example, in conventional ultraviolet, visible, and infrared absorption spectroscopy, the sample is scanned over a range of frequencies. The detector sequentially measures the amount of transmitted light at each frequency. Fourier transformation uses time-domain spectroscopy. The applied rf pulse and the resultant output are acquired as a function of time. The application of energy pulses reduces the amount of time for a single scan so that many scans can be collected in a relatively short period of time. This is because the sum of many sine waves at different frequencies is a single square wave (Fig. 36.3). The squareness of the wave increases with the number of frequencies. By applying a single-square pulse of energy, the sample is exposed to many different frequencies simultaneously.

The output signal is also collected in the time domain, so the entire frequency range is monitored at the same time. The output is recorded as beats, which is the periodicity as the waves go in and out of phase. Many data sets are collected and then ensemble averaging is performed. The signal intensity at each frequency is added together then divided by the number of scans to obtain an average scan. Signals from species of interest appear at precisely the same magnitude and position in each scan. Noise is random and appears at different positions in each data set. The magnitude of signals in the average scan does not change, but the magnitude

$$y = A\left(\sin 2\pi vt + \frac{1}{3}\sin 6\pi v + \frac{1}{5}10\pi vt + \ldots + \frac{1}{n}\sin 2n\pi vt \right)$$

where n is a positive odd number

Fig. 36.3 A Fourier series square waveform or pulse is the sum of simple sine terms (From SKOOG. Principles of Instrumental Analysis, 5E. © 1998 Brooks/Cole, a part of Cengage Learning, Inc. Reproduced by permission. www.cengage.com/permissions)

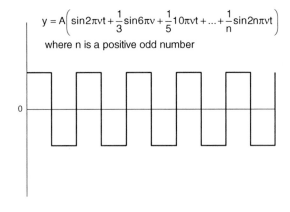

of random noise is significantly reduced. Compared to conventional spectroscopy, Fourier transformation provides higher quality data in a shorter period of time and at a lower cost.

Electron Spin Resonance Spectroscopy

Electron spin resonance (ESR), which is also called electron paramagnetic resonance (EPR), spectroscopy is very similar to ^{1}H NMR. Subatomic particles within the sample are manipulated by electromagnetic radiation while in a strong magnetic field. The difference is that electrons are manipulated, instead of protons, and the manipulating electromagnetic energy is microwave rather than radio frequency. The major difference is that only unpaired, paramagnetic electrons contribute to the ESR signal (Chapter 1). When ESR spectroscopy is used in biochemical applications, signals arising from unpaired electrons of highly reactive radical species are examined. ESR dating of crystalline archaeological or geological materials uses signals from electrons trapped between the valence and conduction bands. Signals from organic radicals present in certain archaeological samples, such as tooth enamel, can interfere with the dating signal; however, ESR measurement conditions can be adjusted to suppress this type of interference.

When placed within the bore of a powerful magnet, paramagnetic electrons behave like small spinning bar magnets. Most of these electrons align in the direction of the magnetic field, which is the low-energy state, but some align against it, which is the high-energy state. Paramagnetic electrons are able to absorb microwave energy applied at a frequency that precisely corresponds to the difference between the high- and the low-energy states; this is electron spin resonance.

Instrumentation

The basic components of an ESR spectrometer are depicted in Fig. 36.4. Microwave radiation is generated by a klystron tube and an attenuator is used to adjust the power (amplitude) of the microwaves. A circulator directs the microwaves to the cavity holding the sample, which is situated in the bore of the magnet. Microwaves emerging from the cavity are directed to the detector by the circulator. The circulator directs any energy deflected by the detector to the load, where it is absorbed.

The microwave (or resonant) cavity is a rectangular metal box; in order for electron resonance to occur, the wavelength (in meters) of the microwave energy must equal the length of the cavity. As described in Chapter 2, since electromagnetic radiation travels at the speed of light, c, wavelength, λ, is inversely related to frequency, $\nu : c = \lambda\nu$. If the cavity is 10 cm (0.1 m) long, it has a corresponding resonant frequency of $2.998 \times 10^9 \text{s}^{-1}$ or 2.998 GHz. A tuning device is used to

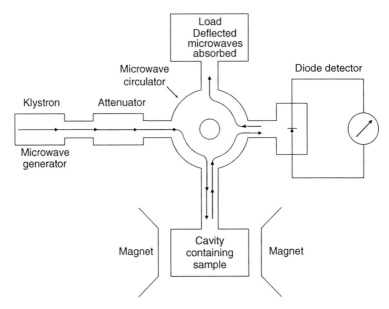

Fig. 36.4 Schematic showing the basic components of an ESR spectrometer (Reproduced with permission, Copyright © 2006, 1989, Tata McGraw-Hill Publishing Company Limited, 7 West Patel Nagar, New Delhi 110 008, India)

adjust the microwave frequency so that it precisely matches the resonant frequency of the cavity.

ESR spectrometers that operate at higher magnetic field strengths are more sensitive. Lowering the temperature of the sample also improves sensitivity, so liquid nitrogen Dewar accessories are available for cavities. Rotating cavities are available for the study of solid samples.

Analysis of the ESR Signal

The frequency at which resonance occurs depends upon the magnetic field strength and the chemical environment of each paramagnetic electron. For this reason, ESR spectroscopy can provide information about the molecular structure of the sample. Since the chemical environment of electrons trapped between the valence and conduction bands in a crystalline sample is similar, they absorb microwave energy at approximately the same field strength. The ESR spectrum consists of one signal or peak in the form of a broad parabola, the size of which provides a measure of the population of trapped electrons. Instead of calculating the area under the peak, the shape of the first derivative of the absorption peak, obtained using calculus, provides a measure of the number of electrons contributing to the signal.

The First Derivative of an Absorption Peak

The central portion of a single absorption peak can be approximated by a simple parabola that opens downward, such as the curve generated by the equation $y = -x^2$ (Fig. 36.5a). Any line that intersects the parabola at only one point is a tangent line and it is useful to describe a particular tangent line by its slope. Slope is defined as the increase in vertical height (rise) over a horizontal distance (run). At the highest point of the parabola ($X = 0$), the tangent line is horizontal and has a slope of zero. Tangent lines to the left of the maximum have positive slopes; those on the right of the maximum have negative slopes. The precise slopes of tangent lines at other points on the X-axis can be obtained by using calculus to obtain the first derivative of the equation, which is $y = -2x$ (Fig. 36.5b).

The flanks of an absorption peak must now be considered. From the leftmost extent of the peak, slopes of tangent lines increase until the curvature changes at the point of inflection. The point of inflection is where the slope of the tangent line reaches its maximum positive value. The opposite occurs on the right flank of the absorption peak, which is the mirror image of the left flank. The point of inflection on the right is where the slope of the tangent line reaches its most negative value. Plots depicting an absorption peak and its first derivative are shown in Fig. 36.6. The vertical distance between the most positive value and the most negative value of the first derivative is used as a measure of the strength of the resonance. Although the advantages of using the first derivative may not be readily apparent for simple spectra, the approach is extremely valuable for the analysis of complex absorption peaks (Fig. 36.7).

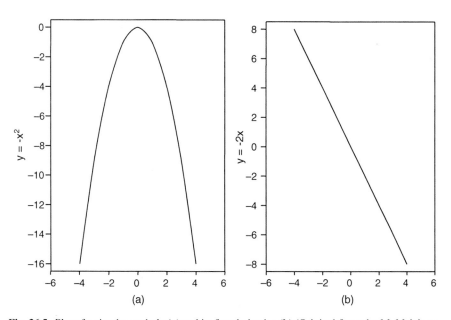

Fig. 36.5 Plot of a simple parabola (**a**) and its first derivative (**b**) (Original figure by M. Malainey and T. Figol)

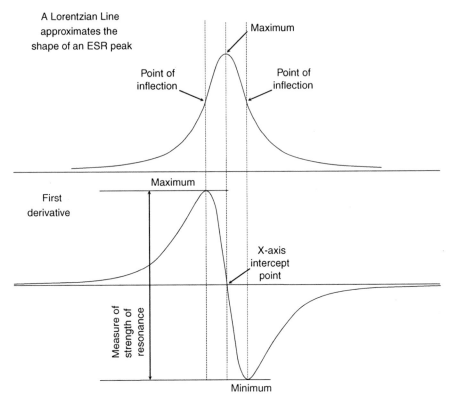

Fig. 36.6 Plot of a hypothetical ESR absorption peak (Lorentzian line) and its first derivative (Reprinted from Walter Gordy, Figure 4.5, "Theory and Application of Electron Spin Resonance", copyright 1980, with permission of John Wiley & Sons, Inc.)

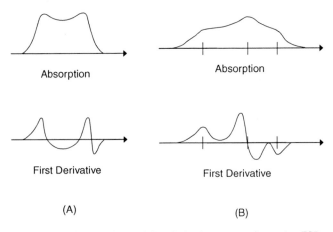

Fig. 36.7 Two examples of absorption and first derivative curves of complex ESR peaks (These figures were published in Structural and Resonance Techniques in Biological Research, edited by Denis L. Rousseau, Electron Spin Resonance by Daniel J. Kosman, pages 96 and 97, Copyright Elsevier 1984.)

Chapter 37
X-Ray and Particle Emission Techniques

The short wavelengths and high energy of X-rays make them a highly versatile form of electromagnetic radiation. Several analytical techniques utilize the interaction of X-rays with a substance, or their emission from excited atoms, to gain information about sample composition. This chapter describes the production of X-rays and analytical techniques that in some way employ the use or analysis of X-rays. Basic descriptions of X-rays and X-ray diffraction and fluorescence are found in physics and chemistry texts (Halliday and Resnick 1981; Halliday et al. 2005; Mortimer 1979, 1986); books on analytical techniques can be consulted for more detailed information (Hammes 2005; Khandpur 2007; Patnaik 2004; Skoog et al. 1998; Verma 2007; Willard et al. 1988). Archaeological applications of these techniques are described by Dran et al. (2000), Elekes et al. (2000), José-Yacamán and Ascencio (2000), Leute (1987), Moens et al. (2000), Rice (1987), Scott (2001), Tite (1992), and Vuorinen (1990).

X-Rays and X-Radiography

X-rays have short wavelengths and high frequencies, 10^{16}–10^{20} Hz, making them highly energetic. X-rays are produced when electrons strike a metal target with great force in a vacuum. Within an X-ray tube, electrons generated by heating a metal wire, or filament, are accelerated by a potential difference toward a metal target (Fig. 37.1). The rate of electron acceleration depends upon the voltage used, but it usually ranges between 30 and 300 kV. The target is the positively charged anode; the force of the impact causes displacement of inner shell electrons of the metal atoms. Electrons fall from outer levels to fill the inner shell vacancies (Chapter 1). The energy released by the rearrangement of electrons due to the inner shell transitions is in the form of X-rays.

The X-rays radiate in geometrically straight beams from the point of impact. They pass through a window in the X-ray tube to the sample, which absorbs them to a greater or lesser degree. Those not absorbed by the sample strike the recording media, which may be a film, screen, or camera. The wavelengths of X-rays emitted

M.E. Malainey, *A Consumer's Guide to Archaeological Science*, Manuals in
Archaeological Method, Theory and Technique, DOI 10.1007/978-1-4419-5704-7_37,
© Springer Science+Business Media, LLC 2011

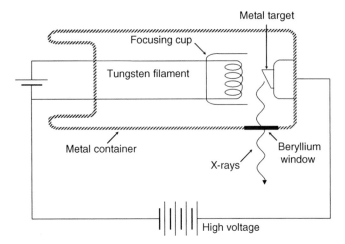

Fig. 37.1 X-ray tube ("Fundamentals of Physics", Second Edition Extended, David Halliday and Robert Resnick, copyright 1981, Reproduced with permission of John Wiley & Sons, Inc.)

are characteristic of the metal used in the target. In general, X-rays produced in a vacuum tube have relatively low energies and longer wavelengths. Intense, short-wave X-rays can be produced by high-energy electron accelerators or synchrotrons.

In X-radiography, X-rays penetrate the sample with more or less absorption until they reach a type of recording media. The amount of absorption largely depends on the thickness of the sample, the atomic numbers of its constituent elements, and the wavelength of the X-rays. Organic samples consist largely of carbon, hydrogen, nitrogen, and oxygen. The highest atomic number among these elements is eight; therefore, most organic materials have a low degree of X-ray absorption. Lead, with an atomic number of 82, has a high degree of X-ray absorption. A sample made up of elements with a wide range of atomic numbers exhibit very good contrast.

X-Ray Diffraction

In 1912, Max von Laue discovered the diffraction of X-rays by crystals. The wavelength of X-rays is approximately the same as the distance between lattice planes in a crystal. The atomic lattice arrangement acts as a diffraction grating for X-rays directed at the sample. Changing the angle with which the X-rays strike the material causes constructive and deconstructive interference of the waves and provides information about its crystalline structure. X-ray diffraction is typically used to characterize minerals; however, X-rays can elucidate the structure of certain organic macromolecules. The structure of DNA and proteins can be examined with X-ray crystallography because strands and fibers within these molecules diffract X-rays.

Diffraction of X-Rays by Crystalline Materials

Crystalline solids, such as minerals, form ionic bonds; unlike the covalent bonds of organic compounds, there is no sharing of electrons (Chapter 1). Instead, an electron from a metal is transferred to a non-metal. The metal becomes a cation, carrying a positive charge; the non-metal becomes an anion, carrying a negative charge. Electrostatic attraction between the oppositely charged ions holds them in a crystal. Each mineral has a unique chemical composition; consequently, the three-dimensional lattice arrangement of atoms in each mineral is also unique. The manner in which a mineral diffracts X-rays is a function of the regular or periodic spacing and arrangement of its constituent atoms.

Diffraction is the scattering of the X-rays by the ordered environment in a crystal. When X-rays strike the sample, some are reflected off the surface while others penetrate and are reflected by planes of atoms within the crystal. Since they have the properties of waves, constructive and deconstructive interference takes place between the reflected and diffracted X-rays.

The path of the X-rays depends on the angle of incidence, θ, which is the angle they strike the surface or inner planes of the crystal. The reflected or exit angle is always identical to incident angle; however, the distance the X-rays travel varies. At certain angles, the extra distance traveled by X-rays diffracted by a subsurface crystal lattice plane equals whole number multiples of wavelength ($n\lambda$). When this occurs, X-rays emerging from within the sample are "in phase" with those reflected by the surface and the waves undergo constructive interference. The amplitude of the X-rays diffracted and reflected by the mineral will be larger than the incident X-rays, producing intensity maxima. At other angles, diffracted X-rays emerge "out of phase" with those reflected by the surface. Under these conditions, deconstructive interference occurs and the amplitude of the X-rays reflected and diffracted by the mineral is smaller than the incident X-rays.

The extra distance the penetrating X-rays travel is related to the interplanar spacing (or distance) of the crystal (Fig. 37.2). The distance between the parallel planes of atoms or molecules within the crystal is a function of its chemical composition. W. L. Bragg described the diffraction of X-rays by crystalline material in the relationship known as Bragg's law or Bragg's equation:

$$n\lambda = 2d \sin \theta$$

where d is the interplanar spacing of the crystal. An X-ray maximum or peak is produced when the difference in path lengths differs by whole numbers of wavelength, $n\lambda$. For X-rays diffracted by the first inner plane of atoms, the extra distance is AP + PC. Right-angle triangles are formed by the points O, A, and P and by the points O, C, and P, with the interplanar spacing, d, as the hypotenuse. We can determine the length of AP (or PC) using geometry:

$$\sin \theta = \frac{\text{opposite}}{\text{hypotensue}} \quad \text{where } \sin \theta = \frac{\text{AP (or PC)}}{d} \quad \text{so that AP} = \text{PC} = d \sin \theta$$

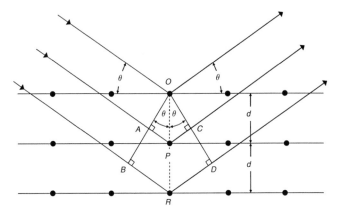

Constructive interference occurs when the distance
$AP + PC = n\lambda$ where n is an integer.

Fig. 37.2 The diffraction of X-rays by a crystal (From SKOOG. Principles of Instrumental Analysis, 5E. © 1998 Brooks/Cole, a part of Cengage Learning, Inc. Reproduced by permission. www.cengage.com/permissions)

The extra distance, $d \sin \theta$, is traveled both before and after diffraction, for a total distance of $2d \sin \theta$. Intensity maxima occur when $2d \sin \theta$ equals 1λ, 2λ, 3λ, and so on. Both the wavelength of the X-rays and the angle at which intensity maxima occur, θ, are known, so the interplanar spacing can be calculated.

The magnitudes of intensity maxima and angles at which they are produced constitutes the diffraction pattern of the crystal. For example, the interplanar spacing of salt, NaCl is 2.52 Å ($\text{Å} = 10^{-10}$ m). If diffraction analysis were performed with X-rays with λ of 1.10 Å, the following pattern of intensity maxima would be observed:

n	1	2	3	4
$\sin \theta / \theta$	0.218/12.6°	0.436/25.8°	0.654/40.8°	0.872/60.7°

Under Bragg's law, there is an underlying assumption that the X-rays will neither gain nor lose energy in the process of being diffracted (elastic scattering). In fact, a very small amount is lost through inelastic scattering, but it is considered negligible.

Sample Analysis

Two methods are used to perform X-ray diffraction, powder photography using the Debye–Scherrer Camera Method and analysis with a diffractometer. In both cases a small portion of the sample must be finely ground or powdered because X-rays only penetrate the uppermost atomic layers of the sample particles. When using either approach there is a danger of preferred orientation. When this occurs, the crystals align themselves in such a way that they never meet Bragg's law. Intensity maxima are not produced so the mineral is undetected.

The wavelengths of X-rays suitable for diffraction analysis range between 0.5–2.5 Å. X-ray tubes with targets composed of copper, molybdenum, iron, or chromium are often used. The various metals produce X-rays that differ with respect to their degree of sample penetration and detectablility by recording media. X-rays are filtered to ensure they have a narrow range of wavelengths (monochromatic).

Powder Photography: Debye–Scherrer Camera Method

Powder photography involves recording intensity maxima of the diffraction pattern on photographic film. A small amount (2–5 mg) of finely ground sample is formed into, or coated onto, a thin rod, about 0.3 mm diameter. This can be accomplished by gently rolling the sample in gum or a vaseline-coated tube or fiber. The goal is to completely coat the exterior of the rod with randomly oriented sample particles. The rod is positioned in the center of a special cylindrical camera with photographic film arranged around the circumference.

Prior to reaching the sample, X-rays are passed through a collimator, consisting of closely spaced metal tubes or plates. Only parallel X-rays pass through the collimator; all others are absorbed. X-rays reflected and diffracted by the rotating sample take the form of concentric cones (Fig. 37.3). The diffraction pattern is recorded on film with intensity maxima appearing as arcs of varying degrees of darkness and thickness. The angle θ is indicated by the position of the arcs on the film. Line intensity is related to quantity of the crystal in the sample and can be measured for quantitative analysis or subjectively assessed. The advantage of powder photography is that it measures a wider range of the total number of diffraction angles.

Diffractometer

Analysis is performed with about 20 mg of powdered sample on a glass slide. The X-ray source is stationary but the sample rotates through angle θ; the detector revolves about the center of the sample through angle 2θ. Diffractometers are fully automated and diffraction patterns can be searched with computerized databases.

X-Ray Fluorescence Spectrometry (XRF)

X-ray fluorescence spectrometry (XRF) is a well-established analytical technique that uses X-rays to excite sample atoms. These primary X-rays cause the release of secondary or fluorescence X-rays, which are analyzed. Different approaches are used to generate the primary X-rays and monitor energy released by sample atoms as they return to ground state. The source of primary X-rays may be an X-ray tube, radioactive isotopes, or a synchrotron. The wavelengths of secondary X-rays can be analyzed using a crystal spectrometer (Bragg reflector) or a lithium-drifted silicon [Si(Li)] detector. Total reflection XRF differs from other types of XRF with respect to the angle at which X-rays strike the sample.

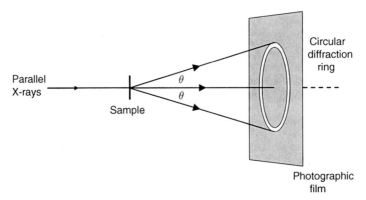

Diffracted X-rays form a cone which appears as
a circle when projected on a vertical surface.

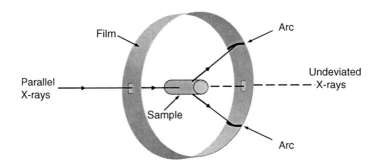

The film intersects only a portion of the cone formed
by the diffracted X-rays so an arc is recorded.

Fig. 37.3 Powder photography: Debye–Scherrer camera method (Reprinted from Harold P. Klug and Leroy E. Alexander, "X-ray diffraction procedures for polycrystalline and amorphous materials" copyright 1974, with permission of John Wiley & Sons, Inc.)

X-rays are a highly energetic form of electromagnetic radiation that penetrate the atom and affect electrons in shells closest to the nucleus. Inner shell electrons absorb the primary X-rays and are ejected. This produces an unstable configuration and electrons from outer levels fill the inner shell vacancies. Energy released by the falling electrons is emitted as secondary or fluorescence X-rays. The energy or wavelengths of the secondary X-rays correspond to the energy differences between the outer shell and the inner shell locations.

All elements, except hydrogen and helium, emit secondary X-rays when exposed to primary X-rays. The energy of primary X-rays used for excitation is constant and the detector measures secondary emissions over a range of wavelengths. An element usually emits a series of secondary X-rays with different wavelengths, so the spectrum for a pure substance can contain multiple peaks. The intensity of secondary X-rays, indicated by the size of the peak, is related to the amount of the element present in the sample. The results are summarized in graphs showing intensity of the peaks (*Y*-axis) plotted against the energy or wavelength (*X*-axis) of the secondary X-rays. Quantitative determinations of elemental concentration are based on X-ray intensities and require the use of a series of calibrations and corrections.

X-ray fluorescence spectrometry of elements with atomic numbers less than 13, ranging from lithium to magnesium on the periodic table, must be performed in a vacuum. The secondary X-rays of these elements have relatively low energies (and longer wavelengths) and are absorbed by air. Consequently, only analysis of elements with atomic number of 13 or higher can be performed in the field using mobile, hand-held X-ray fluorescence devices.

Sample Preparation

The degree of sample preparation required varies with the goal of the analysis. It is possible to non-destructively examine small artifacts that fit inside the sample chamber or a portion of a large, *in situ* structure out in the field. X-ray fluorescence spectrometry is not used for surface analysis but provides information on only the uppermost 100 μm of a sample. Consequently, analysis of an unprepared sample could provide information on a region affected by weathering, corrosion, plating, or other types of contamination. The term "matrix effects" refers to the combined effect of surface texture, sample inhomogeneity, X-ray absorption by the sample, size of particles analyzed, and other factors. Matrix effects are more likely to adversely affect the analysis of unprepared samples. If information about the bulk composition is required, a representative sample should be taken. The sample can be analyzed directly or undergo further preparation. One procedure involves grinding the sample into a fine powder, mixing in a flux to lower its melting point, and heating/baking it so that it fuses into a glass bead.

Even if a sample is not removed, X-ray fluorescence spectrometry may affect an artifact by temporarily or permanently discoloring the specimen. It is not possible to use XRF for point analysis; if required, instruments that use a focused beam, such as an electron microprobe (Chapter 37) or SIMS (Chapter 31), can be used. The precision of analysis varies with the source of primary X-rays, the angle with which the X-rays strike the sample and method used to analyze secondary X-rays.

Instrumentation

Two basic types of X-ray fluorescence instruments are used in archaeological applications: wavelength dispersive (WD) and energy dispersive (ED). These differ with respect to the source of primary X-rays and the analysis of secondary X-rays.

Wavelength-Dispersive XRF (WDXRF or WD-XRF)

Wavelength-dispersive XRF spectrometers generally utilize primary X-rays generated by X-ray tubes operating at 100 kV. These tubes produce a broad beam of

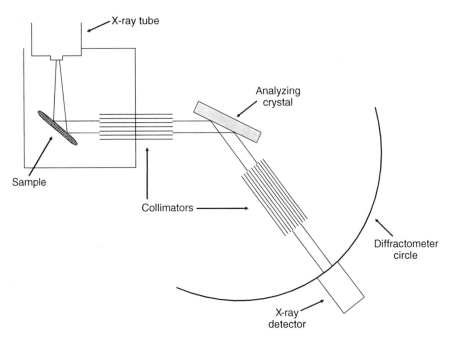

Fig. 37.4 Wavelength-dispersive XRF (Reprinted from E. Ciliberto and G. Spoto, eds., Figure 4.5, "Modern Analytical Methods in Art and Archaeology", copyright 2000, with permission of John Wiley & Sons, Inc.)

X-rays; fine focus tubes can also be used, if desired. The secondary X-rays from the sample are passed through a collimator to ensure that only parallel beams of radiation strike the crystal spectrometer (Fig. 37.4). The diffracted beam is passed into a detector or counter.

A crystal spectrometer, which is also referred to as a Bragg reflector, uses Bragg's law to determine the wavelength of secondary X-rays produced by the sample. The principles are identical to those for X-ray diffraction, described earlier in this chapter. If X-rays of known wavelength (λ) strike the sample at a known angle (θ), the distance between planes of atoms in a crystalline material (d) can be determined. With a crystal spectrometer, the incident angle (θ) and distance between planes of atoms in the crystalline material (d) are known and used to determine the wavelength (λ) of the secondary X-rays. Multiple crystals, each with different atomic lattice arrangements, may be required to analyze all secondary X-rays emitted. Crystal spectrometers perform a sequential analysis of secondary X-rays.

Energy-Dispersive XRF (EDXRF or ED-XRF)

Energy-dispersive XRF spectrometers do not use crystal spectrometers to analyze secondary X-rays; instead, silicon–lithium [Si(Li)] detectors are employed. The temperature of these detectors must be controlled using liquid nitrogen or another process, such as thermoelectrical cooling (Peltier effect). Secondary X-ray photons are converted to electric pulses, which are amplified and simultaneously processed using a multi-channel analyzer. The resolution of these EDXRF instruments is poor compared to WDXRF spectrometers. Regardless, solid-state Si(Li) detectors have enabled the development of hand-held devices that can be used in the field.

A variety of sources can generate primary X-rays for EDXRF. In addition to X-ray tubes, X-rays from radioactive isotope of cadmium (^{109}Cd) and americium (^{241}Am) are utilized. Portable instruments using either primary X-ray source have been developed. The acronym SRXRF refers to XRF using the highly intense X-rays produced by synchrotrons. These primary X-rays have very short wavelengths and improve the sensitivity of the technique. In micro-SRXRF (μ-SRXRF), microbeams improve the spatial resolution so that a small area of the sample can be selected for analysis.

Total Reflection XRF (TXRF)

Total reflection XRF differs from conventional XRF with respect to the interaction of primary X-rays with the sample. The primary X-rays are focused and shaped into a flat, sheet-like beam that strikes the sample at an angle less than 0.1°. At this angle, the incident beam is totally reflected by the surface of the sample carrier (Fig. 37.5). The atoms within the sample are excited and produce secondary X-rays, which are recorded by a detector positioned directly above the sample. Because the sample carrier reflects the primary X-rays, background interferences from it are substantially reduced and the sensitivity of the instrument is improved. Matrix effects are also eliminated when very small samples are analyzed.

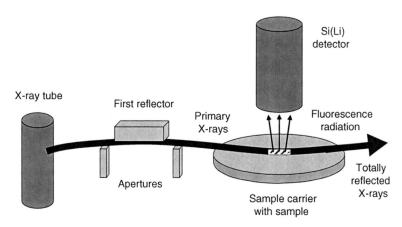

Fig. 37.5 Total reflection XRF (Reprinted from E. Ciliberto and G. Spoto, eds., Figure 4.6, "Modern Analytical Methods in Art and Archaeology", copyright 2000, with permission of John Wiley & Sons, Inc.)

Proton/Particle-Induced X-Ray Emission (PIXE) and Gamma Ray Emission (PIGE)

PIXE is similar to XRF except that heavy-charged particles are used to eject inner shell electrons, rather than primary X-rays. Sample atoms are excited by either protons or alpha particles and produce secondary X-rays. The proton or particle beam can be focused to a diameter less than 1 mm and a large number of elements can be detected and analyzed simultaneously. PIGE is similar but gamma ray emissions from the excited atoms are analyzed instead of X-ray emissions (Fig. 37.6).

As described in Chapters 1 and 3, protons carry a single positive charge and their mass is approximately 1800 times greater than an electron; alpha particles consist of two protons and two neutrons and carry a 2+ charge. Compared to X-rays, protons

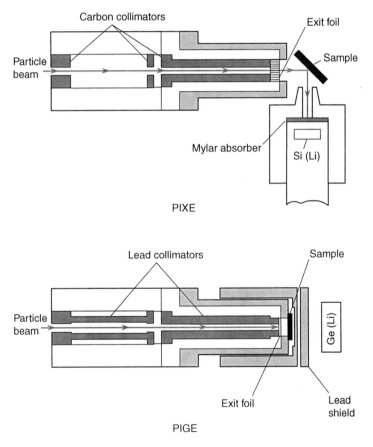

Fig. 37.6 PIXE and PIGE (Reprinted from Journal of Archaeological Science 17(3), Vuorinen, Heikki S., Unto Tapper, and Helena Mussalo-Rauhamaa, and Heavy Metals in Infants, Analysis of Long Bones from Ficana, Italy, 8–6th Century BC., page 240 Copyright 1990, with permission of Elsevier.)

and alpha particles are extremely heavy and large. A particle accelerator operating in the megavolt range is required to produce high-energy beams of either protons or alpha particles for PIXE or PIGE analysis. Lighter elements can be analyzed with a 1 MeV proton beam; heavier elements require a 3 MeV proton beam. Quantitative analysis of elements with atomic numbers ranging between 11 (sodium) and 92 (uranium) is possible using PIXE; however, it is best applied to elements with atomic numbers greater than 15. PIGE can be combined with PIXE to examine the lightest elements in this range, as well as fluorine, with an atomic number of 9.

As with EDXRF, X-ray emissions are usually analyzed using a silicon–lithium [Si(Li)] detector cooled by liquid nitrogen. Filters are required to ensure backscattered particles do not strike the detector. Typically, only X-rays arising from the ejection of electrons from the two innermost levels, K and L, are analyzed (Chapter 1).

Sample Preparation

Due to the size of particles used to induce X-rays, only the surface of the sample is analyzed using PIXE. Analysis can be performed in air or a helium atmosphere, enabling the study of large objects outside a sample chamber with an external ion beam. When used to examine the surface, the technique is non-destructive. A sample must be removed from the object and prepared in order to study its bulk composition. Either a flat cross section of the sample can be analyzed or it can be ground into a powder and formed into a pellet. The quantitative analysis of thin targets is straightforward, while a variety of matrix effects complicate the study of thick targets.

Microbeam analysis typically involves the use of ion beams ranging from about 0.1–1.0 mm, but diameters as small as 20–30 μm have been attained. The risk of damage to the artifact increases with the use of highly focused particle beams.

Surface Characterization Techniques

These techniques are used to study the outer surface of materials. In general, the sample is irradiated with a primary beam, consisting of electrons, ions or neutral molecules; the impact of this beam produces a secondary beam consisting of electrons, ions, neutral molecules, or photons, which is analyzed. With scanning electron microscopy-energy dispersive X-ray spectrometry (SEM-EDS), a beam of electrons is directed at the surface and X-ray emissions from atoms on the surface are analyzed (as with XRF). Electron microprobe analysis is similar but the sample must be prepared as a polished thin section and the diameter of the primary electron beam is between 50 and 200 times smaller, enabling a significantly more precise analysis of the sample surface. Secondary ion mass spectrometry (SIMS) was described in Chapter 31. When using this technique, the surface is bombarded with a beam of positively charged ions formed from argon, cesium, nitrogen, or oxygen. The primary ion beam strips off the surface layer of atoms and occasionally converts them into ions. These secondary ions from the sample are then drawn into a mass spectrometer for analysis.

A common feature of surface characterization techniques is that the primary beam consists of particles, such as ions or electrons. The use of particles ensures that the primary beam does not penetrate the surface and analyze the bulk composition. A 1 keV ion or electron beam will only reach a depth of about 25 Å. A 1 keV photon beam, consisting of highly energetic electromagnetic radiation such as X-rays or gamma rays, can reach a depth of more than 10,000 Å.

Surface Characterization

The aim of many analyses, including provenance studies, is characterization of the bulk composition of a sample. When performing bulk compositional analysis, care must be taken to ensure that a sample representative of the whole is selected. The surface is often avoided because its composition does not necessarily match the bulk.

By definition, the surface of a solid represents a boundary between a solid and an other phase. Under ideal conditions, the surface is not uniform in composition because it is a transition layer that varies continuously with depth. The thickness of the surface is also variable. Under less than ideal conditions, the surface may be altered by reactions with the atmosphere, water, and other substances. These reactions can result in the formation of patinas and hydration rinds; the surface also may be affected by corrosion or rust. The surface may be contaminated by the introduction of another substance into voids. Since it is the area which people see, substances are often intentionally applied to the surface. The surface of an artifact may have been decorated with paints, pigments, slips, or glazes. Substances may have been applied as sealants or adhesives.

Scanning Electron Microscope with ED XRF

Microscopic techniques that merely provide information about the physical nature of the specimen are outside the scope of this text. A scanning electron microscope combined with X-ray fluorescence provides qualitative information about the chemical nature of a sample.

A scanning electron microscope uses a primary beam of electrons to gain information about a sample. The beam is systematically swept across the surface in a raster pattern, which is similar to how images are produced by cathode ray tubes. From its starting position, the beam moves horizontally across the surface of the area selected and returns to its first position. The beam then moves one vertical increment down and scans horizontally and returns to its first position. From there it moves down one additional vertical increment until the area selected is completely scanned.

There is a one-to-one correlation between the signal produced and a corresponding point on the screen where the image is displayed. Magnification is accomplished by reducing the width of the scan lines. By doing this, it is possible to zoom onto a very small portion of the surface. Magnifications of up to 100,000 times are possible when narrow electron beams are used.

Scanning electron microscopy must be performed in a vacuum, with atmospheric pressures of less than 10^{-4} torr. An electron gun is used to produce the beam of electrons. Electromagnetic lenses, one serving as the condenser and another the objective lens, focus the beam down to between 5 and 200 μm. Two pairs of electromagnetic coils inside the objective lens control the position of the beam (Fig. 37.7).

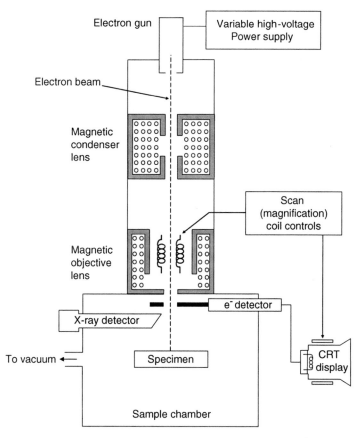

Fig. 37.7 Schematic of a scanning electron microscope (From SKOOG. Principles of Instrumental Analysis, 5E. © 1998 Brooks/Cole, a part of Cengage Learning, Inc. Reproduced by permission. www.cengage.com/permissions)

For scanning electron microscopy, electrons are analyzed using scintillation counters. Bombarding a sample with an electron gun gives rise to two types of electrons that can provide information about its surface: backscattered electrons and secondary electrons. Backscattered electrons are primary electrons that shallowly penetrate the surface then bounce off. Information from the deflecting atoms indicates variations in surface composition, or material contrast, as well as data on surface topography. Secondary electrons are those ejected from the surface by the impact of the primary beam electrons. The number of secondary electrons produced depends upon the angle at which the primary beam strikes the surface, so information about surface topography is generated.

The beam of primary electrons also causes the emission of X-rays. These are analyzed in the same manner as X-ray fluorescence and provide information about the chemical composition of the surface. The scanning electron microscope enables precise selection of the area to be analyzed. The sample is positioned on a stage so that it can be rotated and moved in the X-, Y-, and Z-directions. The data obtained are qualitative and suitable for establishing the presence or absence, but not the quantity or concentration, of a substance.

Sample Requirements

Analysis with scanning electron microscopy with or without energy-dispersive X-ray fluorescence is a non-destructive process. Samples that conduct electricity can be examined without any modification. Conducting samples do not impede the flow of electrons and can conduct heat. A non-conducting material can be examined by coating it with a thin layer of graphite.

Electron Microprobe

Electron microprobe analysis provides both quantitative physical and chemical data about the sample. A narrow, focused beam of electrons is used to stimulate the emission of X-rays from atoms on the surface. These emissions are detected and analyzed in the same manner as X-ray fluorescence.

A vacuum system capable of maintaining pressures of less than 10^{-5} torr is required for the production of the electron beam. Electrons are generated by a heated tungsten cathode and accelerated by an anode to create the beam (Fig. 37.8). Electromagnetic lenses tightly focus the beam on the specimen. Whereas the diameter of an electron beam on the scanning electron microscope ranges between 5 and 200 μm, the beam of an electron microprobe can be focused down to a diameter ranging between 0.1 and 1 μm. An optical microscope is used to select the sample area to be analyzed.

Fluorescent X-rays are collimated and detected by either energy-dispersive or non-dispersive methods. The stage on which the sample is placed can be moved in X- and Y-directions and rotated.

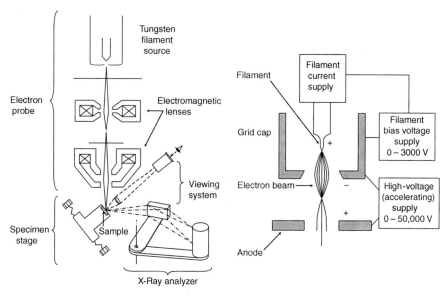

Fig. 37.8 Electron microprobe (From SKOOG. Principles of Instrumental Analysis, 5E. © 1998 Brooks/Cole, a part of Cengage Learning, Inc. Reproduced by permission. www.cengage.com/permissions)

Sample Requirements

Samples must be prepared as polished thin sections prior to analysis. In this respect, it is a destructive process. The sample must be conducting, or coated with conducting material.

Chapter 38
Elemental Analysis

Elemental analysis is the measurement of certain elements within a sample. A variety of techniques used for this purpose, including mass spectrometry, X-ray fluorescence spectrometry, atomic absorption, inductively coupled plasma techniques, and instrumental neutron activation analysis, were described previously. The automatic elemental analyzers discussed in this chapter are typically used to measure elements commonly found in organic and certain inorganic substances, specifically carbon, hydrogen, nitrogen, oxygen, and sulfur. When applied to archaeological materials, carbon–hydrogen–nitrogen (CHN) or carbon/nitrogen (C/N or C:N) ratios can be used to characterize an unknown substance as organic or inorganic or determine if it is a protein. Most textbooks do not explain in detail how these instruments work so the descriptions below are based on product information from Leco Corporation (2007) and PerkinElmer, Inc. (2005). Techniques used for the analysis of data generated by the instruments are outlined in chemistry and instrumentation textbooks such as Pasto and Johnson (1979), Skoog et al. (1998), Solomons (1980), and Solomons and Fryhle (2004).

Instrumentation

The processes performed during elemental analysis vary according to the targeted elements but typically involve sample combustion, reduction to simple gases, and separation of combustion gases and detection. A portion of the substance is carefully weighed into a quartz combustion tube, or a metal vial made of either tin or aluminum, and placed inside the instrument. The system is then completely purged with the carrier gas, usually helium or argon, to remove all interfering gases. Combustion is performed in a stream of oxygen at around 950°C or in the presence of metal oxides, such as cobalt oxide or a mixture of manganese dioxide and tungsten oxide. The combustion gases are then exposed to reagents, usually copper and copper oxide, to convert carbon monoxide, CO, into carbon dioxide, CO_2, and different nitrogen oxides into N_2; sulfur is measured as sulfur dioxide, SO_2, and hydrogen is measured as water vapor, H_2O.

M.E. Malainey, *A Consumer's Guide to Archaeological Science*, Manuals in Archaeological Method, Theory and Technique, DOI 10.1007/978-1-4419-5704-7_38, © Springer Science+Business Media, LLC 2011

As outlined below, oxygen can be indirectly determined as the remainder using mathematical calculations. Direct measurement of oxygen may require a separate instrument module in which the sample undergoes pyrolysis (high-temperature decomposition) with carbon in a helium atmosphere to produce CO. In place of a reduction step, copper oxide is used to oxidize the CO into CO_2, which is then detected and measured.

Prior to analysis of the combustion gases, interfering substances must be removed; what constitutes an interfering substance depends upon the purpose of the analysis. For example, if only carbon and hydrogen are to be analyzed, all other substances must be eliminated, as they would interfere with this measurement. Nitrogen oxides are removed with manganese dioxide; calcium oxides remove sulfur oxides; magnesium oxide eliminates fluorine, and silver wool traps halogens (chlorine, iodine, and bromine) as well as combustion products containing sulfur and phosphorus.

Analysis of the combustion gases involves separation and detection but varies according to system design. Chromatography can be used both to separate the gases and measure the amounts present. In other cases, chromatography is only used to isolate the individual gases; measurements are made simultaneously using dedicated infrared detectors with filters that restrict wavelength.

Determinations of carbon, hydrogen, and nitrogen (CHN) can be made using three pairs of thermal conductivity detectors that monitor changes in the composition of the combustion gases resulting from interactions with absorbent materials. The composition of the combustion gas is determined in the sample side of the first detector; then the gas passes through an absorption trap (containing magnesium perchlorate) that removes the water vapor. The composition of the gas is then re-measured in the reference side of the first detector; the difference represents the amount of water, hence hydrogen, in the sample. The gas is then passed to the second pair of thermal conductivity detectors, which monitor changes in combustion gas composition related to the removal of CO_2 (in a soda–asbestos absorption trap) to determine the amount of carbon in the sample. The third pair of thermal conductivity detectors is used to determine the amount of nitrogen by comparing the combustion gas, which now consists of only nitrogen and helium (the carrier gas), to pure helium.

Elemental analysis of one sample can be completed in less than 10 min; if the instrument is equipped with an autosampler, the next sample is automatically introduced.

Data Analysis

Elemental analysis can be used to establish whether an unknown substance is organic and, if so, determine its empirical formula. Organic compounds contain primarily carbon, hydrogen, oxygen, and nitrogen; when burned in oxygen and reduced, the combustion gases consist of carbon dioxide, water vapor, and nitrogen gas. Levels of these combustion products are determined and the calculated amounts of carbon, hydrogen, and nitrogen are compared to the initial mass of the unknown.

Analysis of compounds containing only carbon and hydrogen (hydrocarbons) or only carbon, hydrogen, and nitrogen (amines) is straightforward. The amount of carbon, hydrogen, and nitrogen in the combustion products fully accounts for the initial mass of the unknown and gives the percentages by weight of these elements. In order to determine the empirical formula, it is necessary to divide the amount of each element by its atomic mass in grams. The empirical formula only gives the simplest ratio of atoms; although C_2H_4, C_4H_8, and C_6H_{12} are different compounds, they share the same empirical formula CH_2. Other analytical techniques, such as mass spectrometry, are required to determine the actual molecular formula of the substance.

Many different compounds, including alcohols, ethers, esters, carboxylic acids, and carbohydrates, contain carbon, hydrogen, and oxygen. Amino acids mainly consist of carbon, hydrogen, nitrogen, and oxygen. Since it is used for combustion, the oxygen composition in the sample must be determined indirectly or measured separately. In order to determine the empirical composition of a compound containing n different elements, the precise amounts of $n-1$ elements need to be determined and the final element is the remainder. Consequently, it is possible to measure oxygen as the difference between the amounts of carbon, hydrogen, and nitrogen produced by combustion and the initial mass of the sample. Some elemental analyzers can also determine the amount of sulfur present in an unknown.

If the empirical formula has been determined and the molecular weight is known, it is possible to determine the molecular formula of a substance. Once this has been obtained, a chemist can further elucidate the structure of the molecule. The number of points of unsaturation can be determined by means of the formula

$$N = \frac{\sum_i n_i(v_i - 2) + 2}{2}$$

where N is the number of sites of unsaturation in a molecule, n_i is the number of atoms of element i and v_i is the absolute value of the valence of element i. The valence of an element is the number of electrons in its outermost shell.

Chapter 39
Mössbauer Spectroscopy

Unlike most other techniques in Part IV, Mössbauer spectroscopy does not regularly appear in general textbooks on chemistry or analytical techniques. Bancroft (1973), Cranshaw et al. (1985), Dickson and Johnson (1984), Edworth et al. (1987), and Verma (2007) offer detailed descriptions of the technique for scientists and an excellent introduction is available from the Royal Society of Chemistry web site (http://www.rsc.org/membership/networking/interestgroups/mossbauerspect/intropart1.asp). Leute (1987) and Rice (1987) describe archaeological applications of the technique.

The technique utilizes the Mössbauer effect, named after the physicist who discovered it, Rudolph Mössbauer. The phenomenon, also known as recoilless gamma ray emission and absorption, is associated with relatively few atomic nuclei. Only about 35 Mössbauer isotopes exist, the most common of which is an iron isotope (^{57}Fe). Other elements with Mössbauer isotopes include ruthenium (Ru), tin (Sn), antimony (Sb), tellurium (Te), iodine (I), europium (Eu), gadolinium (Gd), dysprosium (Dy), erbium (Er), ytterbium (Yb), and neptunium (Np).

The nuclei of Mössbauer isotopes are able to emit and absorb gamma rays, the energy of which corresponds to the difference between their excited and ground states. Resonance is possible because the energy emitted by an excited Mössbauer isotope as it relaxes can be absorbed by the nucleus of another Mössbauer isotope at ground state. The energy of the gamma ray emitted is equal to the energy of the gamma ray absorbed. This unique property makes it possible to investigate the physical and chemical environments of a sample that contains Mössbauer isotopes.

In most other nuclei, a portion of the energy associated with gamma ray emission or absorption is lost to recoil. The recoil associated with emission is akin to that observed when a large cannon is fired. A portion of the energy launches a small, light projectile a great distance in one direction and the remainder causes the large heavy cannon to kickback a small distance in the opposite direction (Fig. 39.1). Likewise, a portion of the absorption energy causes the absorber to move in the same direction of the absorbed energy. The total amount of momentum in the system is conserved:

Total energy (emitted/absorbed) = gamma ray energy + recoil energy

M.E. Malainey, *A Consumer's Guide to Archaeological Science*, Manuals in Archaeological Method, Theory and Technique, DOI 10.1007/978-1-4419-5704-7_39, © Springer Science+Business Media, LLC 2011

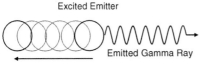

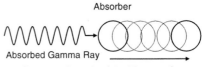

Recoil associated with Emission   **Recoil associated with Absorption**

The gamma ray energy emitted   In order for absorption to occur, a
by the nucleus is less than the   gamma ray must overcome the
difference between the excited   energy loss associated with recoil.
and ground state due to recoil.

Fig. 39.1 Emission and absorption of gamma rays with recoil ([Fig. 1: Recoil of free nuclei in emission or absorption of a gamma-ray, http://www.rsc.org/membership/networking/interestgroups/mossbauerspect/intropart1.asp] – Reproduced by permission of The Royal Society of Chemistry)

Resonance usually cannot occur among these nuclei because the energy of the gamma ray produced does not match the difference between the excited and ground states of that isotope. Rare exceptions occur in less than one in one million cases, when two nuclei are moving in such a way that their respective momentums precisely match the recoil energy and resonance occurs.

If an atom is not chemically bonded to other atoms (i.e., free), the recoil energy associated with emission or absorption causes the particle to be displaced (Fig. 39.1). When an atom bound to others in a crystal lattice emits or absorbs energy, the recoil energy spreads to neighboring particles and the lattice vibrates. These vibrations are called phonons. Mössbauer isotopes within a solid are able to emit or absorb low-energy gamma rays without producing phonons, so the events are, for all intents and purposes, recoilless. Consequently, it is possible for a gamma ray emitted by a Mössbauer isotope fixed in a crystal environment to be absorbed by another of the same species fixed in an identical crystal environment (Fig. 39.2). This "recoil-free resonant absorption of nuclear gamma radiation" is the basis of Mössbauer spectroscopy.

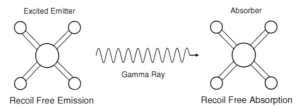

The energy of the gamma ray emitted matches the difference
between the excited and ground state and absorption occurs.

Fig. 39.2 Recoil free emission and absorption of gamma rays ([Fig. 3: Recoil-free emission or absorption of a gamma-ray when the nuclei are in a solid matrix such as a crystal lattice, http://www.rsc.org/membership/networking/interestgroups/mossbauerspect/intropart1.asp] – Reproduced by permission of The Royal Society of Chemistry)

The precise energy of the gamma ray absorbed depends upon the electronic and magnetic environment of a nucleus, which is called its hyperfine interactions. The energy differences associated with these hyperfine interactions are tiny but critical. Absorption only occurs if the local nuclear environment of the emitting Mössbauer isotope matches that of the absorber. By varying the energy of emitted gamma rays slightly, Mössbauer spectroscopy utilizes the energy differences due to hyperfine interactions to gain information about the environment of the absorbing species in a sample.

Instrumentation

Mössbauer spectroscopy is conducted using a specific Mössbauer isotope fixed in a solid matrix in order to target identical Mössbauer isotopes in the solid sample matrix. The energy of the emitted gamma rays directed at the sample is manipulated by slowly moving the gamma ray source material. When the source material moves closer, the effective energies of the gamma rays interacting with the sample increase. When the source material moves away, the effective energies of the gamma rays interacting with the sample decrease. This method of frequency modulation harnesses the Doppler effect.

To visualize the arrangement, picture someone standing by a set of railroad tracks as a train approaches and the engineer blows the whistle. From the perspective of the engineer and everyone else on the train, the sound of the whistle remains constant; the experience of the stationary observer is quite different. As the train approaches, the pitch (i.e., frequency) of the whistle increases to a maximum as it passes by the observer then decreases as the engine moves away. The speed of the train affects the exact sound heard by the observer.

This "frame-of-reference" method of frequency modulation changes the energy of gamma rays experienced by the sample nuclei without altering either the emitter nucleus or the emitted gamma rays. A given Mössbauer isotope in the sample is exposed to a wide range of gamma ray frequencies and absorbs the one consistent with its specific nuclear environment. By systematically changing the speed at which the emitter moves, a variety of energy changes relating to hyperfine interactions between atoms and their environments within a sample can be detected.

The basic components of a Mössbauer spectrometer include (1) a gamma ray source, (2) a frequency modulator that moves the source relative to the sample, (3) a detector able to measure changes in gamma ray energy related to absorption by the sample, and (4) a data processing system (Fig. 39.3). The most widely used gamma ray source is ^{57}Fe, which is the product of the electron capture decay of cobalt-57 (^{57}Co). The radioactive isotope in a matrix of palladium (Pd), platinum (Pt), or rhodium (Rh) has a half-life of 270 days and produces excited ^{57}Fe nuclei. While 91% of all excited ^{57}Fe isotopes decay to produce a 14.4 keV gamma ray, only about 10% of the emissions are recoil free. These emissions are available for absorption by ^{57}Fe in the sample matrix.

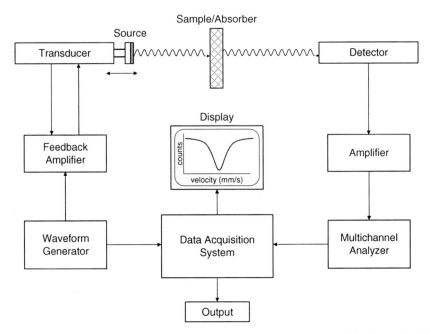

Fig. 39.3 Schematic diagram of a Mossbauer spectrometer (From Atomic and Nuclear Analytical Methods: XRF, Mössbauer, XPS, NAA and Ion-Beam Spectroscopic Techniques, 2007, by H. R. Verma, Figure 4.8, with kind permission of Springer Science and Business Media.)

As noted above, frequency modulation is achieved by moving the source toward and away from the sample; a velocity transducer controls the speed of this movement. Since energy differences related to hyperfine interactions are very small, velocities of only a few millimeters per second (mm/s) are required for the ^{57}Fe source. Instead of using the miniscule frequency changes induced by the back and forth motion, spectra graphically portray absorptions with respect to the speed and direction of the source relative to the sample. Movement toward the sample is given positive values; movement away from the sample is given negative values. If hyperfine interactions are absent from both the isotopes in the source and those in the sample, absorption occurs when the source is at a speed of 0 mm/s (i.e., stationary).

Argon–methane-filled proportional counters are commonly used to monitor gamma rays that pass through the sample; however, scintillation and silicon lithium detectors can also be used. These devices count the number of gamma rays that reach the detector; absorptions are indicated by drops in the gamma ray counts. Samples must be very thin in order to allow the gamma rays to pass through unimpeded. As certain hyperfine interactions are affected by temperature, some systems enable absorptions to be monitored at temperatures that range from –278°C, which is just a few degrees above absolute zero, to more than 1000°C. The detector is able to discriminate the Mössbauer gamma radiation from any external sources in the gamma or X-ray region that may be present.

The data are stored and presented as a plot of counts detected vs. the velocity of the source relative to the sample. A number of velocity sweeps are required to achieve a good signal-to-noise ratio, so a few hours are usually required to complete the analysis. If very little ^{57}Fe is present in the sample, a few days or weeks may be required. Since analysis involves the radioactive decay of the source, the standard deviation is the square root of the number of counts.

Sample Requirements

Mössbauer spectroscopy can only be applied to solid samples that contain a certain minimum quantity of the isotope of interest. The number of Mössbauer nuclei per unit area is related to the sample dimensions and the concentration of nuclei. Although it is the most abundant Mössbauer isotope, only about 2% of iron atoms occur as ^{57}Fe. The sample must be very thin in order to limit non-resonant absorption of gamma radiation by all atoms. A typical sample measures about 0.5 cm thick and has a cross section of about 2 cm^2 (Dickson and Johnson 1984). Ideally, a sample of this size should contain about 10 mg of iron, so that 0.2 mg of the ^{57}Fe isotope is present. Pottery is the most frequently analyzed sample in archaeological applications.

Data Interpretation

As noted above, if the environment of Mössbauer isotopes in the sample is identical to that in the source and does not involve any hyperfine interactions, a single absorption peak will be recorded at 0 mm/s. Deviations from this relate to the specific hyperfine interactions that occur between the nucleus and its surrounding electrons. The three main categories of hyperfine interactions are isomer (or chemical) shifts, quadrupole splitting, and magnetic splitting.

Isomer shifts cause absorptions to occur at speeds other than 0 mm/s. This type of hyperfine interaction arises from differences in the electron charge density around the nucleus. It is of special interest to archaeologists as it can be used to determine the valence state of iron in pottery, which indicates the conditions under which a vessel was fired. A ferrous ion (Fe^{2+}) has six electrons in its 3d shell, while a ferric ion (Fe^{3+}) has only five (see Chapter 1). Occasionally, electrons occupying the spherical 3s orbital are actually farther from the nucleus than those in the 3d orbital. When this occurs, the extra 3d electron in a ferrous ion shields the 3s electrons from the nucleus. The screening effect enables the 3s orbital to expand, which reduces the electron charge density around the nucleus. As a result, ferrous ions have a larger positive isomer shift than ferric ions, so the Fe^{2+} peak occurs to the right of the peak arising from Fe^{3+} ions.

Quadrupole splitting occurs when the charge distribution around the nucleus is non-spherical. When this occurs, the nuclear energy levels are split into a high- and low-spin state. A single peak divides into two, producing a doublet.

Magnetic splitting can arise from the magnetic field generated by unpaired electrons surrounding a nucleus or by an applied magnetic field. Nuclear energy levels are split into different substates, the number of which depends upon the spin state of the nucleus. This type of splitting can result in very complex spectra. In the case of ^{57}Fe, the energy levels of nuclei at ground state are split into two and those of excited nuclei are split into four substates so the peak appears as a sextet (six peaks).

Glossary

Ablation: the process of converting a tiny portion of the surface of a solid into fine particles or vapors, using a spark, arc, or focused laser beam, without altering the chemistry of the sample

Accelerator mass spectrometry (AMS) dating: a method of obtaining a radiocarbon age using accelerator mass spectrometry to measure the relative concentrations of carbon atoms in a sample

Accelerator mass spectrometry: a method of sorting atoms on the basis of mass and measuring their relative concentrations using a particle accelerator and a mass spectrometer.

Accuracy: a term used to describe how close a value, such as one obtained by measurement, is to the actual, true, and correct amount

Acid: a solution with a pH less than 7 (neutral); a solution in which the hydronium ion concentration is higher than the hydroxide ion concentration; a substance that readily donates protons to form hydronium ions

Additive (dose) method: in trapped charge dating, a procedure for extrapolating the equivalent dose, Q, that involves artificially irradiating some portions of the sample prior to measuring its signal then plotting the strengths of the natural signal and natural + artificial signals against the size of the additive doses

Adenine: a purine nitrogen base that occurs in DNA and RNA

Adenosine diphosphate (ADP): a ribonucleic acid (RNA) consisting of adenine bonded to ribose and a highly energetic bond between two inorganic phosphate groups

Adenosine monophosphate (AMP): a ribonucleic acid (RNA) consisting of adenine bonded to ribose and an inorganic phosphate group

Adenosine triphosphate (ATP): a ribonucleic acid (RNA) consisting of adenine bonded to ribose and two highly energetic bonds between three inorganic phosphate groups

M.E. Malainey, *A Consumer's Guide to Archaeological Science*, Manuals in Archaeological Method, Theory and Technique, DOI 10.1007/978-1-4419-5704-7, © Springer Science+Business Media, LLC 2011

Affinity: in chromatography, the degree to which a sample component carried by a mobile phase chemically interacts with the stationary phase

Alcohol: an organic compound containing a hydroxyl group bonded to a carbon atom, named by replacing the -*e* of the corresponding alkane with -*ol*

Aldehyde: an organic compound with an oxygen atom connected to the terminal carbon through a double bond, also known as a carbonyl group; simple aldehydes are named by replacing the -*e* of the corresponding alkane with -*al*

Aliquot: a fractional portion of a larger amount of material

Alkaloid: a special type of amine; a group of compounds that includes nicotine, morphine, and cocaine

Alkane: a hydrocarbon consisting of carbon and hydrogen atoms connected through single covalent bonds

Alkene: a hydrocarbon containing at least one pair of carbon atoms connected through a double bond and named by replacing the -*ane* of the corresponding alkane with -*ene*

Alkyl group: a group formed by the removal of a hydrogen atom from an alkane

Alkyne: a hydrocarbon containing at least one pair of carbon atoms connected through a triple bond and named by replacing the -*ane* of the corresponding alkane with -*yne*

Alpha particle: a subatomic particle consisting of two protons and two neutrons with a charge of 2+ and an atomic mass of 4; a radioactive decay product

Ambient Inhalable Reservoir (AIR):   an international standard used to determine the stable nitrogen composition of a sample; atmospheric nitrogen

Amide: an organic compound that contains a carbonyl group attached to a nitrogen atom

Amine: an organic compound that contains a nitrogen atom bonded to one or more alkyl groups

Amino acid racemization: the reversible process by which the configurations of particular amino acids are converted from their initial L- (or *S*-) form to the D- (or *R*-) form

Amino acid: an organic compound containing a carbon atom bonded to an amino group, a carboxyl group, a hydrogen atom, and 1 of 20 different side chains; the building blocks of proteins; L-amino acids rotate plane-polarized light to the left; D-amino acids rotate plane-polarized light to the right

Amplification: in DNA analysis, use of the polymerase chain reaction to exponentially increase the number of a targeted DNA sequence

Anion: an atom or group of atoms that carries a negative charge because the number of electrons is greater than the number of protons

Annealing: the process of heating material, such as metal and glass, to a certain temperature for a period of time then allowing it to cool to modify its physical properties and relieve internal stress; the second step of PCR where synthetic nucleotide primers attach to sample DNA templates

Annual Dose: in trapped charge dating, the amount of radiation to which the sample was exposed each year

Anode: in an electric field, the positively charged pole toward which negatively charged anions move

Antibody: a globular protein that functions to protect an organism by binding a specific foreign substance, called an antigen

Antibonding orbital: in UV and VIS spectroscopy, the high-energy region in space where electrons shared in a covalent bond have the lowest probability of occurring

Antigen: a potentially harmful substance that is foreign to an organism and can be bound by specific antibodies

Antiserum (-sera): blood serum that carries antibodies for one particular type of protein (monoclonal) or a variety of proteins (polyclonal)

Anti-Stokes line: in Raman spectroscopy, a spectral line arising from an inelastic collision between a photon of light and a molecule that lowers the vibrational state of the molecule

Apatite: a calcium phosphate salt that forms the inorganic component of bones and tooth enamel

Aqueous: water-based

Archaeointensity: the strength of the thermoremanent (heat-induced) magnetization of a sample, which is linearly proportional to intensity of the geomagnetic field in the past

Archaeomagnetic dating: a means of obtaining an age estimate by comparing the direction (and intensity) of magnetization retained in archaeological materials with known variations in the Earth's magnetic field that have occurred over time

Archaeomagnetism: the study of the direction and intensity of magnetization retained in archaeological materials that contain iron-rich minerals and were heated at the time of site occupation

Argon–argon (Ar–Ar): a dating technique applied to volcanic materials involving the measurement of two argon isotopes to determine the amount of argon gas formed from the radioactive decay of potassium-40 and estimate the original amount of potassium in a sample

Aromatic compound: a compound containing an unsaturated, yet highly stable, ring structure, most often consisting of six carbon atoms (a benzene ring)

Arrhenius equation: in chemistry, an equation that describes the exponential relationship between temperature and the rate at which a chemical reaction occurs

Atmospheric pressure chemical ionization (APCI):   a method of ionizing and introducing a sample component separated by high-pressure liquid chromatography into a mass spectrometer; a technique used in LC-MS

Atmospheric pressure photoionization: a method of ionizing and introducing a sample component separated by high-pressure liquid chromatography into a mass spectrometer; a technique used in LC-MS

Atom: a particle consisting of a nucleus containing protons and neutrons surrounded by electrons, except for hydrogen which consists of only one electron and one proton; the smallest portion of an element that retains its fundamental properties

Atomic absorption spectroscopy (AA or AAS):   an analytical technique that determines the presence and concentration of an element on the basis of the amount of characteristic light a sample absorbs when it is atomized in a flame, electrothermal device, or glow discharge device

Atomic mass number: the total number of protons and neutrons in the nucleus of an atom

Atomic number: the number of protons in the nucleus of an atom and the basis of ordering elements in the periodic table; a number shared by all atoms of an element (the number of electrons and neutrons can vary)

Atomic weight: the average mass of the isotopes of an element weighted according to their natural abundance; a value appearing in the periodic table of elements

Atomization: the process of converting the elements of a solid into gaseous atoms and ions using a spark, arc, glow discharge device, or laser

Attenuated total reflectance-IR spectroscopy: an analytical technique that uses energy changes in an IR beam totally reflected by an optically dense crystal to measure the absorption by a sample placed directly on the crystal; a technique that permits IR spectroscopic analysis of liquid or solid samples

Autosampler: a device that automatically loads the next sample into an instrument when analysis of the previous sample is complete

Avogadro's number: 6.022×10^{23}; the number of atoms in 1 mol of any element or the number of molecules in 1 mol of a compound; the number of atoms in exactly 12 g of the most common isotope of carbon, ^{12}C

Bacteriophage: see **Cloning vector**

Base pair rule: in DNA, the purine base, adenine, can only form hydrogen bonds with the pyrimidine base, thymine, and the purine base, guanine, can only form hydrogen bonds with the pyrimidine base, cytosine, which gives rise to two complementary strands

Base: a solution with a pH greater than 7 (neutral); a solution in which the hydroxide ion concentration is higher than the hydronium ion concentration; a substance that readily accepts protons to form hydroxide ions

Basic local alignment search tool (BLAST): a computerized database of reference material for comparisons of nucleotide sequences

Beer's law: the absorbance, A, of electromagnetic energy by an element or molecule is related to its absorptivity, ε, its concentration, c, and the path of the light beam, l; $A = \varepsilon c l$

Beta (β)-sitosterol: a sterol found in plant tissue; a structural lipid present in cell membranes

Beta emission: a subatomic particle with the charge and mass equal to that of an electron; a radioactive decay product

Biogenetic: formed by living organisms

Biomarker: in lipid residue analysis, a molecule associated with a narrow range of substances, or the presence and distribution of certain types of lipid, that enables a residue to be identified with a high degree of precision

Bivariate plot: in compositional analyses, a plot that shows how the concentration of one element varied with respect to another element in samples

Bleaching: in trapped charge dating, the removal of electrons from unstable, shallow traps by exposure to light

Bonding orbital: in UV and VIS spectroscopy, the low-energy region in space where electrons shared in a covalent bond have the highest probability of occurring

Booster PCR: in DNA analysis, a method of increasing the success of PCR amplification by increasing the number of DNA templates

Browser: a ruminant that eats browse (leaves, woody stems, herbs, and fruits)

Buffer: a concentrated solution consisting of a weak acid and its salt that is capable of maintaining its pH at some fairly constant value

C:N (or C/N) ratio: the ratio of carbon atoms to nitrogen atoms in a substance

C_3: the Calvin(–Benson) photosynthetic pathway; plants that follow this process

C_4: the Hatch–Slack photosynthetic pathway; plants that follow this process

Calcite: a mineral consisting only, or mainly, of $CaCO_3$; the principal mineral of limestone and marble

Calibration: in radiocarbon dating, the process of converting a radiocarbon age estimate into calendar years

Calvin (–Benson) cycle: the photosynthetic pathway of plants native to temperate environments, fruits, trees, Near Eastern crops, and aquatic plants that involves intermediate products consisting of three carbon atoms; also called C_3 photosynthesis

Carbohydrate: an organic compound produced by plants that can be broken down into monosaccharides; a group of compounds that includes sugars, starches, and cellulose

Carbon cycle: the movement of organic and inorganic carbon through different parts of the environment

Carbon dioxide: a molecule consisting of one atom of carbon and two atoms of oxygen

Carbon exchange reservoir: the total amount of carbon in the environment

Carbonate: a mineral that contains the carbonate group, CO_3^{2-}; a group of minerals that includes calcite, dolomite, and argonite

Carbonyl group: the functional group common to aldehydes, ketones, and carboxylic acids consisting of an oxygen atom connected to a carbon atom through a double bond

Carboxyl group: the functional group common to carboxylic acids consisting of a hydroxyl group (–OH) attached to the carbon atom of a carbonyl group

Carboxylic acids: an organic compound with a carboxyl group at the end of a carbon chain, named by replacing the *-e* of the corresponding alkane with *-oic acid*

Cathode: in an electric field, the negatively charged pole toward which positively charged cations move

Cation: an atom or group of atoms that carries a positive charge because the number of electrons is less than the number of protons

Cation-ratio dating: a method of inferring the age of archaeological or geological material from the chemical composition of adhering rock (or desert) varnish

Centrifuge: a device that uses centrifugal force to separate material with different densities, such as solids from a solution; the process of rapidly spinning a sample so that the densest material collects at the bottom and least dense (the supernatant) collects at the top

Chemical shift: in NMR spectroscopy, peak position relative to a standard, which is a function of the chemical environment of the nuclei that generated the signal

Chiral: having the property of handedness; a carbon attached to four different groups

Cholesterol: the major sterol in animal tissue; a structural lipid present in cell membranes

Chromatography: the name given to techniques used to separate a sample mixture, carried by a mobile phase, into its constituent components through their differential interactions with a stationary phase

Chromophore: in spectroscopy, a specific grouping of atoms that absorbs light at a particular wavelength

Cloning vector: in DNA analysis, a DNA molecule (plasmid) or a DNA virus (bacteriophage) into which foreign DNA can be inserted for bacterial cloning

Cloning: in DNA analysis, the process of making multiple identical copies (clones) of DNA with bacteria using recombinant DNA technology

Cluster analysis: a type of numerical classification that uses the state or value of attributes that vary significantly between individuals to form groups or clusters of data

Collagen: a fibrous protein that forms most of the organic component of bone

Collimator: a device consisting of closely spaced metal tubes or plates that permit passage of only parallel beams of electromagnetic energy or particles

Colorimetric: techniques that determine the properties of a sample on the basis of color

Complementary: see **Base pair rule**

Compound: a pure substance made up of more than one element in fixed proportions; a substance made up of identical molecules

Compound-specific stable isotope analysis: a technique used to determine the stable isotope values of particular compounds isolated (e.g., separated by gas chromatography) from a sample with a complex composition

Conductor: a substance capable of conducting electricity; a substance with its outermost electrons occurring in a valence band that also serves as a conduction band

Conjugated bonds: in an organic molecule, where adjacent carbons connected by a single covalent bond are each connected to other carbons through double or triple bonds

Consumer: an organism that eats plants (primary consumer) or consumes plant-eating organisms (secondary consumer)

Continuous dynode: in mass spectrometers, a detector consisting of a horn-shaped dynode that converts entering positive ions into electrons then multiplies them as they bounce between the walls of the narrowing dynode prior to exiting

Conventional radiocarbon age (CRA):   an age calculated using Libby's half-life of 5568 years, appropriate modern standards, correction for sample isotopic fractionation, A.D. 1950 as 0 B.P., and the assumption that radiocarbon levels in the different reservoirs have remained constant over time

Cosmic radiation: a type of electromagnetic radiation with very high energy, very high frequency and very short wavelengths compared to visible light; electromagnetic energy from outer space

Covalent bond: a molecular bond that forms between two non-metals, or between a non-metal and hydrogen, involving the mutual sharing of electrons to achieve stable configurations

Crassulacean acid metabolism (CAM) photosynthesis: the pathway of succulent plants that is some combination of C_3 and C_4 processes

Cross polarization (CP):   in ^{13}C NMR spectroscopy, a technique used to reduce the relaxation time of ^{13}C nuclei; a technique that causes the diffusion of excess energy of ^{13}C nuclei to adjacent protons

Crossover (or counter) immuno-electrophoresis (CIEP):   a method of detecting immune response reactions through the formation of precipitin that uses gel electrophoresis to drive the antigen and antisera together

Curie point pyrolysis: a method of introducing an non-volatile solid sample into an instrument for analysis that involves heat-induced decomposition of material applied as a film onto a metal heating element

Curie point pyrolysis-gas chromatography (CuPy-GC/MS):   a technique that involves the analysis of pyrolysis products of a solid sample by gas chromatography-mass spectrometry

Curie point: in instrumental analysis, the maximum temperature to which a metal heating element can be raised in order to induce the decomposition of a solid sample; in archaeomagnetic dating, the temperature at which minerals lose their previously acquired magnetization

Cyclotron: a type of particle accelerator in which particles move outward in a spiral pattern between two oppositely charged D-shaped segments

cytochrome *b* gene: a sequence of mitochondrial DNA commonly targeted for amplification using the polymerase chain reaction

Cytosine: a pyrimidine nitrogen base that occurs in DNA and RNA

De Vries effects: in radiocarbon dating, short-term fluctuations in carbon-14 concentration throughout the Holocene due to natural or cultural activity; see **Suess effects** and **Libby effect**

Deamination: the loss of an amino group; for proteins, a decomposition process that changes an amino acid into a carboxylic acid

Debye–Scherrer camera method: a type of X-ray diffraction analysis that involves recording intensity maxima of the diffraction pattern on photographic film; powder photography

Decarboxylation: the loss of a carboxyl group; for proteins, a decomposition process that changes an amino acid into an amine

Decay constant: a value characteristic for a particular radioactive isotope that can be calculated by dividing its decay rate by the number of radioactive isotopes present

Decay series: the multiple step process involving a radioactive parent isotope that successively decays through many different radioactive daughters until a stable (non-radioactive) product forms

Declination: the angle between magnetic north and true north (the geographic North Pole) at the point of measurement; the horizontal component of the geomagnetic field

Demagnetization: in archaeomagnetic studies, the process of measuring paleointensity by comparing the strength of the original thermoremanent magnetization of a sample to thermoremanent magnetization induced in a laboratory magnetic field

Denaturation: the disruption of the higher level structural organization of a protein through exposure to heat, acids, or bases

Deoxynucleoside 5′ triphosphates (dNTPs): the form in which the four different nucleotides must be present in the environment in order for DNA synthesis to occur

Deoxyribonucleic acid (DNA): a nucleic acid consisting of two complementary strands of nucleotides (containing the sugar, deoxyribose) linked by hydrogen bonds to form a double helix; a molecule that directs cellular functions and contains the instructions for protein manufacture; nuclear DNA occurs in the nucleus of a cell; mitochondrial DNA (mtDNA) occurs in the mitochondria; the acronym of ancient DNA is aDNA

Depleted: in isotopic studies, a term used to describe material that contains a lower concentration of the heavier isotope than other parts of the reservoir due to fractionation processes or radioactive decay

Depth profile: a record of changes in sample composition that occur with increasing depth

Deuterium: the heavy and less abundant isotope of hydrogen with a nucleus consisting of a proton and a neutron; hydrogen-2 or ^2H

Diagenesis: the alteration of the original constituents into new products

Diamagnetic: substances that have only paired electrons occupying orbitals of a subshell; substances that are weakly repelled by magnetic fields

Diastereomers: stereoisomers that are not mirror images of each other

Dideoxynucleoside triphosphates (ddNTPs): molecules containing one of the four nitrogen bases used in the Sanger method of DNA sequencing to terminate DNA replication at different points

Diffraction pattern: in X-ray diffraction, the magnitudes of intensity maxima and angles at which they occur for a substance

Discrete dynode: in mass spectrometers, a detector consisting of several dynodes held at progressively higher voltages, the first of which converts positive ions to electrons then the remaining dynodes successively multiply each striking electron

Discriminant function analysis: a statistical technique which incorporates variables that most significantly account for differences between the groups weighted to maximize the separation between them; a statistical analysis technique used to confirm the existence of discrete groupings and assess the likelihood that a particular sample belongs to a previously established group

Doppler effect: apparent changes in the frequency of energy emitted by a passing object that are perceived by a stationary observer

Dynode: a device used to convert positive ions or photons into electrons then multiplies them exponentially

Early Uptake: in uranium series and ESR dating, a model for the introduction of uranium into a material that assumes uptake occurred soon after deposition and then ceased

Effective diagenetic temperature (EDT): in amino acid racemization dating, an estimate of the temperature to which all bones at a site were exposed, based on the D/L ratio measured in one bone subjected to radiocarbon dating

Effective hydration temperature (EHT): in obsidian hydration dating, a single temperature that approximates the effect of the range of actual temperatures an artifact experienced over the period of time it absorbed water from the environment

Electron impact: in mass spectrometry, a hard ionization technique that involves the destabilization and fragmentation of sample molecules; the most common method of producing ions for mass analysis

Electron microprobe: an instrument that uses a narrow electron beam to stimulate the emission of X-rays that are analyzed to gain information about surface composition

Electron multiplier: in mass spectrometry, a detector that converts positive ions emerging from the mass analyzer into electrons and multiplies them exponentially before recording the signal; see **Discrete dynode** and **Continuous dynode**

Electron paramagnetic spectroscopy: see **Electron spin resonance spectroscopy**

Electron spin resonance (ESR) spectroscopy: an analytical technique that involves placing the sample in a magnet then manipulating paramagnetic electrons trapped at intermediate energy levels with microwave energy, which causes them to resonate; also called electron paramagnetic resonance (EPR)

Electron: a negatively charged subatomic particle that occurs at discrete energy levels around the nucleus of an atom (in the extranuclear space); a subatomic particle with a mass approximately 1836 times smaller than a proton

Electronegativity: a measure of the ability of an element to attract electrons; a property that tends to be highest in atoms with pairs of unbonded electrons

Electronic energy level: in UV and VIS spectroscopy, a quantized energy level that can be occupied by covalently bonded electrons

Electrophoresis: a method of separating components from a sample solution on the basis of their movement through a gel slab to which an electric field is applied

Electrospray ionization: a method of ionizing and introducing a sample component separated by high-pressure liquid chromatography into a mass spectrometer; a technique used in LC-MS

Electrostatic analyzer: a component of a double focusing magnetic sector mass spectrometer; the sector generating an electrostatic field through which sample ions pass to ensure they enter the magnetic field with a narrow range of kinetic energies

Electrothermal vaporization: a process that uses heat to convert a sample solution or solid into vapors for analysis by atomic absorption spectroscopy

Element: a pure substance consisting entirely of atoms with the same number of protons; substances that appear in the Periodic Table of Elements; see Major, Minor, Trace and Rare Earth Elements

Elemental Analyzer: an instrument that can be used to determine the ratios of carbon, hydrogen, nitrogen, oxygen, and/or sulfur in organic and certain inorganic compounds

Eluent: in chromatography, a separated sample component that has emerged from a column

Elution: in chromatography, the movement of sample components through a column; in liquid chromatography, isocratic elution involves the use of a mobile phase consisting of one solvent of constant composition, gradient elution involves the use of a mobile phase consisting of two or more solvents

Empirical formula: the simplest whole number ratio of atoms in a compound containing ionic bonds

Enantiomers: stereoisomers that are mirror images of each other but cannot be superimposed; the two possible spatial arrangements of atoms of a stereoisomer containing one chiral carbon

Endogenous sequences: in DNA analysis, the targeted region amplified by PCR from authentic DNA

Energy-dispersive: in XRF, spectrometers that employ primary X-rays generated by an X-ray tube or radioactive elements and use a silicon–lithium detector to analyze the resulting secondary X-rays

Enriched: in isotopic studies, a term used to describe material that contains a higher concentration of the heavier isotope than other materials or other parts of the reservoir due to fractionation processes or radioactive decay

Enrichment: in isotopic studies, the process that causes the isotopic composition of a substance to be higher (less negative/more positive) than the material from which it was formed; the process that causes the isotopic composition of the tissues of a consumer to be higher than the foods it consumes

Entropy: a measure of randomness or disorder in the universe; a measure that increases during an irreversible process, according to the second law of thermodynamics

Enzyme-linked immunosorbent assay (ELISA): an absorption method of testing for an immune response that uses antibodies conjugated to color-producing enzyme, such as horseradish peroxidase, to enhance the detection of positive reactions

Epimerization: the conversion of a particular type of amino acid with two chiral carbons from its original form to a mixture containing four stereoisomers; the degree to which this process has occurred

Equilibrium: a state where no apparent change is observed in a system, either because all forces acting upon it cancel (net result is zero) or the rate of movement in one direction is equal to the rate of movement in the other

Equivalent dose: in trapped charge dating, an extrapolation of the amount of artificial radiation required to produce a signal identical to the natural signal without a correction for initial supralinear growth

***Escherichia coli* (*E. coli*):** in DNA analysis, bacteria often employed to clone PCR amplification products using recombinant DNA technology

Ester: an organic compound containing a carbonyl group linked to an alkyl group through an oxygen atom; organic compounds synthesized from a carboxylic acid and an alcohol in the presence of water

Ether: an organic compound formed by two alkyl groups linked by an oxygen atom

Evapotranspiration: water loss from a plant through a combination of evaporation and transpiration

Exogenous sequences: in DNA analysis, DNA from contaminants amplified by PCR

Extranuclear space: the area around the nucleus of an atom occupied by electrons

Faraday cup: in mass spectrometry, a detector that collects and neutralizes positive ions emerging from a mass analyzer on a collector electrode and records their impacts as voltage drops across a resistor

Fatty acid methyl esters (FAMES): derivatives of fatty acids suitable for analysis using gas chromatography prepared by treating the total lipid extract with methanol in either an acidic or alkali environment in order to release fatty acids attached to the glycerol backbone and replace the hydrogen (H–) of the carboxyl group with a methyl ($-CH_3$) group

Fatty acid: a carboxylic acid with hydrocarbon chains ranging from 4 to 36 carbon atoms in length; saturated fatty acids consist of carbon atoms connected through single bonds; unsaturated fatty acids contain at least one double bond between carbon atoms

Ferric (Fe^{3+}): the species of iron most abundant in pottery fired in an oxidizing (oxygen-rich) atmosphere; an iron ion with five electrons in its $3d$ orbital

Ferrous (Fe^{2+}): the species of iron most abundant in pottery fired in a reducing (oxygen-poor) atmosphere; an iron ion with six electrons in its $3d$ orbital

Fission track dating: a technique that uses uranium content and the number of fission tracks present in a crystalline substance to estimate the time that has elapsed since the material was free of defects

Fission track: damage to crystalline material caused by the spontaneous splitting of the nucleus of a uranium-238 atom

Fission: splitting apart

Fixation: a process that incorporates a reactant into other compounds; carbon dioxide fixation occurs during photosynthesis; nitrogen-fixation converts atmospheric nitrogen into nitrates and other compounds

Flame ionization detector (FID): in gas chromatography, a device that measures the time required for a separated component to pass through a column (retention time) and its concentration in the mobile phase

Fourier transformation: mathematical processing of the signal detected by an instrument in order to improve the ratio between the signal arising from the species of interest and random noise, called the signal-to-noise (S/N) ratio; the averaging of signals collected as beats in the time domain and their conversion into the frequency domain

Fractionation: the result of a reaction that preferentially favors one isotope of an element over another so the isotopic ratio of the end products differs from the starting material (reactants)

Frequency (v): the number of complete waves that pass a stationary point per unit of time given as a cycles per second or Hertz (Hz); a characteristic of electromagnetic radiation that is directly related to its energy and inversely related to its wavelength

Frequency modulation: in Mössbauer spectroscopy, a method that uses the Doppler effect to modify the apparent energy frequency of a gamma ray source

Functional group: a particular arrangement of atoms common to all members of a family of organic compounds

Gamma rays: a type of electromagnetic radiation with much higher energy and frequency and much shorter wavelengths than visible light; a radioactive decay product; the type of emissions from excited radioactive nuclei used to identify elements in instrumental neutron activation analysis

Gas chromatography (GC): a method of separating individual sample components from a mixture on the basis of volatility and affinity; an analytical technique that involves the vaporization of a liquid sample through progressive heating and transport of vaporized components by a gaseous mobile phase through a fused silica column coated on the interior with a liquid stationary phase

Gas chromatography-combustion-isotope ratio mass spectrometry (GC-C-IRMS): an analytical technique used to measure the ratio between two stable isotopes of an element in components separated from a sample by gas chromatography; an analytical technique that combines gas chromatography with isotope ratio mass spectrometry

Gas chromatography-mass spectrometry (GC-MS): an analytical technique that enables the mass analysis and identification of components separated from a sample by gas chromatography; an analytical technique that combines gas chromatography with mass spectrometry

Geiger(–Müeller) counter: a type of radiation detector; a device that counts positively charged argon ions created by radioactive emissions as electrical pulses that flow through a circuit

Gel diffusion method: see **Ouchterlony**

Geomagnetic field: the Earth's magnetic field

Germanium(–lithium) detector: a device used in INAA to detect and count gamma ray emissions of different frequencies

Gold immunoassay (GIA): an absorption method of testing for an immune response that uses antibodies conjugated to colloidal gold to enhance the detection of positive reactions

Grazer: a ruminant that eats grass

Guanine: a purine nitrogen base that occurs in DNA and RNA

g-value: in ESR spectroscopy, the ratio of absorption frequency in gigahertz, f, to magnetic field strength at resonance in Tesla, H, where, g-value $= f/14H$

Half-life: a value characteristic of a particular radioactive isotope; time required for one-half of the radioactive isotopes present to decay

Haplogroups: groups of related DNA haplotypes that share one or more key markers

Haplotypes: nucleotide changes that occur on the same sequence or chromosome capable of serving as genetic markers

Hatch–Slack pathway: a photosynthetic pathway followed by plants native to hot environments that involves intermediate products consisting of four carbon atoms; C_4 photosynthesis

Hemoglobin (Hb): a globular protein consisting of four heme groups and four amino acid chains (or protomers) linked together by hydrogen bonds.

Hemoglobin crystallization: a technique developed by biologists to identify animal species on the basis of the shapes of hemoglobin crystals that precipitate out of aqueous solutions containing blood

High-pressure (or performance) liquid chromatography (HPLC): a method of separating out individual sample components from a mixture that employs a liquid mobile phase pumped under high pressure through a column consisting of tiny particles made of, or coated with, the stationary phase

Hollow cathode lamp: in atomic absorption spectroscopy, a lamp that emits light characteristic of a particular element with a cup-shaped cathode constructed from, or coated with, that element

Hydration rind: in obsidian hydration dating, the thin outermost portion of the obsidian exterior that has absorbed water from the environment

Hydrogen bond: the attraction between the hydrogen atom of one molecule and a pair of unshared electrons on an atom from a different molecule; a bond between a hydrogen atom on one molecule and an element with a high electronegativity, such as oxygen, nitrogen, and fluorine, occurring on a different molecule; the intermolecular attraction that causes a compound to have higher than expected boiling points, melting points, and viscosities

Hydrolysis: a reaction with water alone, or together with an acid, base, or neutral compound, that causes the decomposition of another reactant; for proteins, the process that breaks bonds between individual amino acids linked together in a polypeptide chain

Hydroxyapatite: see **Apatite**

Hydroxyl group: the oxygen–hydrogen (–OH) functional group that is common to all alcohols

Hyperfine interactions: in Mössbauer spectroscopy, slight differences between the energy of Mössbauer isotopes that provide information about the chemical environment including isomer shifts, quadrupole splitting, and magnetic splitting

Immunoglobulin: protein produced by plasma cells in the lymph system of vertebrates as the body's first line of defense against an attack by a foreign substance, also called gamma (γ) globulin, type G immunoglobulin (IgG), and antibodies

Inclination: the angle between the local horizontal plane and the source of the magnetic field at the Earth's core (the south pole of the dipole) at the point of measurement; the vertical component of the geomagnetic field

Inductively coupled plasma (ICP):   techniques that involve atomizing a sample at very high temperatures in a plasma torch in order to (1) analyze sample elements by the characteristic light they emit by atomic (or optical) emission spectroscopy (ICP-AES or ICP-OES) or (2) analyze the resulting ions by mass spectrometry (ICP-MS)

Infrared energy: a type of electromagnetic radiation with slightly lower energy and frequency and longer wavelengths than visible light; the type of energy used to manipulate molecular bonds in infrared spectroscopy; energy produced by Nernst glowers and Globars for IR spectroscopy

Infrared spectroscopy: an analytical technique that involves the application of infrared light to samples to ascertain the presence of functional groups and types of molecular bonds on the basis of characteristic vibrational absorptions; a technique that uses the absorption of IR energy to increase the vibrational frequency of groups of atoms to the next higher level

Infrared stimulated luminescence (IRSL):   luminescence produced when electrons are evicted from shallow, or bleachable, traps by exposing a sample to monochromatic infrared light; a type of optically stimulated luminescence

Inhibitors: see **PCR inhibitors**

Inorganic: in chemistry, compounds that either do not contain carbon or are not associated with living organisms; not organic

Instrumental neutron activation analysis (INAA):   a technique used to determine the elemental composition of a sample through the analysis of characteristic gamma rays emitted by specific radioactive isotopes artificially created through neutron capture reactions in a nuclear reactor

Insulator: a substance that does not conduct electricity; a substance with a large energy gap between the valence band and the conduction band

Ion trap: a very compact type of mass analyzer that uses constant and oscillating electric fields to control the movement of sample ions; a device that can either sort sample ions according to mass or selectively accumulate a limited number of ions with a desired mass

Ion: an atom that carries a positive or negative charge because the number of electrons does not match the number of protons

Ionic bond: a molecular bond between a metal and a non-metal that involves the transfer of an electron from the metal to the non-metal resulting in electrostatic attraction between the oppositely charged atoms

Ionization: the process of creating ions from electrically neutral atoms or molecules, usually by the removal of electrons

Irradiation: exposure of a substance to radiation

Isoelectric point (pI or IEP): in electrophoresis, the pH at which internal neutralization occurs and a protein becomes electrically neutral

Isomers: compounds with the same molecular formula that differ with respect to how the atoms are joined; structural isomers differ with respect to the order in which atoms are joined; stereoisomers differ with respect to the arrangement of atoms in space but the order in which the atoms are attached is identical

Isothermal: at the same temperature

Isotope: an atom of an element with a specific atomic mass; an atom of a particular element that can differ with respect to the number of neutrons in its nucleus; unstable isotopes are radioactive, stable isotopes are not

Isotopic fractionation: see Fractionation

Ketone: an organic compound with an oxygen atom connected to a carbon through a double bond, also known as a carbonyl group, at some point other than the end of the carbon chain; simple ketones are named by replacing the *-e* of the corresponding alkane with *-one*

Laser ablation (LA): the process of using a laser beam to convert a tiny portion of the surface of a solid into fine particles or vapors; a method of introducing a sample into an instrument for analysis, such as inductively coupled plasma-mass spectrometry (LA-ICP-MS)

Laser beam: a highly focused and intense beam of monochromatic light produced by a laser

Laser microprobe mass spectrometry (or mass analysis) (LAMMS or LMMA): a mass analysis technique that uses a laser for solid sample introduction and ionization

Laser: a device that produces a highly focused and intense beam of monochromatic light; acronym of light amplification by stimulated emission of radiation

Libby effect: also known as atomic bomb or nuclear effect, the increase in the concentration of carbon-14 in the atmosphere due to nuclear weapon tests

(DNA) Ligase: in DNA analysis, an enzyme used to seal a segment of foreign DNA to a severed cloning vector

Ligation: in DNA analysis, the process of inserting foreign DNA into a cloning vector to form recombinant DNA

Linear uptake: in uranium series and ESR dating, a model that assumes uranium absorption by a material occurred at a constant rate starting from the time of deposition

Lipid: an organic compound with a low solubility in water; the group of organic substances that includes fatty acids, triacylglycerols, steroids, waxes, terpenes, phospholipids, glycolipids, and prostaglandins

Liquid chromatography-mass spectrometry (LC-MS): an analytical technique that enables the mass analysis and identification of components separated from a sample by high-pressure liquid chromatography; an analytical technique that combines high-pressure chromatography with mass spectrometry

Liquid scintillation cocktail: a mixture of substances, usually a combination of PPO (2,5-diphenoloxazole) and POPOP [1,4-bis (5-phenyloxazol-2-yl) benzene], that undergoes a fluorescence event (produces a flash of light) when exposed to a beta (β) particle emission

Liquid scintillation counter: in radiocarbon dating, an instrument used to detect beta (β) particle emissions as flashes of light in a liquid scintillation cocktail

Logarithm: in mathematics, the exponent (or power) to which a base must be raised to give a particular number; common logarithms are base 10; natural logarithms are base e, where $e = 2.71828$; in compositional analysis, a data transformation used when comparing major, minor, and trace elements to prevent variables with higher concentrations from having excess weight in calculations of many coefficients of similarity

Luminescence: the emission of light by certain materials

Magic angle spinning (MAS): in ^{13}C NMR spectroscopy, a technique that involves rapidly spinning a sample at 54.7° to reduce line broadening in the spectra of solids

Magnetic sector: a type of mass analysis technique that uses differential deflection of ions in a magnetic field to separate them according to mass

Magnetometer: in archaeomagnetism, a device used to measure the magnetization of a sample

Mahalanobis distance: statistical calculations used to determine the probability that a specimen is a member of any of the groups identified through discriminant analysis

Major element: one of fewer than 10 elements that typically occur in systems at levels greater than 1 wt. % (measured on metal oxide); a category of elements that includes silicon (Si), aluminum (Al), sodium (Na), magnesium (Mg), calcium (Ca), iron (Fe) and oxygen (O)

Mass number: see **Atomic mass number**

Mass spectrometer: an instrument used to produce molecular and elemental ions, sort them according to mass and detect abundances to establish the composition, determine molecular structure or measure isotopic ratios of specific elements

Matrix effects: a term for the wide variety of factors that can affect the compositional analysis of a sample, such as surface texture, sample inhomogeneity, and particle size

Metals: elements that are malleable and ductile, good conductors of heat and electricity and have a characteristic luster; elements mainly appearing on the left and central portions of the periodic table

Metamorphism, -ic: the transformation of sedimentary rock due to exposure to high temperature and pressure; material transformed by exposure to these conditions

Meteoric water: water available for drinking, the source of which is usually precipitation that falls as either rain or snow

Michelson interferometer: in Fourier transform-infrared spectroscopy, an instrument that creates fluctuating interference patterns in the IR energy directed at a sample so that absorption is monitored as a function of time

Micro- (μ-): in the metric system, 1/1000th of the base unit of measurement (i.e., meter, gram, liter); in sample analysis and instrumentation, a prefix commonly used for processes or instruments able to handle very small samples

Microsatellite repeat : see **Simple tandem repeat**

Microwaves: a type of electromagnetic radiation with lower energy and frequency and longer wavelengths than visible light; the type of energy applied to manipulate sample electrons during electron spin resonance spectroscopy

Minor element: one of about seven elements that typically occur in all systems at levels between 0.1 and 1 wt. % (measured on oxide); a category of elements that usually includes hydrogen (H), potassium (K), sulfur (S), carbon (C), phosphorus (P), titanium (Ti), manganese (Mn)

Mobile phase: in chromatography, the liquid or gaseous phase that carries a dissolved sample over or through a stationary phase to facilitate the separation of individual components

Moderator: in instrumental neutron activation analysis, material, such as water, paraffin, or graphite, in a nuclear reactor used to reduce the energy of fast neutrons and convert them to low energy, thermal neutrons

Mole: the quantity of an element or compound with mass equal to its atomic or molecular weight in grams; 6.022×10^{23} atoms of an element or molecules of a compound

Molecular ion: a molecule carrying an electrical charge created in the ion source of a mass spectrometer using a soft ionization technique; an ion created by the addition of a proton or ionized reagent or the removal of a hydrogen atom from a neutral molecule; positively charged molecular ions are also called radical cations

Molecule: a specific combination of elements bonded together to form a particular compound

Monochromator: a detector that can monitor one wavelength of light at a time

Monoclonal: antiserum grown in cell culture containing antibodies for only one specific type of protein

Monosaccharide: a simple carbohydrate that forms the building blocks of di-, tri-, and polysaccharides

Mössbauer effect: a phenomenon associated with only 35 isotopes, called Mössbauer isotopes, that enables energy emitted by an excited nuclei in a crystal environment to be absorbed by another nuclei at ground state in the same crystal environment; recoilless gamma ray emission and absorption

Mössbauer spectroscopy: an analytical technique that uses recoilless gamma ray emission and absorption of Mössbauer isotopes, such as ^{57}Fe, to investigate the chemical and physical environment of the sample

Nd:YAG laser: a type of solid-state laser using a neodymium ion in a host crystal of yttrium aluminum garnet

Nebulizer: a device that converts a sample solution into a fine mist or aerosol

Negative ion chemical ionization (NICI): a soft ionization technique used to produce negatively charged sample molecules for mass analysis

Neutron activation analysis: see instrumental neutron activation analysis

Neutron: an electrically neutral subatomic particle that occurs in the nucleus of an atom; a subatomic particle that is slightly larger than a proton

Nitrogen cycle: the movement of nitrogen through different parts of the environment

Nitrogen fixing: species able to convert molecular nitrogen into nitrates and other compounds

Non-metals: elements that are poor conductors, brittle as solids and lack a characteristic luster; elements appearing on the right side of the periodic table

Nuclear magnetic resonance spectroscopy (NMR): an analytical technique that uses the absorption of radio frequency energy by certain nuclei in a strong magnetic field to obtain information about the chemical environment of nuclei and structure of molecules

Nucleotide: the building blocks of DNA consisting of a phosphate group, a 5-carbon sugar (2-deoxy-D-ribose) and one of the four nitrogen bases: adenine (A), guanine (G), cytosine (C), and thymine (T)

Nucleus: the positively charged dense cluster of protons and neutrons, or a single proton in the case of hydrogen, at the center of an atom

Obligate drinkers: animals unable to obtain all necessary water from their diet and must drink water

Obsidian hydration: the process by which a fresh surface of obsidian slowly absorbs water from its environment causing a hydration rind to form

Obsidian: volcanic glass

Olig-: a prefix used to indicate multiple, but not a large number of items; few

Optical emission spectroscopy (OES): an elemental analysis technique that involves atomizing a sample with heat from a flame, spark or arc so that excited atoms of the elements present emit characteristic light

Optical pumping: the process that occurs in a laser, which is used to create a population inversion of atoms in the lasing material

Optically stimulated luminescence (OSL): luminescence produced when electrons are evicted from shallow, or bleachable, traps by exposing a sample to monochromatic visible light

Orbitals: spherical and non-spherical areas around the nucleus of an atom where electrons have a high probability of occurring; areas in the extranuclear space identified on the basis of their level and shape; see **Bonding orbital** and **Antibonding orbital**

Organic: in chemistry, compounds that contain carbon and are associated with living organisms

Ouchterlony (OCH) method: a means of detecting immune response reactions through the formation of precipitin that involves allowing the antigen and antisera to diffuse through an agar gel plate and come into contact; also called gel diffusion method

Overtone band: in infrared spectroscopy, bands produced when energy absorptions are elevated by more than one vibrational level

Oxidation: the reaction between a substance and oxygen; a reaction that increases the oxygen content or decreases the hydrogen content of an organic molecule

Oxidizing: in pottery analysis, an oxidizing environment refers to firing events that occur with oxygen freely available; in chemistry, an oxidizing agent causes the oxidation of an organic compound and becomes reduced in the process

Paleodose: in trapped charge dating, the total amount of artificial radiation required to produce a signal identical to the natural signal from a sample

Paramagnetic: substances that have unpaired electrons occupying orbitals of a subshell; substances that are drawn into magnetic fields

Particle-induced gamma ray emission (PIGE): an elemental analysis technique that involves the excitation of inner shell electrons through high-energy impacts with alpha particles and analysis of characteristic gamma-rays emitted by sample atoms returning to ground state

Particle-induced X-ray emission (PIXE): an elemental analysis technique that involves the excitation of inner shell electrons through high-energy impacts with alpha particles and analysis of characteristic X-rays emitted by sample atoms returning to ground state

PCR inhibitor: in DNA analysis, a substance whose presence causes the failure of PCR amplification of DNA

Pee Dee Belemnite: a limestone found in Southern Carolina used as the international standard for various compositional (carbon and oxygen isotopic and elemental) analyses

pH scale: shorthand method of expressing the acidity or basicity of a solution with 1 being a strong acid and 14 being a strong base; the negative common logarithm of the concentration of hydronium ions in a solution

Phonon: a vibration of a crystal lattice caused by recoil energy; a vibration that occurs when an atom bound to others in a lattice emits or absorbs energy

Phosphate: a compound that contains the phosphate group, PO_4^{3-}

Photoacoustic spectroscopy: an analytical technique that involves irradiating a sample with pulses of energy in the infrared, ultraviolet, or visible region, the absorption of which induces audible fluctuations in the pressure of the surrounding gas; a technique that permits the spectroscopic analysis of samples that do not transmit light

Photoconducting transducer: in infrared spectroscopy, a detector made of materials that experience temperature-dependent changes in conductivity to measure IR energy

Photodetector: in analytical instruments, any detector that converts light (photons) into an electrical signal

Photodiode: in analytical instruments, a detector consisting of a silicon chip that conducts electricity when struck by photons and records the resulting current

Photoelectric effect: the absorption of electromagnetic radiation by atoms or molecules

Photomultiplier: in analytical instruments, a detector that converts light (photons) into electrons and multiplies them exponentially before recording the signal

Photosynthesis: the process used by green plants and certain microorganisms to transform carbon dioxide and water into sugar and oxygen

Plane-polarized light: light that has passed through a polarizer; light consisting of an electric field that oscillates in only one plane and a magnetic field that oscillates in a perpendicular plane

Plasma emission spectrometer: in inductively coupled plasma emission spectroscopy, a device that sequentially or simultaneously measures the energy and intensity of emissions from elements atomized in a plasma torch

Plasma oxidation: a solid sampling technique involving the use of plasma to remove (burn off) a small portion of organic matter

Plasma torch: in inductively coupled plasma instruments, a device that uses an induced magnetic field to produce high-temperature plasma consisting of argon cations and electrons

Plasma: a gaseous mixture containing significant concentrations of cations and electrons that is able to conduct electricity

Plasmid: see **Cloning vector**

Plateau test: in trapped charge dating, a test used to ensure that the only signals from deep, stable traps are used to calculate the age of a sample

Polar covalent bond: a bond between non-metals with different electronegativities resulting in an unequal sharing of electrons; a covalent bond carrying a partial positive and partial negative charge

Polarity: a measure of the average charge distribution of a molecule; a property that affects solubility in that polar substances more readily dissolve in another polar substance, such as water, and non-polar substances more readily dissolve in non-polar solvents, such as chloroform, benzene, and carbon tetrachloride

Polychromator: a detector that can monitor many wavelengths of light simultaneously

Polyclonal: antiserum that contains antibodies for a variety of proteins typically produced by inoculating live animals

Polymerase chain reaction (PCR): a process that mimics natural enzyme-catalyzed DNA replication; a process that can exponentially increase the number of a targeted segment of DNA if it is present in a sample

Polymerase: a specialized enzyme that catalyzes DNA replication

Polypeptide: sequences of amino acids linked by covalent peptide bonds

Polysaccharide: a complex carbohydrate, the hydrolysis of which yields more than 10 mol of monosaccharides

Portable infrared mineral analyzer (PIMA): a small, transportable, infrared reflectance spectrometer typically used to non-destructively determine the composition of crystalline materials in the field or in a museum setting

Positive ion chemical ionization (PCI): a soft ionization technique used to produce positively charged sample molecules for mass analysis

Positron: a subatomic particle carrying a positive charge and having a mass equal to that of an electron; the radioactive decay product of some artificially produced isotopes

Potassium–argon (K–Ar): a dating technique that uses the amount of argon gas formed from the radioactive decay of potassium-40 to obtain an age estimate for volcanic materials

Precipitin: a three-dimensional lattice produced when antibodies attach to antigens

Precision: a measure of the ability to provide reproducible results when applied to samples processed and analyzed in exactly the same manner

Primer: in DNA replication, a short segment of nucleotides that attaches to a template strand and marks the starting point of DNA synthesis

Primeval (or primordial): relating to the time of the formation of the solar system; a term applied to material synthesized at the time of the formation of the solar system

Principal component analysis: a multivariate analysis technique used to identify clusters in data without forcing the samples into groups; a statistical technique that compresses a large number of variables into a small number of new ones so that relationships between members of a data set can be assessed through two-dimensional scattergrams

Producer: plants that product sugars, cellulose, starches, proteins, and lipids; the food of primary consumers

Prompt radiation: gamma ray emissions emitted by a newly formed radionuclide following a neutron capture reaction

Protein radioimmunoassay (pRIA): see **Radioimmunoassay**

Protein: giant molecules consisting of individual amino acids linked by peptide bonds; fibrous proteins occur as long strands or sheets; globular proteins have precise three-dimensional arrangements

Proton: a positively charged subatomic particle that occurs in the nucleus of an atom; a subatomic particle that is much larger than an electron but slightly smaller than a neutron

Proton-induced X-ray emission (PIXE): an elemental analysis technique that involves the excitation of inner shell electrons through high-energy impacts with protons and analysis of characteristic X-rays emitted by sample atoms returning to ground state

Provenance: the source of raw materials used to manufacture an item or its manufacturing location

Purine: a nitrogen base consisting of two rings that occurs in adenine and guanine; depurination is the loss of this structure from a DNA molecule

Pyrimidine: a nitrogen base consisting of one ring that occurs in thymine and cytosine; depyrimidation is the loss of this structure from a DNA molecule

Pyroelectric transducer: in infrared spectroscopy, a detector made of materials that experience temperature-dependent changes in their residual polarization to measure IR energy

Pyrolysis: heat-induced decomposition; a technique used to introduce samples that do not readily vaporize onto the column of a gas chromatograph

Quadrupole: a type of mass analyzer that filters sample ions according to mass using a variable electric field situated between four parallel rods

Quantized: only able to occur in specific and discontinuous states or levels

Racemization: the degree to which a particular type of amino acid has converted from its original L-form to a 50:50 (racemic) mixture of L-form and D-form enantiomers; the interconversion of L- and D-amino acids

Radio frequency energy: a type of electromagnetic radiation with much lower energy and frequency and longer wavelengths than visible light; the type of energy used to manipulate sample nuclei in nuclear magnetic resonance (NMR) spectroscopy; the type of energy used in the induction coil to form the plasma torch in inductively coupled plasma (ICP) techniques

Radioactive: any isotope that emits alpha particles, beta particles, and/or gamma rays during the transformation from an unstable to a stable nuclear configuration

Radiocarbon dating: the process of determining the age of a sample based on the amount of radioactive carbon (carbon-14) retained

Radiogenic: produced by the radioactive decay of an element

Radioimmunoassay (RIA): an absorption method of testing for an immune response that uses antibodies conjugated to radioactive isotopes to enhance the detection of positive reactions

Radioisotope: an unstable isotope that achieves a stable nuclear configuration through the emission of particles and energy; a radioactive isotope

Radionuclide: a specific radioactive isotope

Raman spectroscopy: an analytical technique that uses energy changes (Raman scattering) caused by inelastic collisions between a photon of light and a molecule to obtain information about its chemical structure

Rare earth element (REE): any one of a group of 17 elements that includes scandium (Sc), yttrium (Y), and those from atomic number 57 to 71: lanthanum (La), cerium (Ce), praseodymium (Pr), neodymium (Nd), promethium (Pm), samarium (Sm), europium (Eu), gadolinium (Gd), terbium (Tb), dysprosium (Dy), holmium (Ho), erbium (Er), thulium (Tm), ytterbium (Yb), and lutetium (Lu)

Rayleigh scattering: in Raman spectroscopy, the energy associated with an elastic collision between a photon of light and a molecule which matches the frequency of the incident beam

Recoilless gamma ray emission and absorption: see **Mössbauer effect**

Recombinant DNA: a cloning vector into which foreign DNA has been inserted prior to bacterial cloning

Reducing: in pottery analysis, a reducing environment refers to firings that occur in the absence of oxygen; in chemistry, a reducing agent causes the reduction of an organic compound and becomes oxidized in the process

Reduction: a reaction that increases hydrogen content or decreases oxygen content of an organic molecule

Regeneration technique: in trapped charge dating, a technique used to determine the supralinearity correction factor for the equivalent dose measurement, obtained through the additive (dose) method, that involves irradiating portions of the sample that have been reset to zero with different, known amounts of radiation

Renaturation: the slow return of denatured proteins to their native state when denaturing conditions cease

Reservoir: in radiocarbon dating, the total amount of carbon in the system (the carbon exchange reservoir); the total amount of carbon in a particular environment (e.g., the marine carbon reservoir)

Resolution: in mass spectrometry, a measure of the ability of an instrument to separate ions of different masses

Restriction endonucleases: in DNA analysis, special enzymes used to precisely cut foreign DNA and the cloning vector in the preparation of recombinant DNA

Retention time: in gas chromatography, the time required for a separated component to emerge from the column

Ribonucleic acid (RNA): a nucleic acid consisting of nucleotides made up of a phosphate group, a 5-carbon sugar (ribose), and one of four different nitrogen bases: adenine, guanine, cytosine, and uracil

Rock (or desert) varnish: a dark, often shiny coating that appears on exposed surfaces in arid environments; the material used for cation-ratio dating

Saponification: alkaline hydrolysis; the process of making soaps (salts of carboxylic acids) by boiling fat and an alkali in water

Scanning electron microscopy: a technique that uses a beam of high-energy electrons directed at a sample to produce backscattered and secondary electrons, which provide information about surface composition and topography

Scanning electron microscopy-energy-dispersive X-ray spectrometry (SEM-EDS): an analytical technique that combines scanning electron microscopy, which provides information about the physical nature of the sample surface, with X-ray fluorescence which provides information about its chemical composition

Secondary ion mass spectrometer: a mass spectrometer that analyzes secondary sample ions generated by a narrow beam of high-energy primary ions

Secular variation: in radiocarbon dating, systemic variability in the concentrations of carbon-14 over time due to processes other than radioactive decay; see **de Vries effects, Suess effects, and Libby effects**

Semiconductors: an insulator that conducts electricity if sufficient energy is applied to move electrons in the valance band across an energy gap into the conduction band

Sensitivity: in compositional analysis, the detection limits of an instrument

Sequencing: the process that uses a device called a sequencer to determine the order of nucleotides in a strand of DNA

Shell: an area surrounding the nucleus of an atom containing electrons with the same energy

Signal splitting: in proton (^{1}H) NMR spectroscopy, the splitting of a peak arising from equivalent nuclei into multiple peaks caused by the magnetic influences of neighboring hydrogen atoms; also called spin–spin splitting

Simple tandem repeat (STR): a short sequence of nucleotides that reoccurs several times and can serve as a genetic marker for a population, also known as a microsatellite repeat

Single aliquot regeneration (SAR): a variation of the regeneration technique used for samples that may not be completely homogeneous to establish the equivalent dose in OSL dating

Spectrometer: an instrument used to monitor the wavelength and intensity of electromagnetic energy emitted or absorbed by a sample

Spectroscopy: the study of energy emitted or absorbed by matter; an analytical technique that involves monitoring the energy absorbed, emitted, or the behavior of molecules in a higher energy or excited state

Spin–spin splitting: see **Signal splitting**

Stable Isotope Ratio Analysis (SIRA) mass spectrometer: an instrument typically used to measure ratios of stable isotopes of carbon, nitrogen, sulfur and oxygen; a magnetic sector mass spectrometer into which sample gas is introduced, often through the use an elemental analyzer

Standard mean ocean water (SMOW): an international standard used in the isotopic analysis of oxygen and hydrogen

Standard: material with a particular and accepted composition used internationally to normalize data obtained on analytical instruments

Stationary phase: in chromatography, a phase consisting of a solid, gel, or liquid over which or through a sample dissolved in a mobile phase passes to facilitate the separation of individual components

Stereochemistry: the arrangement of the atoms of a molecule in space

Sterols: structural lipids containing the perhydrocyclopentanophenanthrene ring system; cholesterol is the major sterol in animal tissues; campesterol, stigmasterol, and sitosterol are sterols found in plant tissue

Stigmasterol: a sterol found in plant tissue; a structural lipid present in cell membranes

Stokes line: in Raman spectroscopy, a spectral line arising from an inelastic collision between a photon of light and a molecule that elevates the vibrational state of the molecule

Suess effects: in radiocarbon dating, the dilution of carbon-14 in the atmosphere due to the burning of fossil fuels; also called industrial effects

Supernatant: the least dense, usually liquid, portion of a sample separated from denser material by a centrifuge

Supralinear: a type of non-linear change where the response is less than expected

Synchrotron radiation XRF (SRXRF): a highly sensitive type of X-ray fluorescence performed using intense primary X-rays generated by a synchrotron

Synchrotron: a type of particle accelerator

Terpene: a lipid built up from two or more five-carbon isoprene units; the prefixes, mono-, sesqui-, di-, and tri-, indicate compounds containing two, three, four, and six isoprene units, respectively

Terpenoid: a terpene containing a hydroxyl group (–OH)

Thermal ionization mass spectrometry (TIMS): a mass analysis technique that uses very high temperatures to evaporate and ionize inorganic samples

Thermal neutrons: in instrumental neutron activation analysis, slow-, low-energy neutrons that undergo neutron capture reactions with sample atoms and convert them into radionuclides

Thermocycler: a programmable device used to automatically control temperature changes for polymerase chain reactions

Thermoluminescence: the emission of light by certain crystalline materials in response to the application of heat; the light produced when electrons, ejected from intermediate energy traps, recombine with holes that are also luminescence centers

Thermoremanent magnetization: heat-induced magnetism retained in iron-rich minerals that corresponds to the direction and intensity of the Earth's magnetic (geomagnetic) field at the time of cooling

Thin layer chromatography: a method of separating out individual sample components from a mixture carried in a liquid mobile phase by capillary action through a gel stationary phase

Thymine: a pyrimidine nitrogen base that occurs in DNA

Time-of-flight (TOF or ToF): a type of mass analyzer where ion sorting is achieved in the absence of magnetic or electric fields; the separation of small packets of ions with the same kinetic energy according to mass as they travel through a 1–2 m long drift tube at different velocities

Total reflection XRF (TXRF): X-ray fluorescence involving primary X-rays that strike the sample and carrier at a very small angle such that those striking the carrier are totally reflected and only secondary X-rays from the sample reach the detector

Touchdown PCR: in DNA analysis, a method of increasing the success of amplification by maintaining a high temperature for several cycles to prevent mispriming

Trace element: one of a large group of elements that typically occur in all systems at levels less than 0.1 wt. % (measured on oxide)

Transmission electron microscopy (TEM): a technique that uses a beam of high-energy electrons that pass through a sample to produce a magnified image and obtain information about material contrast

Trapped charge dating: term applied to techniques that use signals arising from electrons trapped in the crystalline structure of a sample to calculate the time since the traps were empty; thermoluminescence (TL), optically stimulated luminescence (OSL), and electron spin resonance (ESR) dating

Triacylglycerol: a glycerol molecule to which three fatty acids are bonded through ester linkages

Trophic level: a particular link in the food chain; producers are at the bottom of the food chain, followed by primary consumers, secondary consumers, and so on

Ultraviolet (UV) spectroscopy: an analytical technique that involves the application of ultraviolet light to ascertain the presence of functional groups and types of molecular bonds in a sample; the absorption of energy excites low-energy electrons in bonding orbitals to high energy, antibonding orbitals

Ultraviolet light: a type of electromagnetic radiation with slightly higher energy and frequency and shorter wavelengths than visible light; the type of energy applied to manipulate the sample during ultraviolet spectroscopy; light capable of denaturing DNA and used to cleanse equipment and workspaces in ancient DNA (aDNA) laboratories

Uranium decay series: the multiple step process by which radioactive uranium decays through different daughter products to form a stable lead isotope

Uranium series dating: any one of several techniques that use ratios between members of a uranium decay series to obtain an age estimate of a sample

Uranium uptake: the absorption of uranium from the environment into biogenetic material originally free of uranium

Uranium–thorium dating: a technique that uses the ratio between the third and fourth daughter isotopes in the ^{238}U decay series to obtain an age estimate of a sample

Urinalysis test strips: a specific type of colorimetric test for proteins developed to detect blood in urine; also called dip-sticks

Valence band: the location of low-energy, non-conducting electrons of an atom; the uppermost energy level of electrons in insulators

Valence: relating to the outermost shell of an atom

Vector component diagram: in archaeomagnetic studies, a depiction of a three-dimensional magnetism vector in two dimensions; also known as a As-Zijderveld diagram

Vector: a quantity, such as magnetization, that has both direction and magnitude

Vibrational level: in infrared spectroscopy, the quantized vibrational state of a covalent bond

Visible (VIS) spectroscopy: an analytical technique that involves the application of visible light to samples to ascertain the presence of functional groups and molecular bonds; the absorption of energy excites low-energy electrons to high energy (antibonding) states

Visible energy: electromagnetic radiation in the range detectable by the human eye with variations in frequency/wavelength appearing as different colors; the type of energy applied to manipulate the sample during visible spectroscopy

Volatility: a measure of how readily a liquid substance enters the gaseous phase or vaporizes through evaporation or boiling

Wavelength (λ): the distance between two consecutive waves measured at the same point in meters; a property of electromagnetic radiation that is inversely related to the energy and frequency

Wavelength-dispersive: in X-ray fluorescence, spectrometers that typically employ primary X-rays generated by an X-ray tube and use a crystal spectrometer to analyze secondary electrons

Wavenumber: the reciprocal of wavelength

Wax: a lipid consisting of a long-chain alcohol connected to a long-chain fatty acid through an ester linkage

X-radiography: an imaging technique that uses differential absorption of X-rays to detect variations in the constitution of a sample

X-ray diffraction: a technique used to identity crystalline solids on the basis of the constructive and deconstructive interference of X-rays scattered by the parallel planes of atoms within the crystal

X-ray fluorescence: an elemental analysis technique involving the excitation of inner shell electrons with primary X-rays and the analysis of characteristic secondary, fluorescence X-rays emitted as sample atoms return to ground state

X-ray tube: a device used to generate X-rays; a device in which accelerating electrons strike a positively charged metal target, displacing inner shell electrons and causing the release of X-rays as the vacancies are filled

X-rays: a type of electromagnetic radiation with very high energy and frequency and short wavelengths compared to visible light; a type of energy whose absorption, emission, or diffraction by atoms, molecules, or crystal lattices form the basis of a variety of analytical techniques

References

Abbott, David R., Andrew D. Lack, and Gordon Moore
  2008 Chemical Assays of Temper and Clay; Modelling Pottery Production and Exchange in the Uplands North of the Phoenix Basin, Arizona, USA. *Archaeometry* 50(1):48–66.
Addeo, F., L. Barlotti, G. Boffa, A. Di Luccia, A. Malorni, and G. Piccioli
  1979 Costituenti Acidi di una Oleoresina di Conifere Rinvenuta in Anfore Vinarie Durante gli Scavi Archeologici di Oplonti. *Istituto di Industrie Agrarie* 20(6):145–149.
Adderley, W. P., Ian L. Alberts, Ian A. Simpson, and Timothy J. Wess
  2004 Calcium–Iron–Phosphate Features in Archaeological Sediments: Characterization through Microfocus Synchrotron X-ray Scattering Analyses. *Journal of Archaeological Science* 31(9):1215–1224.
Aitken, Martin. J.
  1970 Dating by Archaeomagnetic and Thermoluminescent Methods. *Philosophical Transactions of the Royal Society of London. Series A, Mathematical and Physical Sciences* 269:77–88.
  1989 Luminescence Dating: A Guide for Non-Specialists. *Archaeometry* 31(2):147–159.
  1990 *Science-based Dating in Archaeology*. Longman, London.
  1997 Luminescence Dating. In *Chronometric Dating in Archaeology*, Advances in Archaeological and Museum Science, vol. 2, edited by R. E. Taylor and M. J. Aitken, pp. 183–216. Plenum, New York.
Ambrose, Stanley H.
  1991 Effects of Diet, Climate and Physiology on Nitrogen Isotope Abundances in Terrestrial Foodwebs. *Journal of Archaeological Science* 18(3):293–317.
Ambrose, W. R.
  1976 Intrinsic Hydration Dating of Obsidian. In *Advances in Obsidian Glass Studies: Archaeological and Geochemical Perspectives*, edited by R. E. Taylor, pp. 81–105. Noyes Press, New Jersey.
  2001 Obsidian Hydration Dating. In *Handbook of Archaeological Sciences*, edited by D. R. Brothwell and A. M. Pollard, pp. 81–92. Wiley, Chichester.
Anovitz, Lawrence M., J. M. Elam, Lee R. Riciputi, and David R. Cole
  1999 The Failure of Obsidian Hydration Dating: Sources, Implications, and New Directions. *Journal of Archaeological Science* 26(7):735–752.
Armitage, Ruth A., James E. Brady, Alan Cobb, John R. Southon, and Marvin W. Rowe
  2001 Mass Spectrometric Radiocarbon Dates from Three Rock Paintings of Known Age. *American Antiquity* 66(3):471–480.
Arndt, Allan, Wim Van Neer, Bart Hellemans, Johan Robben, Filip Volckaert, and Marc Waelkens
  2003 Roman Trade Relationships at Sagalassos (Turkey) Elucidated by Ancient DNA of Fish Remains. *Journal of Archaeological Science* 30(9):1095–1105.
Asaro, F., E. Salazar, H. V. Michel, R. L. Burger, and F. H. Stross
  1994 Ecuadorian Obsidian Sources used for Artifact Production and Methods for Provenience Assignments. *Latin American Antiquity* 5(3):257–277.

Attanasio, Donato, D. Capitani, C. Federici, and A. L. Segre
    1995 Electron Spin Resonance Study of Paper Samples Dating from the Fifteenth to the
    Eighteenth Century. *Archaeometry* 37(2):377–384.
Attanasio, Donato, Rosario Platania, and Paolo Rocchi
    2005 White Marbles in Roman Architecture: Electron Paramagnetic Resonance Identification
    and Bootstrap Assessment of the Results. *Journal of Archaeological Science* 32(2):311–319.
Babot, M. D. P., and Maria C. Apella
    2003 Maize and Bone: Residues of Grinding in Northwestern Argentina. *Archaeometry*
    45(1):121–132.
Bacci, Mauro
    2000 UV-VIS-NIR, FT-IR, and FORS Spectroscopies. In *Modern Analytical Methods in Art
    and Archaeology,* Chemical Analysis Series, vol. 155, edited by Enrico Ciliberto and Giuseppe
    Spoto, pp. 321–404. John Wiley and Sons, New York.
Bada, Jeffrey L.
    1985 Amino Acid Racemization Dating of Fossil Bones. In *Annual Review of Earth and
    Planetary Sciences,* vol. 13, edited by George W. Wetherill, Arden L. Albee and Francis G.
    Stehli, pp. 241–268. Annual Reviews Inc., Palo Alto, California.
Badler, Virginia R., Patrick E. McGovern, and Rudolph H. Michel
    1990 Drink And Be Merry! Infrared Spectroscopy and Ancient Near Eastern Wine. *MASCA
    Research Papers in Science and Archaeology* 7:25–36.
Balasse, Marie, Stanley H. Ambrose, Andrew B. Smith, and T. D. Price
    2002 The Seasonal Mobility Model for Prehistoric Herders in the South-western Cape of
    South Africa Assessed by Isotopic Analysis of Sheep Tooth Enamel. *Journal of Archaeological
    Science* 29(9):917–932.
Balasse, Marie, Andrew B. Smith, Stanley H. Ambrose, and Steven R. Leigh
    2003 Determining Sheep Birth Seasonality by Analysis of Tooth Enamel Oxygen Isotope
    Ratios: The Late Stone Age Site of Kasteelberg (South Africa). *Journal of Archaeological
    Science* 30(2):205–215.
Balzer, A., G. Gleixner, G. Grupe, H. L. Schmidt, S. Schramm, and S. Turban-Just
    1997 *In Vitro* Decomposition of Bone Collagen by Soil Bacteria: The Implications for Stable
    Isotope Analysis in Archaeometry. *Archaeometry* 39(2):415–429.
Bamforth, Douglas B.
    1997 Cation-Ratio Dating and Archaeological Research Design: Response to Harry. *American
    Antiquity* 62(1):121–129.
Bancroft, G. M.
    1973 *Mössbauer Spectroscopy: An Introduction for Inorganic Chemists and Geochemists.*
    McGraw-Hill, London.
Banerjee, Monica, and Terence A. Brown
    2004 Non-random DNA Damage Resulting from Heat Treatment: Implications for Sequence
    Analysis of Ancient DNA. *Journal of Archaeological Science* 31(1):59–63.
Barker, James
    1999 *Mass Spectrometry.* 2nd ed. John Wiley and Sons on behalf of Analytical Chemistry by
    Open Learning (ACOL), Chichester, England.
Barnard, H., S. H. Ambrose, D. E. Beehr, M. D. Forster, R. E. Lanehart, M. E. Malainey, R. E.
Parr, M. Rider, C. Solazzo, and R. M. Yohe
    2007 Mixed Results of Seven Methods for Organic Residue Analysis Applied to One Vessel
    with the Residue of a Known Foodstuff. *Journal of Archaeological Science* 34(1):28–37.
Barnes, I. L., J. W. Gramlich, M. G. Diaz, and R. H. Brill
    1978 Possible Change of Lead Isotope Ratios in the Manufacture of Pigments: A Fractionation
    Experiment. In *Archaeological Chemistry,* vol. 2, edited by G. F. Carter, pp. 273–279. American
    Chemical Society, Washington.
Barone, G., V. Crupi, S. Galli, F. Longo, D. Majolino, P. Mazzoleni, and G. Spagnolo
    2004 Archaeometric Analyses on 'Corinthian B' Transport Amphorae found at Gela (Sicily,
    Italy). *Archaeometry* 46(4):553–568.

Bartsiokas, A., and A. P. Middleton
1992 Characterization and Dating of Recent and Fossil Bone by X-Ray Diffraction. *Journal of Archaeological Science* 19(1):63–72.

Baxter, M. J.
1994 *Exploratory Multivariate Analysis of Archaeology.* Edinburgh University, Edinburgh, Scotland.
2001 Multivariate Analysis in Archaeology. In *Handbook of Archaeological Sciences,* edited by D. R. Brothwell and A. M. Pollard, pp. 685–694. Wiley, Chichester.

Baxter, M. J., C. C. Beardah, I. Papageorgiou, M. A. Cau, P. M. Day, and V. Kilikoglou
2008 On Statistical Approaches to the Study of Ceramic Artefacts Using Geochemical and Petrographic Data. *Archaeometry* 50(1):142–157.

Baxter, M. J., C. C. Beardah, and S. Westwood
2000 Sample Size and Related Issues in the Analysis of Lead Isotope Data. *Journal of Archaeological Science* 27(10):973–980.

Baxter, Mike J., and Caitlin E. Buck
2000 Data Handling and Statistical Analysis. In *Modern Analytical Methods in Art and Archaeology,* vol. 155, edited by Enrico Ciliberto and Giuseppe Spoto, pp. 681–746. Wiley, New York.

Baxter, M. J., H. E. M. Cool, and C. M. Jackson
2006 Comparing Glass Compositional Analyses. *Archaeometry* 48(3):399–414.

Beck, Charlotte, and George T. Jones
1994 Dating Surface Assemblages Using Obsidian Hydration. In *Dating in Exposed and Surface Contexts,* edited by Charlotte Beck, pp. 47–76. University of New Mexico, Albuquerque, New Mexico.
2000 Obsidian Hydration Dating, Past and Present. In *It's About Time: A History of Archaeological Dating in North America,* edited by Stephen E. Nash, pp. 124–151. University of Utah Press, Salt Lake City, Utah.

Beck, Curt W.
1986 Spectroscopic Investigations of Amber. *Applied Spectroscopy Reviews* 22(1): 57–110.

Beck, Curt W., Constance A. Fellows, and Edith Mackennan
1974 Nuclear Magnetic Resonance Spectrometry in Archaeology. In *Advances in Chemistry Series No. 138,* edited by Robert F. Gould, pp. 226–235. American Chemical Society, Washington.

Beck, Margaret E., and Hector Neff
2007 Hohokam and Patayan Interaction in Southwestern Arizona: Evidence from Ceramic Compositional Analyses. *Journal of Archaeological Science* 34(2):289–300.

Beck, W., D. J. Donahue, A. J. T. Jull, G. Burr, W. S. Broecker, I. Hajdas, G. Bonani, and E. Malotki
1998 Ambiguities in Direct Dating of Rock Surfaces Using Radiocarbon Measurements. *Science* 280:2132–2135.

Becker, Johanna S.
2007 *Inorganic Mass Spectrometry: Principles and Applications.* Wiley, Chichester, England.

Bednarik, R. G.
1996 Review Article. Only Time Will Tell: A Review of the Methodology of Direct Rock Art Dating. *Archaeometry* 38(1):1–13.

Bellot-Gurlet, Ludovic, Olivier Dorighel, and Gérard Poupeau
2008 Obsidian Provenance Studies in Colombia and Ecuador: Obsidian Sources Revisited. *Journal of Archaeological Science* 35(2):272–289.

Bellot-Gurlet, L., G. Poupeau, O. Dorighel, Th. Calligaro, J.-C. Dran, and J. Salomon
1999 A PIXE/Fission-Track Dating Approach to Sourcing Studies of Obsidian Artefacts in Colombia and Ecuador. *Journal of Archaeological Science* 26(8):855–860.

Benson, L. V., E. M. Hattori, H. E. Taylor, S. R. Poulson, and E. A. Jolie
2006 Isotope Sourcing of Prehistoric Willow and Tule Textiles Recovered from Western

Great Basin Rock Shelters and Caves – Proof of Concept. *Journal of Archaeological Science* 33(11):1588–1599.

Benson, L. V., H. E. Taylor, K. A. Peterson, B. D. Shattuck, C. A. Ramotnik, and J. R. Stein
2008 Development and Evaluation of Geochemical Methods for the Sourcing of Archaeological Maize. *Journal of Archaeological Science* 35(4):912–921.

Bentley, R. A.
2006 Strontium Isotopes from the Earth to the Archaeological Skeleton: A Review. *Journal of Archaeological Method and Theory* 13(3):135–187.

Benzi, Valerio, Laura Abbazzi, Paolo Bartolomei, Massimo Esposito, Cecilia Fasso, Ornella Fonzo, Roberto Giampieri, Francesco Murgia, and Jean-Louis Reyss
2007 Radiocarbon and U-Series Dating of the Endemic Deer Praemegaceros Cazioti (Deperet) from "Grotta Juntu", Sardinia. *Journal of Archaeological Science* 34:790–794.

Berg, Ina
2008 Looking Through Pots: Recent Advances in Ceramics X-Radiography. *Journal of Archaeological Science* 35(5):1177–1188.

Berna, Francesco, Alan Matthews, and Stephen Weiner
2004 Solubilities of Bone Mineral from Archaeological Sites: The Recrystallization Window. *Journal of Archaeological Science* 31(7):867–882.

Beynon, J. H.
1960 *Mass Spectrometry and Its Applications to Organic Chemistry.* Elsevier, Amsterdam.

Bierman, Paul, and Alan Gillespie
1995 Reply to Comment by R. Dorn and C. Harrington and J. Whitney. *Quaternary Research* 43:274–276.

Bishop, Ronald L., Veletta Canouts, Patricia L. Crown, and Suzanne P. De Atley
1990 Sensitivity, Precision, and Accuracy: Their Roles in Ceramic Compositional Data Bases. *American Antiquity* 55(3):537–546.

Bishop, Ronald L., and Hector Neff
1989 Compositional Data Analysis in Archaeology. In *Archaeological Chemistry IV*, edited by Ralph O. Allen, pp. 57–86. American Chemical Society, Washington.

Blau, Soren, Brendan J. Kennedy, and Jean Y. Kim
2002 An Investigation of Possible Fluorosis in Human Dentition Using Synchrotron Radiation. *Journal of Archaeological Science* 29(8):811–817.

Block, B. P., W. H. Powell, and W. C. Fernelius
1990 *Inorganic Chemical Nomenclature, Principles and Practice.* American Chemical Society, Washington.

Bollong, Charles A., Leon Jacobson, Max Peisach, Carlos A. Pineda, and C. G. Sampson
1997 Ordination Versus Clustering of Elemental Data from PIXE Analysis of Herder-Hunter Pottery: A Comparison. *Journal of Archaeological Science* 24(4):319–327.

Bourdon, B., M. Henderson, C. C. Lundstrom, and S. P. Turner (editors)
2003 *Uranium-Series Geochemistry.* Reviews in Mineralogy & Geochemistry vol. 52, Mineralogical Society of America, Washington.

Braun, David P.
1982 Radiographic Analysis of Temper in Ceramic Vessels: Goals and Initial Methods. *Journal of Field Archaeology* 9(2):183–192.

Bronk Ramsey, C.
2008 Radiocarbon Dating: Revolutions in Understanding. *Archaeometry* 50(2):249–275.

Brothwell, D. R. and A. M. Pollard
2001 *Handbook of Archaeological Sciences.* Wiley, Chichester.

Brown, Terence A.
1999 How Ancient DNA May Help in Understanding the Origin and Spread of Agriculture. *Philosophical Transactions: Biological Sciences* 354(1379):89–98.
2001 Ancient DNA. In *Handbook of Archaeological Sciences*, edited by D. R. Brothwell and A. M. Pollard, pp. 301–322. Wiley, Chichester.

Brown, Truman R., and Kâmil Uğurbil
1984 Nuclear Magnetic Resonance. In *Structural and Resonance Techniques in Biological Research,* edited by Denis L. Rousseau, pp. 1–88. Academic, Orlando.

Budd, P., R. Haggerty, A. M. Pollard, B. Scaife, and R. G. Thomas
1996 Rethinking the Quest for Provenance. *Antiquity* 70(267):168–174.

Budd, P., J. Montgomery, A. Cox, P. Krause, B. Barreiro, and R. G. Thomas
1998 The Distribution of Lead within Ancient and Modern Human Teeth: Implications for Long-term and Historical Exposure Monitoring. *Science of the Total Environment* 220 (2–3):121–136.

Buonasera, Tammy
2005 Fatty Acid Analysis of Prehistoric Burned Rocks: A Case Study from Central California. *Journal of Archaeological Science* 32(6):957–965.

Burbidge, C. I., C. M. Batt, S. M. Barnett, and S. J. Dockrill
2001 The Potential for Dating the Old Scatness Site, Shetland, by Optically Stimulated Luminescence. *Archaeometry* 43(4):589–596.

Burton, James H., and T. D. Price
2000 The Use and Abuse of Trace Elements for Paleodietary Research. In *Biogeochemical Approaches to Paleodietary Analysis,* vol. 5, edited by Stanley H. Ambrose and M. A. Katzenberg, pp. 159–171. Kluwer Academic/Plenum Publishers in cooperation with the Society for Archaeological Sciences, New York.

Burton, James H., and Arleyn W. Simon
1993 Acid Extraction as a Simple and Inexpensive Method for Compositional Characterization of Archaeological Ceramics. *American Antiquity* 58(1):45–59.

Bush, Peter, and Ezra B. W. Zubrow
1986 The Art and Science of Eating. *Science and Archaeology* 28:38–43.

Butler, S.
1992 X-Radiography of Archaeological Soil and Sediment Profiles. *Journal of Archaeological Science* 19(2):151–161.

Cackette, M., J. M. D'Auria, and Bryan E. Snow
1987 Examining Earthenware Vessel Function by Elemental Phosphorus Content. *Current Anthropology* 28(1):121–127.

Cannon, Aubrey, and Dongya Y. Yang
2006 Early Storage and Sedentism on the Pacific Northwest Coast: Ancient DNA Analysis of Salmon Remains from Namu, British Colombia. *American Antiquity* 71(1):123–140.

Cano, Raúl
2000 Biomolecular Methods. In *Modern Analytical Methods in Art and Archaeology,* vol. 155, edited by Enrico Ciliberto and Giuseppe Spoto, pp. 241–254. John Wiley and Sons, New York.

Cariati, Franco, and Silvia Bruni
2000 Raman Spectroscopy. In *Modern Analytical Methods in Art and Archaeology,* Chemical Analysis Series, vol. 155, edited by Enrico Ciliberto and Giuseppe Spoto, pp. 255–278. Wiley, New York.

Carr, Christopher
1993 Identifying Individual Vessels with X-Radiography. *American Antiquity* 58(1):96–117.

Carter, Richard J.
1998 Reassessment of Seasonality at the Early Mesolithic Site of Star Carr, Yorkshire Based on Radiographs of Mandibular Tooth Development in Red Deer (*Cervus elaphus*). *Journal of Archaeological Science* 25(9):851–856.

Cassar, M., G. V. Robins, R. A. Fletton, and A. Altsin
1983 Organic Components in Historical Non-Metallic Seals Identified Using ^{13}C NMR Spectroscopy. *Nature* 303(19):238–239.

Cattaneo, C., K. Gelsthorpe, P. Phillips, and R. J. Sokol
1990 Blood in Ancient Human Bone. *Nature* 347(6291):339–339.
1993 Blood Residues on Stone Tools: Indoor and Outdoor Experiments. *World Archaeology* 25(1):29–43.

Charters, S., R. P. Evershed, L. J. Goad, A. Leyden, P. W. Blinkhorn, and V. Denham
1993 Quantification and Distribution of Lipid in Archaeological Ceramics: Implications for Sampling Potsherds for Organic Residue Analysis and the Classification of Vessel Use. *Archaeometry* 35(2):211–223.

Charters, S., R. P. Evershed, A. Quye, P. W. Blinkhorn, and V. Reeves
1997 Simulation Experiments for Determining the Use of Ancient Pottery Vessels: The Behaviour of Epicuticular Leaf Wax During Boiling of a Leafy Vegetable. *Journal of Archaeological Science* 24(1):1–7.

Child, A. M., R. D. Gillard, and A. M. Pollard
1993 Microbially-Induced Promotion of Amino Acid Racemization in Bone: Isolation of the Microorganisms and the Detection of Their Enzymes. *Journal of Archaeological Science* 20(2):159–168.

Christie, W. W.
1982 *Gas Chromatography and Lipids: A Practical Guide.* Oily Press, Ayr, Scotland.

Ciliberto, Enrico and Giuseppe Spoto (editors)
2000 *Modern Analytical Methods in Art and Archaeology.* Chemical Analysis Series, vol. 155. Wiley, New York.

Clark, Robin J. H., and Peter J. Gibbs
1998 Analysis of 16th Century Qazwini Manuscripts by Raman Microscopy and Remote Laser Raman Microscopy. *Journal of Archaeological Science* 25(7):621–629.

Coates, John
1998 A Review of Sampling Methods for Infrared Spectroscopy. In *Applied Spectroscopy: A Compact Reference for Practitioners,* edited by Jerry Workman Jr. and Art W. Springsteen, pp. 50–91. Academic, New York.

Cogswell, James W., Hector Neff, and Michael D. Glascock
1996 The Effect of Firing Temperature on the Elemental Characterization of Pottery. *Journal of Archaeological Science* 23(2):283–287.

Collins, Michael B., Bruce Ellis, and Cathy Dodt-Ellis
1990 Excavations at the Camp Pearl Wheat Site (41KR243) An Early Archaic Campsite on Town Creek Kerr County, Texas. In *Studies in Archaeology 6,* Texas Archaeological Research Laboratory, University of Texas at Austin.

Condamin, J., F. Formenti, M. O. Metais, M. Michel, and P. Blond
1976 The Application of Gas Chromatography to the Tracing of Oil in Ancient Amphorae. *Archaeometry* 18(2):195–201.

Connelly, N. G., T. Damhus, R. M. Hartshorn, and A. T. Hutton
2005 *IUPAC Recommendations 2005:Nomenclature of Inorganic Chemistry.* The Royal Society of Chemistry, Cambridge.

Cook, Andrea C., Jeffrey Wadsworth, John R. Southon, and van der Merwe, Nikolaas J.
2003 AMS Radiocarbon Dating of Rusty Iron. *Journal of Archaeological Science* 30(1): 95–101.

Cook, G. T., C. Bonsall, R. E. M. Hedges, K. McSweeney, V. Boronean, and P. B. Pettitt
2001 A Freshwater Diet-Derived ^{14}C Reservoir Effect at the Stone Age Sites in The Iron Gates Gorge. *Radiocarbon* 43(2A):453–460.

Cooper, Alan, and Hendrick N. Poinar
2000 Ancient DNA: Do It Right or Not At All. *Science* 289(5482):1139–1139.

Cooper, H. K., M. J. M. Duke, Antonio Simonetti, and Guang C. Chen
2008 Trace Element and Pb Isotope Provenance Analyses of Native Copper in Northwestern North America: Results of a Recent Pilot Study Using INAA, ICP-MS, LA-MC-ICP-MS. *Journal of Archaeological Science* 35(6):1732–1747.

Copley, M. S., R. Berstan, S. N. Dudd, G. Docherty, A. J. Mukherjee, V. Straker, S. Payne, and R. P. Evershed
2003 Direct Chemical Evidence for Widespread Dairying in Prehistoric Britain. *Proceedings of the National Academy of Sciences of the United States of America* 100(4):1524–1529.

Copley, M. S., Helen A. Bland, P. Rose, M. Horton, and R. P. Evershed
2005 Gas Chromatographic, Mass Spectrometric and Stable Carbon Isotopic Investigations of Organic Residues of Plant Oils and Animal Fats Employed as Illuminants in Archaeological Lamps from Egypt. *Analyst* 130:860–871.

Copley, Mark S., Pamela J. Rose, Alan Clapham, David N. Edwards, Mark C. Horton, and Richard P. Evershed
2001 Processing Palm Fruits in the Nile Valley: Biomolecular Evidence from Qasr Ibrim. *Antiquity* 75(289):538–542.

Craig, Harmon
1961 Isotopic Variation in Meteoric Waters. *Science* 133(3465):1702–1703.

Craig, N., R. J. Speakman, R. S. Popelka-Filcoff, M. D. Glascock, J. D. Robertson, M. S. Shackley, and M. S. Aldenderfer
2007 Comparison of XRF and PXRF for Analysis of Archaeological Obsidian from Southern Perú. *Journal of Archaeological Science* 34(12):2012–2024.

Craig, Oliver E., and Matthew J. Collins
2000 An Improved Method for the Immunological Detection of Mineral Bound Protein using Hydrofluoric Acid and Direct Capture. *Journal of Immunological Methods* 236(1–2): 89–97.
2002 The Removal of Protein from Mineral Surfaces: Implications for Residue Analysis of Archaeological Materials. *Journal of Archaeological Science* 29(10):1077–1082.

Craig, Oliver E., G. Taylor, J. Mulville, M. J. Collins, and M. Parker Pearson
2005 The Identification of Prehistoric Dairying Activities in the Western Isles of Scotland: An Integrated Biomolecular Approach. *Journal of Archaeological Science* 32(1): 91–103.

Cranshaw, T. E., B. W. Dale, G. O. Longworth, and C. E. Johnson
1985 *Mössbauer spectroscopy and its applications.* Cambridge University, Cambridge.

Curtis, Garniss H.
1975 Improvements in Potassium-Argon Dating: 1962–1975. *World Archaeology* 7(2): 198–209.

Custer, Jay F., John Ilgenfritz, and Keith R. Doms
1988 A Cautionary Note on the Use of Chemstrips for Detection of Blood Residues on Prehistoric Stone Tools. *Journal of Archaeological Science* 15(3):343–345.

Dalrymple, G. B., and Marvin A. Lanphere
1969 *Potassium-Argon Dating: Principles, Techniques and Applications to Geochronology.* W. H. Freeman, San Francisco.

David, Rosalie A., H. G. M. Edwards, D. W. Farwell, and D. L. A. De Faria
2001 Raman Spectroscopic Analysis of Ancient Egyptian Pigments. *Archaeometry* 43(4): 461–473.

Davies, G. R., and A. M. Pollard
1988 Organic Residues in An Anglo-Saxon Grave. In *Science and Archaeology Glasgow 1987. Proceedings at a Conference of the Application of Scientific Techniques to Archaeology. Glasgow, September 1987,* edited by E. A. Slater and J. O. Tate, pp. 391–402. British Archaeological Reports British Series 196(ii), Oxford.

Deal, Michael
1990 Exploratory Analyses of Food Residues from Prehistoric Pottery and other Artifacts from Eastern Canada. *SAS Bulletin* 13(1):6–12.

Deal, Michael, and Peter Silk
1988 Absorption Residues and Vessel Function: A Case Study from the Maine-Maritimes Region. In *A Pot For All Reasons: Ceramic Ecology Revisited,* edited by Charles C. Kolb and Louana M. Lockey, pp. 105–125. Laboratory of Anthropology, Temple University, Philadelphia, Pennsylvania.

Degryse, P., and J. Schneider
2008 Pliny the Elder and Sr-Nd Isotopes: Tracing the Provenance of Raw Materials for Roman Glass Production. *Journal of Archaeological Science* 35(7):1993–2000.

Deguilloux, M. F., L. Bertel, A. Celant, M. H. Pemonge, L. Sadori, D. Magri, and R. J. Petit
2006 Genetic Analysis of Archaeological Wood Remains: First Results and Prospects. *Journal of Archaeological Science* 33(9):1216–1227.

DeNiro, M. J., and S. Epstein
1978 Influence of Diet on the Distribution of Carbon Isotopes in Animals. *Geochimica et Cosmochimica Acta* 42(5):495–506.
1981 Influence of Diet on the Distribution of Nitrogen Isotopes in Animals. *Geochimica et Cosmochimica Acta* 45(3):341–351.

Deo, Jennie N., John O. Stone, and Julie K. Stein
2004 Building Confidence in Shell: Variations in the Marine Radiocarbon Reservoir Correction for the Northwest Coast over the Past 3,000 Years. *American Antiquity* 69(4):771–786.

Derome, Andrew E.
1987 *Modern NMR Techniques for Chemistry Research*. Pergamon, Oxford.

DeVito, Caterina, Vincenzo Ferrini, Silvano Mignardi, Luigi Piccardi, and Rosanna Tuteri
2004 Mineralogical-Petrographic and Geochemical Study to Identify the Provenance of Limestone from Two Archaeological sites in the Sulmona Area (L'Aquila, Italy). *Journal of Archaeological Science* 31(10):1383–1394.

Diaz-Granados, Carol, Marvin W. Rowe, Marian Hyman, James R. Duncan, and John R. Southon
2001 AMS Radiocarbon Dates for Charcoal from Three Missouri Pictographs and their Associated Iconography. *American Antiquity* 66(3):481–492.

Dickson, D. P. E., and C. E. Johnson
1984 Mössbauer Spectroscopy. In *Structural and Resonance Techniques in Biological Research,* edited by Denis L. Rousseau, pp. 245–293. Academic, Orlando.

Dorn, Ronald I.
1983 Cation-ratio Dating: A New Rock Varnish Age-Determination Technique. *Quaternary Research* 20:49–73.
1995 Comment on "Evidence Suggesting That Methods of Rock-Varnish Cation-Ratio Dating are neither Comparable nor Consistently Reliable", by Paul R. Bierman and Alan R. Gillespie. *Quaternary Research* 43:272–273.
1998 Response (to Ambiguities on Direct Dating of Rock Surfaces Using Radiocarbon Measurements by Beck, W., D. J. Donahue, A. J. T. Jull, G. Burr, W. S. Broecker, G. Bonani, I, Hajdas and E. Malotki). *Science* 280:2135–2139.
2007 Rock Varnish. In *Geochemical Sediments and Landscapes,* edited by David J. Nash and Sue J. McLaren, pp. 246–297. Blackwell, London.

Dorn, Ronald I., Margaret Nobbs, and Tom A. Cahill
1988 Cation-Ratio Dating of Rock-Engravings from the Olary Province of Arid South Australia. *Antiquity* 62(237):681–689.

Dostal, J., and C. Elson
1980 General Principles of Neutron Activation Analysis. In *Short Course in Neutron Activation,* edited by G. K. Muekle, pp. 21–42. Mineralogical Association of Canada, Halifax, Nova Scotia.

Downs, Elinor F., and Jerold M. Lowenstein
1995 Identification of Archaeological Blood Proteins: A Cautionary Note. *Journal of Archaeological Science* 22(1):11–16.

Dragovich, D.
2000 Rock Engraving Chronologies and Accelerator Mass Spectrometry Radiocarbon Age of Desert Varnish. *Journal of Archaeological Science* 27(10):871–876.

Dran, Jean-Claude, Thomas Calligaro, and Joseph Salomon
2000 Particle-Induced X-Ray Emission. In *Modern Analytical Methods in Art and Archaeology,* Chemical Analysis Series, vol. 155, edited by Enrico Ciliberto and Giuseppe Spoto, pp. 135–166. Wiley, New York.

Dudd, Stephanie N., and Richard P. Evershed
1999 Evidence for Varying Patterns of Exploitation of Animal Products in Different Prehistoric Pottery Traditions Based on Lipids Preserved in Surface and Absorbed Residues. *Journal of Archaeological Science* 26(12):1473–1482.

Duma, G.
1972 Phosphate Content of Ancient Pots as Indication of Use. *Current Anthropology* 13(1):127–130.

Dungworth, David
1997 Roman Copper Alloys: Analysis of Artefacts from Northern Britain. *Journal of Archaeological Science* 24(10):901–910.

Dunnell, Robert C., and James K. Feathers
1994 Thermoluminescence Dating of Surficial Archaeological Material. In *Dating in Exposed and Surface Contexts,* edited by Charlotte Beck, pp. 115–137. University of New Mexico, Albuquerque.

Dunnell, R. C., and T. L. Hunt
1990 Elemental Composition and Inference of Ceramic Vessel Function. *Current Anthropology* 31(3):330–336.

Duwe, Samuel, and Hector Neff
2007 Glaze and Slip Pigment Analyses of Pueblo IV Period Ceramics from East-Central Arizona Using Time of Flight-Laser Ablation-Inductively Coupled Plasma-Mass Spectrometry (TOF-LA-ICP-MS). *Journal of Archaeological Science* 34(3):403–414.

Dykeman, Douglas D., Ronald H. Towner, and James K. Feathers
2002 Correspondence in Tree-Ring and Thermoluminescence Dating: A Protohistoric Navajo Pilot Study. *American Antiquity* 67(1):145–164.

Ebsworth, E. A. V., David W. H. Rankin, and Stephen Cradock
1987 *Structural Methods in Inorganic Chemistry,* Blackwell Scientific, Boston.

Edwards, H. G. M., M. G. Sibley, and C. Heron
1997 FT-Raman Spectroscopic Study of Organic Residues from 2300-Year-Old Vietnamese Burial Jars. *Spectrochimica Acta Part A Molecular and Biomolecular Spectroscopy* 53(13):2373–2382.

Eighmy, Jeffrey L.
1990 Archaeomagnetic Dating: Practical Problems for the Archaeologist. In *Archaeomagnetic Dating,* edited by Jeffrey L. Eighmy and Robert S. Sternberg, pp. 33–64. University of Arizona, Tucson, Arizona.
2000 Thirty Years of Archaeomagnetic Dating. In *It's About Time: A History of Archaeological Dating in North America,* edited by Stephen E. Nash, pp. 105–123. University of Utah, Salt Lake City, Utah.

Eighmy, Jeffrey L., and Robert S. Sternberg (editors)
1990 *Archaeomagnetic Dating.* University of Arizona, Tucson, Arizona.

Eiland, Murray L., and Quentin Williams
2000 Infra-red Spectroscopy of Ceramics from Tell Brak, Syria. *Journal of Archaeological Science* 27(11):993–1006.

Eisele, J. A., D. D. Fowler, G. Haynes, and R. A. Lewis
1995 Survival and Detection of Blood Residues on Stone Tools. *Antiquity* 69(262):35–46.

Elekes, Z., K. T. Biro, I. Uzonyi, A. Simon, I. Rajta, B. Gratuze, and A. Z. Kiss
2000 *Application of Micro-PIXE and Micro-PIGE Techniques in the Field of Archaeology.* ATOMKI Annual Report 2000. Institute of Nuclear Research of the Hungarian Academy of Sciences, Debrecen, Hungary.

Ellis, L. W.
1997 Hot Rock Technology. In *Hot Rock Cooking on the Greater Edwards Plateau,* vol. 1, edited by S. L. Black, L. W. Ellis, D. G. Creel and G. T. Goode, pp. 79–139. Texas Department of Transportation and Environmental Affair Department, The University of Texas at Austin.

Emerson, Thomas E., Randall E. Hughes, Mary R. Hynes, and Sarah U. Wisseman
2003 The Sourcing and Interpretation of Cahokia-Style Figurines in the Trans-Mississippi South and Southeast. *American Antiquity* 68(2):287–313.

Emery, Kitty F., Lori E. Wright, and Henry Schwarcz
2000 Isotopic Analysis of Ancient Deer Bone: Biotic Stability in Collapse Period Maya Land-use. *Journal of Archaeological Science* 27(6):537–550.

Entwistle, Jane A., and Peter W. Abrahams
1997 Multi-Element Analysis of Soils and Sediments from Scottish Historical Sites. The Potential of Inductively Coupled Plasma-Mass Spectrometry for Rapid Site Investigation. *Journal of Archaeological Science* 24(5):407–416.

Ericson, Jonathon E., Oliver Dersch, and Friedel Rauch
2004 Quartz Hydration Dating. *Journal of Archaeological Science* 31(7):883–902.

Esposito, Massimo, Jean-Louis Reyss, Yaowalak Chaimanee, and Jean-Jacques Jaeger
2002 U-Series Dating of Fossil Teeth and Carbonates from Snake Cave, Thailand. *Journal of Archaeological Science* 29(4):341–349.

Evershed, Richard P.
1992a Mass Spectrometry of Lipids. In *Lipid Analysis: A Practical Approach,* edited by Richard J. Hamilton and Shiela Hamilton, pp. 263–308. IRL Press at Oxford University, Oxford.

1992b Gas Chromatography of Lipids. In *Lipid Analysis A Practical Approach,* edited by Richard J. Hamilton and Shiela Hamilton, pp. 114–151. IRL Press at Oxford University, Oxford.

1993a Biomolecular Archaeology and Lipids. *World Archaeology* 25(1):74–93.

1993b Combined Gas Chromatography-Mass Spectrometry. In *Gas Chromatography: A Practical Approach,* edited by P. J. Baugh, pp. 358–391. IRL Press at Oxford University, Oxford.

1994 Application of Modern Mass Spectrometric Techniques to the Analysis of Lipids. In *Developments in the Analysis of Lipids,* Special Publication No. 160, edited by J. H. P. Tyman and M. H. Gordon, pp. 123–160. Royal Society of Chemistry, Cambridge, UK.

2000 Biomolecular Analysis by Organic Mass Spectrometry. In *Modern Analytical Methods in Art and Archaeology,* Chemical Analysis Series, vol. 155, edited by Enrico Ciliberto and Giuseppe Spoto, pp. 177–239. Wiley, New York.

2008a Organic Residue Analysis in Archaeology: the Archaeological Biomarker Revolution. *Archaeometry* 50(6):895–924.

2008b Experimental Approaches to the Interpretation of Absorbed Organic Residues in Archaeological Reramics. *World Archaeology* 40(1):26–47.

Evershed, R. P., M. S. Copley, L. Dickson, and F. A. Hansel
2008 Experimental Evidence for the Processing of Marine Animal Products and Other Commodities Containing Polyunsaturated Fatty Acids in Pottery Vessels. *Archaeometry* 50(1):101–113.

Evershed, Richard P., Stephanie N. Dudd, Virginia R. Anderson-Stojanovic, and Elizabeth R. Gebhard
2003 New Chemical Evidence for the Use of Combed Ware Pottery Vessels as Beehives in Ancient Greece. *Journal of Archaeological Science* 30(1):1–12.

Evershed, R. P., S. N. Dudd, M. J. Lockheart, and S. Jim
2001 Lipids in Archaeology. In *Handbook of Archaeological Sciences,* edited by D. R. Brothwell and A. M. Pollard, pp. 331–350. Wiley, Chichester.

Evershed, R. P., C. Heron, S. Charters, and L. J. Goad
1992 The Survival of Food Residues: New Methods of Analysis, Interpretation and Application. *Proceedings of the British Academy* 77:187–208.

Evershed, Richard P., Carl Heron, and L. J. Goad
1990 Analysis of Organic Residues of Archaeological Origin by High-Temperature Gas Chromatography and Gas Chromatography Mass Spectrometry. *Analyst* 115(10):1339–1342.

1991 Epicuticular Wax Components Preserved in Potsherds as Chemical Indicators of Leafy Vegetables in Ancient Diets. *Antiquity* 65(248):540–544.

Evershed, R. P., H. R. Mottram, S. N. Dudd, S. Charters, A. W. Stott, G. J. Lawrence, A. M. Gibson, A. Conner, P. W. Blinkhorn, and V. Reeves
1997 New Criteria for the Identification of Animal Fats Preserved in Archaeological Pottery. *Naturwissenschaften* 84(9):402–406.

Evershed, Richard P., Gordon Turner-Walker, Robert E. M. Hedges, Noreen Tuross, and Ann Leyden
1995 Preliminary Results for the Analysis of Lipids in Ancient Bone. *Journal of Archaeological Science* 22(2):277–290.

Evershed, R. P., P. F. van Bergen, T. M. Peakman, E. C. Leigh-Firbank, M. C. Horton, D. Edwards, M. Biddle, B. Kjolbye-Biddle, and P. A. Rowley-Conwy
1997 Archaeological Frankincense. *Nature* 390(18/25 December):667–678.

Feathers, James K.
1997 The Application of Luminescence Dating in American Archaeology. *Journal of Archaeological Method and Theory* 4(1):1–66.
2000 Luminescence Dating and Why It Deserves Wider Application. In *It's About Time: A History of Archaeological Dating in North America,* edited by Stephen E. Nash, pp. 152–166. University of Utah, Salt Lake City, Utah.

Feathers, J. K., M. Berhane, and L. May
1998 Firing Analysis of South-Eastern Missouri Indian Pottery Using Iron Mössbauer Spectroscopy. *Archaeometry* 40(1):59–70.

Feathers, James K., Vance T. Holliday, and David J. Meltzer
2006 Optically Stimulated Luminescence Dating of Southern High Plains Archaeological Sites. *Journal of Archaeological Science* 33(12):1651–1665.

Fenner, Jack N.
2008 The Use of Stable Isotope Ratio Analysis to Distinguish Multiple Prey Kill Events from Mass Kill Events. *Journal of Archaeological Science* 35(3):704–716.

Fiedel, Stuart J.
1996 Blood from Stones? Some Methodological and Interpretive Problems in Blood Residue Analysis. *Journal of Archaeological Science* 23(1):139–147.
1997 Reply to Newman et al. *Journal of Archaeological Science* 24(11):1029–1030.

Fowles, Severin M., Leah Minc, Samuel Duwe, and David V. Hill
2007 Clay, Conflict, and Village Aggregation: Compositional Analyses of Pre-Classic Pottery from Taos, New Mexico. *American Antiquity* 72(1):125–152.

Fralick, Philip W., Stephen A. Kissin, Joe D. Stewart, Neil A. Weir, and Eric Blinman
2000 Paint Composition and Internal Layering of Plaster Ritual Objects from San Lazaro Pueblo (LA92), New Mexico. *Journal of Archaeological Science* 27(11):1039–1053.

Frankel, Edwin N.
1991 Recent Advances in Lipid Oxidation. *Journal of Science of Food Agriculture* 54(4):495–511.

Freestone, I. C.
1982 Applications and Potential of Electron Probe Micro-Analysis in Technological and Provenance Investigations of Ancient Ceramics. *Archaeometry* 24(2):99–116.

Friedman, Elizabeth S., Aaron J. Brody, Marcus L. Young, Jon D. Almer, Carlo U. Segre, and Susan M. Mini
2008 Synchrotron Radiation-Based X-Ray Analysis of Bronze Artifacts from an Iron Age Site in the Judean Hills. *Journal of Archaeological Science* 35(7):1951–1960.

Friedman, Irving, and W. D. Long
1976 Hydration Rate of Obsidian. *Science* 191:347–352.

Friedman, Irving, and Robert L. Smith
1960 A New Dating Method Using Obsidian: Part I, The Development of the Method. *American Antiquity* 25(4):476–493.

Friedman, Irving, Fred W. Trembour, and Richard E. Hughes
    1997 Obsidian Hydration Dating. In *Chronometric Dating in Archaeology*, vol. 3, edited by R. E. Taylor and Martin J. Aitken, pp. 297–321. Plenum, New York.
Frith, J., R. Appleby, R. Stacey, and C. Heron
    2004 Sweetness and Light: Chemical Evidence of Beeswax and Tallow Candles at Fountains Abbey, North Yorkshire. *Medieval Archaeology* 48(1):220–227.
Fuller, B. T., M. P. Richards, S. A. Mays
    2003 Stable Carbon and Nitrogen Isotope Variation in Tooth Dentine Serial Sections from Wharram Percy. *Journal of Archaeological Science* 30(12):1673–1684.
Gale, Noel H.
    1981 Mediterranean Obsidian Source Characterisation by Strontium Isotope Analysis. *Archaeometry* 23(1):41–51.
Gale, Noel H., and Zofia A. Stos-Gale
    1992 Lead Isotope Studies in the Aegean (The British Academy Project). *Proceedings of the British Academy* 77:63–108.
    2000 Lead Isotope Analyses Applied to Provenance Studies. In *Modern Analytical Methods in Art and Archaeology*, Chemical Analysis Series, vol. 155, edited by Enrico Ciliberto and Giuseppe Spoto, pp. 503–584. Wiley, New York.
Gale, N. H., A. P. Woodhead, Z. A. Stos-Gale, A. Walder, and I. Bowen
    1999 Natural Variations Detected in the Isotopic Composition of Copper: Possible Applications to Archaeology and Geochemistry. *International Journal of Mass Spectrometry* 184(1):1–9.
Gallet, Yves, Agnès Genevey, and Vincent Courtillot
    2003 On the Possible Occurrence of "Archaeometric Jerks" in the Geomagnetic Field over the Past Three Millennia. *Earth and Planetary Science Letters* 214:237–242.
Garcia-Heras, M., R. Fernandez-Ruiz, and J. D. Tornero
    1997 Analysis of Archaeological Ceramics by TXRF and Contrasted with NAA. *Journal of Archaeological Science* 24(11):1003–1014.
Garnier, Nicolas, Pascale Richardin, Veronique Cheynier, and Martine Regert
    2003 Characterization of Thermally Assisted Hydrolosis and Methylation Products of Polyphenols from Modern and Archaeological Vine Derivatives Using Gas Chromatography-Mass Spectrometry. *Analytica Chimica Acta* (493):137–157.
Garrigos, J. B., M. A. C. Ontiveros, and V. Kilikoglou
    2003 Chemical Variability in Clays and Pottery from a Traditional Cooking Pot Production Village: Testing Assumptions in Pereruela. *Archaeometry* 45(1):1–17.
Garvie-Lok, Sandra J., Tamara L. Varney, and M. A. Katzenberg
    2004 Preparation of Bone Carbonate for Stable Isotope Analysis: The Effects of Treatment Time and Acid Concentration. *Journal of Archaeological Science* 31(6):763–776.
Gendron, François, David C. Smith, and Aïcha Gendron-Badou
    2002 Discovery of Jadeite-Jade in Guatemala Confirmed by Non-Destructive Raman Microscopy. *Journal of Archaeological Science* 29(8):837–851.
George, Debra, John Southon, and R. E. Taylor
    2005 Resolving an Anomolous Radiocarbon Determination on Mastodon Bone from Monte Verde, Chile. *American Antiquity* 70(4):766–772.
Gerhardt, Klaus O., Scott Searles, and William R. Biers
    1990 Corinthian Figure Vases: Non-Destructive Extraction and Gas Chromatography-Mass Spectrometry. *MASCA Research Papers in Science and Archaeology* 7:41–50.
Geyh, Mebus A., U. Schotterer, and M. Grosjean
    1998 Temporal Changes of the ^{14}C Reservoir Effect in Lakes. *Radiocarbon* 40(2):921–931.
Gibson, J. L., and P. Jagan
    1980 Instrumental Neutron Activation Analysis of Rocks and Minerals. In *Short Course in Neutron Activation*, edited by G. K. Muekle, pp. 109–128. Mineralogical Association of Canada, Halifax, Nova Scotia.

Glascock, M. D.

1994 New World Obsidian: Recent Investigations. In *Archaeometry of Pre-Columbian Sites and Artifacts, Proceedings of the 28th Annual Symposium on Archaeometry,* edited by P. Meyers and D. A. Scott, pp. 113–134. Getty Conservation Institute, Los Angeles.

Glascock, M. D., G. E. Braswell, and R. H. Cobean

1998 A Systematic Approach to Obsidian Source Characterization. In *Archaeological Obsidian Studies: Method and Theory,* edited by M. S. Shackley, pp. 15–66. Springer/Plenum Press, New York.

Goios, Ana, Lourdes Prieto, Antonio Amorim, and Luisa Pereira

2008 Specificity of mtDNA-directed PCR-influence of NUclear mtDNA Insertion (NUMT) Contamination in Routine Samples and Techniques. *International Journal of Legal Medicine* 122(4):341–345.

Goodall, R. A., J. Hall, H. G. M. Edwards, R. J. Sharer, R. Viel, and P. M. Fredericks

2007 Raman Microbe Analysis of Stucco Samples from the Buildings of Maya Classic Copan. *Journal of Archaeological Science* 34(4):666–673.

Gordy, Walter

1980 *Theory and Applications of Electron Spin Resonance.* vol. XV, Wiley, New York.

Gose, Wulf A.

2000 Paleomagnetic Studies of Burned Rocks. *Journal of Archaeological Science* 27(5): 409–421.

Götherström, A., M. J. Collins, A. Angerbjorn, and K. Liden

2002 Bone Preservation and DNA Amplification. *Archaeometry* 44(3):395–404.

Grün, Rainer

1997 Electron Spin Resonance Dating. In *Chronometric Dating in Archaeology,* Advances in Archaeological and Museum Science, vol. 2, edited by R. E. Taylor and M. J. Aitken, pp. 217–260. Plenum, New York.

1998 Reproducibility Measurements for ESR Signal Intensity and Dose Determination: High Precision but Doubtful Accuracy. *Radiation Measurements* 29(2):177–193.

2000 Electron Spin Resonance Dating. In *Modern Analytical Methods in Art and Archaeology,* Chemical Analysis Series, vol. 155, edited by Enrico Ciliberto and Giuseppe Spoto, pp. 641–679. Wiley, New York.

2001 Trapped Charge Dating (ESR, TL, OSL). In *Handbook of Archaeological Sciences,* edited by D. R. Brothwell and A. M. Pollard, pp. 47–62. Wiley, Chichester.

Grün, R., and C. B. Stringer

1991 Electron Spin Resonance Dating and the Evolution of Modern Humans. *Archaeometry* 33(2):153–199.

Grupe, Gisela, Astrid Balzer, and Susanne Turban-Just

2000 Modeling Protein Diagenesis in Ancient Bone: Towards a Validation of Stable Isotope Data. In *Biogeochemical Approaches to Paleodietary Analysis,* vol. 5, edited by Stanley H. Ambrose and M. A. Katzenberg, pp. 173–187. Kluwer Academic/Plenum Publishers in cooperation with the Society for Archaeological Sciences, New York.

Guangyong, Qin, Pan Xianjia, and Li Shi

1989 Mössbauer Firing Study of Terracotta Warriors and Horses of the Qin Dynasty (221 B.C.). *Archaeometry* 31(1):3–12.

Guilderson, Tom P., Paula J. Reimer, and Tom A. Brown

2005 The Boon and Bane of Radiocarbon Dating. *Science* 307(5708):362–364.

Gurfinkel, D. M., and U. M. Franklin

1988 A Study of the Feasibility of Detecting Blood Residue on Artifacts. *Journal of Archaeological Science* 15(1):83–97.

Habicht-Mauche, Judith A., Stephen T. Glenn, Mike P. Schmidt, Rob Franks, Homer Milford, and A. R. Flegal

2002 Stable Lead Isotope Analysis of Rio Grande Glaze Paints and Ores Using ICP-MS: A Comparison of Acid Dissolution and Laser Ablation Techniques. *Journal of Archaeological Science* 29(9):1043–1053.

Hadži, D., and B. Orel
1978 Spectrometrichne Raziskave Izvora Jantaria in Smol iz Prazgodovinskih Najdisc na Slovenskem (Spectrometric Investigations of the Origin of Amber and Pitches from Prehistoric Sites in Slovenia). *Vestnik Slovenskega Kemijskega Drusrustva (Acta Chimica Slovenica)* 25(1):51–63.

Hagelberg, Erika, Lynne S. Bell, Tim Allen, Alan Boyde, Sheila J. Jones, and J. B. Clegg
1991 Analysis of Ancient Bone DNA: Techniques and Applications. *Philosophical Transactions: Biological Sciences* 333(1268):399–407.

Hall, Grant D., Stanley M. Tarka Jr., W. J. Hurst, David Stuart, and Richard E. W. Adams
1990 Cacao Residues in Ancient Maya Vessels from Rio Azul, Guatemala. *American Antiquity* 55(1):138–143.

Halliday, David, and Robert Resnick
1981 *Fundamentals of Physics*. 2nd ed. Wiley, New York.

Halliday, David, Robert Resnick, and J. Walker
2005 *Fundamentals of Physics*. Wiley, Hoboken, NJ.

Hally, D. J.
1983 Use Alteration of Pottery Vessel Surfaces: An Important Source of Evidence of the Identification of Vessel Function. *North American Archaeologist* 4(1):3–26.

Hammes, Gordon G.
2005 *Spectroscopy for the Biological Sciences*. Wiley-Interscience, Hoboken, NJ.

Handt, O., M. Höss, M. Krings, and S. Pääbo
1994 Ancient DNA: Methodological Challenges. *Cellular and Molecular Life Sciences* 50(6):524–529.

Hansel, Fabricio A., Mark S. Copley, Luiz A. S. Madureira, and Richard P. Evershed
2004 Thermally Produced ω-(o-alkylphenyl) Alkanoic Acids Provide Evidence for the Processing of Marine Products in Archaeological Pottery Vessels. *Tetrahedron Letters* 45:2999–3002.

Hansson, Maria C., and Brendan P. Foley
2008 Ancient DNA Fragments Inside Classical Greek Amphoras Reveal Cargo of 2400-Year-Old Shipwreck. *Journal of Archaeological Science* 35(5):1169–1176.

Hare, P. E., Marilyn L. Fogel, Thomas W. Stafford Jr., Alva D. Mitchell, and Thomas C. Hoering
1991 The Isotopic Composition of Carbon and Nitrogen in Individual Amino Acids Isolated from Modern and Fossil Proteins. *Journal of Archaeological Science* 18(3): 277–292.

Hare, P. E., D. W. Von Endt, and J. E. Kokis
1997 Protein and Amino Acid Diagenesis Dating. In *Chronometric Dating in Archaeology,* vol. 3, edited by R. E. Taylor and Martin J. Aitken, pp. 261–296. Plenum, New York.

Harrington, Charles D., and John W. Whitney
1995 Comment on "Evidence Suggesting That Methods of Rock-Varnish Cation-Ratio Dating are neither Comparable nor Consistently Reliable", by Paul R. Bierman and Alan R. Gillespie. *Quaternary Research* 43:268–271.

Harris, Walter E., and Byron Kratochvil
1981 *An Introduction to Chemical Analysis*. Saunders, Philadelphia.

Harry, Karen G.
1995 Cation-Ratio Dating of Varnished Artifacts: Testing the Assumptions. *American Antiquity* 60(1):118–130.

Hart, John P., William A. Lovis, Janet K. Schulenberg, and Gerald R. Urquhart
2007 Paleodietary Implications from Stable Carbon Isotope Analysis of Experimental Cooking Residues. *Journal of Archaeological Science* 34(5):804–813.

Haskin, L. A.
1980 An Overview of Neutron Activation Analysis in Geochemistry. In *Short Course in Neutron Activation,* edited by G. K. Muekle, pp. 1–20. Mineralogical Association of Canada, Halifax, Nova Scotia.

Hastorf, Christine, and Michael J. DeNiro
1985 Reconstruction of Prehistoric Plant Production and Cooking Practices by a New Isotopic Method. *Nature* 315(6):489–491.

Hathaway, J. H., and George J. Krause
1990 The Sun Compass Versus the Magnetic Compass in Archaeomagnetic Sample Collection. In *Archaeomagnetic Dating,* edited by Jeffrey L. Eighmy and Robert S. Sternberg, pp. 139–147. University of Arizona, Tucson, Arizona.

Hatcher, H., M. S. Tite, and J. N. Walsh
1995 A Comparison of Inductively-Coupled Plasma Emission Spectrometry and Atomic Absorption Spectrometry Analysis on Standard Reference Silicate Materials and Ceramics. *Archaeometry* 37(1):83–94.

Haynes, Susan, Jeremy B. Searle, Amanda Bretman, and Keith M. Dobney
2002 Bone Preservation and Ancient DNA: The Application of Screening Methods for Predicting DNA Survival. *Journal of Archaeological Science* 29(6):585–592.

Hedges, Robert E. M.
2000 Radiocarbon Dating. In *Modern Analytical Methods in Art and Archaeology,* vol. 155, edited by Enrico Ciliberto and Giuseppe Spoto, pp. 465–502. Wiley, New York.

Henderson, Julian
2000 *The Science and Archaeology of Materials: An Investigation of Inorganic Materials.* Routledge, London.

Henrickson, Elizabeth F., and Mary M. A. McDonald
1983 Ceramic Form and Function: An Ethnographic Search and an Archaeological Application. *American Anthropologist* 85(3):630–643.

Heron, Carl P.
1989 *The Analysis of Organic Residues from Archaeological Ceramics.* Unpublished Ph. D. dissertation, University of Wales, Cardiff, Wales.

Heron, Carl P., and Richard P. Evershed
1993 The Analysis of Organic Residues and the Study of Pottery Use. In *Archaeological Method and Theory,* vol. 5, edited by Michael B. Schiffer, pp. 247–284. University of Arizona, Tucson.

Heron, C., R. P. Evershed, and L. J. Goad
1991 Effects of Migration of Soil Lipids on Organic Residues Associated with Buried Potsherds. *Journal of Archaeological Science* 18(6):641–659.

Heron, C., and A. M. Pollard
1988 The Analysis of Natural Resinous Materials from Roman Amphoras. In *Science and Archaeology Glasgow 1987. Proceedings at a Conference of the Application of Scientific Techniques to Archaeology. Glasgow, September 1987,* edited by E. A. Slater and J. O. Tate, pp. 429–447. British Archaeological Reports British Series 196 (ii), Oxford.

Herrera, R. S., Hector Neff, Michael D. Glascock, and J. M. Elam
1999 Ceramic Patterns, Social Interactions, and the Olmec: Neutron Activation Analysis of Early Formative Pottery in the Oaxaca Highlands of Mexico. *Journal of Archaeological Science* 26(8):967–987.

Herz, Norman
1990 Stable Isotope Geochemistry Applied to Archaeology. In *Archaeological Geology of North America,* Centennial Special, vol. 4, edited by N. P. Lasca and J. Donahue, pp. 585–595. Geological Society of America, Boulder, Colorado.

Herz, Norman, and Ervan G. Garrison
1998 *Geological Methods for Archaeology.* Oxford University, New York.

Hess, J., and I. Perlman
1974 Mössbauer Spectra of Iron in Ceramics and Their Relation to Pottery Colours. *Archaeometry* 16(2):137–152.

Hesse, Brian, and Paula Wapnish
1985 *Animal Bone Archeology: From Objectives to Analysis.* Taraxacum, Washington.

Higuchi, Russell, Barbara Bowman, Mary Freiberger, Oliver A. Ryder, and Allan C. Wilson
   1984 DNA Sequences from the Quagga, an Extinct Member of the Horse Family. *Nature*
   312:282–284.
Hill, H. E., and J. Evans
   1987 The Identification of Plants Used in Prehistory from Organic Residues. In *Archaeometry:*
   Further Australian Studies, edited by W. R. Ambrose and J. M. J. Mummery, pp.
   90–96. Department of Prehistory, Research School of Pacific Studies, The Australian National
   University, Australian National Gallery, Canberra, Australia.
   1988 Vegeculture in Solomon Islands: Prehistory from Pottery Residues. In *Science and*
   Archaeology Glasgow 1987. Proceedings at a Conference of the Application of Scientific
   Techniques to Archaeology. Glasgow, September 1987, edited by E. A. Slater and J. O. Tate,
   pp. 449–458. British Archaeological Reports British Series 196(ii), Oxford.
   1989 Crops of the Pacific: New Evidence from the Chemical Analysis of Organic Residues in
   Pottery. In *Foraging and Farming: The Evolution of Plant Exploitation,* edited by D. R. Harris
   and G. C. Hillman, pp. 418–425. Unwin Hyman, London.
Hiller, J. C., and T. J. Wess
   2006 The Use of Small-angle X-ray Scattering to Study Archaeological and Experimentally
   Altered Bone. *Journal of Archaeological Science* 33(4):560–572.
Hites, Ronald A.
   1997 Gas Chromatography Mass Spectrometry. In *Handbook of Instrumental Techniques for*
   Analytical Chemistry, edited by Frank A. Settle, pp. 609–626. Prentice Hall, Upper Saddle
   River, NJ.
Hoefs, Jochen
   2009 Stable Isotope Geochemistry. Springer-Verlag, Berlin.
Holliday, Vance T., and William G. Gartner
   2007 Methods of Soil P Analysis in Archaeology. *Journal of Archaeological Science*
   34(2):301–333.
Hortolà, Policarp
   2002 Red Blood Cell Haemotaphonomy of Experimental Human Bloodstains on Techno-
   Prehistoric Lithic Raw Materials. *Journal of Archaeological Science* 29(7):733–739.
Hull, Sharon, Mostafa Fayek, Frances J. Mathien, Phillip Shelley, and Kathy R. Durand
   2008 A New Approach to Determining the Geological Provenance of Turquoise Artifacts
   using Hydrogen and Copper Stable Isotopes. *Journal of Archaeological Science* 35(5):
   1355–1369.
Huntley, Deborah L., Katherine A. Spielmann, Judith A. Habicht-Mauche, Cynthia L. Herhahn,
   and A. R. Flegal
   2007 Local Recipes or Distant Commodities? Lead Isotope and Chemical Compositional
   Analysis of Glaze Paints from the Salinas Pueblos, New Mexico. *Journal of Archaeological*
   Science 34(7):1135–1147.
Hurst, W. J., Robert A. Martin Jr., and Stanley M. Tarka Jr.
   1989 Authentification of Cocoa in Maya Vessels Using High-Performance Liquid
   Chromatographic Techniques. *Journal of Chromatography* 466:279–289.
Hus, J., R. Geeraerts, and S. Spassov
   2003 Archaeomagnetism and Archaeomagnetic Dating. Centre de Physique du Globe, Institut
   Royal Météorologique de Belgique, Dourbes, Belgium.
Hyland, D. C., J. M. Tersak, J. M. Adovasio, and M. I. Siegal
   1990 Identification of the Species of Origin of Residual Blood on Lithic Material. *American*
   Antiquity 55(1):104–112.
Ingo, G. M., L.-I. Manfredi, G. Bultrini, and E. Lo Piccolo
   1997 Quantitative Analysis of Copper-Tin Bronzes by Means of Glow Discharge Optical
   Emission Spectrometry. *Archaeometry* 39(1):59–70.
Ivanovich, M., and R. S. Harmon (editors)
   1992 *Uranium-series Disequilibrium: Applications to Earth, Marine, and Environmental*
   Sciences. 2nd ed. Clarendon, Oxford.

Jaenicke-Després, Viviane R., and Bruce D. Smith
    2006 Ancient DNA and the Integration of Archaeological and Genetic Approaches to the Study
    of Maize Domestication. In *Histories of Maize,* edited by John E. Staller, Robert H. Tykot and
    Bruce F. Benz, pp. 83–95. Academic, Boston.
Jakes, Kathryn A., and John C. Mitchell
    1996 Cold Plasma Ashing Preparation of Plant Phytoliths and their Examination with Scanning
    Electron Microscopy and Energy Dispersive Analysis of X-rays. *Journal of Archaeological
    Science* 23(1):149–156.
Jamieson, R. W., and R. G. V. Hancock
    2004 Neutron Activation Analysis of Colonial Ceramics from Southern Highland Ecuador.
    Archaeometry 46(4):569–583.
Johnson, B. J., and G. H. Miller
    1997 Archaeological Applications of Amino Acid Racemization. *Archaeometry* 39(2):
    265–287.
Jones, George T., Charlotte Beck, Eric E. Jones, and Richard E. Hughes
    2003 Lithic Source Use and Paleoarchaic Foraging Territories in the Great Basin. *American
    Antiquity* 68(1):5–38.
Jones, H. L., W. J. Rink, L. A. Schepartz, S. Miller-Antonio, Huang Weiwen, Hou Yamei, and
    Wang Wei
    2004 Coupled Electron Spin Resonance (ESR)/Uranium-series Dating of Mammalian Tooth
    Enamel at Panxian Dadong, Guizhou Province, China. *Journal of Archaeological Science*
    31(7):965–977.
Jones, M., P. J. Sheppard, and D. G. Sutton
    1997 Soil Temperature and Obsidian Hydration Dating: A Clarification of Variables Affecting
    Accuracy. *Journal of Archaeological Science* 24(6):505–516.
Jones, Martin
    2003 Ancient DNA in Pre-Columbian Archaeology: A Review. *Journal of Archaeological
    Science* 30(5):629–635.
Jones, Terry L., Richard T. Fitzgerald, Douglas J. Kennett, Charles H. Miksicek, John L. Fagan,
    John Sharp, and Jon M. Erlandson
    2002 The Cross Creek Site (CA-SLO-1797) and Its Implication for New World Colonization.
    American Antiquity 67(2):213–230.
José-Yacamán, Miguel, and Jorge A. Ascencio
    2000 Electron Microscopy and Its Application to the Study of Archaeological Materials
    and Art Preservation. In *Modern Analytical Methods in Art and Archaeology,* Chemical
    Analysis Series, vol. 155, edited by Enrico Ciliberto and Giuseppe Spoto, pp. 405–443. Wiley,
    New York.
Kaestle, Frederika A., and K. A. Horsburgh
    2002 Ancient DNA in Anthropology: Methods, Applications, and Ethics. *Yearbook of Physical
    Anthropology* 1999(S35):92–130.
Katzenberg, M. A., and Roman G. Harrison
    1997 What's in a Bone? Recent Advances in Archaeological Bone Chemistry. *Journal of
    Archaeological Research* 5(3):265–293.
Katzenberg, M. Anne and Shelley R. Saunders (editors)
    2008 Biological Anthropology of the Human Skeleton. 2nd ed. Wiley-Liss, Hoboken, NJ.
Keenleyside, A., X. Song, D. R. Chettle, and C. E. Webber
    1996 The Lead Content of Human Bones from the 1845 Franklin Expedition. *Journal of
    Archaeological Science* 23(3):461–465.
Kemp, Brian M., Cara Monroe, and David G. Smith
    2006 Repeat Silica Extraction: A Simple Technique for the Removal of PCR Inhibitors from
    DNA Extracts. *Journal of Archaeological Science* 33(12):1680–1689.

Kennett, Douglas J., Sachiko Sakai, Hector Neff, Richard Gossett, and Daniel O. Larson
    2002 Compositional Characterization of Prehistoric Ceramics: A New Approach. *Journal of Archaeological Science* 29(5):443–455.

Khandpur, R. S.
    2007 *Handbook of Analytical Instruments*. McGraw-Hill, New York.

Kimball, John M.
    1984 *Biology*. 5th ed. Addison Wesley, Reading, Massachusetts.
    1994 *Biology*. Wm. C. Brown, Dubuque, Iowa.

Knudson, Kelly J., Lisa Frink, Brian W. Hoffman, and T. D. Price
    2004 Chemical Characterization of Arctic Soils: Activity Area Analysis in Contemporary Yup'ik Fish Camps using ICP-AES. *Journal of Archaeological Science* 31(4):443–456.

Kohn, Matthew J.
    1996 Predicting animal $\delta^{18}O$; Accounting for diet and physiological adaptation. *Geochimica et Cosmochimica Acta* 60(23):4811–4829.

Kohn, Matthew J., Margaret J. Schoeninger, and John W. Valley
    1996 Herbivore Tooth Oxygen Isotope Compositions: Effects of Diet and Physiology. *Geochimica et Cosmochimica Acta* 60(20):3889–3896.

Koon, H. E. C., R. A. Nicholson, and M. J. Collins
    2003 A Practical Approach to the Identification of Low Temperature Heated Bone using TEM. *Journal of Archaeological Science* 30(11):1393–1399.

Kooyman, Brian, Margaret E. Newman, and Howard Ceri
    1992 Verifying the Reliability of Blood Residue Analysis on Archaeological Tools. *Journal of Archaeological Science* 19(3):265–269.

Kooyman, Brian, Margaret E. Newman, Christine Cluney, Murray Lobb, Shayne Tolman, Paul McNeil, and L. V. Hills
    2001 Identification of Horse Exploitation by Clovis Hunters Based on Protein Analysis. *American Antiquity* 66(4):686–691.

Kosman, Daniel J.
    1984 Electron Spin Resonance. In *Structural and Resonance Techniques in Biological Research,* edited by Denis L. Rousseau, pp. 89–244. Academic, Orlando.

Kovacheva, Mary, Ian Hedley, Neli Jordanova, Maria Kostadinova, and Valentin Gigov
    2004 Archaeomagnetic Dating of Archaeological Sites from Switzerland and Bulgaria. *Journal of Archaeological Science* 31(10):1463–1479.

Kowal, Walter, Owen B. Beattie, Halfdan Baadsgaard, and Peter M. Krahn
    1991 Source Identification of Lead Found in Tissues of Sailors from the Franklin Arctic Expedition of 1845. *Journal of Archaeological Science* 18(2):193–203.

Kuhn, Robert D., and Martha L. Sempowski
    2001 A New Approach to Dating the League of the Iroquois. *American Antiquity* 66(2): 301–314.

Lambert, Joseph B.
    1997 *Traces of the Past: Unraveling the Secrets of Archaeology through Chemistry*. Perseus Publishing. Cambridge, Massachusetts.

Lambert, J. B., C. W. Beck, and J. S. Frye
    1988 Analysis of European Amber by Carbon-13 Nuclear Magnetic Resonance Spectroscopy. *Archaeometry* 30(2):248–263.

Lambert, Joseph B., and James S. Frye
    1982 Carbon Functionalities in Amber. *Science* 217(4554):55–57.

Lambert, J. B., J. S. Frye, and A. Jurkiewicz
    1992 The Provenance and Coal Rank of Jet by Carbon-13 Nuclear Magnetic Resonance Spectroscopy. *Archaeometry* 34(1):121–128.

Lambert, Joseph B., James S. Frye, Thomas A. Lee Jr., Christopher J. Welch, and George O. Poinar Jr.
    1989 Analysis of Mexican Amber by Carbon-13 NMR Spectroscopy. In *Archaeological Chemistry IV,* edited by Ralph O. Allen, pp. 381–388. American Chemical Society, Washington, D. C.

Lambert, Joseph B., James S. Frye, and George O. Poinar
1985 Amber from the Dominican Republic: Analysis by Nuclear Magnetic Resonance Spectroscopy. *Archaeometry* 27(1):43–51.
1990 Analysis of North American Amber by Carbon-13 NMR Spectroscopy. *Geoarchaeology: An International Journal* 5(1):43–52.

Lanteigne, Maurice P.
1991 Cation-Ratio Dating of Rock-Engravings: A Critical Appraisal. *Antiquity* 65(247): 292–295.

Larsen, Clark S.
1997 *Bioarchaeology: Interpreting Behavior from the Human Skeleton.* Cambridge Studies in Biological Anthropology, vol. 21, Cambridge University, New York.

Latham, A. G.
2001 Uranium-Series Dating. In *Handbook of Archaeological Sciences,* edited by D. R. Brothwell and A. M. Pollard, pp. 63–72. John Wiley and Sons, Chichester, England.

Leach, Jeff D.
1998 A Brief Comment on the Immunological Identification of Plant Residues on Prehistoric Stone Tools and Ceramics: Results of a Blind Test. *Journal of Archaeological Science* 25(2):171–175.

Leach, Jeff D., and Raymond P. Mauldin
1995 Additional Comments on Blood Residue Analysis in Archaeology. *Antiquity* 69(266):1020–1022.

Lebo, Susan A., and Kevin T. M. Johnson
2007 Geochemical Sourcing of Rock Specimens and Stone Artifacts from Nihoa and Necker Islands, Hawai'i. *Journal of Archaeological Science* 34(6):858–871.

Leco Corporation
2007 *TruSpec Micro Elemental Series.* Form No. 209–190, R1.56-REV1, Leco Corporation, St. Joseph, Michigan.

Lee, A. P., J. Klinowski, and E. A. Marseglia
1995 Application of Nuclear Magnetic Resonance Spectroscopy to Bone Diagenesis. *Journal of Archaeological Science* 22(2):257–262.

Lee, Gyoung-Ah, Anthony M. Davis, David G. Smith, and John H. McAndrews
2004 Identifying Fossil Wild Rice (*Zizania*) Pollen from Cootes Paradise, Ontario: A New Approach using Scanning Electron Microscopy. *Journal of Archaeological Science* 31(4): 411–421.

Lee, Richard
1969 Chemical Temperature Integration. *Journal of Applied Meteorology* 8(3):423–430.

Lee-Thorp, J. A.
2008 On Isotopes and Old Bones. *Archaeometry* 50(6):925–950.

Lehninger, Albert L.
1970 *Biochemistry: The Molecular Basis of Cell Structure and Function.* Worth, New York.

Lehninger, Albert L., David L. Nelson, and Michael M. Cox
1993 *Principles of Biochemistry.* 2nd. Worth, New York.

Lengyel, Stacey N., and Jeffrey L. Eighmy
2002 A Revision to the U.S. Southwest Archaeomagnetic Master Curve. *Journal of Archaeological Science* 29(12):1423–1433.

Leonard, Jennifer A., Orin Shanks, Michael Hofreiter, Eva Kreuz, Larry Hodges, Walt Ream, Robert K. Wayne, and Robert C. Fleischer
2007 Animal DNA in PCR Reagents Plagues Ancient DNA Research. *Journal of Archaeological Science* 34(9):1361–1366.

Leute, Ulrich
1987 *Archaeometry: An Introduction to Physical Methods in Archaeology and the History of Art.* VCH Verlagsgesellschaft mbH, Weinheim, Germany.

Levine, Ira N.
1978 *Physical Chemistry.* McGraw-Hill, New York.

Levine, Mary A.
   2007 Determining the Provenance of Native Copper from Northeastern North America: Evidence from Instrumental Neutron Activation Analysis. *Journal of Archaeological Science* 34(4):572–587.

Li, Bao-ping, Alan Greig, Jian-Xin Zhao, Kenneth D. Collerson, Kui-shan Quan, Yao-hu Meng, and Zhong-li Ma
   2005 ICP-MS trace element analysis of Song dynasty porcelains from Ding, Jiexiu and Guantai kilns, north China. *Journal of Archaeological Science* 32(2):251–259.

Liden, Kerstin, Cheryl Takahashi, and D. E. Nelson
   1995 The Effects of Lipids in Stable Carbon Isotope Analysis and the Effects of NaOH Treatment on the Composition of Extracted Bone Collagen. *Journal of Archaeological Science* 22(2):321–326.

Lido, David R. (editor)
   2006 *CRC Handbook of Chemistry and Physics.* 87th ed. 2006–2007, CRC Press, Taylor and Francis Group, Boca Raton, Florida.

Linford, Paul
   2006 *Archaeomagnetic Dating: Guidelines on Producing and Interpreting Archaeomagnetic Dates.* English Heritage, Swindon, United Kingdom.

Lipo, Carl P., James K. Feathers, and Robert C. Dunnell
   2005 Temporal Data Requirements, Luminescence Dates, and the Resolution of Chronological Structure of Late Prehistoric Deposits in the Central Mississippi River Valley. *American Antiquity* 70(3):527–544.

Little, Elizabeth A.
   2002 Kautantouwit's Legacy: Calibrated Dates on Prehistoric Maize in New England. *American Antiquity* 67(1):109–118.

Loendorf, Lawrence L.
   1991 Cation-Ratio Varnish Dating and Petroglyph Chronology in Southeastern Colorado. *Antiquity* 65(247):246–255.

Lowe, J. J.
   2001 Quarternary Geochronological Frameworks. In *Handbook of Archaeological Sciences,* edited by D. R. Brothwell and A. M. Pollard, pp. 9–22. John Wiley and Sons, Chichester, England.

Lowenstein, Jerold M.
   1985 Molecular Approaches to the Identification of Species. *American Scientist* 73(6):541–547.

Lowenstein, Jerold M., and Gary Schueunstuhl
   1991 Immunological Methods in Molecular Palaeontology. *Philosophical Transactions of the Royal Society of London. Series B, Biological Sciences* 333(1268):375–380.

Loy, Thomas H.
   1983 Prehistoric Blood Residues: Detection on Tool Surfaces and Identification of Species of Origin. *Science* 220(4603):1269–1271.
   1993 The Artifact as Site: An Example of the Biomolecular Analysis of Organic Residues on Prehistoric Tools. *World Archaeology* 25(1):44–63.
   1994 Residue Analysis of Artifacts and Burned Rock from the Mustang Branch and Barton Sites (41HY209 and 41HY202. In *Archaic and Late Prehistoric Human Ecology in the Middle Onion Creek Valley, Hays County, Texas. Volume 2: Topical Studies* edited by R. A. Ricklis and M. B. Collins, pp. 607–627. University of Texas at Austin, Austin Texas.

Loy, Thomas H., and E. J. Dixon
   1998 Blood Residues on Fluted Points from Eastern Beringia. *American Antiquity* 63(1):21–46.

Loy, T. H., and B. L. Hardy
   1992 Blood Residue Analysis of 90,000-Year-Old Stone Tools from Tabun Cave, Israel. *Antiquity* 66(250):24–35.

Loy, T. H., Rhys Jones, D. E. Nelson, Betty Meehan, John Vogel, John Southon, and Richard Cosgrove

1990 Accelerator Radiocarbon Dating of Human Blood Proteins in Pigments from Late Pleistocene Art Sites in Australia. *Antiquity* 64(242):110–116.

Loy, Thomas H., and Andree R. Wood
1989 Blood Residue Analysis at Çayönü Tepesi, Turkey. *Journal of Field Archaeology* 16(4):451–460.

Lutz, J., and E. Pernicka
1996 Energy Dispersive X-Ray Fluorescence Analysis of Ancient Copper Alloys: Empirical Values for Precision and Accuracy. *Archaeometry* 38(2):313–323.

Lyman, R. L.
1994 *Vertebrate Taphonomy.* Cambridge Manuals in Archaeology, Cambridge University, New York.

Lyons, William H., Michael D. Glascock, and Peter J. Mehringer Jr
2003 Silica from Sources to Site: Ultraviolet Fluorescence and Trace Elements Identify Cherts from Lost Dune, Southeastern Oregon, USA. *Journal of Archaeological Science* 30(9):1139–1159.

MacDougall, I., and T. M. Harrison
1988 *Geochronology and Thermochronology by the Ar-Ar Method.* Oxford Monographs on Geology and Geophysics No. 9, Oxford University, New York.

McGovern, Patrick E., and Rudolph H. Michel
1996 The Analytical and Archaeological Challenge of Detecting Ancient Wine: Two Case Studies from the Ancient Near East. In *The Origins and Ancient History of Wine,* edited by Patrick E. McGovern, Stuart J. Fleming and Solomon H. Katz, pp. 57–65. Gordon and Breach Publishers, Amsterdam.

Malainey, Mary
1997 The Reconstruction and Testing of Subsistence and Settlement Strategies for the Plains, Parkland and Southern Boreal Forest. Unpublished Ph.D. dissertation, University of Manitoba, Winnipeg, Manitoba.
2007 Fatty Acid Analysis of Archaeological Residues: Procedures and Possibilities. In *Theory and Practice of Archaeological Residue Analysis,* edited by H. Barnard and J. W. Eerkens, pp. 77–89. British Archaeological Reports International Series 1650, Oxford.

Malainey, M. E., R. Przybylski, and B. L. Sherriff
1999a The Effects of Thermal and Oxidative Decomposition on the Fatty Acid Composition of Food Plants and Animals of Western Canada: Implications for the Identification of Archaeological Vessel Residues. *Journal of Archaeological Science* 26(1):95–103.
1999b The Fatty Acid Composition of Native Food Plants and Animals of Western Canada. *Journal of Archaeological Science* 26(1):83–94.
1999c Identifying the Former Contents of Late Precontact Period Pottery Vessels from Western Canada using Gas Chromatography. *Journal of Archaeological Science* 26(4):425–438.
2001 One Person's Food: How and Why Fish Avoidance May Affect the Settlement and Subsistence Patterns of Hunter-Gatherers. *American Antiquity* 66(1):141–161.

Mallory-Greenough, Leanne M., John D. Greenough, and J. V. Owen
1998 New Data for Old Pots: Trace Element Characterization of Ancient Egyptian Pottery Using ICP-MS. *Journal of Archaeological Science* 25(1):85–97.

Mannerstrand, Maria, and Lars Lundqvist
2003 Garnet Chemistry from the Slöinge Excavation, Halland and Additional Swedish and Danish Excavations-Comparisons with Garnet Occurring in a Rock Context. *Journal of Archaeological Science* 30(2):169–183.

Marlar, Richard A., Banks L. Leonard, Brian R. Billman, Patricia M. Lambert, and Jennifer E. Marlar
2000 Biochemical Evidence of Cannibalism at a Prehistoric Puebloan Site in Southwestern Colorado. *Nature* 407(6800):74–78.

Marchbanks, Michael L.
1989 *Lipid Analysis in Archaeology: An Initial Study of Ceramics and Subsistence at the George C. Davis Site.* Unpublished Master's thesis, University of Texas at Austin, Austin, TX.

Marchbanks, M. L., and J. M. Quigg
    1990 Appendix G: Organic Residue and Phytolith Analysis. In *Phase II Investigations at Prehistoric and Rock Art Sites, Justiceburg Reservoir, Garza and Kent Counties, Texas,* vol. II, edited by D. K. Boyd, J. T. Abbott, W. A. Bryan, C. M. Garvey, S. A. Tomka and R. C. Fields, pp. 496–519. Prewitt and Associates, Inc., Austin, TX.
Mauk, J. L., and R. G. V. Hancock
    1998 Trace Element Geochemistry of Native Copper from The White Pine Mine, Michigan (USA): Implications for Sourcing Artefacts. *Archaeometry* 40(1):97–107.
Maurer, Joachim, Thomas Mohring, Jürgen Rullkötter, and Arie Nissenbaum
    2002 Plant Lipids and Fossil Hydrocarbons in Embalming Material of Roman Period Mummies from the Dakhleh Oasis, Western Desert, Egypt. *Journal of Archaeological Science* 29(7): 751–762.
Mellars, P. A., L. P. Zhou, and E. A. Marseglia
    1997 Compositional Inhomogeneity of Sediments and its Potential Effects on Dose Rate Estimation for Electron Spin Resonance Dating of Tooth Enamel. *Archaeometry* 39(1): 169–176.
Mills, John S., Raymond White, and Laurence J. Gough
    1984/1985 The Chemical Composition of Baltic Amber. *Chemical Geology* 47(1–2):15–39.
Mishmar, Dan, Eduardo Ruiz-Pesini, Martin Brandon, and Douglas C. Wallace
    2004 Mitochondrial DNA-like Sequences in the Nucleus (NUMTs): Insights into our African origins and the Mechanism of Foreign DNA Integration. *Human Mutation* 23(2):125–133.
Mitchell, Piers D., Eliezer Stern, and Yotam Tepper
    2008 Dysentry in the Crusader Kingdom of Jerusalem: An ELISA Analysis of Two Medieval Latrines in the City of Acre (Israel). *Journal of Archaeological Science* 35(7):1849–1853.
Moens, Luc, Alex von Bohlen, and Peter Vandenabeele
    2000 X-Ray Fluorescence. In *Modern Analytical Methods in Art and Archaeology.* Chemical Analysis Series, vol. 155, edited by Enrico Ciliberto and Giuseppe Spoto, pp. 55–79. Wiley, New York.
Mommsen, H., Th Beier, and A. Hein
    2002 A Complete Chemical Grouping of the Berkeley Neutron Activation Analysis Data on Mycenaean Pottery. *Journal of Archaeological Science* 29(6):613–637.
Monette, Yves, Marc Richer-LaFleche, Marcel Mousette, and Daniel Dufournier
    2007 Compositional Analysis of Local Redwares: Characterizing the Pottery Productions of 16 Workshops Located in Southern Quebec dating from late 17th to late 19th-Century. *Journal of Archaeological Science* 34(1):123–140.
Moorhouse, W. W.
    1959 *The Study of Rocks in Thin Section.* Harper, New York.
Morgan, E. D., L. Titus, R. J. Small, and Corony Edwards
    1983 The Composition of Fatty Materials from a Thule Eskimo Site on Herschel Island. *Arctic* 36(4):356–360.
Morgenstein, M. E., S. Luo, T.-L. Ku, and J. Feathers
    2003 Uranium-Series and Luminescence Dating of Volcanic Lithic Artefacts. *Archaeometry* 45(3):503–518.
Mortimer, Charles E.
    1979 *Chemistry: A Conceptual Approach.* 4th ed. D. Van Nostrand, New York.
    1986 *Chemistry.* Wadsworth, Belmont, California.
Morton, June D.
    1989 *An Investigation of the Use of Stable Isotopic Analysis of Encrustations on Prehistoric Ontario Ceramics and Paleodietary Possibilities.* Unpublished Master's thesis, Department of Geology, McMaster University, Hamilton, Ontario.
Morton, J. D., R. B. Lammers, and H. P. Schwarcz
    1991 Estimation of Palaeodiet: A Model from Stable Isotope Analysis. In *Archaeometry '90,* edited by Ernst Pernick and Günther A. Wagner, pp. 807–820. Birkhauser Verlag Basel, Berlin.

Morton, June D., and Henry P. Schwarcz
  2004 Palaeodietary Implications from Stable Isotopic Analysis of Residues on Prehistoric Ontario Ceramics. *Journal of Archaeological Science* 31(5):503–517.
Mottram, H. R., S. N. Dudd, G. J. Lawrence, A. W. Stott, and R. P. Evershed
  1999 New Chromatographic, Mass Spectrometric and Stable Isotope Approaches to the Classification of Degraded Animal Fats Preserved in Archaeological Pottery. *Journal of Chromatography A* 833(2):209–221.
Mukherjee, Anna J., Alex M. Gibson, and Richard P. Evershed
  2008 Trends in Pig Product Processing at British Neolithic Grooved Ware Sites Traced through Organic Residues in Potsherds. *Journal of Archaeological Science* 35(7):2059–2073.
Müldner, Gundula, and Michael P. Richards
  2005 Fast or Feast: Reconstructing Diet in Later Medieval England by Stable Isotope Analysis. *Journal of Archaeological Science* 32(1):39–48.
Mulligan, Connie J.
  2006 Anthropological Applications of Ancient DNA: Problems and Prospects. *American Antiquity* 71(2):365–380.
Neff, Hector
  1993 Theory, Sampling, and Analytical Techniques in the Archaeological Study of Prehistoric Ceramics. *American Antiquity* 58(1):23–44.
  2000 Neutron Activation Analysis for Provenance Determination in Archaeology. In *Modern Analytical Methods in Art and Archaeology*, Chemical Analysis Series, vol. 155, edited by Enrico Ciliberto and Giuseppe Spoto, pp. 81–134. Wiley, New York.
  2001 Sythesizing Analytical Data – Spatial Results from Pottery Provenance. In *Handbook of Archaeological Sciences*, edited by D. R. Brothwell and A. M. Pollard, pp. 733–748. Wiley, Chichester.
  2003 Analysis of Mesoamerican Plumbate Pottery Surfaces by Laser Ablation-Inductively Coupled Plasma-Mass Spectrometry (LA-ICP-MS). *Journal of Archaeological Science* 30(1):21–35.
Neff, Hector, Ronald L. Bishop, and Edward V. Sayre
  1988 A Simulation Approach to the Problem of Tempering in Compositional Studies of Archaeological Ceramics. *Journal of Archaeological Science* 15(2):159–172.
Nelson, D. E.
  1993 Second Thoughts on a Rock-Art Date. *Antiquity* 67(257):893–895.
Nelson, David L., and Michael M. Cox
  2008 *Lehninger Principles of Biochemistry.* 5th ed. W. H. Freeman, New York.
Newman, Margaret E., Howard Ceri, and Brian Kooyman
  1996 The Use of Immunological Techniques in the Analysis of Archaeological Materials – A Response to Eisele; With Report of Studies at Head-Smashed-In Buffalo Jump. *Antiquity* 70(269):677–682.
Newman, M., and P. Julig
  1989 The Identification of Protein Residues on Lithic Artifacts from a Stratified Boreal Forest Site. *Canadian Journal of Archaeology* 13:119–132.
Newman, Margaret E., Jillian S. Parboosingh, Peter J. Bridge, and Howard Ceri
  2002 Identification of Archaeological Animal Bone by PCR/DNA Analysis. *Journal of Archaeological Science* 29(1):77–84.
Newman, M. E., Robert M. Yohe II, B. Kooyman, and H. Ceri
  1997 "Blood" from Stones? Probably: A Response to Fiedel. *Journal of Archaeological Science* 24(11):1023–1027.
Nir-El, Y., and M. Broshi
  1996 The Red Ink of the Dead Sea Scrolls. *Archaeometry* 38(1):97–102.
O'Connell, T. C., R. E. M. Hedges, M. A. Healey, and Simpson, A. H. R. W.
  2001 Isotopic Comparison of Hair, Nail and Bone: Modern Analyses. *Journal of Archaeological Science* 28(11):1247–1255.

O'Leary, Marion H.
1988 Carbon Isotopes in Photosynthesis. *Bioscience* 38(5):328–336.

O'Neil, J. R., L. J. Roe, E. Reinhard, and R. E. Blake
1994 A Rapid and Precise Method of Oxygen Isotope Analysis of Biogenic Phosphate. *Israel Journal of Earth Sciences* 43:203–212.

O'Rourke, Dennis H., M. G. Hayes, and Shawn W. Carlyle
2000 Ancient DNA Studies in Physical Anthropology. *Annual Review of Anthropology* 29: 217–242.

Oudemans, T. F. M., and J. J. Boon
1991 Molecular Archaeology: Analysis of Charred (Food) Remains from Prehistoric Pottery by Pyrolysis-Gas Chromatography/Mass Spectroscopy. *Journal of Analytical and Applied Pyrolysis* 20:197–227.

Oudemans, T. F. M., J. J. Boon, and R. E. Botto
2007 FTIR and Solid-state ^{13}C CP/MAS NMR Spectroscopy of Charred and Non-Charred Solid Organic Residues Preserved in Roman Iron Age Vessels from the Netherlands. *Archaeometry* 49(3):571–594.

Ownby, Mary F., Charlotte L. Ownby, and Elizabeth J. Miksa
2004 Use of Scanning Electron Microscopy to Characterize Schist as a Temper in Hohokam Pottery. *Journal of Archaeological Science* 31(1):31–38.

Oxenham, Marc F., Cornelia Locher, Nguyen L. Cuong, and Nguyen K. Thuy
2002 Identification of *Areca catechu* (Betel Nut) Residues on the Dentitions of Bronze Age Inhabitants of Nui Nap, Northern Vietnam. *Journal of Archaeological Science* 29(9):909–915.

Pääbo, Svante
1989 Ancient DNA: Extraction, Characterization, Molecular Cloning, and Enzymatic Amplification. *Proceedings of the National Academy of Sciences of the United States of America* 86(6):1939–1943.

Pääbo, Svante, Russel G. Higuchi, and Allan C. Wilson
1989 Ancient DNA and the Polymerase Chain Reaction. *The Journal of Biological Chemistry* 264(17):9709–9712.

Panico, R., W. H. Powell, and W. C. Fernelius (editors)
1993 *A Guide to IUPAC Nomenclature of Organic Compounds, Recommendations 1993.* Blackwell Scientific Publications, Oxford.

Pasto, Daniel J., and Carl R. Johnson
1979 *Laboratory Text for Organic Chemistry.* Prentice-Hall, Englewood Cliffs, NJ.

Pate, F. D.
1994 Bone Chemistry and Paleodiet. *Journal of Archaeological Method and Theory* 1(2): 161–209.

Patnaik, Pradyot
2004 *Dean's Analytical Chemistry Handbook.* 2nd ed. McGraw-Hill, New York.

Patrick, M., A. J. de Koning, and A. B. Smith
1985 Gas Liquid Chromatographic Analysis of Fatty Acids in Food Residues from Ceramics found in the Southwestern Cape, South Africa. *Archaeometry* 27(2):231–236.

Pearson, Charlotte, Sturt W. Manning, Max Coleman, and Kym Jarvis
2005 Can Tree-Ring Chemistry Reveal Absolute Dates for Past Volcanic Eruptions? *Journal of Archaeological Science* 32(8):1265–1274.

Pérez-Arantegui, Josefina, Jaun Á. Paz-Peralta, and Esperanza Ortiz-Palomar
1996 Analysis of the Products Contained in Two Roman Glass *Unguentaria* from the Colony of *Celsa* (Spain). *Journal of Archaeological Science* 23(5):649–655.

Pérez-Arantegui, Josefina, M. I. Urunuela, and Jaun R. Castillo
1996 Roman Glazed Ceramics in the Western Mediterranean: Chemical Characterization by Inductively Coupled Plasma Atomic Emission Spectrometry of Ceramic Bodies. *Journal of Archaeological Science* 23:903–914.

PerkinElmer, Inc.

2005 *FT-IR Spectroscopy: Attenuated Total Reflectance (ATR)*. Technical Note 007024B_01, PerkinElmer, Inc, Shelton, Connecticut. 2005 *Organic Elemental Analysis: 2400 Series II CHNS/O Elemental Analyzer.* Technical Note 007390_01, PerkinElmer, Inc, Shelton, Connecticut.

Petit-Dominguez, Maria D., Rosario Garcia-Giminez, and Maria I. Rucandio

2003 Chemical Characterization of Iberian Amphorae and Tannin Determination as Indicative of Amphora Contents. *Microchimica Acta* 141(1–2):63–68.

Pettitt, P. B., W. Davies, C. S. Gamble, and M. B. Richards

2003 Palaeolithic Radiocarbon Chronology: Quantifying our Confidence Beyond Two Half-lives. *Journal of Archaeological Science* 30(12):1685–1693.

Pfeiffer, Susan, and Tamara L. Varney

2000 Quantifying Histological and Chemical Preservation in Archaeological Bone. In *Biogeochemical Approaches to Paleodietary Analysis,* vol. 5, edited by Stanley H. Ambrose and M. A. Katzenberg, pp. 141–157. Kluwer Academic/Plenum Publishers in cooperation with the Society for Archaeological Sciences, New York.

Pierce, Christopher, Karen R. Adams, and Joe D. Stewart

1998 Determining the Fuel Constituents of Ancient Hearth Ash via ICP-AES Analysis. *Journal of Archaeological Science* 25(6):493–503.

Pierret, A., C. J. Moran, and L. M. Bresson

1996 Calibration and Visualization of Wall-Thickness and Porosity Distributions of Ceramic Using X-Radiography and Image Processing. *Journal of Archaeological Science* 23(3): 419–428.

Poinar, Hendrick N., Matthias Höss, Jeffrey L. Bada, and Svante Pääbo

1996 Amino Acid Racemization and the Preservation of Ancient DNA. *Science* 272(5263):864–866.

Polet, C., and M. A. Katzenberg

2003 Reconstruction of the Diet in a Medieval Monastic Community from the Coast of Belgium. *Journal of Archaeological Science* 30(5):525–533.

Polikreti, K., and Y. Maniatis

2002 A New Methodology for the Provenance of Marble Based on EPR Spectroscopy. *Archaeometry* 44(1):1–21.

Polikreti, K., Y. Maniatis, Y. Bassiakos, N. Kourou, and V. Karageorghis

2004 Provenance of Archaeological Limestone with EPR Spectroscopy: The Case of the Cypriote-type Statuettes. *Journal of Archaeological Science* 31(7):1015–1028.

Pollard, A. M., and Carl Heron

1996 *Archaeological Chemistry.* Royal Society of Chemistry, Cambridge.

Pollard, Mark, Catherine Batt, Ben Stern and Suzanne M. M. Young

2007 *Analytical Chemistry in Archaeology.* Cambridge University, Cambridge.

Pollock, Stephen G., Nathan D. Hamilton, and Richard A. Boisvert

2008 Archaeological Geology of Two Flow-banded Spherulitic Rhyolites in New England, USA: Their History, Exploitation and Criteria for Recognition. *Journal of Archaeological Science* 35(3):688–703.

Popelka-Filcoff, Rachel S., Elizabeth J. Miksa, J. D. Robertson, Michael D. Glascock, and Henry Wallace

2008 Elemental Analysis and Characterization of Ochre Sources from Southern Arizona. *Journal of Archaeological Science* 35(3):752–762.

Poulakakis, N., A. Tselikas, I. Bitsakis, M. Mylonas, and P. Lymberakis

2007 Ancient DNA and the Genetic Signature of Ancient Greek Manuscripts. *Journal of Archaeological Science* 34(5):675–680.

Privat, Karen L., Tamsin C. O'Connell, and Michael P. Richards

2002 Stable Isotope Analysis of Human and Faunal Remains from the Anglo-Saxon Cemetery at Berinsfield, Oxfordshire: Dietary and Social Implications. *Journal of Archaeological Science* 29(7):779–790.

Quigg, J. M., M. E. Malainey, R. Przybylski, and G. Monks
2001 No Bones About It: Using Lipid Analysis of Burned Rock and Groundstone Residues to Examine Late Archaic Subsistence Practices in South Texas. *Plains Anthropologist* 46(177):283–303.

Rafferty, Sean M.
2002 Identification of Nicotine by Gas Chromatography/Mass Spectroscopy Analysis of Smoking Pipe Residue. *Journal of Archaeological Science* 29(8):897–907.
2006 Evidence of Early Tobacco in Northeastern North America? *Journal of Archaeological Science* 33(4):453–458.

Raven, A. M., P. F. van Bergen, A. W. Stott, S. N. Dudd, and R. P. Evershed
1997 Formation of Long-chain Ketones in Archaeological Pottery Vessels by Pyrolysis of Acyl Lipids. *Journal of Analytical and Applied Pyrolysis* 40–41:267–285.

Reber, Eleanora A., and Richard P. Evershed
2004a How Did Mississippians Prepare Maize? The Application of Compound-Specific Carbon Isotope Analysis to Absorbed Pottery Residues from Several Mississippi Valley Sites. *Archaeometry* 46(1):19–33.
2004b Identification of Maize in Absorbed Organic Residues: A Cautionary Tale. *Journal of Archaeological Science* 31(4):399–410.

Reber, Eleanora A., Stephanie N. Dudd, Nikolaas J. van der Merwe, and Richard P. Evershed
2004 Direct Detection of Maize in Pottery Residues Via Compound Specific Stable Carbon Isotope Analysis. *Antiquity* 78(301):682–691.

Rees-Jones, J., and M. S. Tite
1997 Optical Dating Results for British Archaeological Sediments. *Archaeometry* 39(1): 177–187.

Regert, M.
2007 Elucidating Pottery Function using a Multi-step Analytical Methodology Combining Infrared Spectroscopy, Chromatographic Procedures and Mass Spectrometry. In *Theory and Practice of Archaeological Residue Analysis,* edited by H. Barnard and J. W. Eerkens, pp. 60–76. British Archaeological Reports International Series 1650, Oxford.

Regert, M., S. Colinart, L. Degrand, and O. Decavallas
2001 Chemical Alteration and Use of Beeswax Through Time: Accelerated Ageing Tests and Analysis of Archaeological Samples from Various Environmental Contexts. *Archaeometry* 43(4):549–569.

Rehren, Th. and E. Pernicka
2008 Coins, Artefacts and Isotopes—Archaeometallurgy and Archaeometry. *Archaeometry* 50(2):232–248.

Reuther, Joshua D., Jerold M. Lowenstein, S. C. Gerlach, Darden Hood, Gary Scheuenstuhl, and Douglas H. Ubelaker
2006 The Use of an Improved pRIA Technique in the Identification of Protein Residues. *Journal of Archaeological Science* 33(4):531–537.

Reynard, L. M., and R. E. M. Hedges
2008 Stable Hydrogen Isotopes of Bone Collagen in Palaeodietary and Palaeoenvironmental Reconstruction. *Journal of Archaeological Science* 35(7):1934–1942.

Rice, Prudence
1987 *Pottery Analysis: A Sourcebook.* University of Chicago, Chicago.

Richards, M. P., J. A. Pearson, T. I. Molleson, N. Russell, and L. Martin
2003 Stable Isotope Evidence of Diet at Neolithic Çatalhöyük, Turkey. *Journal of Archaeological Science* 30(1):67–76.

Richter, Daniel
2007 Advantages and Limitations of Thermoluminescence Dating of Heated Flint from Paleolithic Sites. *Geoarchaeology: An International Journal* 22(6):671–683.

Rick, Torben C., René L. Vellanoweth, Jon M. Erlandson, and Douglas J. Kennett
2002 On the Antiquity of the Single-Piece Shell Fishhook: AMS Radiocarbon Evidence from the Southern California Coast. *Journal of Archaeological Science* 29(9):933–942.

Rick, Torben C., René L. Vellanoweth, and Jon M. Erlandson
2005 Radiocarbon dating and the "Old Shell" Problem: Direct Dating of Artifacts and Cultural Chronologies in Coastal and Other Aquatic Regions. *Journal of Archaeological Science* 32(11):1641–1648.

Ridings, Rosanna
1991 Obsidian Hydration Dating: The Effects of Mean Exponential Ground Temperature and Depth of Artifact Recovery. *Journal of Field Archaeology* 18(1):77–85.
1996 Where in the World Does Obsidian Hydration Dating Work? *American Antiquity* 61(1):136–148.

Rink, W. J., I. Karavanić, P. B. Pettitt, J. van der Plicht, F. H. Smith, and J. Bartoll
2002 ESR and AMS-based ^{14}C Dating of Mousterian Levels at Mujina Pećina, Dalmatia, Croatia. *Journal of Archaeological Science* 29(9):943–952.

Rink, W. J., D. Richter, H. P. Schwarcz, A. E. Marks, K. Monigal, and D. Kaufman
2003 Age of the Middle Palaeolithic Site of Rosh Ein Mor, Central Negev, Israel: Implications for the Age Range of the Early Levantine Mousterian of the Levantine Corridor. *Journal of Archaeological Science* 30(2):195–204.

Rink, W. J., H. P. Schwarcz, H. K. Lee, V. C. Valdes, F. B. de Quiros, and M. Hoyos
1996 ESR Dating of Tooth Enamel: Comparison with AMS ^{14}C at El Castillo Cave, Spain. *Journal of Archaeological Science* 23(6):945–951.

Rink, W. J., H. P. Schwarcz, A. Ronen, and A. Tsatskin
2004 Confirmation of a Near 400 ka Age for the Yabrudian Industry at Tabun Cave, Israel. *Journal of Archaeological Science* 31(1):15–20.

Roberts, Richard G.
1997 Luminescence Dating in Archaeology: From Origins to Optical. *Radiation Measurements* 27(5/6):819–892.

Robinson, Neil, R. P. Evershed, W. J. Higgs, K. Jerman, and G. Eglinton
1987 Proof of a Pine Wood Origin for Pitch from Tudor (Mary Rose) and Etruscan Shipwrecks: Application of Analytical Organic Chemistry in Archaeology. *Analyst* 112(5): 637–644.

Rogers, Alexander K.
2008 Obsidian Hydration Dating: Accuracy and Resolution Limitations Imposed by Intrinsic Water Variability. *Journal of Archaeological Science* 35(7):2009–2016.

Rottländer, Rolf C. A.
1990 Lipid Analysis in the Identification of Vessel Contents. *MASCA Research Papers in Science and Archaeology* 7:37–40.

Rye, O. S.
1977 Pottery Manufacturing Techniques: X-ray Studies. *Archaeometry* 19(2):205–211.
1981 *Pottery Technology: Principles and Reconstruction*. Taraxacum, Washington.

Rye, O. S., and P. Duerden
1982 Papuan Pottery Sourcing by PIXE: Preliminary Studies. *Archaeometry* 24(1):59–64.

Rypkema, Heather A., Wayne E. Lee, Michael L. Galaty, and Jonathan Haws
2007 Rapid, In-stride Soil Phosphate Measurement in Archaeological survey: A New Method Tested in Loudoun County, Virginia. *Journal of Archaeological Science* 34(11):1859–1867.

Sanders, Jeremy K. M. and Brian K. Hunter
1993 *Modern NMR Spectroscopy: A Guide for Chemists*. Oxford University, Oxford.

Sauter, F., E. W. H. Hayek, W. Moche, and U. Jordis
1987 Betulin aus Archaolgischem Schwelteer (Identification of Betulin in Archaeological Tar). *Verlag der Zeitschrift fur Naturforschung* 42(11–12):1151–1152.

Sayre, Edward

2000 Determination of Provenance. In *Science and Technology in Historic Preservation,* edited by Ray A. Williamson and Paul R. Nickens, pp. 143–169. Kluwer Academic/Plenum, New York.

Schneider, J. S., and P. R. Bierman

1997 Surface Dating Using Rock Varnish. In *Chronometric Dating in Archaeology,* Advances in Archaeological and Museum Science, vol. 2, edited by R. E. Taylor and M. J. Aitken, pp. 357–388. Plenum, New York.

Schoeller, Dale A.

1999 Isotope Fractionation: Why Aren't We What We Eat? *Journal of Archaeological Science* 26(6):667–673.

Schoeninger, Margaret J., Michael J. DeNiro, and Henrik Tauber

1983 Stable Nitrogen Isotope Ratios of Bone Collagen Reflect Marine and Terrestrial Components of Prehistoric Human Diet. *Science* 220(4604):1381–1383.

Schulting, Rick J., and Michael P. Richards

2002 Dogs, Ducks, Deer and Diet: New Stable Isotope Evidence on Early Mesolithic Dogs from the Vale of Pickering, North-east England. *Journal of Archaeological Science* 29(4):327–333.

Schwarcz, Henry P.

1991 Some Theoretical Aspects of Isotope Paleodiet Studies. *Journal of Archaeological Science* 18(3):261–275.

1997 Uranium Series Dating. In *Chronometric Dating in Archaeology,* Advances in Archaeological and Museum Science, vol. 2, edited by R. E. Taylor and M. J. Aitken, pp. 159–182. Plenum, New York.

2000 Some Biochemical Aspects of Carbon Isotopic Paleodiet Studies. In *Biogeochemical Approaches to Paleodietary Analysis,* vol. 5, edited by Stanley H. Ambrose and M. A. Katzenberg, pp. 189–209. Kluwer Academic/Plenum Publishers in cooperation with the Society for Archaeological Sciences, New York.

2006 Stable Carbon Isotope Analysis and Human Diet: A Synthesis. In *Histories of Maize,* edited by John E. Staller, Robert H. Tykot and Bruce F. Benz, pp. 315–322. Academic, Boston.

Schwarcz, H. P., and B. A. Blackwell

1992 Archaeological Applications. In *Uranium-series Disequilibrium: Applications to Earth, Marine, and Environmental Sciences,* edited by M. Ivanovich and R. S. Harmon, pp. 513–552. Claredon, Oxford.

Schwarcz, H. P., J. Melbye, M. A. Katzenberg, and M. Knyf

1985 Stable Isotopes in Human Skeletons of Southern Ontario: Reconstructing Palaeodiet. *Journal of Archaeological Science* 12(3):187–206.

Schwarcz, Henry P., and Margaret J. Schoeninger

1991 Stable Isotope Analyses in Human Nutritional Ecology. *Yearbook of Physical Anthropology* 34:283–321.

Schwedt, A., and H. Mommsen

2004 Clay Paste Mixtures Identified by Neutron Activation Analysis in Pottery of a Roman Workshop in Bonn, Germany. *Journal of Archaeological Science* 31(9):1251–1258.

Scott, D. A.

2001 The Application of Scanning X-Ray Fluorescence Microanalysis in the Examination of Cultural Materials. *Archaeometry* 43(4):475–482.

Scott, D. A., M. Newman, M. Schilling, M. Derrick, and H. P. Khanjian

1996 Blood as a Binding Medium in a Chumash Indian Pigment Cake. *Archaeometry* 38(1):103–112.

Scott, E. M., C. Bryant, I. Carmi, G. Cook, S. Gulliksen, D. Harkness, J. Heinemeier, E. McGee, P. Naysmith, G. Possnert, H. van der Plicht, and M. van Strydonck

2004 Precision and Accuracy in Applied ^{14}C Dating: Some Findings from the Fourth International Radiocarbon Inter-comparison. *Journal of Archaeological Science* 31(9): 1209–1213.

Sealy, Judith
2001 Body Tissue Chemistry and Palaeodiet. In *Handbook of Archaeological Sciences,* edited by D. R. Brothwell and A. M. Pollard, pp. 269–280. Wiley, Chichester.

Sealy, Judith, Richard Armstrong, and Carmel Schrire
1995 Beyond Lifetime Averages: Tracing Life Histories Through Isotopic Analysis of Different Calcified Tissues from Archaeological Human Skeletons. *Antiquity* 69(263):290–300.

Sealy, J. C., N. J. van der Merwe, A. Sillen, F. J. Kruger, and H. W. Krueger
1991 ^{87}Sr/^{86}Sr as a Dietary Indicator in Modern and Archaeological Bone. *Journal of Archaeological Science* 18(3):399–416.

Sempowski, M. L., A. W. Nohe, R. G. V. Hancock, J.-F. Moreau, F. Kwok, S. Aufreiter, K. Karklins, J. Baart, C. Garrad, and I. Kenyon
2001 Chemical Analysis of 17th Century Red Glass Trade Beads from Northeastern North America and Amsterdam. *Archaeometry* 43(4):503–515.

Shackley, M. S.
1998 Gamma Rays, X-Rays and Stone Tools: Some Recent Advances in Archaeological Geochemistry. *Journal of Archaeological Science* 25(3):259–270.
2005 *Obsidian: Geology and Archaeology in the North American Southwest.* University of Arizona Press, Tucson.
2008 Archaeological Petrology and the Archaeometry of Lithic Material. *Archaeometry* 50(2):194–215.

Shackley, Myra
1982 Gas Chromatographic Identification of a Resinous Deposit from a 6th Century Storage Jar and its Possible Identification. *Journal of Archaeological Science* 9(3):305–306.

Shanks, Orin C., Robson Bonnichsen, Anthony T. Vella, and Walt Ream
2001 Recovery of Protein and DNA Trapped in Stone Tool Microcracks. *Journal of Archaeological Science* 28(9):965–972.

Shanks, Orin C., Larry Hodges, Lucas Tilley, Marcel Kornfeld, Mary Lou Larson, and Walt Ream
2005 DNA from Ancient Stone Tools and Bones Excavated at Bugas-Holding, Wyoming. *Journal of Archaeological Science* 32(1):27–38.

Shanks, Orin C., Marcel Kornfeld, and Dee D. Hawk
1999 Protein Analysis of Bugas-Holding Tools: New Trends in Immunological Studies. *Journal of Archaeological Science* 26(9):1183–1191.

Shanks, O. C., M. Kornfeld, and W. Ream
2004 DNA and Protein Recovery from Washed Experimental Stone Tools. *Archaeometry* 46(4):663–672.

Sharp, Zachary
2007 *Principles of Stable Isotope Geochemistry.* Pearson Education, Upper Saddle River.

Sharp, Zachary D., Viorel Atudorei, Héctor O. Panarello, Jorge Fernández, and Chuck Douthitt
2003 Hydrogen Isotope Systematics of Hair: Archeological and Forensic Applications. *Journal of Archaeological Science* 30(12):1709–1716.

Shemesh, Aldo
1990 Crystallinity and Diagenesis of Sedimentary Apatites. *Geochimica et Cosmochimica Acta* 54(9):2433–2438.

Shen, Guanjun, Hai Cheng, and R. L. Edwards
2004 Mass Spectrometric U-series Dating of New Cave at Zhoukoudian, China. *Journal of Archaeological Science* 31(3):337–342.

Shennan, Stephen
1997 *Quantifying Archaeology.* Edinburgh University, Edinburgh, Scotland.

Sherriff, B. L., M. A. Tisdale, B. G. Sayer, H. P. Schwarcz, and M. Knyf
1995 Nuclear Magnetic Resonance Spectroscopic and Isotopic Analysis of Carbonized Residues from Subarctic Canadian Prehistoric Pottery. *Archaeometry* 37(1):95–111.

Shortland, Andrew, Nick Rogers, and Katherine Eremin
  2007 Trace Element Discriminants Between Egyptian and Mesopotamian Late Bronze Age Glasses. *Journal of Archaeological Science* 34(5):781–789.
Sillen, Andrew, Judith C. Sealy, and van der Merwe, Nikolaas J.
  1989 Chemistry and Paleodietary Research: No More Easy Answers. *American Antiquity* 54(3):504–512.
Skibo, J. M.
  1992 *Pottery Function: A Use-Alteration Perspective.* Plenum Press, New York.
Skoog, Douglas A., F. J. Holler, and Timothy A. Nieman
  1998 *Principles of Instrumental Analysis.* 5th ed. Saunders, Philadelphia.
Skoog, Douglas A., and Donald M. West
  1982 *Fundamentals of Analytical Chemistry.* Saunders, Philadelphia.
Smallwood, William L., and Peter Alexander
  1984 *Biology.* Silver Burdett Co., Toronto.
Smith, Gregory D., and Robin J. H. Clark
  2004 Raman Microscopy in Archaeological Science. *Journal of Archaeological Science* 31(8):1137–1160.
Smith Jr., Marion F.
  1983 *The Study of Ceramic Function from Artifact Size and Shape.* Unpublished Ph.D. dissertation, University of Oregon, Eugene, Oregon.
Smith, B. W., E. J. Rhodes, S. Stokes, N. A. Spooner, and M. J. Aitken
  1990 Optical Dating of Sediments: Initial Quartz Results from Oxford. *Archaeometry* 32(1): 19–31.
Smith, P. R., and M. T. Wilson
  1992 Blood Residues on Ancient Stone Tool Surfaces: A Cautionary Note. *Journal of Archaeological Science* 19(3):237–241.
Solomons, T. W. G.
  1980 *Organic Chemistry.* 2nd ed. Wiley, New York.
Solomons, T. W. G., and C. B. Fryhle
  2004 *Organic Chemistry.* Wiley, New York.
Speakman, R. J., and H. Neff (editors)
  2005 *Laser-ablation-ICP-MS in Archaeological Research.* University of New Mexico, Albuquerque.
Speakman, R. J., and M. D. Glascock
  2007 Acknowledging Fifty Years of Neutron Activation Analysis in Archaeology. *Archaeometry* 49(2):179–183.
Speller, Camilla F., Dongya Y. Yang, and Brian Hayden
  2005 Ancient DNA Investigation of Prehistoric Salmon Resource Utilization at Keatley Creek, British Columbia, Canada. *Journal of Archaeological Science* 32(9):1378–1389.
Sponheimer, Matt, and Julia Lee-Thorp
  1999 Oxygen Isotopes in Enamel Carbonate and their Ecological Significance. *Journal of Archaeological Science* 26(6):723–728.
Stacey, R. J., C. R. Cartwright, and C. McEwan
  2006 Chemical Characterization of Ancient Mesoamerican 'Copal' Resins: Preliminary Results. *Archaeometry* 48(2):323–340.
Steelman, Karen L., Marvin W. Rowe, Solveig A. Turpin, Tom Guilderson, and Laura Nightengale
  2004 Nondestructive Radiocarbon Dating: Naturally Mummified Infant Bundle from SW Texas. *American Antiquity* 69(4):741–750.
Stephan, E.
  2000 Oxygen Isotope Analysis of Animal Bone Phosphate: Method Refinement, Influence of Consolidants, and Reconstruction of Paleotemperatures for Holocene Sites. *Journal of Archaeological Science* 27(6):523–535.

Steponaitis, Vincas, James M. Blackman, and Hector Neff
1996 Large-Scale Patterns in the Chemical Composition of Mississippian Pottery. *American Antiquity* 61(3):555–572.

Stern, B., C. Heron, L. Corr, M. Serpico, and J. Bourriau
2003 Compositional Variations in Aged and Heated Pistacia Resin Found in Late Bronze Age Canaanite Amphorae and Bowls from Amarna, Egypt. *Archaeometry* 45(3):457–469.

Stern, B., C. Heron, T. Tellefsen, and M. Serpico
2008 New Investigations in the Uluburun Resin Cargo. *Journal of Archaeological Science* 35(8):2188–2208.

Sternberg, Robert S.
1997 Archaeomagnetic Dating. In *Chronometric Dating in Archaeology,* vol. 3, edited by R. E. Taylor and Martin J. Aitken, pp. 323–356. Plenum, New York.
2001 Magnetic Properties and Archaeomagnetism. In *Handbook of Archaeological Sciences,* edited by D. R. Brothwell and A. M. Pollard, pp. 73–80. Wiley, Chichester.
2008 Archaeomagnetism in *Archaeometry-* A Semi-Centennial Review. *Archaeometry* 50(6):983–998.

Stevenson, Christopher M., Ihab M. Abdelrehim, and Steven W. Novak
2001 Infra-red Photoacoustic and Secondary Ion Mass Spectrometry Measurements of Obsidian Hydration Rims. *Journal of Archaeological Science* 28(1):109–115.
2004 High Precision Measurement of Obsidian Hydration Layers on Artifacts from the Hopewell Site using Secondary Ion Mass Spectrometry. *American Antiquity* 69(3): 555–568.

Stevenson, C. M., J. Carpenter, and B. E. Scheetz
1989 Obsidian Dating: Recent Advances in the Experimental Determination and Application of Hydration Rates. *Archaeometry* 31(2):193–206.

Stoltman, James B.
1989 A Quantitative Approach to the Petrographic Analysis of Ceramic Thin-Sections. *American Antiquity* 54(1):147–160.
1998 The Chaco-Chuska Connection: In Defense of Anna Shepard. In *Pottery and People: A Dynamic Interaction,* edited by J. M. Skibo and G. M. Feinman, pp. 9–24. University of Utah, Salt Lake City, Utah.

Stott, Andrew W., Robert Berstan, and Richard P. Evershed
2003 Direct Dating of Archaeological Pottery by Compound-Specific [14]C Analysis of Preserved Lipids. *Analytical Chemistry* 75(19):5037–5045.

Stott, A. W., R. P. Evershed, S. Jim, V. Jones, J. M. Rogers, N. Tuross, and S. Ambrose
1999 Cholesterol as a New Source of Paleodietary Information: Experimental Approaches and Archaeological Applications. *Journal of Archaeological Science* 26(6):705–716.

Stuiver, Minze, and Thomas F. Braziuna
1993 Modeling Atmospheric [14]C Influences and Radiocarbon Ages of Marine Samples to 10,000 B.C. *Radiocarbon* 35(1):137–189.

Stuiver, Minze, and Henry A. Polach
1977 Discussion: Reporting of [14]C Data. *Radiocarbon* 19(3):355–363.

Takac, Paul R.
2000 Archaeomagnetic Analysis of Archaeological and Experimental Rocks from 41ZP364, Zapata County, Texas. In *Data Recovery at 41ZP364: An Upland Campsite at Falcon Reservoir, Zapata County, Texas,* TRC Technical Report No. 22317, edited by J. M. Quigg and C. Cordova, pp. 298–330. TRC Mariah Associates, Austin.

Tamers, M. A., J. J. Stipp, and J. Collier
1961 High Sensitivity Detection of Naturally Occurring Radiocarbon: Chemistry of the Counting Sample. *Geochimica et Cosmochimica Acta* 24(3–4):266–276.

Tang, C. C., E. J. MacLean, M. A. Roberts, D. T. Clarke, E. Pantos, and Prag, A. J. N. W.
2001 The Study of Attic Black Gloss Sherds using Synchrotron X-ray Diffraction. *Journal of Archaeological Science* 28(10):1015–1024.

Tarling, D. H.
    1975 Archaeomagnetism: The Dating of Archaeological Materials by their Magnetic
    Properties. *World Archaeology* 7(2):185–197.
Taylor, R. E.
    1997 Radiocarbon Dating. In *Chronometric Dating in Archaeology,* vol. 3, edited by R. E.
    Taylor and Martin J. Aitken, pp. 65–96. Plenum, New York.
    2000 The Introduction of Radiocarbon Dating. In *It's About Time: A History of Archaeological
    Dating in North America,* edited by Stephen E. Nash, pp. 84–104. University of Utah, Salt Lake
    City, Utah.
    2001 Radiocarbon Dating. In *Handbook of Archaeological Sciences,* edited by D. R. Brothwell
    and A. M. Pollard, pp. 23–34. Wiley, Chichester.
Taylor, R. E., Minze Stuiver, and Paula J. Reimer
    1996 Development and Extension of the Calibration of the Radiocarbon Time Scale:
    Archaeological Applications. *Quaternary Science Reviews* 15(7):655–668.
Terry, Richard E., Fabian G. Fernández, J. J. Parnell, and Takeshi Inomata
    2004 The Story in the Floors: Chemical Signatures of Ancient and Modern Maya Activities at
    Aguateca, Guatemala. *Journal of Archaeological Science* 31(9):1237–1250.
Thomas, R. H., W. Scaffner, A. C. Wilson, and S. Pääbo
    1989 DNA Phylogeny of the Extinct Marsupial Wolf. *Nature* 340:465–467.
Tieszen, Larry L.
    1991 Natural Variations in the Carbon Isotope Values of Plants: Implications for Archaeology,
    Ecology, and Paleoecology. *Journal of Archaeological Science* 18(3):227–248.
Tite, M. S.
    1992 The Impact of Electron Microscopy on Ceramic Studies. In *New Developments in
    Archaeological Science,* Proceedings of the British Academy, vol. 77, edited by A. M. Pollard,
    pp. 111–131. Oxford University, Oxford.
    1996 In Defence of Lead Isotope Analysis. *Antiquity* 70(270):959–962.
    1999 Pottery Production, Distribution, and Consumption – The Contribution of the Physical
    Sciences. *Journal of Archaeological Method and Theory* 6(3):181–233.
Threadgold, Jayne, and Terence A. Brown
    2003 Degradation of DNA in Artificially Charred Wheat Seeds. *Journal of Archaeological
    Science* 30(8):1067–1076.
Torres, Jesús M., Concepción Borja, and Enrique G. Olivares
    2002 Immunoglobulin G in 1.6 Million-year-old Fossil Bones from Venta Micena (Granada,
    Spain). *Journal of Archaeological Science* 29(2):167–175.
Trembour, Fred W., Irving Friedman, F. J. Jurceka, and F. L. Smith
    1986 A Simple Device for Integrating Temperature, Relative Humidity or Salinity over Time.
    Journal of Atmospheric and Oceanic Technology 3(1):186–190.
Trembour, Fred W., F. L. Smith, and Irving Friedman
    1988 Cells for Integrating Temperature and Humidity over Long Periods of Time. In *Materials
    Issues in Art and Archaeology Materials,* Materials Research Society Symposium, vol. 123,
    edited by Edward Sayre, pp. 245–251.
Triadan, Daniela, Hector Neff, and Michael D. Glascock
    1997 An Evaluation of the Archaeological Relevance of Weak-Acid Extraction ICP: White
    Mountain Redware as a Case Study. *Journal of Archaeological Science* 24(11):997–1002.
Troja, Sebastiano O., and Richard G. Roberts
    2000 Luminescence Dating. In *Modern Analytical Methods in Art and Archaeology,* Chemical
    Analysis Series, vol. 155, edited by Enrico Ciliberto and Giuseppe Spoto, pp. 585–640. Wiley,
    New York.
Trueman, Clive N. G., Anna K. Behrensmeyer, Noreen Tuross, and Steve Weiner
    2004 Mineralogical and Compositional Changes in Bones Exposed on Soil Surfaces in
    Amboseli National Park, Kenya: Diagenetic Mechanisms and the Role of Sediment Pore Fluids.
    Journal of Archaeological Science 31(6):721–739.

Truncer, James
2004 Steatite Vessel Age and Occurrence in Temperate Eastern North America. *American Antiquity* 69(3):487–513.

Truncer, James, M. D. Glascock, and H. Neff
1998 Steatite Source Characterization in Eastern North America: New Results Using Instrumental Neutron Activation Analysis. *Archaeometry* 40(1):23–44.

Tudge, A. P.
1960 A Method of Analysis of Oxygen Isotopes in Orthophosphates – Its use in the Measurement of Paleotemperatures. *Geochimica et Cosmochimica Acta* 18:81–83.

Tuross, Noreen, Ian Barnes, and Richard Potts
1996 Protein Identification of Blood Residues on Experimental Stone Tools. *Journal of Archaeological Science* 23(2):289–296.

Tykot, Robert H.
1997 Characterization of the Monte Arci (Sardinia) Obsidian Sources. *Journal of Archaeological Science* 24(5):467–479.
2006 Isotope Analyses and the Histories of Maize. In *Histories of Maize,* edited by John E. Staller, Robert H. Tykot and Bruce F. Benz, pp. 131–142. Academic, Boston.

van der Merwe, Nikolaas J.
1969 *The Carbon-14 Dating of Iron.* University of Chicago, Chicago, IL.
1982 Carbon Isotopes, Photosynthesis, and Archaeology. *American Scientist* 70:596–606.
1992 Light Stable Isotopes and the Reconstruction of Prehistoric Diets. *Proceedings of the British Academy* 77:247–264.

van der Merwe, Nikolaas J., and Ernesto Medina
1991 The Canopy Effect, Carbon Isotope Ratios and Foodwebs in Amazonia. *Journal of Archaeological Science* 18(3):249–259.

van der Plicht, J., van der Sanden, W. A. B., A. T. Aerts, and H. J. Streurman
2004 Dating Bog Bodies by Means of ^{14}C-AMS. *Journal of Archaeological Science* 31(4): 471–491.

Vanhaeren, Marian, Francesco d'Errico, Isabelle Billy, and Francis Grousset
2004 Tracing the Source of Upper Palaeolithic Shell Beads by Strontium Isotope Dating. *Journal of Archaeological Science* 31(10):1481–1488.

Vaughn, Kevin J., Christina A. Conlee, Hector Neff, and Katharina Schreiber
2006 Ceramic Production in Ancient Nasca: Provenance Analysis of Pottery from the Early Nasca and Tiza Cultures through INAA. *Journal of Archaeological Science* 33(5): 681–689.

Vaughn, Kevin J., and Hector Neff
2004 Tracing the Clay Source of Nasca Polychrome Pottery: Results from a Preliminary Raw Material Survey. *Journal of Archaeological Science* 31(11):1577–1586.

Verma, H. R.
2007 *Atomic and Nuclear Analytical Methods: XRF, Mössbauer, XPS, NAA and Ion-Beam Spectroscopic Techniques.* Springer, Berlin.

Vermilion, Mary R., Mark P. S. Krekeler, and Lawrence H. Keeley
2003 Pigment Identification on Two Moorehead Phase Ramey knives from the Loyd Site, a Prehistoric Mississippian Homestead. *Journal of Archaeological Science* 30(11): 1459–1467.

Vogel, J. C., and van der Merwe, Nikolaas J.
1977 Isotopic Evidence for Early Maize Cultivation in New York State. *American Antiquity* 42(2):238–242.

Vuissoz, Annick, Michael Worobey, Nancy Odegaard, Michael Bunce, Carlos A. Machado, Niels Lynnerup, Elizabeth E. Peacock, and M. T. P. Gilbert
2007 The Survival of PCR-Amplifiable DNA in Cow Leather. *Journal of Archaeological Science* 34(5):823–829.

Vuorinen, Heikki S., Unto Tapper, and Helena Mussalo-Rauhamaa
1990 Trace and Heavy Metals in Infants, Analysis of Long Bones from Ficana, Italy, 8–6th Century B.C. *Journal of Archaeological Science* 17(3):237–254.

Walter, Robert C.
1997 Potassium-Argon/Argon-Argon Dating Methods. In *Chronometric Dating in Archaeology*, vol. 3, edited by R. E. Taylor and Martin J. Aitken, pp. 97–126. Plenum, New York.

Wandsnider, LuAnn
1997 The Roasted and the Boiled: Food Composition and Heat Treatment with Special Emphasis on Pit-hearth Cooking. *Journal of Anthropological Archaeology* 16(1):1–48.

Watson, J. T., and O. D. Sparkman
2007 *Introduction to Mass Spectrometry: Instrumentation, Applications and Strategies for Data Interpretation.* Wiley, Chichester.

Watts, S., and A. M. Pollard
1999 The Organic Geochemistry of Jet: Pyrolysis-gas Chromatography/Mass Spectrometry (Py-GCMS) Applied to Identifying Jet and Similar Black Lithic Materials: Preliminary Results. *Journal of Archaeological Science* 26(8):923–933.

Weber, Andrzej, Hugh G. McKenzie, Roelf Beukens, and Olga I. Goriunova
2005 Evaluation of Radiocarbon Dates from the Middle Holocene Hunter-Gatherer Cemetery Khuzhir-Nuge XIV, Lake Baikal, Siberia. *Journal of Archaeological Science* 32(10): 1481–1500.

Werts, S. P., and A. H. Jahren
2007 Estimation of Temperatures Beneath Archaeological Campfires using Carbon Stable Isotope Composition of Soil Organic Matter. *Journal of Archaeological Science* 34(6):850–857.

White, Christine, Fred J. Longstaffe, and Kimberley R. Law
2004 Exploring the Effects of Environment, Physiology and Diet on Oxygen Isotope Ratios in Ancient Nubian Bones and Teeth. *Journal of Archaeological Science* 31(2):233–250.

Wieser, M. E.
2006 IUPAC Technical Report: Atomic Weights of the Elements 2005. *Pure and Applied Chemist* 78(11):2051–2066.

Weisler, Marshall I., Kenneth D. Collerson, Yue-Xing Feng, Jian-Xin Zhao, and Ke-Fu Yu
2006 Thorium-230 Coral Chronology of a Late Prehistoric Hawaiian Chiefdom. *Journal of Archaeological Science* 33(2):273–282.

Williams, H., F. J. Turner, and C. M. Gilbert
1982 *Petrography: An Introduction to the Study of Rocks in Thin Sections.* W. H. Freeman, San Francisco.

Williams-Thorpe, O.
1995 Obsidian in the Mediterranean and the Near East: A Provenancing Success Story. *Archaeometry* 37(2):217–248.

Williams-Thorpe, O., and R. S. Thorpe
1992 Geochemistry, Sources and Transport of the Stonehenge Bluestones. In *New Developments in Archaeological Science*, Proceedings of the British Academy, vol. 77, edited by A. M. Pollard, pp. 133–161. Oxford University, Oxford.

Williamson, B. S.
2000 Direct Testing of Rock Painting Pigments for Traces of Haemoglobin at Rose Cottage Cave, South Africa. *Journal of Archaeological Science* 27(9):755–762.

Willard, Hobart H., Lynne L. Merritt Jr., John A. Dean, and Frank A. Settle Jr.
1988 *Instrumental Methods of Analysis.* 7th ed. Wadsworth Publishing, Belmont, California.

Wilson, L., and A. M. Pollard
2001 The Provenance Hypothesis. In *Handbook of Archaeological Sciences,* edited by D. R. Brothwell and A. M. Pollard, pp. 507–518. Wiley, Chichester.

Wintle, A. G.
2008 Fifty Years of Luminescence Dating. *Archaeometry* 50(2):276–312.

Wischmann, Hartmut, Susanne Hummel, Markus A. Rothschild, and Bernd Herrmann
2002 Analysis of Nicotine in Archaeological Skeletons from the Early Modern Age and from the Bronze Age. *Ancient Biomolecules* 4(2):47–52.

Workman Jr., Jerry
1998 Optical Spectrometers. In *Applied Spectroscopy: A Compact Reference for Practioners*, edited by Jerry Workman Jr. and Art W. Springsteen, pp. 3–28. Academic Press, San Diego.
1998 Ultraviolet, Visible and Near-Infrared Spectrometry. In *Applied Spectroscopy: A Compact Reference for Practioners*, edited by Jerry Workman Jr. and Art W. Springsteen, pp. 29–48. Academic Press, San Diego.

Workman Jr., Jerry, and Art W. Springsteen (editors)
1998 *Applied Spectroscopy: A Compact Reference for Practitioners.* Academic Press, San Diego.

Wright, Lori E., and Henry P. Schwarcz
1996 Infrared and Isotopic Evidence for Diagenesis of Bone Apatite at Dos Pilas, Guatemala: Paleodietary Implications. *Journal of Archaeological Science* 23(6):933–944.

Yacobaccio, Hugo D., Patricia S. Escola, Fernando X. Pereyra, Marisa Lazzari, and Michael D. Glascock
2004 Quest for Ancient routes: Obsidian Sourcing Research in Northwestern Argentina. *Journal of Archaeological Science* 31(2):193–204.

Yang, Dongya Y., Aubrey Cannon, and Shelley R. Saunders
2004 DNA Species Identification of Archaeological Salmon Bone from the Pacific Northwest Coast of North America. *Journal of Archaeological Science* 31(5):619–631.

Yang, Dongya Y., Barry Eng, John S. Waye, J. C. Dudar, and Shelley R. Saunders
1998 Technical Note: Improved DNA Extraction from Ancient Bones using Silica-Based Spin Columns. *American Journal of Physical Anthropology* 105:539–543.

Yoder, David T.
2008 The Use of "Soft" X-ray Radiography in Determining Hidden Construction Characteristics in Fiber Sandals. *Journal of Archaeological Science* 35(2):316–321.

Yohe II, Robert M., Margaret E. Newman, and Joan S. Schneider
1991 Immunological Identification of Small-Mammal Proteins on Aboriginal Milling Equipment. *American Antiquity* 56(4):659–666.

Young, S. M. M., P. Budd, R. Haggerty, and A. M. Pollard
1997 Inductively Coupled Plasma-Mass Spectrometry for the Analysis of Ancient Metals. *Archaeometry* 39(2):379–392.

Young, Suzanne M. M., and A. M. Pollard
2000 Atomic Spectroscopy and Spectrometry. In *Modern Analytical Methods in Art and Archaeology*, Chemical Analysis Series, vol. 155, edited by Enrico Ciliberto and Giuseppe Spoto, pp. 21–53. Wiley, New York.

Zacharias, N., J. Buxeda i Garrigós, H. Mommsen, A. Schwedt, and V. Kilikoglou
2005 Implications of Burial Alterations on Luminescence dating of Archaeological Ceramics. *Journal of Archaeological Science* 32(1):49–57.

Zhou, L. P., and T. H. van Andel
2000 A Luminescence Dating Study of Open-Air Paleolithic Sites in Western Epirus, Greece. *Journal of Archaeological Science* 27(7):609–620.

Zhu, Jiping, Jie Shan, Ping Qiu, Ying Qin, Changsui Wang, Deliang He, Bo Sun, Peihua Tong, and Shuangcheng Wu
2004 The Multivariate Statistical Analysis and XRD Analysis of Pottery at Xigongqiao Site. *Journal of Archaeological Science* 31(12):1685–1691.

Index

Note: The letters 'f' and 't' following locators refer to figures and tables respectively.

Breinigsville, PA USA
03 November 2010
248464BV00008B/58/P